Grundlehren der mathematischen Wissenschaften 335

A Series of Comprehensive Studies in Mathematics

Colin J. Bushnell · Guy Henniart

The Local Langlands Conjecture for GL(2)

Springer

Colin J. Bushnell
King's College London
Department of Mathematics
Strand, London WC2R 2LS
UK
e-mail: *colin.bushnell@kcl.ac.uk*

Guy Henniart
Université de Paris-Sud et umr 8628 du CNRS
Département de Mathématiques
Bâtiment 425
91405 Orsay
France
e-mail: *guy.henniart@math.u-psud.fr*

Library of Congress Control Number: 2006924564

Mathematics Subject Classification (2000): *Primary:* 11F70, 22E50, 20G05
Secondary: 11D88, 11F27, 22C08

ISSN 0072-7830
ISBN-10 3-540-31486-5 Springer Berlin Heidelberg New York
ISBN-13 978-3-540-31486-8 Springer Berlin Heidelberg New York

Springer is a part of Springer Science+Business Media
springer.com

Printed in The Netherlands

Typesetting: by the author and SPi using a Springer LaTeX macro package
Cover design: *design & production* GmbH, Heidelberg

Printed on acid-free paper SPIN: 11604440 41/SPI 5 4 3 2 1 0

For Elisabeth and Lesley

Foreword

This book gives a complete and self-contained proof of Langlands' conjecture concerning the representations of GL(2) of a non-Archimedean local field. It has been written to be accessible to a doctoral student with a standard grounding in pure mathematics and some extra facility with local fields and representations of finite groups. It had its origins in a lecture course given by the authors at the first Beijing-Zhejiang International Summer School on p-adic methods, held at Zhejiang University Hangzhou in 2004. We hope this is found a fitting response to the efforts of the organizers and the enthusiastic contribution of the student participants.

King's College London and
Université de Paris-Sud.

Contents

Introduction

We work with a non-Archimedean local field F which, we always assume, has finite residue field of characteristic p. Thus F is either a finite extension of the field $\mathbb{Q}_p$ of p-adic numbers or a field $\mathbb{F}_{p^r}((t))$ of formal Laurent series, in one variable, over a finite field. The arithmetic of F is encapsulated in the *Weil group* $\mathcal{W}_F$ of F: this is a topological group, closely related to the Galois group of a separable algebraic closure of F, but with rather more sensitive properties. One investigates the arithmetic via the study of continuous (in the appropriate sense) representations of $\mathcal{W}_F$ over various algebraically closed fields of characteristic zero, such as the complex field $\mathbb{C}$ or the algebraic closure $\overline{\mathbb{Q}}_\ell$ of an ℓ-adic number field.

Sticking to the complex case, the one-dimensional representations of $\mathcal{W}_F$ are the same as the characters (i.e., continuous homomorphisms) $F^\times \to \mathbb{C}^\times$: this is the essence of *local class field theory*. The n-dimensional analogue of a character of $F^\times = \mathrm{GL}_1(F)$ is an irreducible smooth representation of the group $\mathrm{GL}_n(F)$ of invertible $n \times n$ matrices over F. As a specific instance of a wide speculative programme, Langlands [55] proposed, in a precise conjecture, that such representations should parametrize the n-dimensional representations of $\mathcal{W}_F$ in a manner generalizing local class field theory and compatible with parallel global considerations.

The excitement provoked by the local Langlands conjecture, as it came to be known, stimulated a period of intense and widespread activity, reflected in the pages of [8]. The first case, where $n = 2$ and F has characteristic zero, was started in Jacquet-Langlands [46]; many hands contributed but Kutzko, bringing two new ideas to the subject, completed the proof in [52], [53]. Subsequently, the conjecture has been proved in all dimensions, first in positive characteristic by Laumon, Rapoport and Stuhler [58], then in characteristic zero by Harris and Taylor [38], also by Henniart [43] on the basis of an earlier paper of Harris [37].

Throughout the period of this development, the subject has largely remained confined to the research literature. Our aim in this book is to provide a navigable route into the area with a complete and self-contained account of the case $n = 2$, in a tolerable number of pages, relying only on material readily available in standard courses and texts. Apart from a couple of unavoidable caveats concerning Chapter VII, we assume only the standard representation theory for finite groups, the beginnings of the theory of local fields and some very basic notions from topology.

In consequence, our methods are entirely local and elementary. Apart from Chapter I (which could equally serve as the start of a treatise on the representation theory of p-adic reductive groups) and some introductory material in Chapter VII, we eschew all generality. Whenever possible, we exploit special features of GL(2) to abbreviate or simplify the arguments.

The desire to be both compact and complete removes the option of appealing to results derived from harmonic analysis on adèle groups ("base change" [57], [1]) which originally played a determining rôle. This particular constraint has forced us to give the first proof of the conjecture that can claim to be completely local in method.

There is an associated loss, however. The local Langlands Conjecture is just a specific instance of a wide programme, encompassing local and global issues and all connected reductive algebraic groups in one mighty sweep. Beyond the minimal gesture of Chapter XIII, we can give the reader no idea of this. Nor have we mentioned any of the geometric methods currently necessary to prove results in higher dimensions. Fortunately, the published literature contains many fine surveys, from Gelbart's book [32], which still conveys the breadth and excitement of the ideas, to the new directions described in [4].

The approach we take is guided by [46] and [50–53], but we have rearranged matters considerably. We have separated the classification of representations from the functional equation. We have imported ideas of Bernstein and Zelevinsky into the discussion of non-cuspidal representations. While the treatment of cuspidal representations is essentially that of Kutzko, it is heavily informed by hindsight. We have given precedence to the Godement-Jacquet version of the functional equation and so had to treat the Converse Theorem in a novel manner, owing something to ideas of Gérardin and Li. There is also some degree of novelty in our treatment of the Kirillov model and the relation between the functional equation it gives and that of Godement and Jacquet. We have given a quick and explicit proof of the existence of the Langlands correspondence, in the case $p \neq 2$, at an early stage.

The case $p = 2$ has many pages to itself. The method is essentially that of Kutzko, but we have had to bring a new idea to the closing pages (the treatment of the so-called octahedral representations) to avoid an appeal to base change. We regard this case as being particularly important. It remains the one instance of the local conjecture in which the detail is sufficiently complex to be interesting, yet sufficiently visible to illuminate the miracle that is the Langlands correspondence. Even after 25 years, it stands as a sturdy corrective to over-optimistic attitudes to more general problems.

As light relief, we have broadened the picture with some discussion of ℓ-adic representations, since these provide a forum in which the correspondence finds much of its application.

The final Chapter XIII stands outside the main sequence. There, D is the quaternion division algebra over F. The irreducible representations of $D^\times = \mathrm{GL}_1(D)$ can be classified by a method parallel to that used for $\mathrm{GL}_2(F)$. The Jacquet-Langlands correspondence provides a canonical connection between the representation theories of $D^\times$ and $\mathrm{GL}_2(F)$. We include it as an indication of further dimensions in the subject. Given the experience of $\mathrm{GL}_2(F)$, it is a fairly straightforward matter which we have left as a sequence of exercises.

Acknowledgement. The final draft was read by Corinne Blondel, whose acute comments led us to remove a large number of minor errors and obscurities, along with a couple of more significant lapses. It is a pleasure to record our debt to her.

Notation

We list some standard notations which we use repeatedly, without always recalling their meaning.

F = *a non-Archimedean local field;*
$\mathfrak{o}$ = *the discrete valuation ring in F;*
$\mathfrak{p}$ = *the maximal ideal of $\mathfrak{o}$;*
$\boldsymbol{k}$ = $\mathfrak{o}/\mathfrak{p}$; p = *the characteristic of $\boldsymbol{k}$; $q = |\boldsymbol{k}|$;*
U_F = *the group of units of $\mathfrak{o}$; $U_F^n = 1+\mathfrak{p}^n$, $n \geqslant 1$.*

(Thus the characteristic of F is 0 or p: we never need to impose any further restriction.) In addition, $v_F : F^\times \to \mathbb{Z}$ is the normalized (surjective) additive valuation and $\|x\| = q^{-v_F(x)}$. We denote by $\boldsymbol{\mu}_F$ the group of roots of unity in F of order prime to p.

If E/F is a finite field extension, we use the analogous notations $\mathfrak{o}_E$, $\mathfrak{p}_E$, etc. The norm map $E^\times \to F^\times$ is denoted $\mathrm{N}_{E/F}$, and the trace $E \to F$ is $\mathrm{Tr}_{E/F}$. The ramification index and the residue class degree are $e(E|F)$, $f(E|F)$ respectively. The discriminant is $\mathfrak{d}_{E/F} = \mathfrak{p}^{d+1}$, $d = d(E|F)$.

The symbol tr is reserved for the trace of an endomorphism, such as a matrix or a group representation, and det is invariably the determinant.

If R is a ring with 1, $R^\times$ is its group of units and $\mathrm{M}_n(R)$ is the ring of $n\times n$ matrices over R. When R is commutative, $\mathrm{GL}_n(R)$ (resp. $\mathrm{SL}_n(R)$) is the group of $n \times n$ matrices over R which are invertible (resp. of determinant 1). We use the notation B, T, N, Z for the subgroups of GL_2 of matrices of the form

$$\begin{pmatrix} * & * \\ 0 & * \end{pmatrix}, \quad \begin{pmatrix} * & 0 \\ 0 & * \end{pmatrix}, \quad \begin{pmatrix} 1 & * \\ 0 & 1 \end{pmatrix}, \quad \begin{pmatrix} a & 0 \\ 0 & a \end{pmatrix}$$

respectively. Unless otherwise specified, $A = \mathrm{M}_2(F)$ and $G = \mathrm{GL}_2(F)$.

Notes for the reader

Prerequisites. We assume the beginnings of the representation theory of finite groups, including Mackey theory: the first 11 sections of [77] cover it all, bar a couple of results requiring reference to [26]. Of non-Archimedean local fields, we need general structure theory as far as the discriminant and structure of tame extensions, plus behaviour of the norm in tame or quadratic extensions. Practically everything can be found in [30] or the first two parts of [76], while [87] is the source of many of the ideas here. From topology and measure theory, beyond the most elementary concepts, we cover practically everything we need.

All this material is commonly available in many books: we mention only personal favourites.

From Chapter VII onwards, we rely on local class field theory. No detail is involved, so we have been able to take an axiomatic approach. The reader might consult the compact [68] or [74], [76]. More serious is the treatment in §30 of the existence of the Langlands-Deligne local constant. This depends on an interplay between local and global fields using some deep (but classical) theorems. The reader could again take an axiomatic approach. We have included a brief account which is complete modulo the classical background. (The requisite material is in [68] or [54].)

Navigation. Sections are numbered consecutively throughout the book. Each section is divided into (usually) short paragraphs, numbered in the form y.z. A reference y.z Proposition means the (only) proposition in paragraph y.z.

Chapter I stands alone, and could serve as an introduction to much wider areas. Chapter II is elementary, and could be read first. Parts of Chapters VII and X can be read independently. Chapter XIII could be read directly after Chapter VI. Otherwise, the logical dependence is linear and fairly rigid.

Principal series (or non-cuspidal) representations form a distinct subtheme. At a first reading, this could be edited out or pursued exclusively, according to taste. (For a different approach, emphasizing non-cuspidal representations and their importance for L-functions, see Bump's book [10].) Another "short course" option would be to stop at the end of Chapter VIII, by which stage the argument is complete for all but dyadic fields F.

Exercises. A few exercises are scattered through the text. These are intended to illuminate, entertain, or to indicate directions we do not follow. Only the simpler ones ever make a serious contribution to the main argument.

Notes. We have appended brief notes or comments to some chapters, to indicate further reading or wider perspectives. They tend to pre-suppose greater experience than the main text.

History. We have written an account of the subject, not its history: that would be a separate project of comparable scope. We have made no attempt at a complete bibliography. We have cited sources of major importance, and those we have found helpful in the preparation of this volume. We have also mentioned a number of recent works, along with older ones that, in our opinion, remain valuable to one learning the subject.

1

Smooth Representations

This chapter is introductory and foundational in nature. We define a class of topological groups, the *locally profinite groups,* and study their *smooth* representations on complex vector spaces. These representations are often infinite-dimensional, but smoothness imposes a drastic continuity condition. Nontheless, this class of objects is quite wide: it includes, for example, all representations of discrete groups.

We start by recalling some standard facts. We then develop the elementary aspects of smooth representation theory, very much guided by the ordinary representation theory of finite groups. We occasionally turn to non-Archimedean local fields as a source of examples. The topic of Haar measure and integration on topological groups necessarily enters the picture. Since we have only to deal with locally profinite groups, this is a straightforward matter of which we give just as much as we need.

While we will ultimately be concerned only with non-Archimedean local fields F and associated groups like $\mathrm{GL}_2(F)$, there is nothing to be gained from specialization at this stage. Looking beyond the confines of the present book, there is much to be lost. We therefore work, throughout this chapter, in quite extreme generality.

1. Locally Profinite Groups

In this section, we introduce and briefly discuss the notion of a locally profinite group. We concentrate on showing how this framework accommodates the non-Archimedean local fields and some associated groups and rings. We give very few proofs in the first four paragraphs: it is more a case of gathering together the pre-requisite threads.

We conclude the section with a couple of paragraphs about various characters associated with a non-Archimedean local field F. We make unceasing use of this material in the later chapters. More immediately, it gives us some examples to illuminate the general theory of the following sections.

1.1.

Definition. *A* locally profinite group *is a topological group G such that every open neighbourhood of the identity in G contains a compact open subgroup of G.*

For example, any discrete group is locally profinite. A closed subgroup of a locally profinite group is locally profinite. The quotient of a locally profinite group by a closed normal subgroup is locally profinite.

A locally profinite group is locally compact. If it is compact, it is *profinite* in the usual sense, that is, the limit of an inverse system of finite discrete groups. In fact, if G is a compact locally profinite group, it is not hard to show directly that the obvious map

$$G \longrightarrow \varprojlim G/K$$

is a topological isomorphism, where K ranges over the open normal subgroups of G.

In general, any open neighbourhood of 1 in a locally profinite group G contains a compact open subgroup K of G. As we have just seen, K is profinite: the terminology is therefore apt.

Remark. A locally profinite group is locally compact and totally disconnected. In the converse direction, it is known that a compact, totally disconnected topological group is profinite. Likewise, a locally compact, totally disconnected group is locally profinite, but we shall make no use of that fact.

1.2. Let F be a non-Archimedean local field. Thus F is the field of fractions of a discrete valuation ring $\mathfrak{o}$. Let $\mathfrak{p}$ be the maximal ideal of $\mathfrak{o}$ and $\boldsymbol{k} = \mathfrak{o}/\mathfrak{p}$ the residue class field. We will always assume that $\boldsymbol{k}$ is *finite,* and we will generally denote the cardinality $|\boldsymbol{k}|$ by q.

Let ϖ be a prime element of F, that is, an element such that $\varpi\mathfrak{o} = \mathfrak{p}$. Every element $x \in F^\times$ can be written uniquely as $x = u\varpi^n$, for some unit $u \in \mathfrak{o}^\times = U_F$, and some $n \in \mathbb{Z}$. (We use the notation $n = \upsilon_F(x)$.) The field F carries an absolute value

$$\|x\| = q^{-n} = q^{-\upsilon_F(x)}, \quad \|0\| = 0,$$

giving a metric on F, relative to which F is complete. In the metric space topology, F is a topological field. The fractional ideals

$$\mathfrak{p}^n = \varpi^n\mathfrak{o} = \{x \in F : \|x\| \leqslant q^{-n}\}, \quad n \in \mathbb{Z},$$

are open subgroups of F and give a fundamental system of open neighbourhoods of 0 in F.

Combining the definition of the topology with the completeness property, one sees that the canonical map

$$\mathfrak{o} \longrightarrow \varprojlim_{n \geqslant 1} \mathfrak{o}/\mathfrak{p}^n$$

is a topological isomorphism. Since $\boldsymbol{k}$ is finite, each group $\mathfrak{o}/\mathfrak{p}^n$ is finite, and the limit is compact. Each fractional ideal $\mathfrak{p}^n$, $n \in \mathbb{Z}$, is isomorphic to $\mathfrak{o}$ and so is compact. We conclude:

Proposition. *The additive group F is locally profinite, and F is the union of its compact open subgroups.*

1.3. The multiplicative group $F^\times$ is likewise a locally profinite group: the congruence unit groups $U_F^n = 1+\mathfrak{p}^n$, $n \geqslant 1$, are compact open, and give a fundamental system of open neighbourhoods of 1 in $F^\times$.

1.4. Let $n \geqslant 1$ be an integer. The vector space $F^n = F \times \cdots \times F$ carries the product topology, relative to which it is a locally profinite group. As a special case, the matrix ring $\mathrm{M}_n(F)$ is a locally profinite group under addition, in which multiplication of matrices is continuous.

The group $G = \mathrm{GL}_n(F)$ is an open subset of $\mathrm{M}_n(F)$; inversion of matrices is continuous, so G is a topological group. The subgroups

$$K = \mathrm{GL}_n(\mathfrak{o}), \qquad K_j = 1 + \mathfrak{p}^j\mathrm{M}_n(\mathfrak{o}), \quad j \geqslant 1,$$

are compact open, and give a fundamental system of open neighbourhoods of 1 in G. Thus $G = \mathrm{GL}_n(F)$ is a locally profinite group.

More generally, let V be an F-vector space of finite dimension n. The choice of a basis gives an isomorphism $V \cong F^n$, which we use to impose a topology on V. This topology is independent of the choice of basis. The remarks above apply equally to the algebra $\mathrm{End}_F(V)$ and the group $\mathrm{Aut}_F(V)$.

1.5. Let V be an F-vector space of finite dimension n. An $\mathfrak{o}$-*lattice* in V is a finitely generated $\mathfrak{o}$-submodule L of V such that the F-linear span FL of L is V.

Proposition. *Let L be an $\mathfrak{o}$-lattice in V. There is an F-basis $\{x_1, x_2, \dots, x_n\}$ of V such that $L = \sum_{i=1}^n \mathfrak{o}x_i$.*

Proof. By definition, L has a finite $\mathfrak{o}$-generating set. Choose a minimal such set $\{y_1, y_2, \dots, y_m\}$: we show that this is an F-basis of V. It certainly spans V. Suppose it is linearly dependent over F:

$$\sum_{1 \leqslant i \leqslant m} a_i y_i = 0,$$

with $a_i \in F$, not all zero. We can multiply through by an element of $F^\times$ and assume that all $a_i \in \mathfrak{o}$ and that at least one of them, a_j say, is a unit of $\mathfrak{o}$. Thus y_j is an $\mathfrak{o}$-linear combination of the other y_i, contrary to the minimality hypothesis. □

In particular, an $\mathfrak{o}$-lattice L is a compact open subgroup of V. The $\mathfrak{o}$-lattices in V give a fundamental system of open neighbourhoods of 0 in V.

More generally, a *lattice* in V is a compact open subgroup of V. Here we have:

Lemma. *Let L be a subgroup of V; then L is a lattice in V if and only if there exist $\mathfrak{o}$-lattices L_1, L_2 in V such that $L_1 \subset L \subset L_2$.*

Proof. Suppose $L_1 \subset L \subset L_2$, where the L_i are $\mathfrak{o}$-lattices. Since L contains L_1, it is open and hence closed. Since L is contained in L_2, it is compact.

Conversely, if L is a lattice in V, it must contain an $\mathfrak{o}$-lattice (since L is an open neighbourhood of 0), and so $FL = V$. In the opposite direction, we choose a basis $\{x_1, \dots, x_n\}$ of V. The image of L under the obvious projection $V \to Fx_i$ is a compact open subgroup of Fx_i. It is therefore contained in a group of the form $\mathfrak{a}_i x_i$, for some fractional ideal $\mathfrak{a}_i = \mathfrak{p}^{a_i}$ of $\mathfrak{o}$. Thus $L \subset \mathfrak{o}L \subset \sum_i \mathfrak{a}_i x_i$, and this is an $\mathfrak{o}$-lattice. □

1.6. Let G be a locally profinite group.

Proposition. *Let $\psi : G \to \mathbb{C}^\times$ be a group homomorphism. The following are equivalent:*

(1) *ψ is continuous;*
(2) *the kernel of ψ is open.*

If ψ satisfies these conditions and G is the union of its compact open subgroups, then the image of ψ is contained in the unit circle $|z| = 1$ in $\mathbb{C}$.

Proof. Certainly (2) ⇒ (1). Conversely, let $\mathcal{N}$ be an open neighbourhood of 1 in $\mathbb{C}$. Thus $\psi^{-1}(\mathcal{N})$ is open and contains a compact open subgroup K of G. However, if $\mathcal{N}$ is chosen sufficiently small, it contains no non-trivial subgroup of $\mathbb{C}^\times$ and so $K \subset \operatorname{Ker}\psi$.

The unit circle S^1 is the unique maximal compact subgroup of $\mathbb{C}^\times$. If K is a compact subgroup of G, then $\psi(K)$ is compact, and so it is contained in S^1. The final assertion follows. □

We define a *character* of a locally profinite group G to be a continuous homomorphism $G \to \mathbb{C}^\times$. We usually write 1_G, or even just 1, for the trivial (constant) character of G.

We call a character of G *unitary* if its image is contained in the unit circle.

1.7. We will later make frequent use of another property of the local field F.

The set of characters of F is a group under multiplication; we denote it $\widehat{F}$. Since F is the union of its compact open subgroups $\mathfrak{p}^n$, $n \in \mathbb{Z}$, all characters of F are unitary (1.6 Proposition).

If ψ is a character and $\psi \neq 1$, there is a least integer d such that $\mathfrak{p}^d \subset \operatorname{Ker}\psi$.

Definition. *Let $\psi \in \widehat{F}$, $\psi \neq 1$. The* level *of ψ is the least integer d such that $\mathfrak{p}^d \subset \operatorname{Ker}\psi$.*

If we fix d, the set of characters of F of level $\leqslant d$ is the subgroup of $\psi \in \widehat{F}$ such that $\psi \mid \mathfrak{p}^d = 1$.

Proposition ***(Additive duality).*** *Let $\psi \in \widehat{F}$, $\psi \neq 1$, have level d.*

(1) *Let $a \in F$. The map $a\psi : x \mapsto \psi(ax)$ is a character of F. If $a \neq 0$, the character $a\psi$ has level $d - v_F(a)$.*
(2) *The map $a \mapsto a\psi$ is a group isomorphism $F \cong \widehat{F}$.*

Proof. Part (1) is immediate, and $a \mapsto a\psi$ is an injective group homomorphism $F \to \widehat{F}$.

Let $\theta \in \widehat{F}$, $\theta \neq 1$, and let l be the level of θ. Let ϖ be a prime element of F, and $u \in U_F$. The character $u\varpi^{d-l}\psi$ has level l, and so agrees with θ on $\mathfrak{p}^l$. The characters $u\varpi^{d-l}\psi$, $u'\varpi^{d-l}\psi$, $u, u' \in U_F$, agree on $\mathfrak{p}^{l-1}$ if and only if $u \equiv u'$ (mod $\mathfrak{p}$). The group $\mathfrak{p}^{l-1}$ has $q-1$ non-trivial characters which are trivial on $\mathfrak{p}^l$. As u ranges over U_F/U_F^1, the $q-1$ characters $u\varpi^{d-l}\psi \mid \mathfrak{p}^{l-1}$ are distinct, non-trivial, but trivial on $\mathfrak{p}^l$. Therefore one of them, say $u_1\varpi^{d-l}\psi \mid \mathfrak{p}^{l-1}$, equals $\theta \mid \mathfrak{p}^{l-1}$.

Iterating this procedure, we find a sequence of elements $u_n \in U_F$ such that $u_n\varpi^{d-l}\psi$ agrees with θ on $\mathfrak{p}^{l-n}$ and $u_{n+1} \equiv u_n \pmod{\mathfrak{p}^n}$. The Cauchy sequence $\{u_n\}$ converges to some $u \in U_F$ and we have $\theta = u\varpi^{d-l}\psi$. □

Exercise. *Let L be a lattice in F and let χ be a character of L (in the sense of* 1.6*). Show there exists a character ψ of F such that $\psi \mid L = \chi$.*

1.8. We turn to the multiplicative group $F^\times$. Let χ be a character of $F^\times$. By 1.6 Proposition, χ is trivial on U_F^m, for some $m \geqslant 0$.

Definition. *Let χ be a non-trivial character of $F^\times$. The* level *of χ is defined to be the least integer $n \geqslant 0$ such that χ is trivial on U_F^{n+1}.*

We use the same terminology for characters of open subgroups of $F^\times$.

Observe that a character of $F^\times$ need not be unitary: for example, the map $x \mapsto \|x\|$ is a character. Note also that, in a related contrast to the additive case, $F^\times$ has a *unique maximal compact subgroup,* namely U_F.

The structure of the group of characters of $F^\times$ is more subtle than that of $\widehat{F}$. However, we shall make frequent use of a partial description in additive terms. Let m, n be integers, $1 \leqslant m < n \leqslant 2m$. The map $x \mapsto 1{+}x$ gives an isomorphism $\mathfrak{p}^m/\mathfrak{p}^n \cong U_F^m/U_F^n$. This gives an isomorphism of character groups $(\mathfrak{p}^m/\mathfrak{p}^n)\widehat{\ } \cong (U_F^m/U_F^n)\widehat{\ }$, and we can use 1.7 to describe the group $(\mathfrak{p}^m/\mathfrak{p}^n)\widehat{\ }$.

For this purpose, it is convenient to fix a character $\psi_F \in \widehat{F}$ *of level* 1. For $a \in F$, we define a function

$$\psi_{F,a} : F \longrightarrow \mathbb{C}^\times,$$
$$\psi_{F,a}(x) = \psi_F(a(x{-}1)). \tag{1.8.1}$$

Proposition 1.7 then yields:

Proposition. *Let $\psi \in \widehat{F}$ have level* 1. *Let m, n be integers, $0 \leqslant m < n \leqslant 2m{+}1$. The map $a \mapsto \psi_{F,a} \mid U_F^{m+1}$ induces an isomorphism*

$$\mathfrak{p}^{-n}/\mathfrak{p}^{-m} \xrightarrow{\ \approx\ } (U_F^{m+1}/U_F^{n+1})\widehat{\ }.$$

Observe that, viewed as a character of U_F^{m+1}, the function $\psi_{F,a}$ has level $-v_F(a)$. Also, the condition relating m and n can be re-formulated as $\left[\frac{n}{2}\right] \leqslant m < n$, where $x \mapsto [x]$ denotes the greatest integer function.

Terminology. We will use analogues of the notion of *level,* as defined in this paragraph, in many contexts where we study representations of groups with a filtration indexed by the non-negative integers. As we shall see, it is very convenient. From this more general viewpoint, the definition in 1.7 for characters of F appears anomalous. This version is forced on us by a variety of historical conventions: the only point on which these agree is that a character of F of level zero must be trivial on $\mathfrak{o}$ but not on $\mathfrak{p}^{-1}$. This is so firmly established that it would be confusing to change it now.

2. Smooth Representations of Locally Profinite Groups

In this section, we introduce the notion of a *smooth* representation of a locally profinite group G. We develop the basic theory, along lines familiar from the ordinary representation theory of finite groups. New phenomena do arise, but the general outline is very similar to that of the standard theory.

2.1. Let G be a locally profinite group, and let (π, V) be a representation of G. Thus V is a complex vector space and π is a group homomorphism $G \to \mathrm{Aut}_{\mathbb{C}}(V)$. The representation (π, V) is called *smooth* if, for every $v \in V$, there is a compact open subgroup K of G (depending on v) such that $\pi(x)v = v$, for all $x \in K$. Equivalently, if V^K denotes the space of $\pi(K)$-fixed vectors in V, then

$$V = \bigcup_K V^K,$$

where K ranges over the compact open subgroups of G.

In practice, we will usually have to deal with representations of infinite dimension.

A smooth representation (π, V) is called *admissible* if the space V^K is finite-dimensional, for each compact open subgroup K of G.

Let (π, V) be a smooth representation of G; then any G-stable subspace of G provides a further smooth representation of G. Likewise, if U is a G-subspace of V, the natural representation of G on the quotient V/U is smooth. One says that (π, V) is *irreducible* if $V \neq 0$ and V has no G-stable subspace U, $0 \neq U \neq V$.

For smooth representations (π_i, V_i) of G, the set $\mathrm{Hom}_G(\pi_1, \pi_2)$ is just the space of linear maps $f : V_1 \to V_2$ commuting with the G-actions:

$$f \circ \pi_1(g) = \pi_2(g) \circ f, \quad g \in G. \tag{2.1.1}$$

With this definition, the class of smooth representations of G forms a category $\mathrm{Rep}(G)$. We remark that the category $\mathrm{Rep}(G)$ is *abelian.*

We say that two smooth representations (π_1, V_1), (π_2, V_2) of G are *isomorphic,* or *equivalent,* if there exists a $\mathbb{C}$-isomorphism $f : V_1 \to V_2$ satisfying (2.1.1).

Example 1. A character χ of G (1.6) can be viewed as a representation $\chi : G \to \mathbb{C}^\times = \mathrm{Aut}_{\mathbb{C}}(\mathbb{C})$. The representation $(\chi, \mathbb{C})$ is smooth. A one-dimensional representation of G is smooth if and only if it is equivalent to a representation defined by a character of G. Indeed, the set of isomorphism classes of one-dimensional smooth representations of G is in canonical bijection with the group of characters of G.

Example 2. Suppose that G is *compact,* hence profinite. Let (π, V) be an irreducible smooth representation of G. The space V is then *finite-dimensional.* For, if $v \in V$, $v \neq 0$, then $v \in V^K$, for an open subgroup K of G. The index $(G : K)$ is necessarily finite, and the set $\{\pi(g)v : g \in G/K\}$ spans V. Further, if $K' = \bigcap_{g \in G/K} gKg^{-1}$, then K' is an open normal subgroup of G of finite index, acting trivially on V. Thus V is effectively an irreducible representation of the finite discrete group G/K'.

In this more general context, one can still define the group ring $\mathbb{C}[G]$ as the algebra of finite formal linear combinations of elements of G. A smooth representation V of G is then a $\mathbb{C}[G]$-module. However, an arbitrary $\mathbb{C}[G]$-module need not provide a smooth representation of G, and so the group ring is not an effective tool for analyzing smooth representations. For this purpose, one has to replace $\mathbb{C}[G]$ by a different algebra. We discuss this in §4 below.

2.2. We recall a standard concept in the present context.

Proposition. *Let G be a locally profinite group, and let (π, V) be a smooth representation of G. The following conditions are equivalent:*

(1) *V is the sum of its irreducible G-subspaces;*
(2) *V is the direct sum of a family of irreducible G-subspaces;*
(3) *any G-subspace of V has a G-complement in V.*

Proof. We start with the implication (1) $\Rightarrow$ (2). We take a family $\{U_i : i \in I\}$ of irreducible G-subspaces U_i of V such that $V = \sum_{i \in I} U_i$. We consider the set $\mathcal{I}$ of subsets J of I such that the sum $\sum_{i \in J} U_i$ is direct. The set $\mathcal{I}$ is nonempty; we show it is inductively ordered by inclusion. For, suppose we have a totally ordered set $\{J_a : a \in A\}$ of elements of $\mathcal{I}$. Put $J = \bigcup_{a \in A} J_a$. If the sum $\sum_{j \in J} U_j$ is not direct, there is a finite subset S of J for which $\sum_{j \in S} U_j$ is not direct. Since S must be contained in some J_a, this is impossible. Therefore $J \in \mathcal{I}$. We can now apply Zorn's Lemma to get a maximal element J_0 of $\mathcal{I}$. For this set, we have

$$V = \bigoplus_{a \in J_0} U_i,$$

as required for (2).

In (3), let W be a G-subspace of V. By (2), we can assume that $V = \bigoplus_{i \in I} U_i$, for a family (U_i) of irreducible G-subspaces of V. We consider the set $\mathcal{J}$ of subsets J of I for which $W \cap \sum_{i \in J} U_i = 0$. Again, the set $\mathcal{J}$ is nonempty and inductively ordered by inclusion. If J is a maximal element of $\mathcal{J}$, the sum $X = W + \bigoplus_{j \in J} U_j$ is direct. If $X \neq V$, there is $i \in I$ with $U_i \not\subset X$, so the sum $X + U_i$ is direct, and $J \cup \{i\} \in \mathcal{J}$, contrary to hypothesis. Thus (2) $\Rightarrow$ (3).

Suppose now that (3) holds. Let V_0 be the sum of all irreducible G-subspaces of V and write $V = V_0 \oplus W$, for some G-subspace W of V.

Assume for a contradiction that $W \neq 0$. By its definition, the space W has no irreducible G-subspace. However, there is a non-zero G-subspace W_1 of W which is finitely generated over G. By Zorn's Lemma, W_1 has a maximal G-subspace W_0, and then W_1/W_0 is irreducible. We have $V = V_0 \oplus W_0 \oplus U$, for some G-subspace U of V, and hence a G-projection $V \to U$. The image of W_1 in U is an irreducible G-subspace of U, hence an irreducible subspace of V which is not contained in V_0. This is nonsense, so $V = V_0$ and (3) $\Rightarrow$ (1). □

One says that (π, V) is *G-semisimple* if it satisfies the conditions of the proposition. Interesting locally profinite groups G usually have many representations which are not semisimple. However, the property can be employed in a slightly different context:

Lemma. *Let G be a locally profinite group, and let K be a compact open subgroup of G. Let (π, V) be a smooth representation of G. The space V is the sum of its irreducible K-subspaces.*

Proof. Let $v \in V$. As in 2.1 Example 2, v is fixed by an open normal subgroup K' of K, and it generates a finite-dimensional K-space W on which K' acts trivially. Thus W is effectively a finite-dimensional representation of the finite group K/K' and so is the sum of its irreducible K-subspaces. Since $v \in V$ was chosen at random, the lemma follows. □

The lemma says that V is *K-semisimple.*

2.3. Again let G be a locally profinite group and K a compact open subgroup of G.

Let $\widehat{K}$ denote the set of equivalence classes of irreducible smooth representations of K. If $\rho \in \widehat{K}$ and (π, V) is a smooth representation of G, we define V^ρ to be the sum of all irreducible K-subspaces of V of class ρ. We call V^ρ the *ρ-isotypic component of V*. In particular, V^K is the isotypic subspace for the class of the trivial representation 1 of K.

Proposition. *Let (π, V) be a smooth representation of G and let K be a compact open subgroup of G.*

(1) *The space V is the direct sum of its K-isotypic components:*

$$V = \bigoplus_{\rho \in \widehat{K}} V^\rho.$$

(2) *Let (σ, W) be a smooth representation of G. For any G-homomorphism $f : V \to W$ and $\rho \in \widehat{K}$, we have*

$$f(V^\rho) \subset W^\rho \quad \textit{and} \quad W^\rho \cap f(V) = f(V^\rho).$$

Proof. We use 2.2 to write

$$V = \bigoplus_{i \in I} U_i,$$

for a family of irreducible K-subspaces U_i of V. We let $U(\rho)$ be the sum of those U_i of class ρ. We then have

$$V = \bigoplus_{\rho \in \widehat{K}} U(\rho).$$

If W is an irreducible K-subspace of V of class ρ, then $W \subset U(\rho)$: otherwise, there would be a non-zero K-homomorphism $W \to U_i$, for some U_i of some class $\tau \neq \rho$. We deduce that $V^\rho = U(\rho)$ and (1) follows.

In (2), the image of V^ρ is a sum of irreducible K-subspaces of W, all of class ρ and therefore contained in W^ρ. Moreover, $f(V)$ is the sum of the images $f(V^\tau)$, $\tau \in \widehat{K}$, and $f(V^\tau) \subset W^\tau$. Since the sum of the W^τ is direct, $f(V)$ is the direct sum of the $f(V^\tau)$ and the second assertion follows. □

We frequently use part (2) of the Proposition in the following context:

Corollary 1. *Let $a : U \to V$, $b : V \to W$ be G-homomorphisms between smooth representations U, V, W of G. The sequence*

$$U \xrightarrow{a} V \xrightarrow{b} W$$

is exact if and only if

$$U^K \xrightarrow{a} V^K \xrightarrow{b} W^K$$

is exact, for every compact open subgroup K of G.

If H is a subgroup of G, we define

$$V(H) = \text{the linear span of } \{v - \pi(h)v : v \in V,\ h \in H\}. \tag{2.3.1}$$

In particular, $V(H)$ is an H-subspace of V.

Corollary 2. *Let G be a locally profinite group, and let (π, V) be a smooth representation of G. Let K be a compact open subgroup of G. Then*

$$V(K) = \bigoplus_{\substack{\rho \in \widehat{K}, \\ \rho \neq 1}} V^\rho, \qquad V = V^K \oplus V(K),$$

and $V(K)$ is the unique K-complement of V^K in V.

Proof. The sum $W = \bigoplus V^\rho$, with ρ not trivial, is a K-complement of V^K in V. There is therefore a K-surjection $V \to V^K$ with kernel W. Clearly, $V(K)$ is contained in the kernel of any K-homomorphism $V \to V^K$; we conclude that W contains $V(K)$. On the other hand, if U is an irreducible K-space of class $\rho \neq 1$, then $U(K) = U$, so $V^\rho \subset V(K)$. □

Exercises.

(1) *Let* (π, V) *be an* abstract *(i.e., not necessarily smooth) representation of* G. *Define*

$$V^\infty = \bigcup_K V^K,$$

where K *ranges over the compact open subgroups of* G. *Show that* V^∞ *is a* G*-stable subspace of* V. *Define a homomorphism*

$$\pi^\infty : G \longrightarrow \mathrm{Aut}_{\mathbb{C}}(V^\infty)$$

by $\pi^\infty(g) = \pi(g) \mid V^\infty$. *Show that* (π^∞, V^∞) *is a smooth representation of* G.

(2) *Let* (π, V) *be a smooth representation of* G *and* (σ, W) *an abstract representation. Let* $f : V \to W$ *be a* G*-homomorphism. Show that* $f(V) \subset W^\infty$, *and hence* $\mathrm{Hom}_G(V, W) = \mathrm{Hom}_G(V, W^\infty)$.

(3) *Let*

$$0 \to U \xrightarrow{a} V \xrightarrow{b} W \to 0$$

be an exact sequence of G*-homomorphisms of abstract* G*-spaces. Show that the induced sequence*

$$0 \to U^\infty \xrightarrow{a} V^\infty \xrightarrow{b} W^\infty$$

is exact. Show by example that the map $b : V^\infty \to W^\infty$ *need not be surjective.*

2.4. We now consider the notion of an *induced representation.*

Let G be a locally profinite group, and let H be a closed subgroup of G. Thus H is also locally profinite.

Let (σ, W) be a smooth representation of H. We consider the space X of functions $f : G \to W$ which satisfy

(1) $f(hg) = \sigma(h)f(g)$, $h \in H$, $g \in G$;
(2) *there is a compact open subgroup* K *of* G *(depending on* f*) such that* $f(gx) = f(g)$, *for* $g \in G$, $x \in K$.

We define a homomorphism $\Sigma : G \to \mathrm{Aut}_{\mathbb{C}}(X)$ by

$$\Sigma(g)f : x \longmapsto f(xg), \quad g, x \in G.$$

The pair (Σ, X) provides a smooth representation of G. It is called the representation of G *smoothly induced* by σ, and is usually denoted

$$(\Sigma, X) = \mathrm{Ind}_H^G \sigma.$$

The map $\sigma \mapsto \mathrm{Ind}_H^G \sigma$ gives a functor $\mathrm{Rep}(H) \to \mathrm{Rep}(G)$.

There is a canonical H-homomorphism

$$\begin{aligned} \alpha_\sigma : \mathrm{Ind}_H^G \sigma &\longrightarrow W, \\ f &\longmapsto f(1). \end{aligned}$$

The pair $(\mathrm{Ind}_H^G, \alpha)$ has the following fundamental property:

Frobenius Reciprocity. *Let H be a closed subgroup of a locally profinite group G. For a smooth representation (σ, W) of H and a smooth representation (π, V) of G, the canonical map*

$$\begin{aligned} \mathrm{Hom}_G(\pi, \mathrm{Ind}_H^G \sigma) &\longrightarrow \mathrm{Hom}_H(\pi, \sigma), \\ \phi &\longmapsto \alpha_\sigma \circ \phi, \end{aligned}$$

is an isomorphism that is functorial in both variables π, σ.

Proof. Let $f : V \to W$ be an H-homomorphism. We define a G-homomorphism $f_\star : V \to \mathrm{Ind}\,\sigma$ by letting $f_\star(v)$ be the function $g \mapsto f(\pi(g)v)$. The map $f \mapsto f_\star$ is then the inverse of (2.4.2). □

A simple, but very useful, consequence is that $\alpha_\sigma(V) \neq 0$, for any non-zero G-subspace V of $\mathrm{Ind}_H^G \sigma$.

We will also need a less formal property:

Proposition. *The functor* $\mathrm{Ind}_H^G : \mathrm{Rep}(H) \to \mathrm{Rep}(G)$ *is additive and exact.*

Proof. For a smooth representation (σ, W) of H, temporarily let $I(\sigma)$ denote the space of functions $G \to W$ satisfying the first condition $f(hg) = \sigma(h)f(g)$ of the definition above. Thus I is a functor to the category of abstract representations of G; it is clearly additive and exact, while $\mathrm{Ind}_H^G(\sigma) = I(\sigma)^\infty$. Thus Ind_H^G is surely additive, and 2.3 Exercise (3) shows it to be left-exact.

To prove it is right-exact, let (σ, W), (τ, U) be smooth representations of H and let $f : W \to U$ be an H-surjection. Take $\phi \in I(\tau)^\infty$, and choose a compact open subgroup K of G which fixes ϕ. The support of ϕ is a union of cosets HgK, and the value $\phi(g) \in U$ must be fixed by $\tau(H \cap gKg^{-1})$. By 2.3 Corollary 1 (applied to the group H and the trivial representation of its compact open subgroup $H \cap gKg^{-1}$), there exists $w_g \in W$, fixed by $\sigma(H \cap gKg^{-1})$, such that $f(w_g) = \phi(g)$. We define a function $\Phi : G \to W$ to have the same support as ϕ and $\Phi(hgk) = \sigma(h)w_g$, for each $g \in H\backslash \operatorname{supp} \phi/K$. The function Φ is fixed by K, and hence lies in $I(\sigma)^\infty$. Its image in $I(\tau)^\infty$ is ϕ, as required. □

2.5. There is a variation on this theme. With (σ, W) and X as in 2.4, consider the space X_c of functions $f \in X$ which are *compactly supported modulo* H: this means that the image of the support $\operatorname{supp} f$ of f in $H\backslash G$ is compact or, equivalently, $\operatorname{supp} f \subset HC$, for some compact set C in G. The space X_c is stable under the action of G and provides another smooth representation of G. It is denoted $c\text{-}\mathrm{Ind}_H^G \sigma$, and gives a functor

$$c\text{-}\mathrm{Ind}_H^G : \mathrm{Rep}(H) \longrightarrow \mathrm{Rep}(G).$$

One calls it *compact induction,* or *smooth induction with compact supports.*

Exercise 1. *Show that the functor* $c\text{-}\mathrm{Ind}_H^G$ *is additive and exact.*

In all cases, there is a canonical G-embedding $c\text{-}\mathrm{Ind}_H^G \sigma \to \mathrm{Ind}_H^G \sigma$. Put another way, there is a morphism of functors $c\text{-}\mathrm{Ind}_H^G \to \mathrm{Ind}_H^G$. This is an isomorphism if and only if $H\backslash G$ is compact. On the other hand, for specific G, H, σ, the map $c\text{-}\mathrm{Ind}_H^G \sigma \to \mathrm{Ind}_H^G \sigma$ can be an isomorphism even when $H\backslash G$ is not compact. Significant examples of this phenomenon arise in 11.4 below.

This construction is mainly (but not exclusively) of interest when the subgroup H is *open* in G. In this case, there is a canonical H-homomorphism

$$\begin{aligned} \alpha_\sigma^c : W &\longrightarrow c\text{-}\mathrm{Ind}\, \sigma, \\ w &\longmapsto f_w, \end{aligned} \tag{2.5.1}$$

where $f_w \in X_c$ is supported in H and $f_w(h) = \sigma(h)w$, $h \in H$.

Exercise 2. *Suppose H is open in G. Let $\phi : G \to W$ be a function, compactly supported modulo H, such that $\phi(hg) = \sigma(h)\phi(g)$, $h \in H$, $g \in G$. Show that $\phi \in X_c$.*

Lemma. *Let H be an open subgroup of G, and let (σ, W) be a smooth representation of H.*

(1) *The map $\alpha_\sigma^c : w \mapsto f_w$ is an H-isomorphism of W with the space of functions $f \in c\text{-}\mathrm{Ind}_H^G \sigma$ such that $\operatorname{supp} f \subset H$.*
(2) *Let $\mathcal{W}$ be a $\mathbb{C}$-basis of W and $\mathcal{G}$ a set of representatives for G/H. The set $\{gf_w : w \in \mathcal{W}, g \in \mathcal{G}\}$ is a $\mathbb{C}$-basis of $c\text{-}\mathrm{Ind}\, \sigma$.*

Proof. In (1), surely α_σ^c is an H-homomorphism to the space of functions supported in H; the inverse map is $f \mapsto f(1)$.

The support of a function $f \in c\text{-Ind}_H^G \sigma$ is a finite union of cosets Hg^{-1}, for various $g \in \mathcal{G}$, and the restriction of f to any one of these also lies in $c\text{-Ind}\,\sigma$. If $\operatorname{supp} f = Hg^{-1}$, then $g^{-1}f$ has support contained in H, and so is a finite linear combination of functions f_w, $w \in \mathcal{W}$. Clearly, the set of functions gf_w, $w \in \mathcal{W}$, $g \in \mathcal{G}$, is linearly independent, and the proof is complete. □

For open subgroups, compact induction has its own form of Frobenius Reciprocity property:

Proposition. *Let H be an open subgroup of G, let (σ, W) be a smooth representation of H and (π, V) a smooth representation of G. The canonical map*

$$\begin{aligned} \mathrm{Hom}_G(c\text{-Ind}\,\sigma, \pi) &\longrightarrow \mathrm{Hom}_H(\sigma, \pi), \\ f &\longmapsto f \circ \alpha_\sigma^c, \end{aligned}$$

is an isomorphism which is functorial in both variables.

Proof. Let ϕ be an H-homomorphism $W \to V$. There is a unique G-homomorphism $\phi_* : c\text{-Ind}\,\sigma \to V$ such that $\phi_*(f_w) = \phi(w)$, $w \in W$. The map $\phi \mapsto \phi_*$ is then inverse to (2.5.2). □

Remark. Suppose for the moment that G is *finite.* The coincident definitions of induction above are then equivalent to the standard one: this is easily proved directly. Alternatively, the version (2.5.2) of Frobenius Reciprocity is the same as the usual one for finite groups and one can use uniqueness of adjoint functors.

2.6. It is convenient to introduce a technical restriction on the group G. *From now on, we assume that:*

Hypothesis. *For any compact open subgroup K, the set G/K is countable.*

We remark that, if G/K is countable for *one* compact open subgroup K of G, then G/K' is countable for *any* compact open subgroup K' of G. For, $K \cap K'$ is compact, open, and of finite index in K. Thus the surjection $G/(K\cap K') \to G/K$ has finite fibres and $G/(K \cap K')$, hence also G/K', is countable.

Certain of the things we do in this section do not require the property, but it holds for every concrete group in which we shall be interested.

The main effect of the hypothesis is:

Lemma. *Let (π, V) be an irreducible smooth representation of G. The dimension $\dim_{\mathbb{C}} V$ is countable.*

Proof. Let $v \in V$, $v \neq 0$, and choose a compact open subgroup K of G such that $v \in V^K$. Since V is irreducible, the countable set $\{\pi(g)v : g \in G/K\}$ spans V. $\square$

This enables us to generalize a familiar result, as follows.

Schur's Lemma. *If (π, V) is an irreducible smooth representation of G, then* $\mathrm{End}_G(V) = \mathbb{C}$.

Proof. Let $\phi \in \mathrm{End}_G(V)$, $\phi \neq 0$. The image and the kernel of ϕ are both G-subspaces of V, so ϕ is bijective and invertible. Therefore $\mathrm{End}_G(V)$ is a complex division algebra.

If we fix $v \in V$, $v \neq 0$, the G-translates of v span V so an element $\phi \in \mathrm{End}_G(V)$ is determined uniquely by the value $\phi(v)$. We deduce that $\mathrm{End}_G(V)$ has countable dimension. However, any $\phi \in \mathrm{End}_G(V)$, $\phi \notin \mathbb{C}$, is transcendental over $\mathbb{C}$, and generates a field $\mathbb{C}(\phi) \subset \mathrm{End}_G(V)$. The subset $\{(\phi - a)^{-1} : a \in \mathbb{C}\}$ of $\mathbb{C}(\phi)$ is linearly independent over $\mathbb{C}$, so the $\mathbb{C}$-dimension of $\mathbb{C}(\phi)$ is uncountable, and this is impossible. The only conclusion is that $\mathrm{End}_G(V) = \mathbb{C}$, as required. $\square$

Corollary 1. *Let (π, V) be an irreducible smooth representation of G. The centre Z of G then acts on V via a character $\omega_\pi : Z \to \mathbb{C}^\times$, that is, $\pi(z)v = \omega_\pi(z)v$, for $v \in V$ and $z \in Z$.*

Proof. By Schur's Lemma, there is surely a homomorphism $\omega_\pi : Z \to \mathbb{C}^\times$ such that $\pi(z)v = \omega_\pi(z)v$, $z \in Z$, $v \in V$. If K is a compact open subgroup of G such that $V^K \neq 0$, then ω_π is trivial on the compact open subgroup $K \cap Z$ of Z. Thus ω_π is a character of Z. $\square$

One calls ω_π the *central character* of π.

Corollary 2. *If G is abelian, any irreducible smooth representation of G is one-dimensional.*

Remark. If G is compact, the converse of Schur's Lemma holds: a smooth representation (π, V) of G is a direct sum of irreducible representations, so $\mathrm{End}_G(V)$ is one-dimensional if and only if π is irreducible. This is *false* for smooth representations of locally profinite groups in general: see 9.10 below for a significant example.

2.7. We will sometimes need a more general version of the preceding machinery. Before dealing with this, however, it is convenient to interpolate a general lemma.

Lemma. *Let G be a locally profinite group, and let H be an open subgroup of G of finite index.*

(1) *If* (π, V) *is a smooth representation of* G*, then* V *is* G*-semisimple if and only if it is* H*-semisimple.*
(2) *Let* (σ, W) *be a semisimple smooth representation of* H*. The induced representation* $\operatorname{Ind}_H^G \sigma$ *is* G*-semisimple.*

Proof. Suppose that V is H-semisimple, and let U be a G-subspace of V. By hypothesis, there is an H-subspace W of V such that $V = U \oplus W$. Let $f : V \to U$ be the H-projection with kernel W. Consider the map

$$f^G : v \longmapsto (G{:}H)^{-1} \sum_{g \in G/H} \pi(g)\, f(\pi(g)^{-1}v), \quad v \in V.$$

The definition is independent of the choice of coset representatives and it follows that f^G is a G-projection $V \to U$. We then have $V = U \oplus \operatorname{Ker} f^G$ and $\operatorname{Ker} f^G$ is a G-subspace of V. Thus V is G-semisimple (*cf.* 2.2 Proposition).

Conversely, suppose that V is G-semisimple. Thus G is a direct sum of irreducible G-subspaces (2.2), and it is enough to treat the case where V is irreducible over G. As representation of H, the space V is finitely generated and so admits an irreducible H-quotient U. Suppose for the moment that H is a *normal* subgroup of G. By Frobenius Reciprocity (2.4.2), the H-map $V \to U$ gives a non-trivial, hence injective, G-map $V \to \operatorname{Ind}_H^G U$. As representation of H, the induced representation $\operatorname{Ind}_H^G U = c\text{-}\operatorname{Ind}_H^G U$ is a direct sum of G-conjugates of U (*cf.* 2.5 Lemma). These are all irreducible over H, so $\operatorname{Ind} U$ is H-semisimple. Proposition 2.2 then implies that $V \subset \operatorname{Ind} U$ is H-semisimple.

In general, we set $H_0 = \bigcap_{g \in G/H} gHg^{-1}$. This is an open normal subgroup of G of finite index. We have just shown that the G-space V is H_0-semisimple; the first part of the proof shows it is H-semisimple.

This completes the proof of (1), and (2) follows readily from the same arguments. □

We first apply this in the following context. Let Z be the centre of G, and fix a character χ of Z. We consider the class of smooth representations (π, V) of G which *admit* χ *as central character,* that is,

$$\pi(z)v = \chi(z)v, \quad v \in V,\ z \in Z.$$

Proposition. *Let* (π, V) *be a smooth representation of* G*, admitting* χ *as a central character. Let* K *be an open subgroup of* G *such that* KZ/Z *is compact.*

(1) *Let* $v \in V$*. The* KZ*-space spanned by* v *is of finite dimension, and is a sum of irreducible* KZ*-spaces.*
(2) *As representation of* KZ*, the space* V *is semisimple.*

Proof. The vector v is fixed by a compact open subgroup K_0 of K. The set KZ/K_0Z is finite, so the space W spanned by $\pi(KZ)v$ has finite dimension. Surely W is K_0Z-semisimple; the lemma implies it is KZ-semisimple. Since v was chosen arbitrarily, assertion (2) follows. □

In practice, the open subgroup K will contain Z, with K/Z compact. The discussion is equally valid if Z is a closed subgroup of the centre $Z(G)$ of G.

2.8. The notion of *duality,* for smooth representations of a locally profinite group, is both more subtle and more significant than it is for representations of finite groups. We examine it in some detail.

Let (π, V) be a smooth representation of the locally profinite group G. Write $V^* = \mathrm{Hom}_{\mathbb{C}}(V, \mathbb{C})$, and denote by

$$\begin{aligned} V^* \times V &\longrightarrow \mathbb{C}, \\ (v^*, v) &\longmapsto \langle v^*, v\rangle, \end{aligned}$$

the canonical evaluation pairing. The space V^* carries a representation π^* of G defined by

$$\langle \pi^*(g)v^*, v\rangle = \langle v^*, \pi(g^{-1})v\rangle, \quad g \in G,\ v^* \in V^*,\ v \in V.$$

This is not, in general, smooth. We accordingly define

$$\check{V} = (V^*)^\infty = \bigcup_K (V^*)^K,$$

where K ranges over the compact open subgroups of G. Thus (*cf.* 2.3 Exercise (1)) $\check{V}$ is a G-stable subspace of V^*, and provides a smooth representation

$$\check{\pi} = (\pi^*)^\infty : G \longrightarrow \mathrm{Aut}_{\mathbb{C}}(\check{V}).$$

The representation $(\check{\pi}, \check{V})$ is called the *contragredient,* or *smooth dual,* of (π, V). We continue to denote the evaluation pairing $\check{V} \times V \to \mathbb{C}$ by $(\check{v}, v) \mapsto \langle \check{v}, v\rangle$. Therefore

$$\langle \check{\pi}(g)\check{v}, v\rangle = \langle \check{v}, \pi(g^{-1})v\rangle, \quad g \in G,\ \check{v} \in \check{V},\ v \in V. \tag{2.8.1}$$

Let K be a compact open subgroup of G. We recall that V^K has a unique K-complement $V(K)$ in V (2.3). If $\check{v} \in \check{V}$ is fixed under K, we have $\langle \check{v}, V(K)\rangle = 0$, by the definition of $V(K)$. Thus $\check{v} \in \check{V}^K$ is determined by its effect on V^K.

Proposition. *Restriction to* V^K *induces an isomorphism* $\check{V}^K \cong (V^K)^*$.

Proof. One can extend a linear functional on V^K to an element of $\check{V}^K$ by deeming that it be trivial on $V(K)$. □

Corollary. *Let (π, V) be a smooth representation of G, and let $v \in V$, $v \neq 0$. There exists $\check{v} \in \check{V}$ such that $\langle \check{v}, v \rangle \neq 0$.*

Remark. Let (π, V) be a smooth representation of G. Although undeclared in the notation, the subspace $\check{V}$ of V^* does depend on G, in the following sense. Let H be a closed subgroup of G, and let $\widetilde{V}$ denote the space of H-smooth vectors in V^*. Certainly $\check{V} \subset \widetilde{V}$, but it is easy to produce examples where $\widetilde{V} \neq \check{V}$.

2.9. Let (π, V) be a smooth representation of the locally profinite group G. We can form the smooth dual $(\check{\check{\pi}}, \check{\check{V}})$ of $(\check{\pi}, \check{V})$. There is a canonical G-map $\delta : V \to \check{\check{V}}$ given (in the obvious notation) by

$$\langle \delta(v), \check{v} \rangle_{\check{V}} = \langle \check{v}, v \rangle_V, \quad v \in V, \ \check{v} \in \check{V}.$$

It is injective (2.8 Corollary).

Proposition. *Let (π, V) be a smooth representation of a locally profinite group G. The canonical map $\delta : V \to \check{\check{V}}$ is an isomorphism if and only if π is admissible.*

Proof. The map δ induces a map $\delta^K : V^K \to \check{\check{V}}^K$ for each compact open subgroup K of G, and δ is surjective if and only if δ^K is surjective for all K (2.3 Corollary). However, δ^K is the canonical map $V^K \to (V^K)^{**}$, which is surjective if and only if $\dim V^K < \infty$. □

2.10. Let (π, V), (σ, W) be smooth representations of G, and let $f : V \to W$ be a G-map. We define a map $\check{f} : \check{W} \to \check{V}$ by the relation

$$\langle \check{f}(\check{w}), v \rangle = \langle \check{w}, f(v) \rangle, \quad \check{w} \in \check{W}, \ v \in V.$$

The map $\check{f}$ is a G-homomorphism, and $(\pi, V) \mapsto (\check{\pi}, \check{V})$ gives a contravariant functor of $\mathrm{Rep}(G)$ to itself.

Lemma. *The contravariant functor*

$$\begin{aligned} \mathrm{Rep}(G) &\longrightarrow \mathrm{Rep}(G), \\ (\pi, V) &\longmapsto (\check{\pi}, \check{V}), \end{aligned}$$

is exact.

Proof. If we have an exact sequence of smooth representations (π_i, V_i) of G:

$$0 \to V_1 \longrightarrow V_2 \longrightarrow V_3 \to 0,$$

the sequence

$$0 \to V_1^K \longrightarrow V_2^K \longrightarrow V_3^K \to 0$$

is exact (2.3). The sequence of dual spaces $(V_i^K)^*$ is then exact, that is,

$$0 \to \check{V}_1^K \longrightarrow \check{V}_2^K \longrightarrow \check{V}_3^K \to 0$$

is exact. The result now follows from 2.3 Corollary 1. □

We deduce:

Proposition. *Let (π, V) be an admissible representation of G. Then (π, V) is irreducible if and only if $(\check{\pi}, \check{V})$ is irreducible.*

Remark. For certain locally profinite groups G, there do exist examples of irreducible smooth representations π for which $\check{\pi}$ is not irreducible: see 8.2 Remark for an example.

Exercise. *Let (π, V), (σ, W) be smooth representations of G. Let $\wp(\pi, \sigma)$ be the space of G-invariant bilinear pairings $V \times W \to \mathbb{C}$. Show that there are canonical isomorphisms*

$$\mathrm{Hom}_G(\pi, \check{\sigma}) \cong \wp(\pi, \sigma) \cong \mathrm{Hom}_G(\sigma, \check{\pi}).$$

3. Measures and Duality

We now give a quick, but essentially complete, account of some invariant measures on a locally profinite group and associated homogeneous spaces. Locally profinite groups are such that their measure theory is effectively algebraic in nature, and can be treated as an episode in their representation theory.

3.1. Let G be a locally profinite group. Let $C_c^\infty(G)$ be the space of functions $f : G \to \mathbb{C}$ which are locally constant and of compact support.

Let $f \in C_c^\infty(G)$. Local constancy and compactness of support together imply that there exist compact open subgroups K_1, K_2 of G such that $f(k_1 g) = f(g) = f(g k_2)$, for all $g \in G$, $k_i \in K_i$. Taking $K = K_1 \cap K_2$, one sees that f is a finite linear combination of characteristic functions of double cosets KgK.

The group G acts on $C_c^\infty(G)$ by *left translation* λ and by *right translation* ρ:

$$\begin{aligned} \lambda_g f &: x \longmapsto f(g^{-1}x), \\ \rho_g f &: x \longmapsto f(xg), \end{aligned} \qquad x, g \in G,\ f \in C_c^\infty(G). \tag{3.1.1}$$

Both of the G-representations $(C_c^\infty(G), \lambda)$, $(C_c^\infty(G), \rho)$ are smooth.

Definition. *A* right Haar integral *on G is a non-zero linear functional*

$$I : C_c^\infty(G) \longrightarrow \mathbb{C}$$

such that

(1) $I(\rho_g f) = I(f)$, $g \in G$, $f \in C_c^\infty(G)$, *and*
(2) $I(f) \geqslant 0$ *for any* $f \in C_c^\infty(g)$, $f \geqslant 0$.

One defines a *left Haar integral* similarly, using left translation λ instead of right translation. We now show that G possesses a right Haar integral, and essentially only one.

Proposition. *There exists a right Haar integral $I : C_c^\infty(G) \to \mathbb{C}$. Moreover, a linear functional $I' : C_c^\infty(G) \to \mathbb{C}$ is a right Haar integral if and only if $I' = cI$, for some constant $c > 0$.*

Proof. Let K be a compact open subgroup of G; we denote by ${}^K C_c^\infty(G)$ the space of functions in $C_c^\infty(G)$ fixed by $\lambda(K)$. We view ${}^K C_c^\infty(G)$ as G-space via right translation. It is then identical to the representation of G compactly induced from the trivial representation 1_K of K:

$${}^K C_c^\infty(G) = c\text{-}\mathrm{Ind}_K^G \, 1_K.$$

Lemma. *Viewing $\mathbb{C}$ as the trivial G-space, we have*

$$\dim_{\mathbb{C}} \mathrm{Hom}_G({}^K C_c^\infty(G), \mathbb{C}) = 1.$$

There exists a non-zero element $I_K \in \mathrm{Hom}_G({}^K C_c^\infty(G), \mathbb{C})$ such that $I_K(f) \geqslant 0$ whenever $f \geqslant 0$. If h_K is the characteristic function of K, then $I_K(h_K) > 0$.

Proof. The first assertion is given by 2.5 Proposition. For $g \in G$, let f_g denote the characteristic function of Kg. The set of functions f_g, $g \in K\backslash G$, then forms a $\mathbb{C}$-basis of the space ${}^K C_c^\infty(G)$ (2.5 Lemma). The functional $I_K : f_g \mapsto 1$ has the required properties, noting that $h_K = f_1$. □

We choose a descending sequence $\{K_n\}_{n \geqslant 1}$ of compact open subgroups K_n of G such that $\bigcap_n K_n = 1$. We then have

$$C_c^\infty(G) = \bigcup_{n \geqslant 1} {}^{K_n} C_c^\infty(G).$$

For each $n \geqslant 1$, there is a unique right G-invariant functional I_n on ${}^{K_n} C_c^\infty(G)$ which maps the characteristic function of K_n to $(K_1 : K_n)^{-1}$. We have $I_{n+1} \mid {}^{K_n} C_c^\infty(G) = I_n$, and so the family $\{I_n\}$ gives a functional on $C_c^\infty(G)$ of the required kind. The uniqueness statement is immediate. □

Remark. The lemma also implies that, if we view $C_c^\infty(G)$ as a smooth representation of G under right translation, then

$$\dim \mathrm{Hom}_G(C_c^\infty(G), \mathbb{C}) = 1. \tag{3.1.2}$$

The functional I of the Proposition is a right Haar integral on G. One can produce a left Haar integral in exactly the same way. Alternatively, one can proceed as follows.

Corollary. *For $f \in C_c^\infty(G)$, define $\check{f} \in C_c^\infty(G)$ by $\check{f}(g) = f(g^{-1})$, $g \in G$. The functional*

$$I' : C_c^\infty(G) \longrightarrow \mathbb{C},$$
$$I'(f) = I(\check{f}),$$

is a left Haar integral on G. Moreover, any left Haar integral on G is of the form cI', for some $c > 0$.

The uniqueness statement follows from the proposition on observing that, if J is a left Haar integral, then $f \mapsto J(\check{f})$ is a right Haar integral.

Let I be a *left* Haar integral on G. Let $S \neq \emptyset$ be a compact open subset of G and let Γ_S be its characteristic function. We define

$$\mu_G(S) = I(\Gamma_S).$$

Then $\mu_G(S) > 0$ and the measure μ_G satisfies $\mu_G(gS) = \mu_G(S)$, $g \in G$. One refers to μ_G as a *left Haar measure* on G. The relation with the integral is expressed via the traditional notation

$$I(f) = \int_G f(g)\, d\mu_G(g), \quad f \in C_c^\infty(G).$$

Further traditional abbreviations are frequently permitted, in particular,

$$\int_G \Gamma_S(x) f(x)\, d\mu_G(x) = \int_S f(x)\, d\mu_G(x).$$

Definition. *The group G is* unimodular *if any left Haar integral on G is a right Haar integral.*

3.2. One can extend the domain of Haar integration, much as in the classical case of the Lebesgue measure. We outline a few examples of what we will need.

First, one can integrate more general functions. For example, let f be a function on G invariant under left translation by a compact open subgroup

K of G. Let μ_G be a left Haar measure on G. If the series

$$\sum_{g\in K\backslash G} \int_{Kg} |f(x)|\, d\mu_G(x)$$

converges, so does the series without the absolute value, and we put

$$\int_G f(x)\, d\mu_G(x) = \sum_{g\in K\backslash G} \int_{Kg} f(x)\, d\mu_G(x).$$

The result does not depend on the choice of K, and this extended Haar integral has the translation invariance property of the original.

Next, let G_1, G_2 be locally profinite groups, and set $G = G_1 \times G_2$. Then G is locally profinite. An element $\sum_{1\leqslant i\leqslant r} f_i^1 \otimes f_i^2$ of the tensor product $C_c^\infty(G_1)\otimes C_c^\infty(G_2)$ gives a function on G by

$$\Phi(g_1, g_2) = \sum_i f_i^1(g_1) f_i^2(g_2).$$

Then $\Phi \in C_c^\infty(G)$, and this process gives an isomorphism $C_c^\infty(G_1)\otimes C_c^\infty(G_2) \to C_c^\infty(G)$. Let μ_j be a left Haar measure on G_j, $j = 1, 2$. There is then a unique left Haar measure μ_G on G such that

$$\int_G f_1 \otimes f_2(g)\, d\mu_G(g) = \int_{G_1} f_1(g_1)\, d\mu_1(g_1) \int_{G_2} f_2(g_2)\, d\mu_2(g_2),$$

for $f_i \in C_c^\infty(G_i)$. One writes $\mu_G = \mu_1 \otimes \mu_2$.

For a general $f \in C_c^\infty(G)$, the function

$$f_1(g_1) = \int_{G_2} f(g_1, g_2)\, d\mu_2(g_2)$$

lies in $C_c^\infty(G_1)$. We have

$$\int_G f(g)\, d\mu_G(g) = \int_{G_1} f_1(g_1)\, d\mu_1(g_1),$$

and symmetrically: if f is of the form $f_1\otimes f_2$, this is obvious, and such functions span $C_c^\infty(G)$.

Next, let V be a complex vector space, and consider the space $C_c^\infty(G; V)$ of locally constant, compactly supported functions $f : G \to V$. This space is isomorphic to $C_c^\infty(G) \otimes V$ in the obvious way: a tensor $\sum_i f_i \otimes v_i$ gives the function

$$g \longmapsto \sum_i f_i(g) v_i.$$

If μ_G is a left Haar measure on G, there is a unique linear map $I_V : C_c^\infty(G; V) \to V$ such that

$$I_V(f \otimes v) = \int_G f(g)\, d\mu_G(g) \cdot v.$$

We write

$$I_V(\phi) = \int_G \phi(g)\, d\mu_G(g), \quad \phi \in C_c^\infty(G; V).$$

This has the same invariance properties as the Haar integral on scalar-valued functions.

3.3. Let μ_G be a left Haar measure on G. For $g \in G$, consider the functional

$$\begin{aligned} C_c^\infty(G) &\longrightarrow \mathbb{C}, \\ f &\longmapsto \int_G f(xg)\, d\mu_G(x). \end{aligned}$$

This is a left Haar integral on G, so there is a unique $\delta_G(g) \in \mathbb{R}_+^\times$ such that

$$\delta_G(g) \int_G f(xg)\, d\mu_G(x) = \int_G f(x)\, d\mu_G(x),$$

for all $f \in C_c^\infty(G)$. The function δ_G is a homomorphism $G \to \mathbb{R}_+^\times$. It is trivial if G is abelian; more generally, δ_G is trivial if and only if G is unimodular.

Taking f to be the characteristic function of a compact open subgroup K of G, we see that δ_G is trivial on K and hence a character of G. One calls δ_G the *module* of G.

The functional

$$f \longmapsto \int_G \delta_G(x)^{-1} f(x)\, d\mu_G(x), \quad f \in C_c^\infty(G),$$

is a right Haar integal on G.

The mnemonic

$$d\mu_G(xg) = \delta_G(g)\, d\mu_G(x)$$

may be found helpful.

Remark. We have already observed that δ_G is trivial on any compact open subgroup of G. In particular, if G is compact, then $\delta_G = 1$ and G is unimodular. In the general case, *any* character $G \to \mathbb{R}_+^\times$ is trivial on compact subgroups, since $\mathbb{R}_+^\times$ has only the trivial compact subgroup.

3.4 Let H be a closed subgroup of G, with module δ_H. Let $\theta : H \to \mathbb{C}^\times$ be a character of H. We consider the space of functions $f : G \to \mathbb{C}$ which are G-smooth under right translation, compactly supported modulo H, and satisfy

$$f(hg) = \theta(h)f(g), \quad h \in H, \ g \in G.$$

We call this space $C_c^\infty(H\backslash G, \theta)$, and view it as a smooth G-space via right translation ρ. (Note that $C_c^\infty(H\backslash G, \theta) = c\text{-Ind}_H^G\,\theta$, but this characterization is not helpful.)

Proposition. *Let $\theta : H \to \mathbb{C}^\times$ be a character of H. The following are equivalent:*

(1) *There exists a non-zero linear functional $I_\theta : C_c^\infty(H\backslash G, \theta) \to \mathbb{C}$ such that $I_\theta(\rho_g f) = I_\theta(f)$, for all $g \in G$.*
(2) $\theta\delta_H = \delta_G \mid H$.

When these conditions hold, the functional I_θ is uniquely determined up to constant factor.

Proof. Let μ_G, μ_H be left Haar measures on G, H respectively. We define a G-map $C_c^\infty(G) \to C_c^\infty(H\backslash G, \theta)$, denoted $f \mapsto \widetilde{f}$, by

$$\widetilde{f}(g) = \int_H \theta\delta_H(h)^{-1} f(hg)\, d\mu_H(h).$$

This map satisfies

$$\widetilde{\lambda_k f} = \delta_H\theta(k)^{-1}\widetilde{f},$$

for $k \in H$ and $f \in C_c^\infty(G)$. We prove it is surjective.

If K is a compact open subgroup of G, the space $C_c^\infty(G)^K$ is spanned by the characteristic functions of cosets gK, $g \in G/K$. On the other hand, each coset HgK supports, at most, a one-dimensional space of functions in $C_c^\infty(H\backslash G, \theta)^K$. These subspaces span $C_c^\infty(H\backslash G, \theta)^K$, so the map $f \mapsto \widetilde{f}$ is surjective on K-fixed functions. It is therefore surjective.

Suppose that the space $C_c^\infty(H\backslash G, \theta)$ admits a functional I_θ of the required kind. The map $f \mapsto I_\theta(\widetilde{f})$ is then a non-trivial G-homomorphism

$$(C_c^\infty(G), \rho) \longrightarrow \mathbb{C}.$$

However, the space $\mathrm{Hom}_G(C_c^\infty(G), \mathbb{C})$ has dimension 1 (3.1.2), and is spanned by a right Haar integral. The condition (1) holds, therefore, if and only if the right Haar integral $C_c^\infty(G) \to \mathbb{C}$ factors through the quotient map $C_c^\infty(G) \to C_c^\infty(H\backslash G, \theta)$. When it does hold, I_θ is determined up to a constant factor.

The kernel of the map $f \mapsto \widetilde{f}$ contains all functions $\lambda_h f - \delta_H\theta(h)^{-1}f$, for $h \in H$ and $f \in C_c^\infty(G)$. Applying the right Haar integral on G to such functions, we get

$$\int_G (\lambda_h f(g) - \delta_H\theta(h^{-1})\ f(g))\delta_G(g)^{-1}\,d\mu_G(g)$$
$$= \big(\delta_G(h^{-1}) - \delta_H\theta(h^{-1})\big)\int_G f(g)\delta_G(g)^{-1}\,d\mu_G(g).$$

This vanishes identically if and only if $\delta_G(h^{-1}) = \delta_H\theta(h^{-1})$, for all $h \in H$. Thus (1) $\Rightarrow$ (2).

For the converse, we take a function $f \in C_c^\infty(G)$ such that $\widetilde{f} = 0$. The function f is fixed by a compact open subgroup K, and it is enough to treat the case where $\operatorname{supp} f \subset HgK$, for some $g \in K$. Thus f is a finite linear combination of the characteristic function of cosets h_igK, $h_i \in H$. The condition $\widetilde{f} = 0$ then amounts to

$$\mu_H(H \cap gKg^{-1}) \sum_i \theta\delta_H(h_i)^{-1} f(h_i g) = 0,$$

since $\theta\delta_H$ is trivial on the compact subgroup $H \cap gKg^{-1}$ of H. On the other hand,

$$\int_G f(x)\delta_G(x)^{-1}\,d\mu_G(x) = \mu_G(K)\delta_G(g)^{-1} \sum_i \delta_G(h_i)^{-1} f(h_i g).$$

If (2) holds, this surely vanishes as required. □

When the conditions of the proposition hold, the character θ takes only positive real values. Let $f \in C_c^\infty(G)$ satisfy $f(g) \geqslant 0$, for all $g \in G$; we then have $\widetilde{f}(g) \geqslant 0$ for all g. Consequently:

Corollary. *Suppose that the conditions of the proposition hold. There is then a non-zero linear functional I_θ on $C_c^\infty(H\backslash G, \theta)$ such that:*

(1) $I_\theta(\rho_g f) = I_\theta(f)$, for $f \in C_c^\infty(H\backslash G, \theta)$, $g \in G$;
(2) $I_\theta(f) \geqslant 0$, for $f \in C_c^\infty(H\backslash G, \theta)$, $f \geqslant 0$.

These conditions determine I_θ uniquely, up to a positive constant factor.

One habitually uses a notation like

$$I_\theta(f) = \int_{H\backslash G} f(g)\,d\mu_{H\backslash G}(g), \quad f \in C_c^\infty(H\backslash G, \theta),$$

and calls $\mu_{H\backslash G}$ a positive semi-invariant measure on $H\backslash G$. (Since $\theta = \delta_H^{-1}\delta_G \mid H$ is uniquely determined, there is no real need to refer to it again.)

3.5. Let G be a locally profinite group and H a closed subgroup of G. Put

$$\delta_{H\backslash G} = \delta_H^{-1}\delta_G \mid H : H \longrightarrow \mathbb{R}_+^\times.$$

Duality Theorem. *Let $\dot\mu$ be a positive semi-invariant measure on $H\backslash G$. Let (σ, W) be a smooth representation of H. There is a natural isomorphism*

$$\big(c\text{-}\mathrm{Ind}_H^G\,\sigma\big)^\vee \cong \mathrm{Ind}_H^G\,\delta_{H\backslash G}\otimes\check\sigma,$$

depending only on the choice of $\dot\mu$.

Proof. We view $\delta_{H\backslash G}\otimes\check\sigma$ as acting on the same space $\check W$ as $\check\sigma$. Let $(\check w, w)\mapsto \langle \check w, w\rangle$ be the evaluation pairing $\check W\times W\to\mathbb{C}$. Let $\phi\in c\text{-}\mathrm{Ind}\,\sigma$, $\Phi\in\mathrm{Ind}\,\delta_{H\backslash G}\otimes\check\sigma$, and consider the function

$$f : g\longmapsto \langle\Phi(g),\phi(g)\rangle,\quad g\in G.$$

This lies in $C_c^\infty(H\backslash G, \delta_{H\backslash G})$, so we have a G-invariant pairing

$$\mathrm{Ind}\,\delta_{H\backslash G}\otimes\check\sigma\times c\text{-}\mathrm{Ind}\,\sigma\longrightarrow\mathbb{C},$$
$$(\Phi,\phi)\longmapsto\int_{H\backslash G}\langle\Phi(g),\phi(g)\rangle\,d\dot\mu(g).$$

This induces a G-homomorphism $\mathrm{Ind}\,\delta_{H\backslash G}\otimes\check\sigma\to\big(c\text{-}\mathrm{Ind}\,\sigma\big)^\vee$. It is clearly natural in σ; we show it is an isomorphism.

To do this, we take a compact open subgroup K of G and describe the space $(c\text{-}\mathrm{Ind}\,\sigma)^K$ of K-fixed functions in $c\text{-}\mathrm{Ind}\,\sigma$. The support of such a function, f say, is a finite union of cosets HgK. The value $f(g)\in W$ is then fixed by the compact open subgroup $H\cap gKg^{-1}$ of H. We choose a set $\mathcal{G}$ of representatives for the coset space $H\backslash G/K$. For each $g\in\mathcal{G}$, we choose a basis $\mathcal{W}_g$ of the space $W^{H\cap gKg^{-1}}$. For each $g\in\mathcal{G}$ and each $w\in\mathcal{W}_g$, there is a unique function $f_{g,w}\in(c\text{-}\mathrm{Ind}\,\sigma)^K$, supported on HgK, such that $f_{g,w}(g)=w$. The set

$$\{f_{g,w} : g\in\mathcal{G}, w\in\mathcal{W}_g\}$$

is then a basis of $(c\text{-}\mathrm{Ind}\,\sigma)^K$. That is:

Lemma 1. *The space $(c\text{-}\mathrm{Ind}\,\sigma)^K$ consists of the finite linear combinations of functions $f_{g,w}$, $g\in\mathcal{G}$, $w\in\mathcal{W}_g$.*

Let $\mathcal{W}_g^*$ denote the basis of $\check W_g = \mathrm{Hom}_{\mathbb{C}}(W^{H\cap gKg^{-1}},\mathbb{C})$ dual to $\mathcal{W}_g$. For $g\in\mathcal{G}$, $\check w\in\mathcal{W}_g^*$, we define $f_{g,\check w}\in(\mathrm{Ind}\,\delta_{H\backslash G}\otimes\check\sigma)^K$ in the same way as before. We then have:

Lemma 2. *The space $(\mathrm{Ind}\,\delta_{H\backslash G}\otimes\check\sigma)^K$ consists of all functions f such that, for any $g\in\mathcal{G}$, the restriction $f\mid HgK$ is a finite linear combination of functions $f_{g,\check w}$, $\check w\in\mathcal{W}_g^*$.*

For $g_1, g_2 \in \mathcal{G}$, $w \in \mathcal{W}_{g_1}$, $\check{w} \in \mathcal{W}^*_{g_2}$, we have

$$\langle f_{g_2,\check{w}}, f_{g_1,w}\rangle = \begin{cases} \dot{\mu}(Hg_1K)\langle \check{w}, w\rangle & \text{if } Hg_1K = Hg_2K, \\ 0 \quad \text{otherwise,} \end{cases}$$

for some $\dot{\mu}(Hg_1K) > 0$. The pairing thus identifies $(\operatorname{Ind} \delta_{H\backslash G} \otimes \check{\sigma})^K$ with the linear dual of $(c\text{-}\operatorname{Ind} \sigma)^K$, and hence $\operatorname{Ind} \delta_{H\backslash G} \otimes \check{\sigma}$ with $(c\text{-}\operatorname{Ind} \sigma)^\vee$, as required. □

4. The Hecke Algebra

If G is a finite group, the concept of a representation of G is essentially identical to that of a module over the group algebra $\mathbb{C}[G]$. This relation extends to smooth representations of a locally profinite group, but only with a suitable definition of 'group algebra'.

4.1. To avoid pointless technical complications, we impose the following:

Hypothesis. *Unless otherwise stated, we assume that G is unimodular.*

We fix a Haar measure μ on G. For $f_1, f_2 \in C_c^\infty(G)$, we define

$$f_1 * f_2(g) = \int_G f_1(x) f_2(x^{-1}g)\, d\mu(x).$$

The function $(x, g) \mapsto f_1(x) f_2(x^{-1}g)$ lies in $C_c^\infty(G \times G)$ so (*cf.* 3.2) $f_1 * f_2 \in C_c^\infty(G)$. Similarly, for $f_i \in C_c^\infty(G)$, the integral expressing $f_1 * (f_2 * f_3)(g)$ is that of a function from $C_c^\infty(G \times G)$, so we can manipulate formally:

$$\begin{aligned} f_1 * (f_2 * f_3)(g) &= \iint f_1(x) f_2(y) f_3(y^{-1}x^{-1}g)\, d\mu(y) d\mu(x) \\ &= \iint f_1(x) f_2(x^{-1}y) f_3(y^{-1}g)\, d\mu(y) d\mu(x) \\ &= \iint f_1(x) f_2(x^{-1}y) f_3(y^{-1}g)\, d\mu(x) d\mu(y) \\ &= (f_1 * f_2) * f_3(g). \end{aligned}$$

The binary operation $*$, called *convolution*, is thus associative. The pair

$$\mathcal{H}(G) = (C_c^\infty(G), *)$$

is an associative $\mathbb{C}$-algebra called the *Hecke algebra* of G.

In general, $\mathcal{H}(G)$ has no unit element; it is commutative if G is commutative.

Remark 1. The algebra structure on $\mathcal{H}(G)$, that is, the convolution operation, clearly depends on the choice of Haar measure μ. However, suppose we have two Haar measures μ, ν, giving rise to algebra structures $\mathcal{H}_\mu(G)$, $\mathcal{H}_\nu(G)$ on $C_c^\infty(G)$. There is a constant $c > 0$ such that $\nu = c\mu$. The map $f \mapsto c^{-1}f$ is then an algebra isomorphism $\mathcal{H}_\mu(G) \to \mathcal{H}_\nu(G)$.

Remark 2. Suppose that G is *discrete.* For Haar integral, we may take

$$\int_G f(g)\, d\mu(g) = \sum_{g\in G} f(g).$$

The map

$$f \longmapsto \sum_{g\in G} f(g)g$$

is then an algebra isomorphism of $\mathcal{H}(G)$ with the group algebra $\mathbb{C}[G]$. Observe that in this case (and only this case — exercise) $\mathcal{H}(G)$ has a unit element.

While, in general, $\mathcal{H}(G)$ has no unit element, it does have a copious supply of *idempotent* elements. For example, let K be a compact open subgroup of G, and define a function $e_K \in \mathcal{H}(G)$ by

$$e_K(x) = \begin{cases} \mu(K)^{-1} & \text{if } x \in K, \\ 0 & \text{if } x \notin K. \end{cases} \tag{4.1.1}$$

Proposition.

(1) *The function e_K satisfies $e_K * e_K = e_K$.*
(2) *A function $f \in \mathcal{H}(G)$ satisfies $e_K * f = f$ if and only if $f(kg) = f(g)$, for all $k \in K$, $g \in G$.*
(3) *The space $e_K * \mathcal{H}(G) * e_K$ is a sub-algebra of $\mathcal{H}(G)$, with unit element e_K.*

Proof. First consider the integral

$$e_K * e_K(g) = \int_G e_K(x)\, e_K(x^{-1}g)\, d\mu(x).$$

If $g \notin K$, the integrand is identically zero. If $g \in K$, it vanishes for $x \notin K$ while, for $x \in K$, it takes the value $\mu(K)^{-2}$. Integrating, we get the first part.

Taking $f \in \mathcal{H}(G)$, $k \in K$, $g \in G$, we have

$$\begin{aligned} e_K * f(kg) &= \int_G e_K(x) f(x^{-1}kg)\, d\mu(x) \\ &= \int_G e_K(kx) f(x^{-1}g)\, d\mu(x) \\ &= \int_G e_K(x) f(x^{-1}g)\, d\mu(x) = e_K * f(g). \end{aligned}$$

If f is left K-invariant, the last integral reduces to $f(g)$, while the function $e_K * f$ is visibly left K-invariant.

Part (3) is now obvious. □

We note that $e_K * \mathcal{H}(G) * e_K$ is the space of $f \in \mathcal{H}(G)$ satisfying $f(k_1 g k_2) = f(g)$, for $g \in G$, $k_1, k_2 \in K$. We often write $\mathcal{H}(G, K) = e_K * \mathcal{H}(G) * e_K$.

4.2. Let M be a left $\mathcal{H}(G)$-module: it will be convenient to denote the module action by $(f, m) \mapsto f * m$, for $f \in \mathcal{H}(G)$, $m \in M$. We say that M is *smooth* if $\mathcal{H}(G) * M = M$. Since $\mathcal{H}(G)$ is the union of its sub-algebras $e_K * \mathcal{H}(G) * e_K$, the module M is smooth if and only if, for every $m \in M$, there is a compact open subgroup K of G such that $e_K * m = m$.

Let M_1, M_2 be smooth $\mathcal{H}(G)$-modules; we define $\mathrm{Hom}_{\mathcal{H}(G)}(M_1 M_2)$ to be the space of all $\mathcal{H}(G)$-homomorphisms $M_1 \to M_2$. With this definition, the class of smooth $\mathcal{H}(G)$-modules forms a category, which we denote $\mathcal{H}(G)$-Mod. Let (π, V) be a smooth representation of G. For $f \in \mathcal{H}(G)$, $v \in V$, we set

$$\pi(f)v = \int_G f(g)\pi(g)v \, d\mu(g). \tag{4.2.1}$$

The integrand lies in $C_c^\infty(G; V)$ (notation of 3.2), so the integral (4.2.1) defines an element of V. Alternatively, we can choose a compact open subgroup K of G which fixes v and f (under right translation); we may then interpret the integral as the finite sum

$$\pi(f)v = \mu(K) \sum_{g \in G/K} f(g)\pi(g)v. \tag{4.2.2}$$

Immediately from (4.2.2), we get

$$\pi(e_K)v = v, \quad v \in V^K.$$

Proposition 1. *Let (π, V) be a smooth representation of G. The operation*

$$(f, v) \longmapsto \pi(f)v, \quad v \in V, \ f \in \mathcal{H}(G),$$

gives V the structure of a smooth $\mathcal{H}(G)$-module. If (π', V') is another smooth representation of G and $\phi : V \to V'$ is a G-homomorphism, then ϕ is also an $\mathcal{H}(G)$-homomorphism:

$$\phi \circ \pi(f) = \pi'(f) \circ \phi, \quad f \in \mathcal{H}(G). \tag{4.2.3}$$

Proof. Let $v \in V$, $f_1, f_2 \in \mathcal{H}(G)$; as in 4.1, we can compute formally:

$$\begin{aligned}
\pi(f_1 * f_2)v &= \int_G f_1 * f_2(g)\pi(g)v\, d\mu(g) \\
&= \iint f_1(h) f_2(h^{-1}g)\pi(g)v\, d\mu(h)d\mu(g) \\
&= \iint f_1(h) f_2(h^{-1}g)\pi(g)v\, d\mu(g)d\mu(h) \\
&= \iint f_1(h) f_2(g)\pi(hg)v\, d\mu(g)d\mu(h) \\
&= \int f_1(h)\pi(h) \int f_2(g)\pi(g)v\, d\mu(g)\, d\mu(h) \\
&= \int f_1(h)\pi(h)\, \pi(f_2)v\, d\mu(h) \\
&= \pi(f_1)\pi(f_2)v.
\end{aligned}$$

So, in $\mathrm{End}_{\mathbb{C}}(V)$, we have $\pi(f_1 * f_2) = \pi(f_1)\,\pi(f_2)$. Thus $f \mapsto \pi(f)$ is a homomorphism of $\mathbb{C}$-algebras $\mathcal{H}(G) \to \mathrm{End}_{\mathbb{C}}(V)$, and gives V the structure of an $\mathcal{H}(G)$-module. If K is a compact open subgroup of G, we have $\pi(e_K)v = v$, $v \in V^K$, so V has become a smooth $\mathcal{H}(G)$-module. The relation (4.2.3) follows from (4.2.2). □

Example. Take $(\pi, V) = (\lambda, \mathcal{H}(G))$ (notation of (3.1.1)). For $\phi, f \in \mathcal{H}(G)$, $\lambda(\phi)f = \phi * f$. If, on the other hand, $(\pi, V) = (\mathcal{H}(G), \rho)$, then $\rho(\phi)f = f * \check{\phi}$, where $\check{\phi}$ is the function $g \mapsto \phi(g^{-1})$.

One can work the process in the other direction. Let M be a smooth $\mathcal{H}(G)$-module, and denote the action of $\mathcal{H}(G)$ on M by

$$(f, m) \longmapsto f * m, \quad f \in \mathcal{H}(G),\ m \in M.$$

With this notation:

Proposition 2. *There is a unique G-homomorphism $\pi : G \to \mathrm{Aut}_{\mathbb{C}}(M)$ such that (π, M) is a smooth representation of G and*

$$\pi(f)m = f * m, \quad f \in \mathcal{H}(G),\ m \in M. \tag{4.2.4}$$

Moreover, if M' is a smooth $\mathcal{H}(G)$-module with associated G-representation (π', M'), then any $\mathcal{H}(G)$-homomorphism $M \to M'$ is a G-homomorphism $\pi \to \pi'$.

Proof. Consider the canonical map

$$\mathcal{H}(G) \otimes_{\mathcal{H}(G)} M \longrightarrow M. \tag{4.2.5}$$

This is surjective, by the definition of smoothness. Let $\sum_{1\leqslant i\leqslant r} f_i \otimes m_i$ lie in the kernel. We choose a compact open subgroup K of G which fixes each f_i on either side and such that $m_i \in e_K * M$ for all i. In particular, $e_K * m_i = m_i$ for all i. We then have

$$\sum_{i=1}^{r} f_i \otimes m_i = e_K \otimes \sum_i f_i * m_i = 0,$$

so the map (4.2.5) is injective.

Thus (4.2.5) is an $\mathcal{H}(G)$-isomorphism $\mathcal{H}(G) \otimes_{\mathcal{H}(G)} M \cong M$. The tensor product carries a smooth representation of G, via left translation on the first factor. The isomorphism transfers the G-action to M, to give a smooth representation (π, M), say.

One can write this action more explicitly. Take $m \in M$, and choose a compact open subgroup K of G such that $e_K * m = m$. For $g \in G$, let f denote the characteristic function of gK. We then have

$$\pi(g)m = \mu(K)^{-1} f * m. \tag{4.2.6}$$

The property (4.2.4) now follows from the example above. The final assertion is clear from (4.2.6). □

The concepts "smooth representation of G" and "smooth $\mathcal{H}(G)$-module" are thus identical and interchangeable. For example, if (π, V) is a smooth representation of G, the G-subspaces of V are just the $\mathcal{H}(G)$-submodules of V, the G-homomorphisms $V \to V'$ are the $\mathcal{H}(G)$-homomorphisms $V \to V'$, and so on. In particular, the categories $\mathrm{Rep}(G)$, $\mathcal{H}(G)$-Mod are equivalent (but this statement is weaker than what we have just proved).

4.3. There are some associated algebra structures of interest. We fix a compact open subgroup K of G, and define $e_K \in \mathcal{H}(G)$ as in (4.1.1). We consider the sub-algebra $\mathcal{H}(G,K) = e_K * \mathcal{H}(G) * e_K$ of $\mathcal{H}(G)$. Since e_K is idempotent, this algebra has a unit element, namely e_K. Using the notation of 2.3, we have:

Lemma. *Let (π, V) be a smooth representation of G. The operator $\pi(e_K)$ is the K-projection $V \to V^K$ with kernel $V(K)$. The space V^K is an $\mathcal{H}(G,K)$-module on which e_K acts as the identity.*

Proof. For $v \in V$ and $k \in K$, we have (by a simple calculation)

$$\pi(k)\,\pi(e_K)\,v = \pi(e_K)\,\pi(k)\,v = \pi(e_K)\,v.$$

Thus $\pi(e_K)$ is a K-map $V \to V^K$. As $\pi(e_K)v = v$ for $v \in V^K$, the image of $\pi(e_K)$ is V^K. Since e_K is idempotent, so is $\pi(e_K)$. Certainly $\pi(e_K)$ annihilates the unique K-complement $V(K)$ of V^K, so $\pi(e_K)$ is the unique K-projection $V \to V^K$. □

The $\mathcal{H}(G,K)$-module V^K encapsulates significant information about the representation (π, V).

Proposition.

(1) *Let (π, V) be an irreducible smooth representation of G. The space V^K is either zero or a simple $\mathcal{H}(G,K)$-module.*

(2) *The process $(\pi, V) \mapsto V^K$ induces a bijection between the following sets of objects:*

  (a) *equivalence classes of irreducible smooth representations (π, V) of G such that $V^K \neq 0$;*

  (b) *isomorphism classes of simple $\mathcal{H}(G,K)$-modules.*

Proof. Let (π, V) be an irreducible smooth representation of G; suppose that $V^K \neq 0$, and let M be a non-zero $\mathcal{H}(G,K)$-submodule of V^K. The space $\pi(\mathcal{H}(G))M$ is a non-zero G-subspace of V, hence equal to V itself. We therefore have

$$V^K = \pi(e_K)V = \pi(e_K)\pi(\mathcal{H}(G))M = \pi(\mathcal{H}(G,K))M = M.$$

This proves (1).

So, $V \mapsto V^K$ gives a map from equivalence classes of irreducible representations (π, V) with $V^K \neq 0$ to isomorphism classes of simple $\mathcal{H}(G,K)$-modules. In the opposite direction, let M be a simple $\mathcal{H}(G,K)$-module and consider the G-space $U = \mathcal{H}(G) \otimes_{\mathcal{H}(G,K)} M$, where G acts on the first tensor factor by left translation. The space U satisfies

$$U^K = e_K * \mathcal{H}(G) \otimes_{\mathcal{H}(G,K)} M = e_K \otimes M \cong M.$$

Zorn's Lemma gives the existence of a G-subspace X of U which is maximal for the property $X \cap e_K \otimes M = 0$ or, equivalently, $X^K = 0$. Given any other G-subspace Y of U with $Y^K = 0$, we have $(X+Y)^K = X^K + Y^K = 0$. So, by the maximality of X, we have $Y \subset X$ and X is the unique G-subspace maximal for the property $X^K = X \cap e_K \otimes M = 0$.

A G-subspace W of U strictly containing X must meet, and therefore contain, the simple $\mathcal{H}(G,K)$-module $e_K \otimes M$, with the result that $W = U$. Thus X is a maximal G-subspace of U and $V = U/X$ is irreducible. It satisfies $V^K \cong M$ as $\mathcal{H}(G,K)$-module.

It remains to show that the isomorphism class of V depends only on that of M. Let $f : M \to M'$ be an isomorphism of $\mathcal{H}(G,K)$-modules. The map f extends uniquely to an $\mathcal{H}(G)$-isomorphism

$$\begin{aligned} f : U = \mathcal{H}(G) \otimes_{\mathcal{H}(G,K)} M &\longrightarrow \mathcal{H}(G) \otimes_{\mathcal{H}(G,K)} M' = U', \\ \phi \otimes m &\longmapsto \phi \otimes f(m). \end{aligned}$$

The image $X' = f(X)$ is then the unique maximal $\mathcal{H}(G)$-subspace of U' such that $X'^K = 0$. The map f therefore induces an isomorphism $U/X \cong U'/X'$, as required. □

Corollary. *Let (π, V) be a smooth representation of G, $V \neq 0$; then (π, V) is irreducible if and only if, for any compact open subgroup K of G, the space V^K is either zero or $\mathcal{H}(G,K)$-simple.*

Proof. One implication is given by the proposition. Conversely, suppose (π, V) is not irreducible, and let $U \subsetneq V$ be a non-zero G-subspace. Set $W = V/U$. There exists a compact open subgroup K of G such that both spaces U^K, W^K are non-zero. The sequence

$$0 \to U^K \longrightarrow V^K \longrightarrow W^K \to 0$$

is exact, and is an exact sequence of $\mathcal{H}(G,K)$-modules. Thus V^K is non-zero and not simple over $\mathcal{H}(G,K)$. □

4.4. There is a more general version of 4.3 which we outline. We start with a compact open subgroup K of G and $\rho \in \widehat{K}$. We consider the function $e_\rho \in \mathcal{H}(G)$, with support contained in K, and given by

$$e_\rho(x) = \frac{\dim \rho}{\mu(K)} \operatorname{tr} \rho(x^{-1}), \quad x \in K. \tag{4.4.1}$$

If K' is the kernel of ρ, then e_ρ is constant on cosets gK' and $K'g'$, so $e_{K'} * e_\rho = e_\rho * e_{K'} = e_\rho$. Thus e_ρ lies in the subalgebra $\mathcal{H}(K,K')$ of $\mathcal{H}(G,K')$. However, $e_{K'} \mapsto 1$ induces an algebra isomorphism $\mathcal{H}(K,K') \to \mathbb{C}[K/K']$. This isomorphism takes e_ρ to the idempotent of the group algebra corresponding to the irreducible representation ρ of the finite group K/K'. Therefore:

Proposition.

(1) *The function $e_\rho \in \mathcal{H}(G)$ is idempotent.*
(2) *If (π, V) is a smooth representation of G, then $\pi(e_\rho)$ is the K-projection $V \to V^\rho$.*

In particular, $V^\rho = \pi(e_\rho)V$, and the isotypic space V^ρ is a module over the subalgebra $e_\rho * \mathcal{H}(G) * e_\rho$ of $\mathcal{H}(G)$. The exact analogue of 4.3 Proposition holds, with e_ρ replacing e_K and V^ρ replacing V^K.

Example. Suppose that $\dim \rho = 1$, that is, ρ is a character of K. The algebra $e_\rho * \mathcal{H}(G) * e_\rho$ is then the space of $f \in \mathcal{H}(G)$ such that $f(kgk') = \rho(kk')^{-1} f(g)$, $k, k' \in K$, $g \in G$.

4.5. If G is a finite group, a key property of the group algebra $\mathbb{C}[G]$ is that it is *semisimple*. Since $\mathbb{C}[G]$ has finite dimension over $\mathbb{C}$, this is equivalent to the Jacobson radical of $\mathbb{C}[G]$ being trivial or, again, to the property that if $x \in \mathbb{C}[G]$, $x \neq 0$, then there exists a simple $\mathbb{C}[G]$-module M such that $xM \neq 0$.

For a locally profinite group G satisfying 2.6 Hypothesis, there is a corresponding property for the Hecke algebra $\mathcal{H}(G)$. Although we will not use this result, we include a proof to give a satisfying completeness to the generalization.

Separation property. *Let $f \in \mathcal{H}(G)$, $f \neq 0$. There exists an irreducible smooth representation (π, V) of G such that $\pi(f) \neq 0$.*

Proof. We shall need:

Lemma. *Let $\mathcal{A}$ be a $\mathbb{C}$-algebra with 1, of countable dimension. Let $a \in \mathcal{A}$ be non-nilpotent.*

(1) *There exists $\lambda \in \mathbb{C}^\times$ such that $a-\lambda 1$ is not invertible in $\mathcal{A}$.*
(2) *There exists a simple $\mathcal{A}$-module M such that $aM \neq 0$.*

Proof. Suppose, for a contradiction, that $a-\mu 1 \in \mathcal{A}^\times$, for all $\mu \in \mathbb{C}^\times$. In particular, $a \neq \mu 1$, for any $\mu \in \mathbb{C}^\times$. The set $\{(a-\mu 1)^{-1} : \mu \in \mathbb{C}\}$ is uncountable, and so linearly dependent over $\mathbb{C}$. A non-trivial linear dependence relation

$$\sum_{i=1}^{n} \alpha_i \, (a-\mu_i 1)^{-1} = 0$$

yields a non-trivial polynomial $\phi(t) \in \mathbb{C}[t]$ such that

$$\phi(a) = \prod_{j=1}^{r} (a-\lambda_j 1) = 0,$$

for some $r \geqslant 1$ and certain $\lambda_j \in \mathbb{C}$. Not all λ_j are zero, since a is not nilpotent. If $\lambda_j \neq 0$, then $b = a-\lambda_j 1$ is a divisor of zero in $\mathcal{A}$ and so not invertible. Consequently, $\mathcal{A}b \neq \mathcal{A}$ and there exists a maximal left ideal $\mathcal{I}$ of $\mathcal{A}$ such that $b \in \mathcal{I}$. The $\mathcal{A}$-module $\mathcal{A}/\mathcal{I}$ is then simple, and $a(1+\mathcal{I}) = \lambda_j + \mathcal{I} \neq 0$. □

Let $f \in \mathcal{H}(G)$, $f \neq 0$. Define $f^* \in \mathcal{H}(G)$ by $f^*(g) = \overline{f(g^{-1})}$, where the bar denotes complex conjugation. We set $h = f^* * f$. The function h satisfies $h^* = h$ and $h(1_G) > 0$. In particular, $h \neq 0$. Applying the same argument to h, we have $h*h = h^* * h \neq 0$. Inductively, we conclude that h is not nilpotent. There exists a compact open subgroup K of G such that $h \in \mathcal{H}(G, K)$, and $\mathcal{H}(G, K)$ is a $\mathbb{C}$-algebra with 1, of countable dimension. By the lemma, there is a simple $\mathcal{H}(G, K)$-module M such that $hM \neq 0$. By 4.3 Proposition, there is an irreducible smooth representation (π, V) of G such that V^K is $\mathcal{H}(G, K)$-isomorphic to M. We have $\pi(h) \neq 0$, and so $\pi(f) \neq 0$, as required. □

Further reading. For foundational matters concerning locally profinite groups, or locally compact, totally disconnected groups, see [63]. The various topologies associated with local fields are discussed systematically in [87]. Measure theory is available in many texts but, particularly as regards Haar measure, [84] remains a classic.

2

Finite Fields

In this brief chapter, we work out the irreducible representations of the group $\mathrm{GL}_2(\boldsymbol{k})$ of invertible 2×2 matrices over a finite field $\boldsymbol{k}$. This anticipates some of the phenomena which arise in the representation theory of $\mathrm{GL}_2(F)$, where F is a non-Archimedean local field. Indeed, if $\boldsymbol{k}$ is the residue field of the non-Archimedean local field F, the representation theory of $\mathrm{GL}_2(\boldsymbol{k})$ actually governs a part of that of $\mathrm{GL}_2(F)$: see 11.5 below for a first example of this.

5. Linear Groups

In this section only, F denotes an *arbitrary* field. We briefly recall some basic facts about the group $G = \mathrm{GL}_2(F)$ of 2×2 invertible matrices over F. The aim is to fix some notation and terminology for use in later sections.

We give no proofs: these are at the level of elementary linear algebra.

5.1. The group G possesses some particularly important algebraic subgroups:

$$\begin{gathered} B = \{\left(\begin{smallmatrix} a & b \\ 0 & c\end{smallmatrix}\right) \in G\}, \quad N = \{\left(\begin{smallmatrix} 1 & b \\ 0 & 1\end{smallmatrix}\right) \in G\}. \\ T = \{\left(\begin{smallmatrix} a & 0 \\ 0 & b\end{smallmatrix}\right) \in G\}. \end{gathered} \tag{5.1.1}$$

The group B is called the standard Borel subgroup of $G = \mathrm{GL}_2(F)$, and N is the unipotent radical of B. The group T is the standard split maximal torus in G. We have a semi-direct product decomposition $B = T \ltimes N$. We also put

$$Z = \{\left(\begin{smallmatrix} a & 0 \\ 0 & a\end{smallmatrix}\right) \in G\}.$$

Thus Z is the centre of G, and Z is canonically isomorphic to $F^\times$.

5.2. In the notation (5.1.1), elementary methods of linear algebra give the *Bruhat decomposition*

$$G = B \cup BwB,$$

where w denotes the permutation matrix

$$w = \begin{pmatrix} 0 & 1 \\ 1 & 0 \end{pmatrix}.$$

That is, $\{1, w\}$ is a set of representatives for the coset space $B\backslash G/B$.

The coset BwB consists of the matrices with non-zero $(2,1)$-entry. Since $B = NT = TN$ and w normalizes T, we have $BwB = NwB = BwN$. Moreover, the map

$$B \times N \longrightarrow BwN,$$
$$(b, n) \longmapsto bwn,$$

is bijective.

5.3. We recall the conjugacy class structure of the group G. For $g \in G$, let $\mathrm{ch}_g(t) = \det(tI - g)$ denote the characteristic polynomial of g. Thus $\mathrm{ch}_g(t)$ is a monic quadratic polynomial with coefficients in F, and $\mathrm{ch}_g(0) = \det g \neq 0$.

Proposition. *Let $g \in G$ and set $f(t) = \mathrm{ch}_g(t)$.*

(1) *Suppose that $f(t)$ is irreducible over F. The sub-algebra $F[g]$ of $A = \mathrm{M}_2(F)$ is a field and the G-centralizer of g is $F[g]^\times$. If $f(t) = t^2 + at + b$, then g is G-conjugate to the matrix*

$$\begin{pmatrix} 0 & -b \\ 1 & -a \end{pmatrix}.$$

An element $h \in G$ is G-conjugate to g if and only if $\mathrm{ch}_h(t) = f(t)$.

(2) *Suppose $f(t)$ has distinct roots $a, b \in F^\times$. Then g is G-conjugate to the matrix*

$$\begin{pmatrix} a & 0 \\ 0 & b \end{pmatrix}.$$

The G-centralizer of g is $F[g]^\times$.

(3) *Suppose that $f(t)$ has a repeated root $a \in F^\times$. Then g is conjugate to exactly one of the matrices*

$$\begin{pmatrix} a & 0 \\ 0 & a \end{pmatrix}, \quad \begin{pmatrix} a & 1 \\ 0 & a \end{pmatrix}.$$

In the first case, g lies in the centre of G. In the second, the G-centralizer of g is $F[g]^\times$.

6. Representations of Finite Linear Groups

In this section, $\boldsymbol{k}$ denotes a finite field with q elements. We classify the irreducible (complex) representations of the finite group $G = \mathrm{GL}_2(\boldsymbol{k})$.

We use the notation B, N, T, Z , as in §5.

6.1. The group G has order $(q^2-1)(q^2-q)$. Further:

Lemma. *The group G has exactly q^2-1 conjugacy classes.*

Proof. The field $\boldsymbol{k}$ has exactly $(q^2-q)/2$ irreducible, monic polynomials of degree 2. Thus there are $(q^2-q)/2$ conjugacy classes as in 5.3 Proposition (1). The second case of 5.3 Proposition gives $(q-1)(q-2)/2$ classes, while the third case gives $2(q-1)$. The lemma follows immediately. □

Thus G has precisely q^2-1 irreducible representations, up to isomorphism.

6.2. The group N of upper triangular unipotent matrices in G is isomorphic to the additive group of $\boldsymbol{k}$, via the map $x \mapsto \left(\begin{smallmatrix} 1 & x \\ 0 & 1 \end{smallmatrix}\right)$.

If we fix a non-trivial character ψ of $\boldsymbol{k}$, the function $a\psi : x \mapsto \psi(ax)$, $x \in \boldsymbol{k}$, ranges over the characters of $\boldsymbol{k}$ as a ranges over $\boldsymbol{k}$. Under the natural action of T (or B) on N, the characters of N thus fall into two conjugacy classes, consisting respectively of the trivial character and all of the non-trivial ones.

6.3. We first consider a straightforward method of constructing irreducible representations of G.

Let χ_1, χ_2 be characters of $\boldsymbol{k}^\times$. We form the character

$$\chi = \chi_1 \otimes \chi_2 : \begin{pmatrix} a & 0 \\ 0 & b \end{pmatrix} \longmapsto \chi_1(a)\chi_2(b)$$

of T: we regard this as a character of B, trivial on N, via the quotient $B \to B/N \cong T$. We form the induced representation $\mathrm{Ind}_B^G \chi$ of G, and consider its irreducible components. These are characterized independently as follows:

Lemma. *Let π be an irreducible representation of G. The following conditions are equivalent:*

(1) *π is equivalent to a G-subspace of $\mathrm{Ind}_B^G \chi$, for some character χ of T;*
(2) *π contains the trivial character of N.*

Proof. The representation π contains the trivial character of N if and only if it contains an irreducible representation σ of B containing the trivial character of N. This condition is equivalent to σ being the inflation of a character of T. The lemma now follows from Frobenius Reciprocity. □

We have to analyze the induced representation $\mathrm{Ind}_B^G \chi$. To do this, we form the character

$$\chi^w : \begin{pmatrix} a & 0 \\ 0 & b \end{pmatrix} \longmapsto \chi_2(a)\chi_1(b)$$

of T. We abuse notation and view this as a character of B trivial on N. We then have:

Proposition. *Let χ, ξ be characters of T, viewed as characters of B which are trivial on N.*

(1) *The space $\mathrm{Hom}_G(\mathrm{Ind}_B^G(\chi), \mathrm{Ind}_B^G(\xi))$ is trivial unless $\chi = \xi$ or $\chi = \xi^w$.*
(2) *The spaces*

$$\mathrm{Hom}_G(\mathrm{Ind}_B^G(\chi), \mathrm{Ind}_B^G(\chi)), \quad \mathrm{Hom}_G(\mathrm{Ind}_B^G(\chi), \mathrm{Ind}_B^G(\chi^w)),$$

have the same dimension. This dimension is 2 if $\chi = \chi^w$, and otherwise is 1.

Proof. We use the canonical isomorphism of Frobenius Reciprocity

$$\mathrm{Hom}_G(\mathrm{Ind}_B^G(\chi), \mathrm{Ind}_B^G(\xi)) \cong \mathrm{Hom}_B(\mathrm{r}_B^G \mathrm{Ind}_B^G(\chi), \xi) \tag{6.3.1}$$

where r_B^G denotes the functor of restriction of representations from G to B. The restriction-induction formula of Mackey gives

$$\mathrm{r}_B^G \mathrm{Ind}_B^G(\chi) \cong \sum_{y \in B\backslash G/B} \mathrm{Ind}_{B\cap B^y}^B(\mathrm{r}_{B\cap B^y}^{B^y}(\chi^y)), \tag{6.3.2}$$

where χ^y denotes the character $b \mapsto \chi(yby^{-1})$ of the group $B^y = y^{-1}By$. By 5.2, we only have to consider the cases $y = 1$ and $y = w$. The term in (6.3.2) corresponding to $y = 1$ is just χ, and so contributes a factor $\mathrm{Hom}_T(\chi, \xi)$ to (6.3.1). The term corresponding to w contributes

$$\mathrm{Hom}_B(\mathrm{Ind}_T^B(\chi^w), \xi) \cong \mathrm{Hom}_T(\chi^w, \xi).$$

Altogether, (6.3.1) reduces to

$$\mathrm{Hom}_B(\mathrm{Ind}_B^G(\chi), \mathrm{Ind}_B^G(\xi)) \cong \mathrm{Hom}_T(\chi, \xi) \oplus \mathrm{Hom}_T(\chi^w, \xi),$$

whence follows the result. □

We deduce:

Corollary 1. *Let χ be a character of T, viewed as a character of B which is trivial on N.*

(1) *The representation $\mathrm{Ind}_B^G \chi$ is irreducible if and only if $\chi \neq \chi^w$.*
(2) *If $\chi = \chi^w$, the representation $\mathrm{Ind}_B^G \chi$ has length 2, with distinct composition factors.*

Example. Consider the trivial character 1_B of B. The trivial character 1_G occurs in $\mathrm{Ind}_B^G 1_B$, so

$$\mathrm{Ind}_B^G 1_B = 1_G \oplus \mathrm{St}_G,$$

for a unique irreducible representation St_G, called the *Steinberg representation*. Clearly, $\dim \mathrm{St}_G = q$.

The representations $\mathrm{Ind}_B^G \chi$, for characters χ of T, are called the *principal series*. The terminology is often extended to include their irreducible components, that is, all irreducible representations of G containing the trivial character of N. Counting, we get:

Corollary 2. *Up to isomorphism, the group G has precisely $\frac{1}{2}(q^2+q) - 1$ irreducible representations which contain the trivial character of N.*

6.4. An irreducible representation of G not containing the trivial character of N is called *cuspidal*. Such a representation σ must contain some non-trivial character of N. Any two non-trivial characters of N are B-conjugate (6.2), so σ contains all non-trivial characters of N. The cuspidal representations cannot be constructed directly by induction.

Let $\boldsymbol{l}/\boldsymbol{k}$ be a quadratic field extension: thus $\boldsymbol{l}$ is uniquely determined, up to $\boldsymbol{k}$-isomorphism. The non-trivial $\boldsymbol{k}$-automorphism of the field $\boldsymbol{l}$ is $x \mapsto x^q$. A character θ of $\boldsymbol{l}^\times$ is called *regular* if $\theta^q \neq \theta$.

Choosing a $\boldsymbol{k}$-basis of $\boldsymbol{l}$ identifies $\boldsymbol{l}$ with the vector space $\boldsymbol{k} \oplus \boldsymbol{k}$ and G with $\mathrm{Aut}_{\boldsymbol{k}}(\boldsymbol{l})$. The natural action of $\boldsymbol{l}^\times$ on $\boldsymbol{l}$ thus gives an embedding of $\boldsymbol{l}^\times$ in G, the G-conjugacy class of which is uniquely determined (5.3 Proposition (1)). We henceforward identify $\boldsymbol{l}^\times$ with a subgroup E of G. Note that any element of G with irreducible characteristic polynomial is conjugate to an element of E.

Let θ be a regular character of E and ψ a non-trivial character of N. We define a character θ_ψ of ZN by

$$\theta_\psi : \left(\begin{smallmatrix} a & 0 \\ 0 & a \end{smallmatrix}\right) u \longmapsto \theta(a)\psi(u), \quad a \in \boldsymbol{k}^\times,\ u \in N.$$

We observe that, by (6.2), the representation $\mathrm{Ind}_{ZN}^G \theta_\psi$ is, up to equivalence, independent of the choice of ψ.

Theorem. *Let θ be a regular character of E and ψ a non-trivial character of N.*

(1) *The virtual representation*

$$\pi_\theta = \mathrm{Ind}_{ZN}^G \theta_\psi - \mathrm{Ind}_E^G \theta$$

is an irreducible cuspidal representation of G, of dimension $q-1$.

(2) *Let θ_1, θ_2 be regular characters of E; then $\pi_{\theta_1} \cong \pi_{\theta_2}$ if and only if $\theta_2 = \theta_1$ or $\theta_2 = \theta_1^q$.*

(3) *Every irreducible cuspidal representation of G is of the form π_θ, for some regular character θ of E.*

Proof. We observe that part (3) follows from parts (1) and (2) on counting up the dimensions. One may prove the first two parts directly, by computing characters. One finds:

$$\begin{aligned} \operatorname{tr}\pi_\theta(z) &= (q-1)\,\theta(z), \quad z \in Z; \\ \operatorname{tr}\pi_\theta(zu) &= -\theta(z), \quad z \in Z,\ u \in N,\ u \neq 1; \\ \operatorname{tr}\pi_\theta(y) &= -(\theta(y)+\theta^q(y)), \quad y \in E \smallsetminus Z. \end{aligned} \tag{6.4.1}$$

The character value $\operatorname{tr}\pi_\theta(g)$ is zero if g is not conjugate to an element of $ZN \cup E$. The character table (6.4.1) gives part (2) straightaway. One checks that

$$\frac{1}{|G|}\sum_{g\in G} |\operatorname{tr}\pi_\theta(g)|^2 = 1,$$

so π_θ is an irreducible representation of G. Finally,

$$\sum_{u\in N} \operatorname{tr}\pi_\theta(u) = 0,$$

so $\pi_\theta \mid N$ does not contain the trivial character of N, and π_θ is therefore cuspidal. □

Exercise. *Let ψ be a non-trivial character of N, and consider the representation*

$$\mathcal{F} = \operatorname{Ind}_N^G \psi.$$

Let σ be an irreducible representation of G. Show that:

(1) *If $\sigma = \phi \circ \det$, for a character ϕ of $\mathbf{k}^\times$, then σ does not occur in $\mathcal{F}$.*
(2) *Otherwise, σ occurs in $\mathcal{F}$ with multiplicity one.*

Further reading. The complex representation theory of the group $\mathrm{GL}_n(k)$, for a finite field k, was originally worked out by J.A. Green [36]. For a systematic development, emphasizing the related combinatorics, see [60]. The representation theory of reductive algebraic groups over finite fields has been intensively studied: see [22], [23], [29] for introductions.

3

Induced Representations of Linear Groups

From now on, we concentrate on the group $G = \mathrm{GL}_2(F)$ over a non-Archimedean local field F. The group G inherits a locally profinite topology from the base field F, as in 1.4. Our objective is the classification of the irreducible smooth representations of G, although we shall not achieve it until the end of Chapter IV.

In this chapter, the algebraic subgroups B, N, T of G (as in 5.1) play a pivotal rôle. In parallel with the representation theory of the finite group $\mathrm{GL}_2(\boldsymbol{k})$ worked out in §6, the irreducible smooth representations of G fall into two broad classes. First, there is a "principal series" of representations: these are the composition factors of representations obtained from characters of T by a process of inflation to B and then induction to G. Frobenius Reciprocity characterizes them, among the irreducible smooth representations of G, as those admitting a non-trivial quotient on which N acts trivially. The irreducible smooth representations not obtainable this way are called "cuspidal". The main result of this chapter (9.11) gives a complete classification of the principal series representations.

There is a further subgroup of G which plays a surprisingly important part in the classification process. This is the "mirabolic subgroup" M of matrices $(x_{ij}) \in B$ with $x_{22} = 1$. The group M has a very simple representation theory: besides an obvious family of characters, it has a unique irreducible smooth representation. Further, irreducible representations of G decompose very little when restricted to M. This is the basis of our detailed analysis of the principal series representations of G.

In this chapter, we make only rather general remarks about the irreducible cuspidal representations of G. We give a characterization of them more helpful than that of not being in the principal series, and a speculative method for constructing them. This prepares the ground for the analysis in Chapter IV.

Some of the arguments and results here, particularly in §11, apply to quite arbitrary locally profinite groups: we point these out as they arise. Most of the time, we work exclusively with $\mathrm{GL}_2(F)$ or its subgroups, and exploit this restriction as much as we can to simplify and abbreviate the treatment. We are rarely unwilling to substitute an explicit matrix calculation for a more general abstract argument.

7. Linear Groups over Local Fields

As noted in §1, the group $G = \mathrm{GL}_2(F)$ has many compact open subgroups, of which a small number are of particular importance. This is expressed first via various coset decompositions of G, beyond the universal Bruhat decomposition of 5.2. Using these decompositions, one can turn the general measure theory of §3 into an effective computational tool, necessary for handling the integrals arising within the representation theory of G.

This section thus amounts to a course of calculus on $\mathrm{GL}_2(F)$, which can be skimmed at first reading and referred back to at need. The only result to which we will return is the Duality Theorem at the end, but the general techniques developed here are used frequently.

7.1. For this section, we set $V = F \oplus F$, and think of it as the space of column vectors with G acting on the left. The standard subgroups B, N, T, Z are as in §5. These are all closed subgroups of G. The group isomorphisms $B/N \cong T$ and $B \cong T \ltimes N$ are homeomorphisms.

7.2. Reflecting the special nature of the base field F, the group G admits decompositions besides the Bruhat decomposition of 5.2. The first of these is:

(7.2.1) ***Iwasawa decomposition.*** *Let B be the standard Borel subgroup of G and set $K = \mathrm{GL}_2(\mathfrak{o})$; then $G = BK$.*

Proof. Take $g \in G$; if the $(2,1)$-entry of g is zero, then $g \in B$. Otherwise, post-multiplying by the permutation matrix $w \in K$ if necessary, we can assume $v_F(g_{21}) \geqslant v_F(g_{22})$. We can then post-multiply by a lower triangular matrix in K to achieve $g_{21} = 0$. □

Consequently, the quotient space $B\backslash G$ is a continuous image of the compact group $K = \mathrm{GL}_2(\mathfrak{o})$, and so:

Corollary. *The quotient space $B\backslash G$ is compact.*

Continuing with the notation $K = \mathrm{GL}_2(\mathfrak{o})$, we also have:

(7.2.2) *Cartan decomposition.* *Let ϖ be a prime element of F. The matrices*

$$\begin{pmatrix} \varpi^a & 0 \\ 0 & \varpi^b \end{pmatrix}, \quad a, b \in \mathbb{Z}, \ a \leqslant b,$$

form a set of representatives for the coset space $K\backslash G/K$.

Proof. Permuting rows and columns, using the permutation matrix in K, we can arrange for the largest entry of g (in absolute value) to be in the $1,1$ place. Multiplying by elementary matrices from K, we can then arrange for g to be diagonal, and unit factors can be absorbed into K. This gives

$$G = \bigcup_{a \leqslant b} K \begin{pmatrix} \varpi^a & 0 \\ 0 & \varpi^b \end{pmatrix} K.$$

We have to prove this union is disjoint. That is, we have to recover the integers a, b from the coset KgK, where

$$g = \begin{pmatrix} \varpi^a & 0 \\ 0 & \varpi^b \end{pmatrix}.$$

First, we have $a+b = v_F(\det h)$, for any $h \in KgK$. Next, the group index $(K : K \cap hKh^{-1})$ depends only on the coset KhK, and $(K : K \cap gKg^{-1}) = 1$ if $b = a$, or $(q+1)q^{b-a-1}$ if $b > a$. □

Corollary. *If K is a compact open subgroup of G, the set G/K is countable.*

Proof. As observed in 2.6, it is enough to show that G/K is countable for one choice of K: we take $K = \mathrm{GL}_2(\mathfrak{o})$. The space $K\backslash G/K$ is certainly countable, and each double coset KgK contains only finitely many cosets $g'K$. □

That is, G satisfies the countability hypothesis of 2.6.

Exercise. *Let K be a compact subgroup of G. Show that $gKg^{-1} \subset \mathrm{GL}_2(\mathfrak{o})$, for some $g \in G$. Deduce that, up to G-conjugacy, $\mathrm{GL}_2(\mathfrak{o})$ is the unique maximal compact subgroup of G.*

Hint. There are two steps. One first shows that there exists a K-stable $\mathfrak{o}$-lattice in V: consider the $\mathfrak{o}$-span of KL, for a randomly chosen $\mathfrak{o}$-lattice L. The second consists of showing that the only $\mathrm{GL}_2(\mathfrak{o})$-stable lattices in V are the obvious ones $\mathfrak{p}^j \oplus \mathfrak{p}^j$, $j \in \mathbb{Z}$.

7.3. The *standard Iwahori subgroup* of G is the compact open subgroup

$$I = \left\{ \begin{pmatrix} a & b \\ c & d \end{pmatrix} : a, d \in U_F, \ b \in \mathfrak{o}, \ c \in \mathfrak{p} \right\}.$$

Let $N' = N^w$ denote the group of lower triangular unipotent matrices in G.

(7.3.1) ***Iwahori decomposition.*** *We have* $I = (I\cap N')(I\cap T)(I\cap N)$. *More precisely, the product map*

$$I\cap N' \times I\cap T \times I\cap N \longrightarrow I$$

is bijective, and a homeomorphism, for any ordering of the factors on the left hand side.

Proof. The product map is certainly continuous. It is elementary to write down its inverse and observe that it is continuous. □

Set $K = \mathrm{GL}_2(\mathfrak{o})$; under the canonical surjection $K \to \mathrm{GL}_2(\boldsymbol{k})$, the image of I is the standard Borel subgroup of $\mathrm{GL}_2(\boldsymbol{k})$. The Bruhat decomposition for $\mathrm{GL}_2(\boldsymbol{k})$ implies

$$K = I \cup IwI. \tag{7.3.2}$$

Combining (7.3.2) with the Iwasawa decomposition (7.2.1) for G, we obtain the more symmetric double-coset decomposition

$$G = BI \cup BwI = B(I\cap N') \cup Bw(I\cap N). \tag{7.3.3}$$

The cosets BwI, BI are both open in G.

Remark. Let $L = \mathfrak{o}\oplus\mathfrak{o}$, $L' = \mathfrak{o}\oplus\mathfrak{p}$. The Iwahori subgroup I is then the common G-stabilizer of the two lattices L, L'.

7.4. We now describe the Haar measures attached to the various locally profinite groups under discussion. We start with the basic example of the field F itself.

Lemma. *The vector space* $C_c^\infty(F)$ *is spanned by the characteristic functions of sets* $a+\mathfrak{p}^m$, $a \in F$, $m \in \mathbb{Z}$.

Proof. Surely the characteristic function of $a+\mathfrak{p}^m$ lies in $C_c^\infty(F)$. Conversely, let $\Phi \in C_c^\infty(F)$. Since Φ has compact support, there exists $n \in \mathbb{Z}$ such that $\operatorname{supp}\Phi \subset \mathfrak{p}^n$. Also, Φ is fixed under translation by a compact open subgroup of F, hence by $\mathfrak{p}^m$, for some $m \in \mathbb{Z}$. Thus Φ is a linear combination of characteristic functions of sets $a+\mathfrak{p}^m$, $a \in \mathfrak{p}^n/\mathfrak{p}^m$. □

If Φ_0 denotes the characteristic function of $\mathfrak{o}$ and μ is a Haar measure on F, we have

$$\int_F \Phi_0(x)\,d\mu(x) = c_0,$$

for some $c_0 > 0$. If Φ_1 is the characteristic function of a coset $a+\mathfrak{p}^b$, $a \in F$, $b \in \mathbb{Z}$, then

$$\int_F \Phi_1(x)\,d\mu(x) = c_0 q^{-b}.$$

This identity suffices for integrating any function $\Phi \in C_c^\infty(F)$.

Now take $\Phi \in C_c^\infty(F)$ and $y \in F^\times$. Using the identity above, we find

$$\int_F \Phi(xy)\, d\mu(x) = \|y\|^{-1} \int_F \Phi(x)\, d\mu(x),$$

where, we recall, $\|y\| = q^{-v_F(y)}$. We accordingly define a measure $\mu^\times$ on $F^\times$ by $d\mu^\times(x) = d\mu(x)/\|x\|$, meaning the following. If $\Phi \in C_c^\infty(F^\times)$, the function $x \mapsto \|x\|^{-1}\Phi(x)$ (vanishing at 0) lies in $C_c^\infty(F)$, so we can put

$$\int_{F^\times} \Phi(x)\, d\mu^\times(x) = \int_F \Phi(x)\|x\|^{-1}\, d\mu(x), \quad \Phi \in C_c^\infty(F^\times). \tag{7.4.1}$$

A simple manipulation shows that (7.4.1) defines a Haar integral on $F^\times$.

7.5. The matrix ring $A = \mathrm{M}_2(F)$ is (as additive group) a product of 4 copies of F and a Haar measure is obtained by taking a (tensor) product of 4 copies of a Haar measure on F.

Proposition. *Let μ be a Haar measure on A. For $\Phi \in C_c^\infty(G)$, the function $x \mapsto \Phi(x)\|\det x\|^{-2}$ (vanishing on $A \smallsetminus G$) lies in $C_c^\infty(A)$. The functional*

$$\Phi \mapsto \int_A \Phi(x)\|\det x\|^{-2}\, d\mu(x), \quad \Phi \in C_c^\infty(G),$$

is a left and right Haar integral on G. In particular, G is unimodular.

Proof. Let $g \in G$ and consider the functionals

$$\Phi \longmapsto \begin{cases} \displaystyle\int_A \Phi(gx)\, d\mu(x), \\ \displaystyle\int_A \Phi(xg)\, d\mu(x), \end{cases} \qquad \Phi \in C_c^\infty(A).$$

Each is a Haar integral on A and differs from the initial one by a positive constant (depending on g). To evaluate this constant, we take Φ to be the characteristic function of $\mathfrak{m} = \mathrm{M}_2(\mathfrak{o})$. In the first instance, the function $x \mapsto \Phi(gx)$ is the characteristic function of the lattice $\mathfrak{m}' = g^{-1}\mathfrak{m}$. Thus

$$\int_A \Phi(gx)\, d\mu(x) = \mu(\mathfrak{m}') = \mu(\mathfrak{m})\,(\mathfrak{m}' : \mathfrak{m} \cap \mathfrak{m}')/(\mathfrak{m} : \mathfrak{m} \cap \mathfrak{m}').$$

This quotient of indices depends only on the double coset KgK, $K = \mathrm{GL}_2(\mathfrak{o})$. Taking g in diagonal form (7.2.2), one gets

$$\int_A \Phi(gx)\, d\mu(x) = \|\det g\|^{-2} \int_A \Phi(x)\, d\mu(x).$$

The second instance is treated in same way to get

$$\int_A \Phi(xg)\, d\mu(x) = \|\det g\|^{-2} \int_A \Phi(x)\, d\mu(x).$$

The proposition then follows from simple manipulations. For example, if $\Phi \in C_c^\infty(G)$,

$$\begin{aligned}\int_A \Phi(xg)\|\det x\|^{-2}\, d\mu(x) &= \|\det g\|^{-2} \int_A \Phi(x)\|\det xg^{-1}\|^{-2}\, d\mu(x)\\ &= \int_A \Phi(x)\|\det x\|^{-2}\, d\mu(x),\end{aligned}$$

as required. □

7.6. We turn to the subgroups B, N, T of G. Since $N \cong F$ and $T \cong F^\times \times F^\times$, there is nothing more to say about them. We have $B = T \ltimes N$; we define a linear functional on the space $C_c^\infty(B) = C_c^\infty(T) \otimes C_c^\infty(N)$ by

$$\Phi \longmapsto \int_T \int_N \Phi(tn)\, d\mu_T(t) d\mu_N(n), \quad \Phi \in C_c^\infty(B),$$

where μ_T, μ_N are Haar measures on T, N respectively. One verifies immediately that this functional is left B-invariant, so it is a left Haar integral on B. We are so justified in denoting it

$$\Phi \longmapsto \int_B \Phi(b)\, d\mu_B(b).$$

The Haar measure μ_B may be thought of as the tensor product, $\mu_B = \mu_T \otimes \mu_N$, but the two factors do not commute. This reflects the fact that the group B is not unimodular. In the language of 3.3:

Proposition. *The module δ_B of the group B is given by*

$$\delta_B : tn \longmapsto \|t_2/t_1\|, \qquad n \in N,\ t = \begin{pmatrix} t_1 & 0 \\ 0 & t_2 \end{pmatrix} \in T. \tag{7.6.1}$$

Proof. Setting

$$c = sm, \quad m \in N,\ s = \begin{pmatrix} s_1 & 0 \\ 0 & s_2 \end{pmatrix} \in T,$$

we get

$$\int_B \Phi(bc)\, d\mu_B(b) = \int_T \int_N \Phi(ts\, s^{-1}ns\, m)\, d\mu_T(t) d\mu_N(n).$$

We use the obvious isomorphism $N \to F$ to identify μ_N with a certain Haar measure μ_F on F. For $\phi \in C_c^\infty(N)$, we then have

$$\begin{aligned}\int_N \phi(s^{-1}ns)\, d\mu_N(n) &= \int_F \phi\begin{pmatrix} 1 & s_1^{-1}xs_2 \\ 0 & 1 \end{pmatrix} d\mu_F(x)\\ &= \|s_1 s_2^{-1}\| \textstyle\int_N \phi(n)\, d\mu_N(n).\end{aligned}$$

By definition,

$$\int_B \Phi(bc)\, d\mu_B(b) = \delta_B(c)^{-1} \int_B \Phi(b)\, d\mu_B(b),$$

and the result follows. □

In the notation of 3.4, we now have:

Corollary. *The space $C_c^\infty(B\backslash G, \delta_B^{-1})$ admits a positive semi-invariant measure $\dot{\mu}$, where δ_B is given by (7.6.1). If $K = \mathrm{GL}_2(\mathfrak{o})$, there is a Haar measure μ_K on K such that*

$$\int_{B\backslash G} f(g)\, d\dot{\mu}(g) = \int_K f(k)\, d\mu_K(k),$$

for $f \in C_c^\infty(B\backslash G, \delta_B^{-1})$.

Proof. The character δ_B is trivial on the compact group $K \cap B$. Restriction of functions is an isomorphism $C_c^\infty(B\backslash G, \delta_B^{-1}) \to C_c^\infty(K \cap B\backslash K, 1)$, where 1 denotes the trivial character of $K \cap B$. The semi-invariant measure $\dot{\mu}$ thus restricts to a semi-invariant measure on $C_c^\infty(K \cap B\backslash K, 1)$, but so does any Haar measure on K. □

We observe that μ_K is effectively just the restriction of a Haar measure μ_G on G. Comparing with the proof of 3.4 Proposition, there is a left Haar measure μ_B on B such that

$$\int_G \phi(g)\, d\mu_G(g) = \int_K \int_B \phi(bk)\, d\mu_B(b) d\mu_G(k), \quad \phi \in C_c^\infty(G). \tag{7.6.2}$$

Exercises.

(1) *Let I be the standard Iwahori subgroup; let dn', dt, dn be Haar measures on the groups $I \cap N'$, $I \cap T$, $I \cap N$ respectively. Show that the functional*

$$f \longmapsto \iiint f(n'tn)\, dn'dt\, dn, \quad f \in C_c^\infty(I),$$

is a Haar integral on I.

(2) *Let $C = N'TN$. Show that C is open and dense in G, and that the product map $N' \times T \times N \to C$ is a homeomorphism.*

(3) *Let dg be a Haar measure on G. Show that there are Haar measures dn', dt, dn on N', T, N such that*

$$\int_G f(g)\, dg = \iiint f(n'tn)\delta_B(t)^{-1}\, dn'dt\, dn, \quad f \in C_c^\infty(G).$$

7.7. Let σ be a smooth representation of T, viewed as representation of B trivial on N. Corollary 7.2 implies that the canonical inclusion map

$$c\text{-Ind}_B^G \sigma \longrightarrow \text{Ind}_B^G \sigma$$

is an isomorphism. We can therefore apply the Duality Theorem of 3.5 to get:

Duality Theorem. *Let σ be a smooth representation of T, viewed as representation of B trivial on N, and fix a positive semi-invariant measure $\dot{\mu}$ on $C_c^\infty(B\backslash G, \delta_B^{-1})$. There is a canonical isomorphism*

$$(\text{Ind}_B^G \sigma)^\vee \cong \text{Ind}_B^G \delta_B^{-1} \otimes \check{\sigma},$$

depending only on the choice of $\dot{\mu}$.

8. Representations of the Mirabolic Group

Before starting on the representation theory of the group $G = \text{GL}_2(F)$, we study the representations of a certain subgroup of G, the so-called *mirabolic subgroup*

$$M = \left\{ \left(\begin{smallmatrix} a & x \\ 0 & 1 \end{smallmatrix}\right) : a \in F^\times, x \in F \right\}.$$

Thus M is the semi-direct product of N by the group $S = T \cap M \cong F^\times$.

8.1. To start with, let (π, V) be a smooth representation of N and let ϑ be a character of N. We denote by $V(\vartheta)$ the linear subspace of V spanned by the vectors $\pi(n)v - \vartheta(n)v$, $n \in N$, $v \in V$. We set $V_\vartheta = V/V(\vartheta)$: this is the unique maximal N-quotient of V on which N acts via the character ϑ.

If ϑ_0 is the trivial character of N, we have $V(\vartheta_0) = V(N)$ (notation of 2.3) and we write $V_{\vartheta_0} = V_N$.

Lemma. *Let μ_N be a Haar measure on N and ϑ a character of N.*

(1) *Let (π, V) be a smooth representation of N and $v \in V$. The vector v lies in $V(\vartheta)$ if and only if there is a compact open subgroup N_0 of N such that*

$$\int_{N_0} \vartheta(n)^{-1} \pi(n) v \, d\mu_N(n) = 0. \tag{8.1.1}$$

(2) *The process $(\pi, V) \mapsto V_\vartheta$ is an exact functor from* $\text{Rep}(N)$ *to the category of complex vector spaces.*

Proof. We assume first that ϑ is the trivial character of N.

The group $N \cong F$ is the union of an ascending sequence of compact open subgroups. So, if

$$v = \sum_{i=1}^{r} v_i - \pi(n_i) v_i \in V(N),$$

there is a compact open subgroup N_0 of N containing all n_i. The relation (8.1.1) then holds for this choice of N_0.

Conversely, let $v \in V$ and suppose (8.1.1) holds. There is an open normal subgroup N_1 of N_0 such that $v \in V^{N_1}$. The space V^{N_1} carries a representation of the finite group N_0/N_1. Therefore, in the obvious notation, $V^{N_1} = V^{N_1}(N_0/N_1) \oplus V^{N_0}$ (*cf.* 2.3) and the map

$$w \longmapsto \mu_N(N_0)^{-1} \int_{N_0} \pi(n) w \, d\mu_N(n), \quad w \in V^{N_1},$$

is the N_0-projection $V^{N_1} \to V^{N_0}$. This has kernel $V^{N_1}(N_0/N_1) \subset V(N)$ and we have proved (1) for the trivial character of N.

Now let ϑ be an arbitrary character of N, and consider the representation (π', V') of N, where $V' = V$ and $\pi'(n) = \vartheta(n)^{-1}\pi(n)$. We then have $V(\vartheta) = V'(N)$ and so (1) follows in general.

Part (2) is an immediate consequence of (1). □

We mention some simple consequences of the lemma. If (π, V) is a smooth representation of N (or of M), then $V(N)$ is an N- (or M-) subspace of V. The exact sequence

$$0 \to V(N) \longrightarrow V \longrightarrow V_N \to 0$$

gives an exact sequence

$$0 \to V(N)_N \longrightarrow V_N \longrightarrow V_N \to 0$$

in which the map $V_N \to V_N$ is the identity. Therefore

$$V(N)_N = 0 \quad \text{and} \quad V(N)(N) = V(N). \tag{8.1.2}$$

Suppose that $\vartheta \neq 1$. As N acts trivially on $V/V(N)$, we have $(V/V(N))_\vartheta = 0$ and so the inclusion $V(N) \to V$ induces an isomorphism

$$V(N)_\vartheta \cong V_\vartheta. \tag{8.1.3}$$

Proposition. *Let (π, V) be a smooth representation of N, and let $v \in V$, $v \neq 0$. There exists a character ϑ of N such that $v \notin V(\vartheta)$.*

Proof. We write

$$N_j = \left(\begin{smallmatrix} 1 & \mathfrak{p}^j \\ 0 & 1 \end{smallmatrix}\right), \quad j \in \mathbb{Z}. \tag{8.1.4}$$

Take $v \in V$, $v \neq 0$. We choose $j_0 \in \mathbb{Z}$ such that N_{j_0} fixes v. For $j \leqslant j_0$, let V_j denote the N_j-space generated by v. This is the direct sum of isotypic components V_j^η, as η ranges over the characters of N_j trivial on N_{j_0}. For each $j \leqslant j_0$, there exists η_j such that $V_j^{\eta_j} \neq 0$; by definition, we have

$$\int_{N_j} \eta_j(n)^{-1}\pi(n)v\,dn \neq 0.$$

The N_{j-1}-space generated by $V_j^{\eta_j}$ is contained in V_{j-1}, so we may choose η_{j-1} such that $\eta_{j-1} \mid N_j = \eta_j$. It follows (compare the argument in 1.7) that there exists a character ϑ of N such that, for all $j \leqslant j_0$, we have

$$\int_{N_j} \vartheta(n)^{-1}\pi(n)v\,dn \neq 0.$$

Therefore $v \notin V(\vartheta)$, as required. □

Corollary 1. *Let (π, V) be a smooth representation of N. If $V_\vartheta = 0$ for all characters ϑ of N, then $V = 0$.*

Now let (π, V) be a smooth representation of M. The space $V(N)$ is then an M-subspace of V and V_N carries a natural representation of $M/N = S$. On the other hand, $\pi(S)$ permutes the subspaces $V(\vartheta)$, $\vartheta \neq 1$, transitively. Explicitly, for $s \in S$, $\pi(s)V(\vartheta) = V(\vartheta')$, where $\vartheta'(n) = \vartheta(s^{-1}ns)$. We can therefore sharpen Corollary 1 for representations of M:

Corollary 2. *Let (π, V) be a smooth representation of M. Suppose that $V_N = 0$ and that $V_\vartheta = 0$ for some non-trivial character ϑ of N. Then $V = 0$.*

8.2. We now fix a non-trivial character ϑ of N, and consider the two M-spaces $\mathrm{Ind}_N^M \vartheta$, c-$\mathrm{Ind}_N^M \vartheta$. Observe that, if ϑ' is some other non-trivial character of N, then $\mathrm{Ind}_N^M \vartheta'$ is M-isomorphic to $\mathrm{Ind}_N^M \vartheta$ and similarly for the compactly induced representations.

Proposition. *Let ϑ be a non-trivial character of N and set $\mathcal{W} = \mathrm{Ind}_N^M \vartheta$, $\mathcal{W}^c =$ c-$\mathrm{Ind}_N^M \vartheta$. Let $\alpha : \mathcal{W} \to \mathbb{C}$ denote the canonical map $f \mapsto f(1)$.*

(1) *We have $\mathcal{W}(N) = \mathcal{W}^c(N) = \mathcal{W}^c$ and $(\mathcal{W}/\mathcal{W}^c)(N) = 0$.*
(2) *The map α induces isomorphisms $\mathcal{W}_\vartheta \cong \mathbb{C}$ and $\mathcal{W}^c_\vartheta \cong \mathbb{C}$.*

Proof. Let $f \in \mathcal{W}$ and $n \in N$. For $a \in S$, we have $f(an) = \vartheta(ana^{-1})f(a)$. As there is an integer j such that N_j (as in (8.1.4)) fixes f, we see that $f(a) = 0$ if $\|\det a\|$ is sufficiently large. On the other hand, $f(an) = f(a)$ if $\|\det a\|$ is sufficiently small, so $nf - f$ vanishes at a if $\|\det a\|$ is sufficiently small. This implies that $nf - f \in \mathcal{W}^c$. Thus $\mathcal{W}(N) \subset \mathcal{W}^c$ and N acts trivially on $\mathcal{W}/\mathcal{W}^c$.

We next prove that $\mathcal{W}^c(N) = \mathcal{W}^c$. Let ψ be the character of F defined by

$$\psi(x) = \vartheta\begin{pmatrix} 1 & x \\ 0 & 1 \end{pmatrix}.$$

Take $a \in F^\times$, $j \in \mathbb{Z}$, $j \geqslant 1$. Let $f_{a,j} \in \mathcal{W}^c$ be the function such that

$$f_{a,j} : \begin{pmatrix} 1 & x \\ 0 & 1 \end{pmatrix}\begin{pmatrix} au & 0 \\ 0 & 1 \end{pmatrix} \longmapsto \psi(x), \quad u \in U_F^j,$$

and which vanishes elsewhere. The various functions $f_{a,j}$ span $\mathcal{W}^c$ over $\mathbb{C}$. We have

$$\begin{pmatrix} 1 & x \\ 0 & 1 \end{pmatrix} f_{a,j}\begin{pmatrix} b & 0 \\ 0 & 1 \end{pmatrix} = \psi(bx) f_{a,j}\begin{pmatrix} b & 0 \\ 0 & 1 \end{pmatrix}.$$

We deduce that $nf_{a,j}$ has the same support as $f_{a,j}$, $n \in N$. We can certainly find $x \in F$ such that the function $u \mapsto \psi(aux)$, $u \in U_F^j$, is constant, equal to c say, with $c \neq 1$. If $n = \left(\begin{smallmatrix} 1 & x \\ 0 & 1 \end{smallmatrix}\right)$, then $nf_{a,j} - f_{a,j} = (c-1)f_{a,j}$, whence $f_{a,j} \in \mathcal{W}^c(N)$ and so $\mathcal{W}^c(N) = \mathcal{W}^c$. This implies $\mathcal{W}(N) = \mathcal{W}^c$ also, and we have proved (1).

The map α induces a surjection $\mathcal{W}_\vartheta \to \mathbb{C}$. On the other hand, since N acts trivially on $\mathcal{W}/\mathcal{W}^c$, the inclusion $\mathcal{W}^c \hookrightarrow \mathcal{W}$ induces an isomorphism $\mathcal{W}^c_\vartheta \cong \mathcal{W}_\vartheta$. To prove (2), therefore, it is enough to show that any $f \in \mathcal{W}^c$ with $f(1) = 0$ belongs to $\mathcal{W}^c(\vartheta)$. A function f, vanishing at 1, is a finite linear combination of functions $f_{a,j}$ with $a \notin U_F^j$, so it is enough to treat such functions. However, as $a \notin U_F^j$, there exists $x \in F$ such that the function $u \mapsto \psi(aux) - \psi(x)$, $u \in U_F^j$, is a non-zero constant. Taking $n = \left(\begin{smallmatrix} 1 & x \\ 0 & 1 \end{smallmatrix}\right)$, the same calculation as before shows that $nf_{a,j} - \vartheta(n)f_{a,j}$ is a non-zero constant multiple of $f_{a,j}$, whence $f_{a,j} \in \mathcal{W}^c(\vartheta)$ and $\mathcal{W}^c(\vartheta) = \mathcal{W}^c$, as required. □

Corollary. *The representation* $c\text{-Ind}_N^M \vartheta$ *is irreducible over* M.

Proof. Let V be an M-subspace of $\mathcal{W}^c$. As $\mathcal{W}^c_N = 0$, the spaces V_N, $(\mathcal{W}^c/V)_N$ are both zero. The sequence

$$0 \to V_\vartheta \longrightarrow \mathcal{W}^c_\vartheta \longrightarrow (\mathcal{W}^c/V)_\vartheta \to 0$$

is exact. As $\dim \mathcal{W}^c_\vartheta = 1$, we conclude that $\dim V_\vartheta$ is 0 or 1. In the first case, $V = 0$ by 8.1 Corollary 2. In the second, $(\mathcal{W}^c/V)_\vartheta = 0$ and so $\mathcal{W}^c = V$. □

We display some of the remarks made in the course of the proof of the proposition:

Gloss.

(1) *A function* $f \in \mathcal{W}$ *is determined by its restriction to* $S \cong F^\times$. *The restriction* $f \mid F^\times$ *is a smooth function on* $F^\times$.

(2) *A smooth function* ϕ *on* $F^\times$ *is of the form* $\phi = f \mid F^\times$, *for some* $f \in \mathcal{W}$, *if and only if there exists* $c > 0$ *such that* $\phi(x) = 0$ *for all* x *satisfying* $\|x\| > c$.

(3) *A function* $f \in \mathcal{W}$ *lies in* $\mathcal{W}^c$ *if and only if* $f \mid F^\times \in C_c^\infty(F^\times)$.

Remark. Part (3) implies that the representation $\mathrm{Ind}_N^M \vartheta$ is never irreducible, for any non-trivial character ϑ of N. The Duality Theorem 3.5 implies

$$(c\text{-}\mathrm{Ind}_N^M \vartheta)^\vee \cong \mathrm{Ind}_N^M \check{\vartheta}.$$

Thus $c\text{-}\mathrm{Ind}_N^M \vartheta$ provides an example of an irreducible smooth representation with *reducible* contragredient (*cf.* 2.10).

8.3. Again let ϑ be a non-trivial character of N. Let (π, V) be a smooth representation of M. Frobenius Reciprocity (2.4) gives a canonical isomorphism

$$\mathrm{Hom}_N(V, V_\vartheta) \cong \mathrm{Hom}_M(V, \mathrm{Ind}_N^M V_\vartheta).$$

Let $q : V \to V_\vartheta$ denote the quotient map and let $q_\star$ be the map $V \to \mathrm{Ind}_N^M V_\vartheta$ corresponding to q under this isomorphism. Explicitly, for $v \in V$, $q_\star(v)$ is the function $m \mapsto q(\pi(m)v)$.

Theorem. *Let* (π, V) *be a smooth representation of* M. *The* M*-homomorphism* $q_\star : V \to \mathrm{Ind}_N^M V_\vartheta$ *induces an isomorphism* $V(N) \cong c\text{-}\mathrm{Ind}_N^M V_\vartheta$.

Proof. The N-space V_ϑ is a direct sum of copies of ϑ. Therefore $\mathrm{Ind}_N^M V_\vartheta$ is a direct sum of copies of $\mathrm{Ind}_N^M \vartheta$. Proposition 8.2 so yields

$$(\mathrm{Ind}_N^M V_\vartheta)(N) = c\text{-}\mathrm{Ind}_N^M V_\vartheta = (c\text{-}\mathrm{Ind}_N^M V_\vartheta)(N). \tag{8.3.1}$$

The M-homomorphism $q_\star : V \to \mathrm{Ind}_N^M V_\vartheta$ surely maps $V(N)$ to $(\mathrm{Ind}_N^M V_\vartheta)(N) = c\text{-}\mathrm{Ind}_N^M V_\vartheta$.

Let $W = \mathrm{Ker}\, q_\star \cap V(N)$ and $C = c\text{-}\mathrm{Ind}_N^M V_\vartheta / q_\star(V(N))$. The natural map $W_N \to V(N)_N$ is injective, by 8.1 Lemma (2), so $W_N = 0$. Likewise, $(c\text{-}\mathrm{Ind}\, V_\vartheta)_N$ is zero, so $C_N = 0$.

The map $q_\star : V \to c\text{-}\mathrm{Ind}\, V_\vartheta$ induces a map

$$q_{\star,\vartheta} : V_\vartheta = V(N)_\vartheta \longrightarrow (c\text{-}\mathrm{Ind}\, V_\vartheta)_\vartheta.$$

By 8.2 Proposition (2), the canonical N-map $\mathrm{Ind}\, V_\vartheta \to V_\vartheta$ induces an isomorphism $\alpha_\vartheta : (c\text{-}\mathrm{Ind}\, V_\vartheta)_\vartheta \to V_\vartheta$. The composite map $\alpha_\vartheta \circ q_{\star,\vartheta} : V_\vartheta \to V_\vartheta$ is the identity. However, the kernel of this map is W_ϑ and its cokernel is isomorphic to C_ϑ. We have shown that the spaces W_N, W_ϑ, C_N, C_ϑ are all zero. By 8.1 Corollary 2, therefore, both W and C are trivial and $q_\star : V(N) \to c\text{-}\mathrm{Ind}\, V_\vartheta$ is an isomorphism, as desired. □

Corollary. *Let (π, V) be an irreducible smooth representation of M. Either:*

(1) $\dim V = 1$ *and* π *is the inflation of a character of* $M/N \cong F^\times$, *or*
(2) $\dim V$ *is infinite and* $\pi \cong c\text{-Ind}_N^M \vartheta$, *for any character* $\vartheta \neq 1$ *of* N.

In case (1), $\dim V_N = 1$ *and* $V_\vartheta = 0$ *for* $\vartheta \neq 1$. *In case* (2), $V_N = 0$ *and* $\dim V_\vartheta = 1$ *for all* $\vartheta \neq 1$.

Proof. If $V(N) = 0$, then N acts trivially on V. The group M/N is abelian, so Schur's Lemma 2.6 implies $\dim V = 1$ and we are in case (1).

If $V(N) \neq 0$, then $V(N) = V$ and $V_N = 0$. Therefore $V_\vartheta \neq 0$ for all characters $\vartheta \neq 1$ of N, and so $\dim V$ is infinite. The theorem implies that $V = V(N)$ is M-isomorphic to $c\text{-Ind}_N^M V_\vartheta$. The N-space V_ϑ is a direct sum of copies of ϑ, so V is a direct sum of copies of $c\text{-Ind}_N^M \vartheta$ and, since it is irreducible, $V \cong c\text{-Ind}_N^M \vartheta$. □

9. Jacquet Modules and Induced Representations

We start the process of classifying the irreducible smooth representations of the locally profinite group $G = \mathrm{GL}_2(F)$. In this section, we deal completely with those irreducible smooth representations (π, V) of G (the "principal series") for which $V_N \neq 0$.

9.1. Let (π, V) be a smooth representation of G. As in 8.1, $V(N)$ denotes the subspace of V spanned by the vectors $v - \pi(x)v$, for $v \in V$ and $x \in N$. The space $V_N = V/V(N)$ inherits a representation π_N of $B/N = T$, which is smooth. We call (π_N, V_N) the *Jacquet module* of (π, V) at N.

In particular, *the Jacquet functor*

$$\begin{aligned} \mathrm{Rep}(G) &\longrightarrow \mathrm{Rep}(T), \\ (\pi, V) &\longmapsto (\pi_N, V_N), \end{aligned} \tag{9.1.1}$$

is exact and additive.

Let (σ, W) be a smooth representation of T. We view σ as a smooth representation of B which is trivial on N, and form the smooth induced representation $\mathrm{Ind}_B^G \sigma$. (We sometimes abbreviate $\mathrm{Ind}_B^G \sigma = \mathrm{Ind}\, \sigma$, since B and G are the only groups involved for most of the time.)

If (π, V) is a smooth representation of G, Frobenius Reciprocity (2.4) gives an isomorphism

$$\mathrm{Hom}_G(\pi, \mathrm{Ind}\, \sigma) \cong \mathrm{Hom}_B(\pi, \sigma).$$

However, σ is trivial on N so any B-homomorphism $\pi \to \sigma$ factors through the quotient map $\pi \to \pi_N$. We deduce

$$\mathrm{Hom}_G(\pi, \mathrm{Ind}\, \sigma) \cong \mathrm{Hom}_T(\pi_N, \sigma). \tag{9.1.2}$$

This has the following consequence:

Proposition. *Let (π, V) be an irreducible smooth representation of G. The following are equivalent:*

(1) *The Jacquet module V_N is non-zero.*
(2) *The representation π is isomorphic to a G-subspace of a representation $\mathrm{Ind}_B^G \chi$, for some character χ of T.*

Proof. Suppose (2) holds. From (9.1.2) we get

$$\mathrm{Hom}_T(\pi_N, \chi) \cong \mathrm{Hom}_G(\pi, \mathrm{Ind}\,\chi) \neq 0,$$

so surely $\pi_N \neq 0$.

To prove (1)⇒(2), we have to show that the Jacquet module V_N admits an irreducible T- (or B-) quotient.

Choose $v \in V$, $V \neq 0$. Since V is irreducible over G, any element of V is a finite linear combination of translates $\pi(g)v$ of v, for various $g \in G$. Write $K = \mathrm{GL}_2(\mathfrak{o})$. The vector v is fixed by a subgroup K' of K of finite index; let $\{v_1, v_2, \dots, v_r\}$ be the distinct elements of the form $\pi(k)v$, $k \in K$. In particular, $r \leqslant (K{:}K')$. Since $G = BK$, the elements $v_1, \dots, v_r$ generate V over B, and their images generate V_N over T.

Thus V_N is finitely generated as a representation of T. We choose a minimal generating set $\{u_1, \dots, u_t\}$, $t \geqslant 1$, say. A standard Zorn's Lemma argument shows that V_N has a T-subspace U, containing $u_1, \dots, u_{t-1}$, and maximal for the property $u_t \notin U$. Then U is a maximal T-subspace of V_N and V_N/U is an irreducible representation of T, hence a character (2.6 Corollary 2). □

An irreducible smooth representation (π, V) of G is called *cuspidal* if V_N is zero. In the literature, cuspidal representations are usually called *supercuspidal* or *absolutely cuspidal.* On the other hand, if $V_N \neq 0$, one says that π is in the *principal series.*

9.2. In the case of a finite field $\boldsymbol{k}$, we divided the irreducible representations according to whether or not they contained the trivial character of N. For $\mathrm{GL}_2(F)$ we use the existence of an N-trivial quotient, for the following reason:

Exercise 1. *Let (π, V) be an irreducible smooth representation of G with a non-trivial $\pi(N)$-fixed vector. Show that $\pi = \phi \circ \det$, for some character ϕ of $F^\times$.*

Hint. If $v \in V$ is fixed by N, it is fixed by the subgroup H of G generated by N and some open subgroup K. Show that, since H contains a lower triangular unipotent matrix, it also contains $\mathrm{SL}_2(F)$.

Exercise 2. *Let (π, V) be an irreducible smooth representation of G such that $\dim V$ is finite. Show that V has a non-zero $\pi(N)$-fixed vector. Deduce that $\dim V = 1$ and π is of the form $\phi \circ \det$, for some character ϕ of $F^\times$.*

These exercises help to explain the direction we take, although they play no part in the argument to follow. (We will, however, need Exercise 1 at a later stage.)

In this connection, we note:

Proposition. *Any character of G is of the form $\phi \circ \det$, for some character ϕ of $F^\times$.*

Proof. If χ is a character of G, its kernel contains the commutator subgroup of G. Since F is infinite, this commutator subgroup is $\mathrm{SL}_2(F)$, so $\chi = \phi \circ \det$, for some homomorphism $\phi : F^\times \to \mathbb{C}^\times$. The determinant map is surjective and open, so ϕ is a character. □

9.3. An important fact concerns the structure of the Jacquet module $(\mathrm{Ind}_B^G \chi)_N$ of an induced representation. Here, it is no more difficult to give a very general result.

Let μ_N be a Haar measure on N and let $t \in T$. The measure $S \mapsto \mu_N(t^{-1}St)$ is the Haar measure $\delta_B(t)\mu_N$, for δ_B as in (7.6.1):

$$\int_N f(txt^{-1})\, d\mu_N(x) = \delta_B(t) \int_N f(x)\, d\mu_N(x), \quad f \in C_c^\infty(N).$$

As before, let w denote the permutation matrix

$$w = \begin{pmatrix} 0 & 1 \\ 1 & 0 \end{pmatrix}.$$

If σ is a smooth representation of T, we can form the representation $\sigma^w : t \mapsto \sigma(wtw^{-1})$, and view it as a representation of B which is trivial on N.

As in 2.4, α_σ denotes the canonical B-map $\mathrm{Ind}\,\sigma \to \sigma$ given by $f \mapsto f(1)$. It induces a canonical T-map $(\mathrm{Ind}\,\sigma)_N \to \sigma$, which we continue to denote α_σ.

Restriction-Induction Lemma. *Let (σ, U) be a smooth representation of T and set $(\Sigma, X) = \mathrm{Ind}_B^G \sigma$. There is an exact sequence of representations of T:*

$$0 \to \sigma^w \otimes \delta_B^{-1} \longrightarrow \Sigma_N \xrightarrow{\alpha_\sigma} \sigma \to 0.$$

Proof. By definition, X is the space of G-smooth functions $f : G \to U$ such that $f(bg) = \sigma(b)f(g)$, $b \in B$, $g \in G$. The canonical map $\alpha_\sigma : X \to U$ amounts to restriction of functions to B. Set $V = \mathrm{Ker}\,\alpha_\sigma$. Thus V provides a smooth representation of B and there is an exact sequence

$$0 \to V_N \longrightarrow X_N \longrightarrow U \to 0.$$

We have to identify the T-representation V_N.

We recall that $G = B \cup BwN$. A function $f \in X$ thus lies in V if and only if $\operatorname{supp} f \subset BwN$. More precisely:

Lemma. *Let $f \in X$; then $f \in V$ if and only if there is a compact open subgroup N_0 of N (depending on f) such that* $\operatorname{supp} f \subset BwN_0$.

Proof. A function $f \in X$ lies in V if and only if $f(1) = 0$. Since f is G-smooth, such a function f vanishes on a set BN_0', where N_0' is some compact open subgroup of N'. The identity (for $x \neq 0$)

$$\begin{pmatrix} 1 & 0 \\ x & 1 \end{pmatrix} = \begin{pmatrix} 1 & x^{-1} \\ 0 & 1 \end{pmatrix} \begin{pmatrix} -x^{-1} & 0 \\ 0 & x \end{pmatrix} w \begin{pmatrix} 1 & x^{-1} \\ 0 & 1 \end{pmatrix} \in Bw \begin{pmatrix} 1 & x^{-1} \\ 0 & 1 \end{pmatrix}$$

implies that $\operatorname{supp} f \subset BwN_0$, for some compact open subgroup N_0 of N, as required. □

Let $f \in V$; in view of the lemma, we can define a function $f_N : T \to U$ by

$$f_N(x) = \int_N f(xwn)\, dn = \sigma(x)\, f_N(1), \quad x \in T.$$

By 8.1 Lemma, the kernel of the map $f \mapsto f_N$ is $V(N)$, and so $f \mapsto f_N(1)$ gives a bijective map $V_N \to U$. Taking $t \in T$ and $f \in V$, we have

$$\begin{aligned} (tf)_N(x) &= \int_N f(xwnt)\, dn \\ &= \delta_B(t^{-1}) \int_N f(xt^w wn)\, dn = \delta_B^{-1}(t)\, (t^w f_N)(x). \end{aligned}$$

Thus $f \mapsto f_N(1)$ is a B-homomorphism $V \to \sigma^w \otimes \delta_B^{-1}$ inducing a T-isomorphism $V_N \cong \sigma^w \otimes \delta_B^{-1}$. □

9.4. The irreducible representations of G exhibit a helpful finiteness property:

Proposition. *Let (π, V) be an irreducible smooth representation of G which is not cuspidal. The representation π is admissible.*

Proof. By definition, $V_N \neq 0$. By 9.1 Proposition, π is equivalent to a sub-representation of $\operatorname{Ind}_B^G \chi$, for some character χ of T. It is enough, therefore, to prove that $\operatorname{Ind} \chi$ is admissible.

We fix a compact open subgroup K of G; shrinking it if necessary, we may assume that $K \subset K_0 = \mathrm{GL}_2(\mathfrak{o})$. The space X^K of K-fixed points in $\operatorname{Ind} \chi$ consists of the functions $f : G \to \mathbb{C}$ satisfying

$$f(bgk) = \chi(b) f(g), \quad b \in B,\ g \in G,\ k \in K. \tag{9.4.1}$$

We have $G = BK_0$, so the set $B\backslash G/K$ is finite, and each double coset BgK supports, at most, a one-dimensional space of functions satisfying (9.4.1) (*cf.* 3.5). Thus X^K is finite-dimensional, as required. □

Remark. The irreducible cuspidal representations of G are likewise admissible, but the proof requires different techniques: see 10.2 Corollary below.

9.5. We introduce another notation. If (π, V) is a smooth representation of G and ϕ is a character of $F^\times$, we define a smooth representation $(\phi\pi, V)$ of G by setting

$$\phi\pi(g) = \phi(\det g)\,\pi(g), \quad g \in G. \tag{9.5.1}$$

One calls $\phi\pi$ the *twist* of π by ϕ.

Similarly for characters of T: if $\chi = \chi_1 \otimes \chi_2$ is a character of T and ϕ is a character of $F^\times$, then we put $\phi \cdot \chi = \phi\chi_1 \otimes \phi\chi_2$. If we inflate $\phi \cdot \chi$ to a representation of B trivial on N, we get $\phi \cdot \chi = (\phi \circ \det \mid B) \otimes \chi$. It follows immediately that

$$\mathrm{Ind}_B^G(\phi \cdot \chi) \cong \phi\, \mathrm{Ind}_B^G \chi. \tag{9.5.2}$$

This allows us to make convenient adjustments to the character χ without changing the essential structure of the induced representation.

9.6. We aim to give a precise account of the structure of representations of the form $\mathrm{Ind}_B^G \chi$. The main step is:

Irreducibility Criterion. *Let $\chi = \chi_1 \otimes \chi_2$ be a character of T, and set $(\Sigma, X) = \mathrm{Ind}_B^G \chi$.*

(1) *The representation (Σ, X) is reducible if and only if $\chi_1\chi_2^{-1}$ is either the trivial character or the character $x \mapsto \|x\|^2$ of $F^\times$.*
(2) *Suppose that (Σ, X) is reducible. Then:*
   (a) *the G-composition length of X is 2;*
   (b) *one composition factor of X has dimension 1, the other is of infinite dimension;*
   (c) *X has a 1-dimensional G-subspace if and only if $\chi_1\chi_2^{-1} = 1$;*
   (d) *X has a 1-dimensional G-quotient if and only if $\chi_1\chi_2^{-1}(x) = \|x\|^2$, $x \in F^\times$.*

We will refine this to a classification of the irreducible principal series representations in 9.11 below. The proof of the theorem occupies paragraphs 9.7–9.9 to follow.

9.7. We use the notation of 9.6. Let

$$V = \{f \in X : f(1) = 0\}.$$

This is a B-subspace of X and we have an exact sequence

$$0 \to V \longrightarrow X \longrightarrow \mathbb{C} \to 0,$$

where the one-dimensional space $\mathbb{C} = X/V$ carries the character χ of T. By the Restriction-Induction Lemma (9.3), $V_N \cong \delta_B^{-1}\chi^w$.

Proposition. *Let W be the kernel $V(N)$ of the canonical map $V \to V_N$. The space W is irreducible over B.*

Proof. We shall actually prove that W is irreducible as a representation of the mirabolic group M of §8. We observe that, by (8.1.2), $W_N = 0$ and $W = W(N)$.

Lemma. *For $f \in V$, define a function $f_N \in C_c^\infty(N)$ by $f_N(n) = f(wn)$, $n \in N$. The map*

$$\begin{aligned} V &\longrightarrow C_c^\infty(N), \\ f &\longmapsto f_N, \end{aligned} \tag{9.7.1}$$

is an N-isomorphism.

Proof. As in 9.3 Lemma, the support of $f \in V$ is contained in $Bw\mathcal{N}$, for some compact open subgroup $\mathcal{N}$ of N. The assertions follow immediately (observing that the notation f_N here is not the same as that in 9.3). □

For $\phi \in C_c^\infty(N)$ and $a \in F^\times$, we define $a\phi \in C_c^\infty(N)$ by

$$a\phi\begin{pmatrix} 1 & x \\ 0 & 1 \end{pmatrix} = \chi_2(a)\,\phi\begin{pmatrix} 1 & a^{-1}x \\ 0 & 1 \end{pmatrix}.$$

This gives an action of $F^\times$ on $C_c^\infty(N)$ which we regard as an action of the group S of matrices $\left(\begin{smallmatrix} a & 0 \\ 0 & 1 \end{smallmatrix}\right)$. We combine this with the natural action of N to give $C_c^\infty(N)$ the structure of a smooth representation of M. With this structure, the map (9.7.1) is a M-isomorphism.

Let ϑ be a non-trivial character of N. The map $f \mapsto \vartheta f$ is a linear automorphism of $V = C_c^\infty(N)$ carrying $V(N)$ to $V(\vartheta)$. Since $V/V(N)$ has dimension 1, we deduce that $\dim V_\vartheta = 1$ also. However, since N acts trivially on V_N, the inclusion $W \to V$ induces an isomorphism $W_\vartheta \cong V_\vartheta$, whence $W_\vartheta \cong \vartheta$. We apply 8.3 Theorem to get $W = W(N) \cong c\text{-}\mathrm{Ind}_N^M \vartheta$ which, by 8.2 Corollary, is irreducible. □

As a direct consequence of the Proposition, we have:

Corollary. *As a representation of B or of M, $\mathrm{Ind}_B^G \chi$ has composition length 3. Two of the composition factors have dimension one, and the third is of infinite dimension. In particular, the G-composition length of the representation $\mathrm{Ind}_B^G \chi$ is at most 3.*

9.8. We continue with the same notation and observe:

Proposition. *The following are equivalent:*

(1) $\chi_1 = \chi_2$;
(2) *X has a one-dimensional N-subspace.*

When these conditions hold,

(3) *X has a unique one-dimensional N-subspace X_0;*
(4) *X_0 is a G-subspace of X, and it is not contained in V.*

Proof. If (1) holds, we may as well take $\chi_1 = \chi_2 = 1$ (*cf.* (9.5.2)). The (non-zero) constant function then spans a one-dimensional G-subspace of X.

Conversely, let $f \in X$ span an N-stable subspace of dimension 1. Thus N acts on f (by right translation) as a character. The support of f is left-invariant under B, so $\operatorname{supp} f$ is either G or BwN. The second case is impossible: if $f(1) = 0$, then the support of f is confined to BwN_0, for some compact open subgroup N_0 of N (9.3 Lemma). Thus $\operatorname{supp} f = G$ and f vanishes nowhere. In particular $f(1) \neq 0$, and $f \notin V$. The canonical N-map $X \to \mathbb{C} = X/V$ identifies the N-space $\mathbb{C}f$ with the trivial N-space $\mathbb{C}$. It follows that N fixes f under right translation. Take $x \in F^\times$, and consider the identity

$$w\begin{pmatrix} 1 & x \\ 0 & 1 \end{pmatrix} = \begin{pmatrix} 1 & x^{-1} \\ 0 & 1 \end{pmatrix}\begin{pmatrix} -x^{-1} & 0 \\ 0 & x \end{pmatrix}\begin{pmatrix} 1 & 0 \\ x^{-1} & 1 \end{pmatrix}.$$

If $\|x\|$ is sufficiently large, then f is fixed under right translation by $\left(\begin{smallmatrix} 1 & 0 \\ x^{-1} & 1 \end{smallmatrix}\right)$. So, as f is fixed by N, we have

$$f(w) = \chi_1(-1)\,\chi_1^{-1}\chi_2(x)\,f(1),$$

for all $x \in F^\times$ of sufficiently large absolute value. Thus $\chi_1 = \chi_2 = \phi$, say, and $f(g) = \phi(\det g)f(1)$. We have proved (1) $\Leftrightarrow$ (2), and that the one-dimensional N-subspace is uniquely determined. We have already shown that it is not contained in V. □

9.9. We now finish the proof of the Irreducibility Criterion 9.6. Assume X is reducible. Its G-length is 2 or 3, and it has either a finite-dimensional G-subspace or a finite-dimensional G-quotient (9.7 Corollary).

Assume the first alternative. Thus X has a one-dimensional N-subspace, and we are in the situation of 9.8: X has a one-dimensional G-subspace L, and $\chi_1 = \chi_2 = \phi$, say. Moreover, G acts on L as the character $\phi \circ \det$ and $L \cap V = 0$ (notation of 9.7). The quotient $Y = X/L$ is therefore B-isomorphic to V. If X has G-length 3, then Y has G-length 2. However, V has B-length 2 and a unique B-quotient, which is of dimension 1. This gives a G-quotient of Y on which G must act as a character $\phi' \circ \det$ (9.2). This would force $\phi' \otimes \phi'$ to appear as a factor in the Jacquet module $Y_N \cong \phi \cdot \delta_B^{-1}$, which it cannot. Thus X has G-length 2 and we are in case (2)(c) of 9.6.

In the other alternative, X has a finite-dimensional G-quotient. The representation $\check{X}$ therefore has a finite-dimensional G-subspace, and we are back with the first alternative. By the Duality Theorem of 7.7, $\check{X} \cong \operatorname{Ind}_B^G \delta_B^{-1}\check{\chi}$, so we are in the case (2)(d) of 9.6.

Thus, in part (1) of the theorem, we have shown that X reducible implies χ has the stated form. The converse is given by 9.8 and the dual case. We have also proved statements (a)–(d). □

9.10. To get a classification of the irreducible, non-cuspidal representations of G, we need to investigate the homomorphisms between induced representations:

Proposition. *Let* χ, ξ *be characters of* T. *The space* $\mathrm{Hom}_G(\mathrm{Ind}_B^G \chi, \mathrm{Ind}_B^G \xi)$ *has dimension* 1 *if* $\xi = \chi$ *or* $\chi^w \delta_B^{-1}$, *zero otherwise.*

Proof. We use Frobenius Reciprocity (9.1.2):

$$\mathrm{Hom}_G(\mathrm{Ind}_B^G \chi, \mathrm{Ind}_B^G \xi) \cong \mathrm{Hom}_T((\mathrm{Ind}\,\chi)_N, \xi).$$

The Jacquet module $(\mathrm{Ind}\,\chi)_N$ fits into the exact sequence

$$0 \to \chi^w \delta_B^{-1} \longrightarrow (\mathrm{Ind}\,\chi)_N \longrightarrow \chi \to 0.$$

If we assume that $\chi \neq \chi^w \delta_B^{-1}$, then this sequence splits and the result follows immediately. The equation $\chi = \chi^w \delta_B^{-1}$ amounts to $\chi_1(x) = \|x\|\chi_2(x)$, $x \in F^\times$. In this case, $\mathrm{Ind}\,\chi$ is irreducible and the result again follows. □

By way of some examples, we examine in more detail the case where $\mathrm{Ind}_B^G \chi$ is *reducible.* Thus there is a character ϕ of $F^\times$ such that $\chi = \phi \cdot 1_T$ or $\chi = \phi \cdot \delta_B^{-1}$. Twisting does not affect the situation materially, so we assume $\phi = 1$.

Consider first the case $\mathrm{Ind}_B^G 1_T$. The irreducible G-quotient of $\mathrm{Ind}_B^G 1_T$ is called the *Steinberg representation* of G, and is denoted St_G:

$$0 \to 1_G \longrightarrow \mathrm{Ind}_B^G 1_T \longrightarrow \mathrm{St}_G \to 0. \qquad (9.10.3)$$

Its dimension is infinite and $(\mathrm{St}_G)_N \cong \delta_B^{-1}$. Likewise, if ϕ is a character of $F^\times$, we have an exact sequence

$$0 \to \phi_G \longrightarrow \mathrm{Ind}_B^G \phi_T \longrightarrow \phi \cdot \mathrm{St}_G \to 0,$$

where $\phi_G = \phi \circ \det$ and $\phi_T = \phi \otimes \phi$. (Representations of the form $\phi \cdot \mathrm{St}_G$ are sometimes called *special.*)

Taking the smooth dual of (9.10.3), we get an exact sequence

$$0 \to \mathrm{St}_G^\vee \longrightarrow \mathrm{Ind}_B^G \delta_B^{-1} \longrightarrow 1_G \to 0. \qquad (9.10.4)$$

The proposition implies

$$\mathrm{St}_G \cong \mathrm{St}_G^\vee. \qquad (9.10.5)$$

Remark. The proposition also implies that the space $\mathrm{End}_G(\mathrm{Ind}\,1_T)$ has dimension 1, while $\mathrm{Ind}\,1_T$ is not irreducible. Thus the converse of Schur's Lemma fails in this context, as remarked in 2.6.

Observe also the imperfect parallelism between the proposition above and the corresponding result (6.3) for the finite field case.

9.11. We introduce a new notation. If σ is a smooth representation of T, we define

$$\iota_B^G\,\sigma = \mathrm{Ind}_B^G(\delta_B^{-1/2}\otimes\sigma). \tag{9.11.1}$$

This provides another exact functor $\mathrm{Rep}(T)\to\mathrm{Rep}(G)$, known as *normalized* or *unitary* smooth induction. It gives rise to more convenient combinatorics, for example:

$$\left(\iota_B^G\,\sigma\right)^\vee \cong \iota_B^G\,\check{\sigma}. \tag{9.11.2}$$

In this language, the Irreducibility Criterion (9.6) and 9.10 Proposition say:

Lemma.

(1) *Let $\chi=\chi_1\otimes\chi_2$ be a character of T. The representation $\iota_B^G\,\chi$ is reducible if and only if $\chi_1\chi_2^{-1}$ is one of the characters $x\mapsto\|x\|^{\pm1}$ of $F^\times$ or, equivalently, $\chi=\phi\cdot\delta_B^{\pm1/2}$ for some character ϕ of $F^\times$.*
(2) *Let χ, ξ be characters of T. The space $\mathrm{Hom}_G(\iota_B^G\,\chi,\iota_B^G\,\xi)$ is not zero if and only if $\xi=\chi$ or $\xi=\chi^w$.*

Gathering up our earlier arguments and results, we get:

Classification Theorem. *The following is a complete list of the isomorphism classes of irreducible, non-cuspidal representations of G:*

(1) *the irreducible induced representations $\iota_B^G\,\chi$, where $\chi\neq\phi\cdot\delta_B^{\pm1/2}$ for any character ϕ of $F^\times$;*
(2) *the one-dimensional representations $\phi\circ\det$, where ϕ ranges over the characters of $F^\times$;*
(3) *the special representations $\phi\cdot\mathrm{St}_G$, where ϕ ranges over the characters of $F^\times$.*

The classes in this list are all distinct except that, in (1), we have $\iota_B^G\,\chi\cong\iota_B^G\,\chi^w$.

Proof. The examples in 9.10 show that every irreducible, non-cuspidal representation of G appears in this list. The relations between the irreducibly induced ones are given by 9.10 Proposition. The same result also implies that no special representation $\phi\cdot\mathrm{St}_G$ can be equivalent to $\phi'\cdot\mathrm{St}_G$, $\phi'\neq\phi$, or any irreducibly induced representation. □

10. Cuspidal Representations and Coefficients

In 9.1, we defined an irreducible smooth representation (π,V) of $\mathrm{GL}_2(F)$ to be cuspidal if its Jacquet module (π_N,V_N) is trivial or, equivalently, if it is not isomorphic to a composition factor of an induced representation $\mathrm{Ind}_B^G\,\chi$, for any character χ of T. Such a negative and exclusive approach yields essentially no information about this particularly important class of representations. We now give an alternative definition, valid in a much wider context, and show it is equivalent to the original one. It provides the starting point for a method of constructing cuspidal representations.

From this new viewpoint, irreducible cuspidal representations have a striking algebraic property: they are projective objects in the appropriate subcategory of $\mathrm{Rep}(G)$. We have no direct need for this result, but we have included it as an appendix.

10.1. Let (π, V) be a smooth representation of $G = \mathrm{GL}_2(F)$; from vectors $v \in V$, $\check{v} \in \check{V}$, we get a smooth function on G by

$$\gamma_{\check{v}\otimes v} : g \longmapsto \langle \check{v}, \pi(g)v \rangle.$$

We let $\mathcal{C}(\pi)$ be the vector space spanned by the functions $\gamma_{\check{v}\otimes v}$, $\check{v}\otimes v \in \check{V}\otimes V$. The functions $f \in \mathcal{C}(\pi)$ are called the *(matrix) coefficients* of π.

The space $\check{V}\otimes V$ carries a smooth representation of the group $G\times G$, while $G \times G$ acts on the function space $\mathcal{C}(\pi)$ by translation: the first factor acts by left translation and the second by right translation. The map $\check{v}\otimes v \mapsto \gamma_{\check{v}\otimes v}$ is then a surjective $G \times G$-homomorphism $\check{V} \otimes V \to \mathcal{C}(\pi)$.

The space $\mathcal{C}(\pi)$ is primarily, but not exclusively, of interest in the case where π is irreducible. When π is irreducible, the centre Z of G acts on V via the central character ω_π of π and

$$\gamma(zg) = \omega_\pi(z)\gamma(g), \quad z \in Z, g \in G, \gamma \in \mathcal{C}(\pi).$$

The support of a coefficient is therefore invariant under translation by Z.

Definition. *Let (π, V) be an irreducible smooth representation of G; one says that π is* γ-cuspidal *if every $\gamma \in \mathcal{C}(\pi)$ is compactly supported modulo Z.*

The term "γ-cuspidal" is a convenient, but temporary, expedient.

Convention. To save adjectives, if a representation is described as cuspidal or γ-cuspidal, it is implicitly assumed to be smooth.

We first achieve some technical control:

Proposition.

(1) *If (π, V) is an irreducible γ-cuspidal representation of G, then π is admissible.*
(2) *Let (π, V) be an irreducible admissible representation of G, and suppose that some non-zero coefficient of π is compactly supported modulo Z; then π is γ-cuspidal.*

Proof. In part (1), we suppose for a contradiction that π is not admissible. We choose a compact open subgroup K such that V^K has infinite dimension. This dimension, we note, is countable (2.6). The dimension of $\check{V}^K \cong \mathrm{Hom}_{\mathbb{C}}(V^K, \mathbb{C})$ is therefore uncountable.

We fix a non-zero $v \in V^K$ and consider the map $\Gamma_v : \check{V}^K \to \mathcal{C}(\pi)$ given by $\check{v} \mapsto \gamma_{\check{v}\otimes v}$. Since the translates gv, $g \in G$, span V, the map Γ_v is injective. Its image is a space of functions f on G, satisfying

$$f(zkgk') = \omega_\pi(z)f(g), \quad g \in G, \ z \in Z, \ k, k' \in K, \tag{10.1.1}$$

and supported on a finite union of cosets $ZKgK$. The dimension of $\Gamma_v(\check{V}^K)$ is therefore at most countable, while Γ_v is injective and $\dim \check{V}^K$ is uncountable. This gives the desired contradiction.

We turn to part (2). The smooth dual $(\check{\pi}, \check{V})$ is irreducible and admissible (2.10). We view the space $\check{V} \otimes V$ as a smooth representation of $G \times G$, and hence as a smooth module over $\mathcal{H}(G \times G) = \mathcal{H}(G) \otimes \mathcal{H}(G)$.

If K is a compact open subgroup of G, we have

$$(\check{V} \otimes V)^{K\times K} = (e_K \otimes e_K) * (\check{V} \otimes V) = \check{V}^K \otimes V^K.$$

If K is sufficiently small, the spaces V^K, $\check{V}^K$ are finite-dimensional simple modules over $\mathcal{H}(G, K)$. The Jacobson Density Theorem implies that $\check{V}^K \otimes V^K$ is a simple module over $\mathcal{H}(G, K) \otimes \mathcal{H}(G, K) \cong \mathcal{H}(G \times G, K \times K)$. This holds for all sufficiently small K, so $\check{V} \otimes V$ is an irreducible admissible $G \times G$-space (4.3 Corollary).

The surjective $G \times G$-homomorphism $\gamma : \check{V} \otimes V \to \mathcal{C}(\pi)$ is therefore an isomorphism and $\mathcal{C}(\pi)$ is irreducible over $G \times G$. If $\gamma \in \mathcal{C}(\pi)$ is non-zero, any $\gamma' \in \mathcal{C}(\pi)$ is a finite linear combination of functions $(g, h)\gamma$, $(g, h) \in G \times G$. If γ is compactly supported modulo Z, then so is γ'. □

Remark 1. All of the preceding definitions and arguments apply in the general case, where G is a unimodular locally profinite group satisfying 2.6 Hypothesis. Indeed, 2.6 is only used at one point, in the proof of part (1) of the proposition. Even this can be avoided by noting that the dual of a vector space W has dimension strictly greater than $\dim W$ except when $\dim W$ is finite.

Remark 2. In the general context of Remark 1, part (2) of the proposition *fails* when the irreducible smooth representation (π, V) is not admissible. An example is given by the representation $c\text{-}\mathrm{Ind}_N^M \vartheta$ considered in 8.2: see especially 8.2 Remark.

10.2. The reason for introducing the notion of γ-cuspidality is explained by:

Theorem. *Let (π, V) be an irreducible smooth representation of G; then π is cuspidal if and only if it is γ-cuspidal.*

Proof. We first assume that π is cuspidal, and show it is γ-cuspidal. Let ϖ be a prime element of F, and put

$$t = \begin{pmatrix} \varpi & 0 \\ 0 & 1 \end{pmatrix}.$$

The set T^+ of powers t^n, $n \geqslant 0$, then provides a family of representatives for $ZK\backslash G/K$, $K = \mathrm{GL}_2(\mathfrak{o})$ (7.2.2).

Lemma. *Let $v \in V$, $\check{v} \in \check{V}$. There exists $m \geqslant 0$ such that $\gamma_{\check{v}\otimes v}(t^n) = 0$, for all $n \geqslant m$.*

Proof. We choose a compact open subgroup N_1 of N which fixes $\check{v}$. Since $V_N = 0$, we have $v \in V(N)$ and (8.1) there is a compact open subgroup N_2 of N such that

$$\int_{N_2} \pi(x)v\,dx = 0.$$

We then have

$$\int_{N_0} \pi(x)v\,dx = 0,$$

for any compact open subgroup N_0 of N containing N_2. However, there exists $m \geqslant 0$ such that $t^a N_2 t^{-a} \subset N_1$ for all $a \geqslant m$. For such a we have (for certain positive constants k_1, k_2)

$$\begin{aligned}
\langle \check{v}, \pi(t^a)v\rangle &= k_1 \int_{N_1} \langle \check{\pi}(x^{-1})\check{v}, \pi(t^a)v\rangle\,dx \\
&= k_1 \int_{N_1} \langle \check{\pi}(t^{-a})\check{v}, \pi(t^{-a}xt^a)v\rangle\,dx \\
&= k_2 \int_{t^{-a}N_1t^a} \langle \check{\pi}(t^{-a})\check{v}, \pi(x)v\rangle\,dx \\
&= 0,
\end{aligned}$$

since $t^{-a}N_1t^a \supset N_2$. $\square$

Continuing with the proof of the theorem, we fix a non-zero coefficient $f = \gamma_{\check{v}\otimes v}$ of π. We write $K = \mathrm{GL}_2(\mathfrak{o})$ and let K' be an open normal subgroup of K fixing both $\check{v}$ and v. We let $k_1, k_2, \dots, k_r$ be a set of coset representatives for K/K'. Thus, if $g \in G$, there exists $n \geqslant 0$ such that

$$ZKgK = ZKt^nK = \bigcup_{i,j} ZK'k_i^{-1}t^nk_jK'.$$

It follows that

$$\operatorname{supp} f \subset \bigcup_{1\leqslant i,j\leqslant r} ZK'\,(\operatorname{supp} f_{ij} \cap T^+)\,K',$$

where f_{ij} denotes the coefficient function $x \mapsto f(k_i x k_j^{-1})$. This set is compact modulo Z, by the lemma. It follows that all coefficients $\gamma_{\check{v}\otimes v}$ of π are compactly supported modulo Z, and π is therefore γ-cuspidal.

Combining this argument with 9.4 Proposition and 10.1 Proposition, we have shown:

Corollary. *Every irreducible smooth representation of $G = \mathrm{GL}_2(F)$ is admissible.*

We now prove the converse statement in the theorem. Let (π, V) be an irreducible γ-cuspidal representation of G. In particular, (π, V) is admissible. By 2.10 Proposition, the dual $(\check{\pi}, \check{V})$ is irreducible and admissible. Let K_n denote the group $1+\mathfrak{p}^n\mathrm{M}_2(\mathfrak{o})$, $n \geqslant 1$. We take $v \in V$ and choose $n \geqslant 1$ so that v is fixed by $\pi(K_n)$. We take t as before.

For $\check{v} \in \check{V}^{K_n}$, the function $g \mapsto \langle \check{v}, \pi(g)v\rangle$ is compactly supported modulo Z; we deduce that $\langle \check{v}, \pi(t^a)v\rangle = 0$ for all $a \in \mathbb{Z}$ sufficiently large. Since $\check{V}^{K_n}$ is of finite dimension there is a constant c such that $\langle \check{v}, \pi(t^a)v\rangle = 0$ for all $\check{v} \in \check{V}^{K_n}$ and all $a \geqslant c$. This implies $\pi(e_{K_n})\pi(t^a)v = 0$ for $a \geqslant c$. We write, for $j \in \mathbb{Z}$,

$$N_j = \begin{pmatrix} 1 & \mathfrak{p}^j \\ 0 & 1 \end{pmatrix}, \quad N'_j = \begin{pmatrix} 1 & 0 \\ \mathfrak{p}^j & 1 \end{pmatrix}, \quad T_n = K_n \cap T,$$

so that $K_n = N_n T_n N'_n$. Set $K_n^{(a)} = t^{-a}K_n t^a = N_{n-a}T_n N'_{n+a}$; we then have, for $a \geqslant c$,

$$\begin{aligned} 0 = \pi(e_{K_n})\pi(t^a)v &= \pi(t^a)\pi(e_{K_n^{(a)}})v \\ &= \pi(t^a) \sum_{x \in N_{n-a}/N_n} \pi(x)\pi(e_{K_n^{(a)}\cap K_n})v. \end{aligned}$$

However, v is fixed by $K_n^{(a)} \cap K_n$, so this equation reduces to

$$0 = k\,\pi(t^a) \int_{N_{n-a}} \pi(x)v\,dx,$$

for a constant $k > 0$ depending on the choice of a Haar measure dx on N. We deduce that $v \in V(N)$ (8.1 Lemma). This applies to all $v \in V$, so π is cuspidal, as required. □

10a. Appendix: Projectivity Theorem

We give another property of γ-cuspidal representations. We will not use this result, but we have included it for its power and beauty. It also holds in a very general context.

10a.1. Let G be a locally profinite group with centre Z. Let χ be a character of Z and let (π, V) be a smooth representation of G. We recall that π *admits* χ *as central character* if $\pi(z)v = \chi(z)v$, for $z \in Z$, $v \in V$.

Projectivity Theorem. *Let G be a unimodular locally profinite group, satisfying* (2.6) *and with centre Z. Assume that any character $\chi : Z \to \mathbb{R}_+^\times$ extends to a character $G \to \mathbb{R}_+^\times$.*

Let (π, V) be an irreducible γ-cuspidal representation of G, and let (τ, U) be a smooth representation of G admitting ω_π as central character. Let $f : U \to V$ be a surjective G-homomorphism. There exists a G-homomorphism $\phi : V \to U$ such that $f \circ \phi = 1_V$.

10a.2. The group G/Z is locally profinite. One sees easily that it is unimodular. Indeed, let μ_G, μ_Z be Haar measures on G, Z respectively. By 3.4 Proposition, there is a unique right Haar measure $\dot{\mu}$ on G/Z such that

$$\int_G f(g)\, d\mu_G(g) = \int_{G/Z} \int_Z f(zg)\, d\mu_Z(z)\, d\dot{\mu}(g), \quad f \in C_c^\infty(G).$$

Symmetrically, $\dot{\mu}$ is also a left Haar measure on G/Z.

Schur's orthogonality relation. *Let $d\dot{g}$ be a Haar measure on G/Z, and let $v_1, v_2 \in V$, $\check{v}_1, \check{v}_2 \in \check{V}$. The function*

$$g \longmapsto \langle \check{\pi}(g)\check{v}_1, v_1\rangle \langle \check{v}_2, \pi(g)v_2\rangle, \quad g \in G,$$

is invariant under translation by Z and

$$\int_{G/Z} \langle \check{\pi}(g)\check{v}_1, v_1\rangle \langle \check{v}_2, \pi(g)v_2\rangle\, d\dot{g} = d(\pi)^{-1} \langle \check{v}_1, v_2\rangle \langle \check{v}_2, v_1\rangle,$$

for a constant $d(\pi) > 0$ depending only on π and the measure $d\dot{g}$.

Proof. Since π is γ-cuspidal, the integrand has compact support in G/Z and the integral converges. If we fix, say, $\check{v}_1$ and v_2, the integral determines a G-invariant pairing $\check{V} \times V \to \mathbb{C}$. Such a pairing is given by a G-homomorphism $\Theta : \check{V} \to \check{V}$ (*cf.* Exercise 2.10). Since V is admissible (10.1 Proposition) and irreducible, the same applies to $\check{V}$ (2.10) and Schur's Lemma (2.6) implies that any G-invariant pairing $\check{V} \times V \to \mathbb{C}$ is a scalar multiple of the standard one. Therefore, there is a constant $c_{\check{v}_1, v_2}$ such that

$$\int_{G/Z} \langle \check{\pi}(g)\check{v}_1, v_1\rangle \langle \check{v}_2, \pi(g)v_2\rangle\, d\dot{g} = c_{\check{v}_1, v_2} \langle \check{v}_2, v_1\rangle.$$

The function $(\check{v}_1, v_2) \mapsto c_{\check{v}_1, v_2}$ is again a G-invariant bilinear pairing $\check{V} \times V \to \mathbb{C}$, so

$$c_{\check{v}_1, v_2} = c_\pi \langle \check{v}_1, v_2\rangle,$$

for a constant c_π.

It remains only to prove that $c_\pi > 0$. The assumption on G allows us to replace π by a twist and assume that $|\omega_\pi| = 1$. The space V then admits a positive definite, G-invariant Hermitian form h, constructed as follows. One chooses a nonzero element $\check{v} \in \check{V}$ and sets

$$h(v_1, v_2) = \int_{G/Z} \overline{\langle \check{v}, \pi(g)v_1\rangle} \langle \check{v}, \pi(g)v_2\rangle \, d\dot{g}.$$

There is then a complex anti-linear G-isomorphism $\Theta : (\pi, V) \to (\check{\pi}, \check{V})$ such that $h(v_1, v_2) = \langle \Theta v_1, v_2\rangle$. Schur's Lemma again implies that h is the unique G-invariant, positive definite Hermitian form on V, up to a positive constant factor.

Going through the same argument, one sees that

$$\int_{G/Z} h(\pi(g)v_1, v_2)\, h(v_3, \pi(g)v_4)\, d\dot{g} = b_\pi\, h(v_1, v_4)\, h(v_3, v_2),$$

for a constant b_π. On taking $v_1 = v_2 = v_3 = v_4 \neq 0$, one sees that $b_\pi > 0$. On the other hand,

$$\begin{aligned}\int_{G/Z} h(\pi(g)v_1, v_2)\, h(v_3, \pi(g)v_4)\, d\dot{g} &= \int_{G/Z} \langle \check{\pi}(g)\Theta(v_1), v_2\rangle \langle \Theta(v_3), \pi(g)v_4\rangle \, d\dot{g} \\ &= c_\pi \, \langle \Theta(v_1), v_4\rangle \langle \Theta(v_3), v_2\rangle \\ &= c_\pi \, h(v_1, v_4)\, h(v_3, v_2).\end{aligned}$$

Therefore $c_\pi = b_\pi > 0$, as required. □

Remark. Let (π, V) be an irreducible smooth representation of G such that $|\omega_\pi| = 1$. One says that π is *square-integrable modulo* Z if

$$\int_{G/Z} |\langle \check{v}, \pi(g)v\rangle|^2 \, d\dot{g} < \infty$$

for all $\check{v} \in \check{V}$, $v \in V$. The orthogonality relation then holds for π, with exactly the same proof. The positive constant $d(\pi)$ is called the *formal degree* of π, relative to the measure $d\dot{g}$. (For a full discussion of square-integrable representations of $\mathrm{GL}_2(F)$, see 17.4 *et seq.* below.)

10a.3. We now prove the Projectivity Theorem. First, we need to generalize the constructions of 4.1, 4.2. Let χ be a character of $F^\times$. Let $\mathcal{H}_\chi(G)$ be the space of locally constant functions $f : G \to \mathbb{C}$, which are compactly supported modulo Z, such that $f(zg) = \chi(z)^{-1} f(g)$, $z \in Z$, $g \in G$. Using a Haar measure on G/Z, we define convolution on $\mathcal{H}_\chi(G)$ as in (4.1). If (σ, W) is a smooth representation of G admitting χ as central character, we extend the action of G on W to one of $\mathcal{H}_\chi(G)$, just as in 4.2.

We let (π, V) be an irreducible γ-cuspidal representation of G, as in the theorem. We abbreviate $\omega = \omega_\pi$. We take a smooth representation (τ, U) of G, admitting ω as central character, and a G-surjection $f : U \to V$. If $u \in U$ satisfies $f(u) \neq 0$, the restriction of f to the G-space $\tau(\mathcal{H}_\omega(G))u$ generated by u is still surjective. Composing it with the obvious G-surjection $\mathcal{H}_\omega(G) \to \tau(\mathcal{H}_\omega(G))u$, we get a G-surjection

$$\begin{aligned} \Pi : \mathcal{H}_\omega(G) &\longrightarrow V, \\ \phi \longmapsto f(\tau(\phi)u) &= \pi(\phi)v_0, \end{aligned}$$

where $v_0 = f(u)$. It is enough to show that Π splits over G.

We choose a vector $\check{v}_0 \in \check{V}$ such that $d(\pi)^{-1}\langle \check{v}_0, v_0 \rangle = 1$. The function $\phi_v : g \mapsto \langle \check{\pi}(g)\check{v}_0, v \rangle$ lies in $\mathcal{H}_\omega(G)$ and the map

$$\begin{aligned} \Phi : V &\longrightarrow \mathcal{H}_\omega(G), \\ v &\longmapsto \phi_v, \end{aligned}$$

is a G-homomorphism. The composite map $\Pi \circ \Phi$ is given by

$$w \longmapsto \pi(\phi_w)v_0 = \int_{G/Z} \pi(g)\phi_w(g)v_0 \, d\dot{g}, \quad w \in V.$$

For $\check{w} \in \check{V}$, this gives

$$\langle \check{w}, \Pi\Phi(w) \rangle = \int_{G/Z} \langle \check{w}, \check{\pi}(g)v_0 \rangle \langle \check{\pi}(g)\check{v}_0, w \rangle \, d\dot{g} = \langle \check{w}, w \rangle,$$

whence $\Pi\Phi(w) = w$, as required. □

11. Intertwining, Compact Induction and Cuspidal Representations

We describe a method for constructing irreducible cuspidal representations of $G = \mathrm{GL}_2(F)$, using compact induction from open subgroups. At this stage, it is a purely formal matter: it is not clear that the necessary hypotheses are satisfied sufficiently often to give useful results. Such issues are the subject of the next chapter. Here, we have to be content with one interesting example.

11.1. We start with general considerations so, for the time being, G is a unimodular locally profinite group with the countability property 2.6. Throughout, Z denotes the centre of G. If K is a compact open subgroup of G, we write $\widehat{K}$ for the set of isomorphism classes of irreducible smooth representations of K.

Definition 1. *For $i = 1, 2$, let K_i be a compact open subgroup of G and let $\rho_i \in \widehat{K}_i$. Let $g \in G$. The element g* intertwines *ρ_1* with *ρ_2 if*

$$\mathrm{Hom}_{K_1^g \cap K_2}(\rho_1^g, \rho_2) \neq 0,$$

where ρ_1^g denotes the representation $x \mapsto \rho_1(gxg^{-1})$ of the group $K_1^g = g^{-1}K_1g$.

As a property of g, this depends only on the double coset K_1gK_2.

The definition applies equally if the K_i are just closed subgroups of G; we will often need it in the case where the K_i are open and compact modulo the centre of G.

Definition 2. *Let K be a compact open subgroup of G, and let (π, V) be a smooth representation of G. We say that π* contains *ρ, or ρ* occurs in *π, if $\mathrm{Hom}_K(\rho, \pi) \neq 0$.*

Again, we can use the same definition in more general contexts, for example, if K is open and compact modulo the centre of G and π admits a central character (see 2.7). We also use it when G is compact and K is a closed subgroup of G.

Remaining with the compact open case for the time being, the significance of the concept of intertwining is first indicated by the following.

Proposition 1. *For $i = 1, 2$, let K_i be a compact open subgroup of G and let $\rho_i \in \widehat{K}_i$. Let (π, V) be an irreducible smooth representation of G which contains both ρ_1 and ρ_2. There then exists $g \in G$ which intertwines ρ_1 with ρ_2.*

Proof. For each i, we have the decomposition of V into K_i-isotypic components (2.3 Proposition). The hypothesis is equivalent to $V^{\rho_i} \neq 0$, $i = 1, 2$.

Let e_2 denote the K_2-projection $V \to V^{\rho_2}$. Since π is irreducible and $V^{\rho_1} \neq 0$, the spaces $\pi(g^{-1})V^{\rho_1} = V^{\rho_1^g}$, $g \in G$, span V. We can therefore choose $g \in G$ such that $e_2 \circ \pi(g^{-1})$ induces a non-zero map $V^{\rho_1} \to V^{\rho_2}$: this is the required element g. □

Take (K_i, ρ_i) as in the Proposition. The representations ρ_1^g, ρ_2 of $K_1^g \cap K_2$ are semisimple, so the spaces

$$\mathrm{Hom}_{K_1^g \cap K_2}(\rho_1^g, \rho_2), \qquad \mathrm{Hom}_{K_1^g \cap K_2}(\rho_2, \rho_1^g) \cong \mathrm{Hom}_{K_1 \cap K_2^{g^{-1}}}(\rho_2^{g^{-1}}, \rho_1)$$

have the same dimension. Therefore g intertwines ρ_1 with ρ_2 if and only if g^{-1} intertwines ρ_2 with ρ_1.

Language.

(1) We say that the ρ_i *intertwine in* G if there exists $g \in G$ which intertwines ρ_1 with ρ_2. The relation of G-intertwining between pairs (K_i, ρ_i) is therefore symmetric and reflexive; it is not transitive.
(2) If we have a single pair (K, ρ), we say that g intertwines ρ if it intertwines ρ with itself.

Remark. We will often wish to use this approach when K is not compact, but only an open subgroup of G which is *compact modulo* Z. One cannot, in general, decompose a smooth representation (π, V) of G into a direct sum of K-isotypic components. Such a decomposition does exist (2.7) if (π, V) admits a central character ω_π, in particular, if π is irreducible. With this caveat, we can treat open, compact modulo centre subgroups of G in the same way as compact open subgroups.

We will later (in Chapter VI) need another intertwining criterion. (We use the notation of 4.4 here.)

Proposition 2. *Let K be a compact open subgroup of G, let $g \in G$, and $\rho \in \widehat{K}$. The following are equivalent:*

(1) *there exists $f \in e_\rho * \mathcal{H}(G) * e_\rho$ such that $f \mid KgK \neq 0$;*
(2) *g intertwines ρ.*

Proof. Consider the space $C^\infty(KgK)$ of G-smooth functions on the coset KgK. This carries a smooth representation of $K \times K$ by

$$(k_1, k_2)f : x \longmapsto f(k_1^{-1} x k_2).$$

Let H denote the group of pairs $(k, g^{-1}kg) \in K \times K$, $k \in K \cap gKg^{-1}$. The map $f \mapsto f(g)$ is then an H-homomorphism $C^\infty(KgK) \to \mathbb{C}$ (with H acting trivially). By Frobenius Reciprocity (2.4), this induces a $K \times K$-homomorphism

$$C^\infty(KgK) \longrightarrow \mathrm{Ind}_H^{K\times K}(1_H). \tag{11.1.1}$$

We show this is an isomorphism.

The space $V = \mathrm{Ind}_H^{K\times K}(1_H)$ consists, by definition, of smooth functions $\phi : K \times K \to \mathbb{C}$ such that $\phi(hk_1, g^{-1}hgk_2) = \phi(k_1, k_2)$, $k_i \in K$, $h \in K \cap gKg^{-1}$. Given such a function ϕ, we can define $f_\phi \in C^\infty(KgK)$ by setting $f_\phi(k_1 g k_2) = \phi(k_1^{-1}, k_2)$; the map $\phi \mapsto f_\phi$ is the inverse of the map (11.1.1).

In these terms, condition (1) amounts to $e_\rho * C^\infty(KgK) * e_\rho \cong V^{\rho\otimes\check{\rho}} \neq 0$. Equivalently,

$$\mathrm{Hom}_{K\times K}(\rho \otimes \check{\rho}, V) \cong \mathrm{Hom}_H(\rho \otimes \check{\rho}, 1_H) \neq 0.$$

The last relation is equivalent to the representation $k \mapsto \rho(k) \otimes \check{\rho}(g^{-1}kg)$ of $K \cap gKg^{-1}$ having a fixed vector, that is, $\mathrm{Hom}_{K\cap gKg^{-1}}(\rho, \rho^{g^{-1}}) \neq 0$, as required. □

11.2. Let K be an open subgroup of G, containing and compact modulo the centre Z of G. Let (ρ, W) be an irreducible smooth representation of K. Let $\mathcal{H}(G,\rho)$ be the space of functions $f : G \to \mathrm{End}_{\mathbb{C}}(W)$ which are compactly supported modulo Z and satisfy

$$f(k_1 g k_2) = \rho(k_1) f(g) \rho(k_2), \quad k_i \in K,\ g \in G.$$

Observe that the support of any $f \in \mathcal{H}(G,\rho)$ is a finite union of double cosets KgK.

Let $\dot\mu$ be a Haar measure on G/Z. For $\phi_1, \phi_2 \in \mathcal{H}(G,\rho)$, we set

$$\phi_1 * \phi_2(g) = \int_{G/Z} \phi_1(x)\phi_2(x^{-1}g)\, d\dot\mu(x), \quad g \in G.$$

The function $\phi_1 * \phi_2$ lies in $\mathcal{H}(G,\rho)$ and, under this operation of convolution, the space $\mathcal{H}(G,\rho)$ is an associative $\mathbb{C}$-algebra with 1.

Remark. The algebra $\mathcal{H}(G,\rho)$ is called the *ρ-spherical Hecke algebra* of G, or the *intertwining algebra* of ρ in G. It is closely related to the algebra $e_\rho * \mathcal{H}(G) * e_\rho$: there is a canonical algebra isomorphism $e_\rho * \mathcal{H}(G) * e_\rho \cong \mathcal{H}(G,\rho) \otimes \mathrm{End}_{\mathbb{C}}(W)$. (In the literature, the algebra we have defined is sometimes denoted $\mathcal{H}(G,\check\rho)$.)

Lemma. *Let $g \in G$; there exists $\phi \in \mathcal{H}(G,\rho)$ with support KgK if and only if g intertwines ρ.*

Proof. Let $f \in \mathrm{End}_{\mathbb{C}}(W)$; for a fixed $g \in G$, the assignment $kgk' \mapsto \rho(k) f \rho(k')$, $k, k' \in K$, gives an element of $\mathcal{H}(G,\rho)$ if and only if, for $k \in K^g \cap K$, we have $f \circ \rho(k) = \rho^g(k) \circ f$. That is, if and only if $f \in \mathrm{Hom}_{K^g\cap K}(\rho, \rho^g)$. The representations ρ, ρ^g of $K \cap K^g$ are semisimple, so the spaces $\mathrm{Hom}_{K^g\cap K}(\rho,\rho^g)$, $\mathrm{Hom}_{K^g\cap K}(\rho^g,\rho)$ have the same dimension. The Lemma now follows. □

We have actually shown that the space of functions $f \in \mathcal{H}(G,\rho)$ supported on KgK is canonically isomorphic to $\mathrm{Hom}_{K^g\cap K}(\rho,\rho^g)$.

11.3. With (K,ρ) as in 11.2, we consider the compactly induced representation $c\text{-}\mathrm{Ind}_K^G\,\rho$, as in 2.5. The space X underlying this representation consists of the functions $f : G \to W$, which are compactly supported modulo Z, and satisfy $f(kg) = \rho(k) f(g)$, $k \in K$, $g \in G$. The group G acts by right translation. (All functions $f \in X$ are G-smooth for this action, since K is open: see 2.5 Exercise 2.)

For $\phi \in \mathcal{H}(G,\rho)$ and $f \in c\text{-}\mathrm{Ind}\,\rho$, we define

$$\phi * f(g) = \int_{G/Z} \phi(x) f(x^{-1}g)\, d\dot\mu(x), \quad g \in G.$$

Clearly, $\phi * f \in X$, and this action gives a homomorphism of $\mathbb{C}$-algebras

$$\mathcal{H}(G,\rho) \longrightarrow \mathrm{End}_G(c\text{-}\mathrm{Ind}\,\rho). \tag{11.3.1}$$

Proposition. *The map* (11.3.1) *is an isomorphism of* $\mathbb{C}$*-algebras.*

Proof. We use the relation $\mathrm{End}_G(c\text{-Ind}\,\rho) \cong \mathrm{Hom}_K(\rho, c\text{-Ind}\,\rho)$ of (2.5.2). Let $\phi^0 : w \to \phi^0_w$ be the canonical map $W \to c\text{-Ind}\,\rho$, corresponding to the identity endomorphism of $c\text{-Ind}\,\rho$: the function ϕ^0_w has support K and $\phi^0_w(k) = \rho(k)w$. The isomorphism $\mathrm{End}_G(c\text{-Ind}\,\rho) \to \mathrm{Hom}_K(\rho, c\text{-Ind}\,\rho)$ is composition with ϕ^0. Composing (11.3.1) with ϕ^0, we get a map $\mathcal{H}(G,\rho) \to \mathrm{Hom}_K(W, c\text{-Ind}\,\rho)$. We write down its inverse. Let

$$\begin{aligned} \phi : W &\longrightarrow c\text{-Ind}\,\rho, \\ w &\longmapsto \phi_w, \end{aligned}$$

be a K-homomorphism. We define a function $\Phi : G \to \mathrm{End}_{\mathbb{C}}(W)$ by

$$\Phi(g) : w \longmapsto \phi_w(g).$$

For $k \in K$, we have $\Phi(kg) : w \mapsto \phi_w(kg) = \rho(k)\phi_w(g)$, so $\Phi(kg) = \rho(k)\Phi(g)$. Also, $\Phi(gk) : w \mapsto \phi_w(gk) = \phi_{\rho(k)w}(g)$, since ϕ is a K-map. Therefore $\Phi \in \mathcal{H}(G,\rho)$ and $\phi \mapsto \dot{\mu}(K/Z)^{-1}\Phi$ is the required inverse map. □

11.4. The central result of this section is:

Theorem. *Let K be an open subgroup of $G = \mathrm{GL}_2(F)$, containing and compact modulo Z. Let (ρ, W) be an irreducible smooth representation of K and suppose that an element $g \in G$ intertwines ρ if and only if $g \in K$. The compactly induced representation $c\text{-Ind}_K^G\,\rho$ is then irreducible and cuspidal.*

Proof. We write $(\Sigma, X) = c\text{-Ind}_K^G\,\rho$. We first show that the representation Σ has a non-zero coefficient which is compactly supported modulo Z. To see this, we use the canonical K-embedding $\phi^0 : W \to X$ of the preceding proof, which identifies W with the space of functions in X that are supported in K (2.5 Lemma).

The groups K, G are unimodular, so the Duality Theorem of 3.5 implies that $\check{X} \cong \mathrm{Ind}_K^G\,\check{\rho}$. The induced representation $\mathrm{Ind}_K^G\,\check{\rho}$ contains $c\text{-Ind}_K^G\,\check{\rho}$ as G-subspace. The canonical K-embedding $\check{W} \to c\text{-Ind}_K^G\,\check{\rho}$ identifies $\check{W}$ with the space of functions in $\check{X}$ with support contained in K. We take non-zero functions $\check{w} \in \check{W} \subset \check{X}$ and $w \in W \subset X$: the coefficient $\gamma_{\check{w}\otimes w}$ is then non-zero and supported in K.

Consequently, we need only prove that X is irreducible: it is then admissible (10.2 Corollary) and we can apply 10.1 Proposition (2) to show it is γ-cuspidal, hence cuspidal.

The centre Z of G acts on X via the character ω_ρ, where $\rho(z)w = \omega_\rho(z)w$, $z \in Z$, $w \in W$. Therefore X is the direct sum of its K-isotypic components (2.7). Any K-map $W \to X$ has image contained in X^ρ, so:

$$\mathrm{Hom}_K(W, X^\rho) = \mathrm{Hom}_K(W, X) \cong \mathrm{End}_G(X) \cong \mathcal{H}(G, \rho).$$

However, the intertwining condition implies that $\dim \mathcal{H}(G, \rho) = 1$. The space $\mathrm{Hom}_K(W, X^\rho)$ therefore has dimension 1, and we conclude that $W = X^\rho$.

Let Y be a non-zero G-subspace of X. Therefore

$$0 \neq \mathrm{Hom}_G(Y, X) \subset \mathrm{Hom}_G(Y, \mathrm{Ind}_K^G \rho) \cong \mathrm{Hom}_K(Y, \rho).$$

Since Y is semisimple over K (2.7), we have $Y^\rho \neq 0$. Thus $Y \supset W = X^\rho$, since W is irreducible over K. As W generates X over G, we conclude that $Y = X$. Thus X is irreducible, as required. □

Remark 1. The theorem holds (with the conclusion that c-$\mathrm{Ind}\,\rho$ is γ-cuspidal), with the same proof, in considerable generality. It is valid for a unimodular locally profinite group G, satisfying 2.6 Hypothesis, and such that any irreducible smooth representation of G is admissible.

Remark 2. The converse of the theorem also holds. If ρ is intertwined by some $g \in G \smallsetminus K$, then $\mathcal{H}(G, \rho) \cong \mathrm{End}_G(c\text{-}\mathrm{Ind}\,\rho)$ has dimension > 1. Thus c-$\mathrm{Ind}\,\rho$ has a non-scalar endomorphism and cannot be irreducible.

Remark 3. In the situation of the theorem, the smooth dual $(c\text{-}\mathrm{Ind}\,\rho)^\vee$ is irreducible. It is, however, isomorphic to $\mathrm{Ind}\,\check{\rho}$. We deduce that $\mathrm{Ind}\,\check{\rho} = c\text{-}\mathrm{Ind}\,\check{\rho} \cong (c\text{-}\mathrm{Ind}\,\rho)^\vee$. Since these representations are all admissible, we can dualize again to get c-$\mathrm{Ind}\,\rho = \mathrm{Ind}\,\rho$.

11.5. We give an example illustrative of the above procedures. Let $G = \mathrm{GL}_2(F)$, $K = \mathrm{GL}_2(\mathfrak{o})$ and $K_1 = 1{+}\mathfrak{p}\mathrm{M}_2(\mathfrak{o})$. Thus K_1 is an open normal subgroup of K and $K/K_1 \cong \mathrm{GL}_2(\boldsymbol{k})$. We also let I_1 denote the group of matrices

$$I_1 = 1 + \begin{pmatrix} \mathfrak{p} & \mathfrak{o} \\ \mathfrak{p} & \mathfrak{p} \end{pmatrix}.$$

Thus I_1 is the inverse image in K of the standard group $N(\boldsymbol{k})$ of upper triangular unipotent matrices in $\mathrm{GL}_2(\boldsymbol{k})$.

Theorem. *Let (π, V) be an irreducible smooth representation of G, and suppose that π contains the trivial character of K_1. Exactly one of the following holds:*

(1) *π contains a representation λ of K, inflated from an irreducible cuspidal representation $\tilde{\lambda}$ of $\mathrm{GL}_2(\boldsymbol{k})$;*
(2) *π contains the trivial character of I_1.*

In the first case, π is cuspidal, and there exists a representation Λ of ZK such that $\Lambda \mid K \cong \lambda$ and

$$\pi \cong c\text{-}\mathrm{Ind}_{ZK}^G \Lambda.$$

Proof. The group K stabilizes the space V^{K_1}, which is therefore a direct sum of irreducible representations of K which are trivial on K_1, that is, they are inflated from $\mathrm{GL}_2(\boldsymbol{k})$. Let λ be one of these, inflated from $\tilde{\lambda}$. Either $\tilde{\lambda}$ is cuspidal (in the sense of §6), or it is not. In the latter case, it contains the trivial character of $N(\boldsymbol{k})$, whence λ contains the trivial character of I_1.

We have to show that the two cases cannot occur together. To do this, we interpolate a useful general lemma.

Lemma. *For $i = 1, 2$, let $\tilde{\rho}_i$ be an irreducible representation of $\mathrm{GL}_2(\boldsymbol{k})$, and let ρ_i denote the inflation of $\tilde{\rho}_i$ to a representation of K. Suppose that $\tilde{\rho}_1$ is cuspidal.*

(1) *The representations ρ_i intertwine in G if and only if $\tilde{\rho}_1 \cong \tilde{\rho}_2$.*
(2) *An element $g \in G$ intertwines ρ_1 if and only if $g \in ZK$.*

Proof. Let $g \in G$ intertwine ρ_2 with ρ_1. It is only the coset $KgZK$ which intervenes, so we can take g of the form

$$g = \begin{pmatrix} \varpi^a & 0 \\ 0 & 1 \end{pmatrix},$$

for some $a \geqslant 0$. If $a = 0$, we have $g = 1$ and there is nothing to do. We therefore assume $a \geqslant 1$. The group $K_1^g \cap K$ contains the group

$$N_0 = \begin{pmatrix} 1 & \mathfrak{o} \\ 0 & 1 \end{pmatrix} \subset \begin{pmatrix} 1 & \mathfrak{p} \\ 0 & 1 \end{pmatrix}^g$$

on which ρ_2^g is trivial. Since $\tilde{\rho}_1$ is cuspidal, ρ_1 does not contain the trivial character of N_0, so g cannot intertwine the ρ_i. All assertions now follow. □

It follows from 11.1 Proposition 1 that, in the theorem, the two cases cannot occur together. We now assume that $\tilde{\lambda}$ is cuspidal. Surely π contains some representation Λ of ZK extending λ. Thus we have a non-trivial ZK-homomorphism $\Lambda \to \pi$, giving a non-trivial G-homomorphism $c\text{-}\mathrm{Ind}_{ZK}^G \Lambda \to \pi$. However, by part (2) of the lemma and 11.4 Theorem, the representation $c\text{-}\mathrm{Ind}\,\Lambda$ is irreducible, so $\pi \cong c\text{-}\mathrm{Ind}\,\Lambda$, as desired. □

Remark. We will eventually see (14.5) that the theorem has a kind of converse. If (π, V) is an irreducible representation of G containing the trivial character of K_1, then it is cuspidal if and only if it satisfies condition (1) of the theorem.

Further reading.

Although we have focused exclusively on $G = \mathrm{GL}_2(F)$, many elements reflect the much more general discussions in the papers [5,6] of Bernstein and Zelevinsky. These apply in the context of connected reductive algebraic groups over

F and centre on general versions of the Restriction-Induction Lemma (9.3) and the homomorphism theorem in the form 9.11 Lemma 2. That programme culminates in a classification of the non-cuspidal representations of $\mathrm{GL}_n(F)$, [90]. Rodier's report [71] is a helpful introduction. The eternal pre-print [25] is also written in these terms, from a slightly different point of view. Only in very few cases, however, does one have a good command of the non-cuspidal representations of groups besides $\mathrm{GL}_n(F)$.

The initial analysis of cuspidal representations in this chapter is quite general in tone, and holds very widely. Even 11.5 and its converse have close analogues for completely general reductive groups [67], [64].

4

Cuspidal Representations

The objective of this chapter is to classify the irreducible cuspidal representations of $G = \mathrm{GL}_2(F)$.

The method is different from that of Chapter III, being based on the process of restricting representations of G to certain, rather special, compact open subgroups. The maximal compact subgroup $K = \mathrm{GL}_2(\mathfrak{o})$ and the standard Iwahori subgroup I of G have already appeared in incidental rôles, but they now move centre-stage. Each of these groups has a canonical filtration by open normal subgroups $\{K_n\}_{n\geqslant 0}$, $\{I_n\}_{n\geqslant 0}$ respectively. Given an irreducible smooth representation (π, V) of G, one looks for the largest proper subgroup in the filtration (there is some subtlety as to whether one chooses K or I) for which V has a fixed vector. The representations of the next largest filtration subgroup occurring in π encapsulate a significant amount of information concerning π.

The example of 11.5 gives an idea of the method. There, one knows that π has a fixed vector for the group K_1. The natural representation of K on the space of K_1-fixed vectors in π reveals whether π is cuspidal or not. If it is cuspidal, one rapidly gets an explicit description of it as an induced representation.

The general case is slightly more involved. It is essential to operate, for some of the time, in an intrinsic manner but we reduce to matrix calculations whenever possible. The groups K, I and their filtrations have ring-theoretic descriptions which are particularly convenient. We start the chapter by developing these. We then have to examine the restrictions of representations of G to the filtration subgroups and interpret our findings. This results in a clear characterization of the cuspidal representations, which we then refine to a classification.

The final section looks in a different direction. Irreducible representations of G containing the trivial character of the Iwahori subgroup I play a special rôle in representation theory: this is true in a much wider context than the present one. We need to consider them for two reasons. First, we have to exclude the possibility that any of them is cuspidal. Second, we have to give an account of the non-cuspidal, square-integrable representations (10a.2 Remark) to facilitate some calculations in Chapter VI. Representations with Iwahori-fixed vector are pivotal in this matter.

12. Chain Orders and Fundamental Strata

In this section, we construct a family of compact open subgroups of G, and then a family of characters of these subgroups. The key point is that an irreducible smooth representation of G (not already covered by 11.5) contains one, and essentially only one, of these characters.

12.1. We need a special family of compact open subgroups of $G = \mathrm{GL}_2(F)$. These are groups of units of certain rings, which we now define.

Temporarily write $V = F \oplus F$, so that $G = \mathrm{Aut}_F(V)$ and $A = \mathrm{End}_F(V)$. An $\mathfrak{o}$-*lattice chain in* V is a non-empty set $\mathcal{L}$ of $\mathfrak{o}$-lattices in V, linearly ordered under inclusion and stable under multiplication by $F^\times$.

If $\mathcal{L}$ is an $\mathfrak{o}$-lattice chain in V, we can number its elements:

$$\mathcal{L} = \{L_i : i \in \mathbb{Z}\}, \quad L_i \supsetneq L_{i+1}.$$

Stability under translation by $F^\times$ implies the existence of a positive integer $e = e_{\mathcal{L}}$ such that

$$xL_i = L_{i+ev_F(x)},$$

for all $i \in \mathbb{Z}$ and $x \in F^\times$.

Lattice chains are easy to describe completely:

Proposition. *Let* $\mathcal{L} = \{L_i\}_{i\in\mathbb{Z}}$ *be an* $\mathfrak{o}$*-lattice chain in* $V = F \oplus F$*. Then* $e_{\mathcal{L}} = 1$ *or* 2*. Moreover:*

(1) *Suppose* $e_{\mathcal{L}} = 1$*. There exists* $g \in G$ *such that*

$$gL_i = \mathfrak{p}^i \oplus \mathfrak{p}^i, \quad i \in \mathbb{Z}.$$

(2) *Suppose that* $e_{\mathcal{L}} = 2$. *There exists* $g \in G$ *such that*

$$\begin{aligned} gL_{2i} &= \mathfrak{p}^i \oplus \mathfrak{p}^i, \\ gL_{2i+1} &= \mathfrak{p}^i \oplus \mathfrak{p}^{i+1}, \quad i \in \mathbb{Z}. \end{aligned} \tag{12.1.1}$$

Proof. Set $e_{\mathcal{L}} = e$. The quotient $L_0/L_e = L_0/\mathfrak{p}L_0$ is a vector space over $\boldsymbol{k} = \mathfrak{o}/\mathfrak{p}$ of dimension 2. The quotients L_i/L_e, $0 \leqslant i \leqslant e$, form a flag of subspaces of L_0/L_e, whence $1 \leqslant e \leqslant 2$, as required.

The lattice L_0 is the $\mathfrak{o}$-span of an F-basis of V (1.5 Proposition), so there exists $g \in G$ such that $gL_0 = \mathfrak{o} \oplus \mathfrak{o}$. Hence $gL_{ei} = \mathfrak{p}^i \oplus \mathfrak{p}^i$, $i \in \mathbb{Z}$, and this deals completely with the case $e = 1$.

If $e = 2$, we have $\mathfrak{o} \oplus \mathfrak{o} \supset gL_1 \supset \mathfrak{p} \oplus \mathfrak{p}$, and $U = gL_1/(\mathfrak{p} \oplus \mathfrak{p})$ is a one-dimensional $\boldsymbol{k}$-subspace of $\boldsymbol{k} \oplus \boldsymbol{k} = (\mathfrak{o} \oplus \mathfrak{o})/(\mathfrak{p} \oplus \mathfrak{p})$. There exists $\bar{h} \in \mathrm{GL}_2(\boldsymbol{k})$ such that $\bar{h}U = \boldsymbol{k} \oplus 0$. If $h \in \mathrm{GL}_2(\mathfrak{o})$ has image $\bar{h}$ in $\mathrm{GL}_2(\boldsymbol{k})$, then $hgL_1 = \mathfrak{o} \oplus \mathfrak{p}$, and the result follows. □

Let $\mathcal{L} = \{L_i\}_{i \in \mathbb{Z}}$ be a lattice chain in V, and define

$$\begin{aligned} \mathfrak{A}_{\mathcal{L}} &= \bigcap_{i \in \mathbb{Z}} \mathrm{End}_{\mathfrak{o}}(L_i) = \bigcap_{0 \leqslant i \leqslant e-1} \mathrm{End}_{\mathfrak{o}}(L_i) \\ &= \{x \in A : xL_i \subset L_i, i \in \mathbb{Z}\}. \end{aligned}$$

Thus $\mathfrak{A}_{\mathcal{L}}$ is a ring (with 1) and also an $\mathfrak{o}$-lattice in A. (Such rings are called $\mathfrak{o}$-*orders* in A.) The $\mathfrak{o}$-lattices $L \in \mathcal{L}$ are $\mathfrak{A}_{\mathcal{L}}$-modules. Interpreting the proposition in terms of $\mathfrak{A}_{\mathcal{L}}$, we get:

Corollary. *Let* $\mathcal{L}$ *be an* $\mathfrak{o}$*-lattice chain in* V, *and set* $\mathfrak{A} = \mathfrak{A}_{\mathcal{L}}$. *There exists* $g \in G$ *such that*

$$g\mathfrak{A}g^{-1} = \begin{cases} \mathfrak{M} = \begin{pmatrix} \mathfrak{o} & \mathfrak{o} \\ \mathfrak{o} & \mathfrak{o} \end{pmatrix} & \text{if } e_{\mathcal{L}} = 1, \\ \mathfrak{J} = \begin{pmatrix} \mathfrak{o} & \mathfrak{o} \\ \mathfrak{p} & \mathfrak{o} \end{pmatrix} & \text{if } e_{\mathcal{L}} = 2. \end{cases} \tag{12.1.2}$$

12.2. An $\mathfrak{A}_{\mathcal{L}}$-*lattice* in V is, by definition, an $\mathfrak{o}$-lattice in V which is also an $\mathfrak{A}_{\mathcal{L}}$-module.

Proposition 1. *Let* $\mathcal{L}$ *be a lattice chain in* V *and let* L *be an* $\mathfrak{A}_{\mathcal{L}}$*-lattice in* V. *Then* $L \in \mathcal{L}$.

Proof. It is enough to treat the cases (12.1.2). We start with $\mathfrak{A}_{\mathcal{L}} = \mathfrak{J}$. The order $\mathfrak{J}$ contains the idempotent matrices

$$e_1 = \begin{pmatrix} 1 & 0 \\ 0 & 0 \end{pmatrix}, \quad e_2 = \begin{pmatrix} 0 & 0 \\ 0 & 1 \end{pmatrix}.$$

It follows that $L \supset e_1L + e_2L$, whence $L = e_1L \oplus e_2L$. That is, $L = \mathfrak{p}^a \oplus \mathfrak{p}^b$, for certain integers a, b. The lattice L is invariant under the group of upper triangular unipotent matrices in $\mathfrak{J}$, so $a \leqslant b$. It is also invariant under the group of lower triangular unipotent matrices in $\mathfrak{J}$, so $b \leqslant a+1$. The only possibilities, therefore, are $L = \mathfrak{p}^a \oplus \mathfrak{p}^a$ or $L = \mathfrak{p}^a \oplus \mathfrak{p}^{a+1}$, for some $a \in \mathbb{Z}$. All of these lie in $\mathcal{L}$. The case $\mathfrak{A}_{\mathcal{L}} = \mathfrak{M}$ is similar but easier, so we omit the details. □

In other words, $\mathcal{L}$ is the set of all $\mathfrak{A}_{\mathcal{L}}$-lattices in V, and so we can recover $\mathcal{L}$ (up to its numbering) from the ring $\mathfrak{A}_{\mathcal{L}}$.

It is usually more convenient to work directly with the ring $\mathfrak{A}_{\mathcal{L}}$ than via the lattice chain $\mathcal{L}$. We therefore introduce the following terminology.

A *chain order in* $A = \mathrm{M}_2(F)$ is a ring of the form $\mathfrak{A} = \mathfrak{A}_{\mathcal{L}}$, for some $\mathfrak{o}$-lattice chain $\mathcal{L}$ in V. We set $e_{\mathfrak{A}} = e_{\mathcal{L}}$.

A special rôle is played by the Jacobson radical $\mathfrak{P}$ of the chain order $\mathfrak{A}$: we use the notation $\mathfrak{P} = \operatorname{rad}\mathfrak{A}$.

Proposition 2. *Let $\mathfrak{A}$ be a chain order in A with $\mathfrak{P} = \operatorname{rad}\mathfrak{A}$, and set $e = e_{\mathfrak{A}}$. Then $\mathfrak{P}^e = \mathfrak{p}\mathfrak{A}$, and there exists $\Pi \in G$ such that*

$$\mathfrak{P} = \Pi\mathfrak{A} = \mathfrak{A}\Pi.$$

Proof. It is enough to treat the standard examples of (12.1.2). For these, we have:

$$\operatorname{rad}\mathfrak{M} = \varpi\,\mathfrak{M}, \quad \operatorname{rad}\mathfrak{J} = \begin{pmatrix} 0 & 1 \\ \varpi & 0 \end{pmatrix} \mathfrak{J}, \tag{12.2.1}$$

for any prime element ϖ of F. □

Such an element Π is called a *prime element of* $\mathfrak{A}$. We have

$$\mathfrak{P}^n = \Pi^n\mathfrak{A} = \mathfrak{A}\Pi^n, \quad n \geqslant 0.$$

We take this as the definition of $\mathfrak{P}^n$ when $n \leqslant -1$. Observe that if $\mathfrak{A} = \mathfrak{A}_{\mathcal{L}}$, for a lattice chain $\mathcal{L} = \{L_i\}$, then $\Pi L_i = L_{i+1}$ for all i and:

$$\mathfrak{P}^n = \bigcap_{i\in\mathbb{Z}} \operatorname{Hom}_{\mathfrak{o}}(L_i, L_{i+n}) = \{x \in A : xL_i \subset L_{i+n}, i \in \mathbb{Z}\}, \quad n \in \mathbb{Z}.$$

Remark. There is a purely ring-theoretic description of the class of rings we have called "chain orders". The chain orders in A are conventionally known as the *hereditary* $\mathfrak{o}$-orders in A.

12.3. Let $\mathfrak{A}$ be a chain order in A, with $\mathfrak{P} = \operatorname{rad}\mathfrak{A}$. We put

$$U_{\mathfrak{A}}^0 = U_{\mathfrak{A}} = \mathfrak{A}^\times,$$
$$U_{\mathfrak{A}}^n = 1+\mathfrak{P}^n, \quad n \geqslant 1.$$

For example, observe that $U_{\mathfrak{M}} = \mathrm{GL}_2(\mathfrak{o})$ and $U_{\mathfrak{J}} = I$, the standard Iwahori subgroup of G (7.3).

In general, the groups $U_{\mathfrak{A}}^n$, $n \geqslant 0$, are compact open subgroups of G, and each $U_{\mathfrak{A}}^n$ is a normal subgroup of $U_{\mathfrak{A}}$. If $2m \geqslant n > m \geqslant 1$, the map $x \mapsto 1+x$ induces an isomorphism

$$\mathfrak{P}^m/\mathfrak{P}^n \xrightarrow{\approx} U_{\mathfrak{A}}^m/U_{\mathfrak{A}}^n. \tag{12.3.1}$$

We shall also need the group

$$\mathcal{K}_{\mathfrak{A}} = \{g \in G : g\mathfrak{A}g^{-1} = \mathfrak{A}\}. \tag{12.3.2}$$

If $\mathfrak{A} = \mathfrak{A}_{\mathcal{L}}$, for some lattice chain $\mathcal{L}$, we also have

$$\mathcal{K}_{\mathfrak{A}} = \mathrm{Aut}_{\mathfrak{o}}(\mathcal{L}) = \{g \in G : gL \in \mathcal{L}, \forall L \in \mathcal{L}\}.$$

Either way, $\mathcal{K}_{\mathfrak{A}}$ is the semi-direct product $\langle \Pi \rangle \ltimes U_{\mathfrak{A}}$ of $U_{\mathfrak{A}}$ and the cyclic group generated by a prime element Π of $\mathfrak{A}$. In particular, $\mathcal{K}_{\mathfrak{A}}$ is open in G. It contains the centre Z of G, and $\mathcal{K}_{\mathfrak{A}}/Z$ is compact. It normalizes all of the groups $U_{\mathfrak{A}}^j$, $j \geqslant 0$.

Exercises.

(1) *Let $g \in G$; show that $g \in \mathcal{K}_{\mathfrak{A}}$ if and only if $g\mathfrak{A} = \mathfrak{P}^m$, for some $m \in \mathbb{Z}$.*
(2) *Show that $\mathcal{K}_{\mathfrak{A}}$ is the normalizer $N_G(U_{\mathfrak{A}})$ of $U_{\mathfrak{A}}$ in G.*
(3) *Show that $\mathcal{K}_{\mathfrak{A}}$ is the G-normalizer of $U_{\mathfrak{A}}^m$, for any $m \geqslant 0$.*

Hint. In (2), the $\mathfrak{o}$-algebra generated by $U_{\mathfrak{A}}$ is $\mathfrak{A}$ *except* in the case $e = 2$ and $q = 2$.

12.4 Example. There is one important abstract context in which chain orders arise. Let E be an F-subalgebra of A which is a quadratic field extension of F. Thus $V = F \oplus F$ is an E-vector space of dimension one.

Proposition. *Let E be an F-subalgebra of A such that E/F is a quadratic field extension.*

(1) *The set of $\mathfrak{o}_E$-lattices in V forms an $\mathfrak{o}$-lattice chain $\mathcal{L}$, with the property $e_{\mathcal{L}} = e(E|F)$. Further, $\mathcal{L}$ is the unique lattice chain in V which is stable under translation by $E^\times$.*
(2) *The order $\mathfrak{A} = \mathfrak{A}_{\mathcal{L}}$ is the unique chain order in A such that $E^\times \subset \mathcal{K}_{\mathfrak{A}}$.*
(3) *If $\mathfrak{P} = \mathrm{rad}\,\mathfrak{A}$, then $x\mathfrak{A} = \mathfrak{P}^{v_E(x)}$, $x \in E^\times$, and $\mathcal{K}_{\mathfrak{A}} = E^\times U_{\mathfrak{A}}$.*

Proof. Let $v \in V$, $v \neq 0$; then $\mathcal{L} = \{\mathfrak{p}_E^j v : j \in \mathbb{Z}\}$ is the set of all $\mathfrak{o}_E$-lattices in V, and is defined independently of v. We have $\mathcal{L} = E^\times L = \{xL : x \in E^\times\}$, for any $L \in \mathcal{L}$.

If $\mathcal{L}'$ is a lattice chain stable under $E^\times$, each $L \in \mathcal{L}'$ is stable under the action of U_E; we have $\mathfrak{o}_E = \mathfrak{o}[U_E]$, so L is an $\mathfrak{o}_E$-lattice and $\mathcal{L}' \subset \mathcal{L}$. But $\mathcal{L} = E^\times L \subset \mathcal{L}'$, for any $L \in \mathcal{L}' \subset \mathcal{L}$, so $\mathcal{L}' = \mathcal{L}$. The remaining assertions follow readily. $\square$

12.5. We take a character $\psi \in \widehat{F}$, $\psi \neq 1$. Let $\widehat{A}$ denote the group of characters of A and define $\psi_A \in \widehat{A}$ by

$$\psi_A(x) = \psi(\mathrm{tr}_A\, x),$$

where tr_A denotes the trace map $A \to F$. For $a \in A$, we can define a character $a\psi_A$ of A by

$$a\psi_A(x) = \psi_A(ax) = \psi_A(xa), \quad x \in A.$$

Lemma. *The map $a \mapsto a\psi_A$ gives an isomorphism $A \cong \widehat{A}$.*

Proof. Working explicitly, $A = \mathrm{M}_2(F) \cong F \oplus F \oplus F \oplus F$ (as additive group), so $\widehat{A} = \widehat{F} \times \widehat{F} \times \widehat{F} \times \widehat{F} \cong F \oplus F \oplus F \oplus F$, via the isomorphism $F \cong \widehat{F}$ of 1.7. A character of A is therefore of the form

$$\begin{pmatrix} a & b \\ c & d \end{pmatrix} \longmapsto \psi(\alpha a)\psi(\beta b)\psi(\gamma c)\psi(\delta d),$$

for uniquely determined elements $\alpha, \beta, \gamma, \delta \in F$. This expression, however, is just

$$\psi_A\left(\begin{pmatrix} \alpha & \gamma \\ \beta & \delta \end{pmatrix}\begin{pmatrix} a & b \\ c & d \end{pmatrix}\right),$$

and the lemma follows. □

It is now convenient to *fix $\psi \in \widehat{F}$ of level* 1 (*cf.* 1.7). Let L be some $\mathfrak{o}$-lattice in A; we define

$$L^* = \{x \in A : \psi_A(xy) = 1, y \in L\}.$$

If $\{a_i\}_{1 \leqslant i \leqslant 4}$ is an $\mathfrak{o}$-basis of L, then $L^* = \sum \mathfrak{p} a_i'$, where $\{a_i'\}$ is the dual basis of A defined by $\mathrm{tr}_A(a_i a_j') = \delta_{ij}$ (Kronecker delta).

Thus L^* is an $\mathfrak{o}$-lattice and $L^{**} = L$. Because of this "bi-duality" property, the obvious relation $(L_1 + L_2)^* = L_1^* \cap L_2^*$ implies that $(L_1 \cap L_2)^* = L_1^* + L_2^*$, for $\mathfrak{o}$-lattices L_i in A.

Via the isomorphism $A \cong \widehat{A}$ given by ψ_A, we can identify L^* with the group of characters of A which are trivial on L. Moreover, if we have $\mathfrak{o}$-lattices $L_1 \subset L_2$, then L_1^*/L_2^* is identified with the group of characters of L_2 which are trivial on L_1.

Proposition. *Let $\mathfrak{A}$ be a chain order in A with radical $\mathfrak{P}$, and let $\psi \in \widehat{F}$ have level one.*

(1) *For $n \in \mathbb{Z}$, we have $(\mathfrak{P}^n)^* = \mathfrak{P}^{1-n}$.*
(2) *Let m, n be integers such that $2m+1 \geqslant n > m \geqslant 0$. Let $(U_{\mathfrak{A}}^{m+1}/U_{\mathfrak{A}}^{n+1})^\wedge$ denote the group of characters of the finite abelian group $U_{\mathfrak{A}}^{m+1}/U_{\mathfrak{A}}^{n+1}$. The map*

$$\begin{aligned} \mathfrak{P}^{-n}/\mathfrak{P}^{-m} &\longrightarrow (U_{\mathfrak{A}}^{m+1}/U_{\mathfrak{A}}^{n+1})^\wedge, \\ a+\mathfrak{P}^{-m} &\longmapsto \psi_{A,a} \mid U_{\mathfrak{A}}^{m+1}, \end{aligned}$$

is an isomorphism, where $\psi_{A,a}$ denotes the function $x \mapsto \psi_A(a(x-1))$.

Proof. The explicit form (12.1.2) gives $\mathfrak{A}^* = \mathfrak{P}$. If L is an $\mathfrak{o}$-lattice in A and $g \in G$, then $(gL)^* = L^*g^{-1}$. So, if Π is a prime element of $\mathfrak{A}$, then $(\mathfrak{P}^n)^* = (\Pi^n\mathfrak{A})^* = \mathfrak{A}^*\Pi^{-n} = \mathfrak{P}^{1-n}$, as required for (1). The second assertion then follows from (12.3.1). □

We normally abbreviate $\psi_{A,a} = \psi_a$.

Remark 1. Let $\eta \in \widehat{F}$, $\eta \neq 1$. We may again form the character $\eta_A = \eta \circ \mathrm{tr}_A \in \widehat{A}$ and the function $\eta_a : x \mapsto \eta_A(a(x-1))$. There exists $c \in F^\times$ such that $\eta = c\psi$; the level of η is $1-v_F(c)$ and $\eta_a = \psi_{ca}$. The map $a \mapsto \eta_a$ induces an isomorphism $c^{-1}\mathfrak{P}^{-n}/c^{-1}\mathfrak{P}^{-m} \cong \big(U_{\mathfrak{A}}^{m+1}/U_{\mathfrak{A}}^{n+1}\big)^\wedge$. Taking our basic character to have level one thus gives the simplest numerology.

Remark 2. The focus of interest here is on the characters χ of the finite groups $U_{\mathfrak{A}}^{m+1}/U_{\mathfrak{A}}^{n+1}$, particularly their intertwining properties. If we choose $\psi \in \widehat{F}$ (of level 1, for convenience) and write $\chi = \psi_a$, these properties of χ are reflected in properties of the coset $a+\mathfrak{P}^{-m}$, and are hence describable in terms of matrices.

As always, there are disadvantages to making a non-canonical "choice of coordinates" to describe a mathematical object. However, it is easy to work out the effect of a change of coordinates: replacing ψ by another character $\psi' = u^{-1}\psi$ of level one, gives $\chi = \psi_a = \psi'_{ua}$, for some $u \in U_F$. The relevant properties of cosets $a+\mathfrak{P}^{-m}$ will be unaffected by such a change, and there is no apparent way of proceeding otherwise.

12.6. We start to apply these concepts to the analysis of representations of G.

Let (π, V) be an irreducible smooth representation of G. Let $\mathcal{S}(\pi)$ denote the set of pairs $(\mathfrak{A}, n)$, where $\mathfrak{A}$ is a chain order in A and $n \geqslant 0$ is an integer, subject to the condition that π contains the trivial character of $U_{\mathfrak{A}}^{n+1}$. We define the *normalized level* $\ell(\pi)$ of π by

$$\ell(\pi) = \min\{n/e_{\mathfrak{A}} : (\mathfrak{A}, n) \in \mathcal{S}(\pi)\}.$$

Clearly, $2\ell(\pi)$ is a non-negative integer.

In the notation (12.1.2), we have $U_{\mathfrak{M}}^1 \subset U_{\mathfrak{J}}^1$, so

Proposition. *Let π be an irreducible smooth representation of G; then $\ell(\pi) = 0$ if and only if π contains the trivial character of $U_{\mathfrak{M}}^1$.*

We have given a preliminary analysis of such representations in 11.5 (using the notation $U_{\mathfrak{M}}^1 = K_1$).

12.7. To deal with the representations π for which $\ell(\pi) > 0$, we introduce a new concept. *For the remainder of this chapter, we work relative to a fixed choice of character $\psi \in \widehat{F}$ of level one.*

A *stratum* in A is a triple $(\mathfrak{A}, n, a)$, where $\mathfrak{A}$ is a chain order in A (with radical $\mathfrak{P}$, say), n is an integer and $a \in \mathfrak{P}^{-n}$.

We say that strata $(\mathfrak{A}, n, a_1)$, $(\mathfrak{A}, n, a_2)$ are *equivalent* if $a_1 \equiv a_2 \pmod{\mathfrak{P}^{1-n}}$.

If $n \geqslant 1$, we can associate to a stratum $(\mathfrak{A}, n, a)$ the character ψ_a of $U_{\mathfrak{A}}^n$, which is trivial on $U_{\mathfrak{A}}^{n+1}$. This character depends only on the equivalence class of the stratum (and the choice of ψ).

Proposition. *Let* $(\mathfrak{A}_i, n_i, a_i)$, $i = 1, 2$, *be strata in* A, *let* $\mathfrak{P}_i = \operatorname{rad} \mathfrak{A}_i$, *and let* $g \in G$. *Assume that* $n_i \geqslant 1$, $i = 1, 2$. *The following are equivalent:*

(1) *The element* g *intertwines the character* ψ_{a_1} *of* $U_{\mathfrak{A}_1}^{n_1}$ *with the character* ψ_{a_2} *of* $U_{\mathfrak{A}_2}^{n_2}$.
(2) *The intersection* $g^{-1}(a_1 + \mathfrak{P}_1^{1-n_1})g \cap (a_2 + \mathfrak{P}_2^{1-n_2})$ *is non-empty.*

Proof. Consider the chain order $\mathfrak{A}_3 = g^{-1}\mathfrak{A}_1 g$; this has radical

$$\operatorname{rad} A_3 = g^{-1}\mathfrak{P}_1 g.$$

The character $(\psi_{a_1})^g$ of the group $(U_{\mathfrak{A}_1}^{n_1})^g = U_{\mathfrak{A}_3}^{n_1}$ is associated to the stratum $(\mathfrak{A}_3, n_1, g^{-1}a_1 g)$. It is therefore enough to consider the case $g = 1$.

If (2) holds, we take an element a of the intersection. We then have $\psi_a = \psi_{a_i}$ on $U_{\mathfrak{A}_i}^{n_i}$, so $\psi_{a_1} = \psi_{a_2} = \psi_a$ on $U_{\mathfrak{A}_1}^{n_1} \cap U_{\mathfrak{A}_2}^{n_2}$.

Conversely, suppose that the ψ_{a_i} agree on $U_{\mathfrak{A}_1}^{n_1} \cap U_{\mathfrak{A}_2}^{n_2}$:

$$\psi_A(a_1 x) = \psi_A(a_2 x), \quad x \in \mathfrak{P}_1^{n_1} \cap \mathfrak{P}_2^{n_2}.$$

In the notation of 12.5, we have

$$\left(\mathfrak{P}_1^{n_1} \cap \mathfrak{P}_2^{n_2}\right)^* = \left(\mathfrak{P}_1^{n_1}\right)^* + \left(\mathfrak{P}_2^{n_2}\right)^* = \mathfrak{P}_1^{1-n_1} + \mathfrak{P}_2^{1-n_2},$$

and so

$$a_1 \equiv a_2 \pmod{\mathfrak{P}_1^{1-n_1} + \mathfrak{P}_2^{1-n_2}}.$$

That is, there exist $x_i \in \mathfrak{P}_i^{1-n_i}$ such that $a_2 = a_1 + x_1 + x_2$, or

$$a_2 - x_2 = a_1 + x_1 \in (a_1 + \mathfrak{P}_1^{1-n_1}) \cap (a_2 + \mathfrak{P}_2^{1-n_2}),$$

as required. □

When the element g satisfies condition (2) of the Proposition, we say that it *intertwines* $(\mathfrak{A}_1, n_1, a_1)$ *with* $(\mathfrak{A}_2, n_2, a_2)$.

12.8. Not all strata are of equal interest: we have to distinguish a particular class of them.

Definition. *Let* $\mathfrak{A}$ *be a chain order in* A, *and set* $\mathfrak{P} = \operatorname{rad} \mathfrak{A}$. *A stratum* $(\mathfrak{A}, n, a)$ *in* A *is called* fundamental *if the coset* $a + \mathfrak{P}^{1-n}$ *contains no nilpotent element of* A.

This property depends only on the equivalence class of the stratum. It is indeed a property of the character ψ_a of $U_{\mathfrak{A}}^n$ (assuming $n \geqslant 1$): changing our choice of ψ changes $a+\mathfrak{P}^{1-n}$ to $ua+\mathfrak{P}^{1-n}$, for some $u \in U_F$, and this does not affect the defining property.

We first need an effective method of recognizing fundamental strata:

Proposition. *Let $(\mathfrak{A}, n, a)$ be a stratum in A, and put $\mathfrak{P} = \operatorname{rad}\mathfrak{A}$. The following are equivalent:*

(1) *The coset $a+\mathfrak{P}^{1-n}$ contains a nilpotent element of A.*
(2) *There is an integer $r \geqslant 1$ such that $a^r \in \mathfrak{P}^{1-rn}$.*

Proof. Condition (2) is equivalent to saying that $x_1x_2\ldots x_r \in \mathfrak{P}^{1-rn}$, for any choice of elements $x_1, x_2, \ldots, x_r \in a+\mathfrak{P}^{1-n}$. Thus condition (2) is a property of the coset $a+\mathfrak{P}^{1-n}$, rather than just the element a. That said, the implication (1) $\Rightarrow$ (2) is clear.

We assume (2) holds. Nothing is changed if we replace $(\mathfrak{A}, n, a)$ by a G-conjugate. This reduces us to the cases $\mathfrak{A} = \mathfrak{M}$ or $\mathfrak{J}$ (12.1.2). Likewise nothing changes if we replace $(\mathfrak{A}, n, a)$ by $(\mathfrak{A}, n-e, \varpi a)$, for a prime element ϖ of F, $e = e_{\mathfrak{A}}$. This reduces us to the cases $(\mathfrak{A}, n) = (\mathfrak{M}, 0), (\mathfrak{J}, 0)$ or $(\mathfrak{J}, -1)$.

In the first case, we have $a \in \mathfrak{M}$; let $\tilde{a}$ be the image of a in $\mathrm{M}_2(\boldsymbol{k})$. Condition (2) holds if and only if $\tilde{a}^r = 0$, for some $r \geqslant 1$. This is equivalent to $\tilde{a}$ being zero (that is, $0 \in a+\varpi\mathfrak{M}$) or $\mathrm{GL}_2(\boldsymbol{k})$-conjugate to $\left(\begin{smallmatrix} 0 & 1 \\ 0 & 0 \end{smallmatrix}\right)$. In the latter case, $a+\varpi\mathfrak{M}$ contains a $\mathrm{GL}_2(\mathfrak{o})$-conjugate of $\left(\begin{smallmatrix} 0 & 1 \\ 0 & 0 \end{smallmatrix}\right)$ and (1) holds.

Now let $\mathfrak{P} = \operatorname{rad}\mathfrak{J}$; let $m = 0, 1$. We then have

$$a+\mathfrak{P}^{m+1} = \begin{pmatrix} 0 & 1 \\ \varpi & 0 \end{pmatrix}^m \begin{pmatrix} a_1 & 0 \\ 0 & a_2 \end{pmatrix} + \mathfrak{P}^{m+1},$$

for $a_i \in \mathfrak{o}$. The coset $a+\mathfrak{P}^{m+1}$ only determines the a_i modulo $\mathfrak{p}$. If $m = 0$, condition (2) holds if and only if $a_1 \equiv a_2 \equiv 0 \pmod{\mathfrak{p}}$. This clearly implies (1). In the other case $m = 1$, (2) is equivalent to $a_1a_2 \equiv 0 \pmod{\mathfrak{p}}$ and again (1) follows. □

Using the calculations in the last proof, we can list the equivalence classes of non-fundamental strata, up to conjugation in G. Call a stratum $(\mathfrak{A}, n, a)$ *trivial* if $a \in \mathfrak{P}^{1-n}$, where $\mathfrak{P} = \operatorname{rad}\mathfrak{A}$.

Gloss. *Let ϖ be a prime element of F. A non-trivial, non-fundamental stratum in A is equivalent to a G-conjugate of one of the following:*

$$(\mathfrak{M}, n, \varpi^{-n}a), \quad a = \left(\begin{smallmatrix} 0 & 1 \\ 0 & 0 \end{smallmatrix}\right), \tag{12.8.1}$$

$$(\mathfrak{J}, 2n-1, \varpi^{-n}\alpha), \quad \alpha = \left(\begin{smallmatrix} 0 & 1 \\ 0 & 0 \end{smallmatrix}\right), \tag{12.8.2}$$

for some $n \in \mathbb{Z}$.

12.9. Let (π, V) be an irreducible smooth representation of G. We say that π *contains the stratum* $(\mathfrak{A}, n, a)$ if $n \geqslant 1$ and π contains the character ψ_a of $U_{\mathfrak{A}}^n$. Observe that, if this is the case, then $n/e_{\mathfrak{A}} \geqslant \ell(\pi)$ by definition.

The main result here is:

Theorem. *Let π be an irreducible smooth representation of G and let $(\mathfrak{A}, n, a)$ be a stratum in A, contained in π. The following are equivalent:*

(1) *$(\mathfrak{A}, n, a)$ is fundamental;*
(2) *$\ell(\pi) = n/e_{\mathfrak{A}}$.*

In particular, π contains a fundamental stratum if and only if $\ell(\pi) > 0$.

Proof. The first step is:

Lemma 1.

(1) *Let $(\mathfrak{A}, n, a)$ be a non-fundamental stratum in A, and let $\mathfrak{P}$ be the radical of $\mathfrak{A}$. There is a chain order $\mathfrak{A}_1$ in A, with radical $\mathfrak{P}_1$, and an integer n_1, such that*

$$a+\mathfrak{P}^{1-n} \subset \mathfrak{P}_1^{-n_1}, \quad \text{and} \quad n_1/e_{\mathfrak{A}_1} < n/e_{\mathfrak{A}}.$$

(2) *Let π be an irreducible smooth representation of G, containing a non-fundamental stratum $(\mathfrak{A}, n, a)$. We then have $\ell(\pi) < n/e_{\mathfrak{A}}$.*

Proof. Part (1) is trivial if the stratum $(\mathfrak{A}, n, a)$ is trivial, so we assume otherwise. The issue is unchanged if we replace $(\mathfrak{A}, n, a)$ by a G-conjugate. Similarly if we replace $(\mathfrak{A}, n, a)$ by $(\mathfrak{A}, n-e_{\mathfrak{A}}r, \varpi^r a)$, for an integer r and a prime element ϖ of F. This reduces us to the cases (12.8.1), (12.8.2) with $n = 0$. In the first of these, we have

$$\begin{pmatrix} 0 & 1 \\ 0 & 0 \end{pmatrix} + \mathfrak{P}_{\mathfrak{M}} \subset \mathfrak{P}_{\mathfrak{I}},$$

and in the second

$$\begin{pmatrix} 0 & 1 \\ 0 & 0 \end{pmatrix} + \mathfrak{P}_{\mathfrak{I}}^2 \subset \mathfrak{P}_1,$$

where $\mathfrak{P}_1 = \operatorname{rad} \mathfrak{A}_1$ and

$$\mathfrak{A}_1 = \begin{pmatrix} \mathfrak{o} & \mathfrak{p}^{-1} \\ \mathfrak{p} & \mathfrak{o} \end{pmatrix},$$

which is conjugate to $\mathfrak{M}$.

For the second part, we apply (1); the relation $a+\mathfrak{P}^{1-n} \subset \mathfrak{P}_1^{-n_1}$ implies $\mathfrak{P}^{1-n} \subset \mathfrak{P}_1^{-n_1}$ and, dualizing, $\mathfrak{P}_1^{n_1+1} \subset \mathfrak{P}^n$. This further gives $U_{\mathfrak{A}_1}^{n_1+1} \subset U_{\mathfrak{A}}^n$, and the character ψ_a of $U_{\mathfrak{A}}^n$ is trivial on $U_{\mathfrak{A}_1}^{n_1+1}$. Therefore $\ell(\pi) \leqslant n_1/e_{\mathfrak{A}_1} < n/e_{\mathfrak{A}}$, as required. □

If $\ell(\pi) > 0$ then, by definition, π contains a stratum $(\mathfrak{A}, n, a)$ with $n/e_{\mathfrak{A}} = \ell(\pi)$. By Lemma 1, this stratum must be fundamental. If π contains another stratum $(\mathfrak{B}, m, b)$, then (12.7, 11.1) it must intertwine with $(\mathfrak{A}, n, a)$.

Lemma 2. *Let $(\mathfrak{A}, n, a)$ be a fundamental stratum in A. Let $(\mathfrak{B}, m, b)$ be a stratum in A which intertwines with $(\mathfrak{A}, n, a)$. Then $m/e_{\mathfrak{B}} \geqslant n/e_{\mathfrak{A}}$, with equality if and only if $(\mathfrak{B}, m, b)$ is fundamental.*

Proof. Let $\mathfrak{P} = \operatorname{rad}\mathfrak{A}$, $\mathfrak{Q} = \operatorname{rad}\mathfrak{B}$. The property in question depends only on the coset $b+\mathfrak{Q}^{1-m}$; replacing $(\mathfrak{B}, m, b)$ by a G-conjugate, we can arrange that $b \in a+\mathfrak{P}^{1-n}$. Assume, for a contradiction, that $m/e_{\mathfrak{B}} < n/e_{\mathfrak{A}}$. Thus there is an integer $r \geqslant 1$ such that

$$\mathfrak{p}^{-rme_{\mathfrak{A}}}\mathfrak{B} \subset \mathfrak{p}^{1-rne_{\mathfrak{B}}}\mathfrak{A}.$$

Consider the element $x = b^{re_{\mathfrak{A}}e_{\mathfrak{B}}}$. Since $(\mathfrak{A}, n, a)$ is fundamental, we have $x \notin \mathfrak{p}^{1-rne_{\mathfrak{B}}}\mathfrak{A}$, but certainly $x \in \mathfrak{p}^{-rme_{\mathfrak{A}}}\mathfrak{B}$. This contradiction proves $m/e_{\mathfrak{B}} \geqslant n/e_{\mathfrak{A}}$.

If $(\mathfrak{B}, m, b)$ is fundamental, symmetry implies $m/e_{\mathfrak{B}} \leqslant n/e_{\mathfrak{A}}$, and hence $m/e_{\mathfrak{B}} = n/e_{\mathfrak{A}}$. Conversely, suppose that $m/e_{\mathfrak{B}} = n/e_{\mathfrak{A}}$ but that $(\mathfrak{B}, m, b)$ is not fundamental. We have $b^{2r} \in \mathfrak{p}^{-2rn/e_{\mathfrak{A}}}\mathfrak{A} \smallsetminus \mathfrak{p}^{-2rn/e_{\mathfrak{A}}}\mathfrak{P}$ for all integers $r \geqslant 1$, while $b^{2r} \in \mathfrak{p}^{-2rm/e_{\mathfrak{B}}+j_r}\mathfrak{B}$, for a sequence of integers j_r tending to infinity as $r \to \infty$. For r sufficiently large, therefore, $\mathfrak{p}^{j_r}\mathfrak{B} \subset \mathfrak{P}$ which implies $b^{2r} \in \mathfrak{p}^{-2rn/e_{\mathfrak{A}}}\mathfrak{P}$. This is impossible, so $(\mathfrak{B}, m, b)$ is fundamental. □

In the case $\ell(\pi) > 0$, the proof of the theorem is complete. It remains only to show that, if $\ell(\pi) = 0$, then π cannot contain a fundamental stratum $(\mathfrak{A}, n, a)$, $n \geqslant 1$. However, by definition, π contains the trivial character of $U^1_{\mathfrak{M}}$. It therefore contains some character of $U^1_{\mathfrak{J}}$ which is trivial on $U^1_{\mathfrak{M}} \supset U^2_{\mathfrak{J}}$. That is, π contains a stratum $(\mathfrak{J}, 1, a)$ in which the element a takes the form

$$a \equiv \begin{pmatrix} 0 & 0 \\ a_0 & 0 \end{pmatrix} \pmod{\mathfrak{J}},$$

for some $a_0 \in \mathfrak{o}$. Thus $(\mathfrak{J}, 1, a)$ is non-fundamental. If π contains a stratum $(\mathfrak{A}, n, b)$, then $n/e_{\mathfrak{A}} \geqslant 1/2$; since this stratum must intertwine with $(\mathfrak{J}, 1, a)$, it cannot be fundamental by Lemma 2. □

13. Classification of Fundamental Strata

Given that there are, in effect, only two chain orders in A (12.1.2), it is not hard to describe the fundamental strata completely, up to G-conjugacy. We carry out this process, to obtain a preliminary classification of the irreducible representations of G in terms of strata. We continue to work relative to a fixed $\psi \in \widehat{F}$ of level one.

13.1. The first case is:

Proposition 1.

(1) *Let* $(\mathfrak{A}, n, a)$ *be a stratum in which* $e_{\mathfrak{A}} = 2$ *and* n *is odd. Let* $\mathfrak{P} = \operatorname{rad}\mathfrak{A}$. *The stratum* $(\mathfrak{A}, n, a)$ *is fundamental if and only if* $a\mathfrak{A} = \mathfrak{P}^{-n}$ *or, equivalently,* $a \in \Pi^{-n}U_{\mathfrak{A}}$ *for a prime element* Π *of* $\mathfrak{A}$.

(2) *Let* π *be an irreducible smooth representation of* G *with* $\ell(\pi) > 0$. *If* $\ell(\pi) = n/2 \notin \mathbb{Z}$, *then* π *contains a fundamental stratum* $(\mathfrak{I}, n, a)$.

Proof. The first assertion concerns only the conjugacy class of the stratum, so we can take $\mathfrak{A} = \mathfrak{I}$. The result then follows from 12.8 Gloss. In (2), 12.9 Theorem says that π contains a fundamental stratum; since $\ell(\pi) \notin \mathbb{Z}$, it must be conjugate to one of the form $(\mathfrak{I}, n, a)$. □

Definition. *A* ramified simple stratum *is a fundamental stratum* $(\mathfrak{A}, n, \alpha)$ *in which* $e_{\mathfrak{A}} = 2$ *and* n *is odd.*

Proposition 2.

If $0 < \ell(\pi) = n \in \mathbb{Z}$, *then* π *contains a fundamental stratum of the form* $(\mathfrak{M}, n, \alpha)$.

Proof. This follows from 12.9 Theorem and the observation that

$$U_{\mathfrak{M}}^{n+1} \subset U_{\mathfrak{I}}^{2n+1} \subset U_{\mathfrak{I}}^{2n} \subset U_{\mathfrak{M}}^{n}. \qquad \square \tag{13.1.1}$$

For this reason, there is no need to consider fundamental strata of the form $(\mathfrak{I}, 2n, a)$, $n \in \mathbb{Z}$. More precisely,

Corollary. *Let* π *be an irreducible smooth representation of* G *with* $\ell(\pi) > 0$. *Then* π *contains a fundamental stratum* $(\mathfrak{A}, n, a)$ *such that* $\gcd(n, e_{\mathfrak{A}}) = 1$.

13.2. Consider a stratum $(\mathfrak{A}, n, \alpha)$ in which $e_{\mathfrak{A}} = 1$. We may write $\alpha = \varpi^{-n}\alpha_0$, for some $\alpha_0 \in \mathfrak{A}$. Let $f_\alpha(t) \in \mathfrak{o}[t]$ be the characteristic polynomial of α_0, and let $\tilde{f}_\alpha(t) \in \boldsymbol{k}[t]$ be its reduction modulo $\mathfrak{p}$: thus $\tilde{f}_\alpha(t)$ is the characteristic polynomial of the image of α_0 in $\mathfrak{A}/\mathfrak{P} \cong \mathrm{M}_2(\boldsymbol{k})$, where $\mathfrak{P} = \operatorname{rad}\mathfrak{A}$.

If we regard the prime element ϖ as fixed, the polynomial $\tilde{f}_\alpha(t)$ depends only on the equivalence class of the stratum $(\mathfrak{A}, n, \alpha)$, and is indeed an invariant of the G-conjugacy class of the stratum.

Observe that the stratum $(\mathfrak{A}, n, \alpha)$ is fundamental if and only if $\tilde{f}_\alpha(t) \neq t^2$ (12.8 Proposition).

Definition. *Let* $(\mathfrak{A}, n, \alpha)$ *be a fundamental stratum with* $e_{\mathfrak{A}} = 1$. *We say that* $(\mathfrak{A}, n, \alpha)$ *is*

$$\left.\begin{array}{r}\text{unramified simple}\\ \text{split}\\ \text{essentially scalar}\end{array}\right\} \quad \textit{if} \quad \left\{\begin{array}{l}\tilde{f}_\alpha(t) \text{ is irreducible in } \boldsymbol{k}[t],\\ \tilde{f}_\alpha(t) \text{ has distinct roots in } \boldsymbol{k},\\ \tilde{f}_\alpha(t) \text{ has a repeated root in } \boldsymbol{k}^\times.\end{array}\right.$$

A stratum $(\mathfrak{A}, n, \alpha)$ is called *simple* if it is either ramified or unramified simple.

Remark. All of the qualities assigned to fundamental strata $(\mathfrak{A}, n, \alpha)$ in the preceding two definitions are qualities of the associated character ψ_α of $U_{\mathfrak{A}}^n$.

Proposition.

(1) *A ramified simple stratum cannot intertwine with any fundamental stratum of the form $(\mathfrak{M}, n, \alpha)$.*
(2) *Let $(\mathfrak{M}, n, \alpha)$, $(\mathfrak{M}, n, \beta)$ be fundamental strata which intertwine. We then have $\tilde{f}_\alpha(t) = \tilde{f}_\beta(t)$.*

Proof. (1) follows from 12.8 Lemma 2. In (2), there exists $g \in G$ and $\beta' \in \beta+\mathfrak{P}_{\mathfrak{M}}^{1-n}$ such that $g^{-1}\beta' g \in \alpha+\mathfrak{P}_{\mathfrak{M}}^{1-n}$. The characteristic polynomial of the element $\varpi^n g^{-1}\beta' g$ is the same as that of $\varpi^n\beta'$, so, when we reduce it modulo $\mathfrak{p}$, we get $\tilde{f}_\beta(t)$. On the other hand, since $g^{-1}\beta' g \in \alpha+\mathfrak{P}_{\mathfrak{M}}^{1-n}$, this reduction is also $\tilde{f}_\alpha(t)$. Thus $\tilde{f}_\beta = \tilde{f}_\alpha$, as required. □

13.3. One of the cases of 13.2 Definition is easily explained. If π is an irreducible smooth representation of G and if χ is a character of $F^\times$, we recall that $\chi\pi$ denotes the representation $g \mapsto \chi(\det g)\pi(g)$.

Theorem. *Let π be an irreducible smooth representation of G with $\ell(\pi) > 0$. The following are equivalent:*

(1) *The representation π contains an essentially scalar stratum $(\mathfrak{M}, n, \alpha)$.*
(2) *There is a character χ of $F^\times$ such that $\ell(\chi\pi) < \ell(\pi)$.*

Proof. Suppose that $(\mathfrak{M}, n, \alpha)$ is an essentially scalar stratum occurring in π. Replacing α by a $U_{\mathfrak{M}}$-conjugate, we can assume

$$\alpha \equiv \varpi^{-n}\begin{pmatrix} a & b \\ 0 & a \end{pmatrix} \pmod{\mathfrak{p}^{1-n}\mathfrak{M}}, \tag{13.3.1}$$

for a prime element ϖ of F and $a \in U_F$, $b \in \mathfrak{o}$. Let χ be a character of $F^\times$, of level n, such that $\chi(1+x) = \psi(-a\varpi^{-n}x)$, $x \in \mathfrak{p}^n$ (*cf.* 1.8). We then have $\chi \circ \det \mid U_{\mathfrak{M}}^n = \psi_{-a\varpi^{-n}}$, so the representation $\chi\pi$ contains the stratum $(\mathfrak{M}, n, \beta)$, with

$$\beta \equiv \varpi^{-n}\begin{pmatrix} 0 & b \\ 0 & 0 \end{pmatrix} \pmod{\mathfrak{p}^{1-n}\mathfrak{M}}.$$

This is not fundamental and so $\ell(\chi\pi) < n$.

For the converse, we change notation and take for π an irreducible smooth representation of G with $\ell(\pi) < n$. Thus π contains the trivial character of $U_{\mathfrak{M}}^n$ (13.1.1). Let ϕ be a character of $F^\times$, of level n, with $\phi(1+x) = \psi(a\varpi^{-n}x)$, $x \in \mathfrak{p}^n$ and some $a \in U_F$. The representation $\phi\pi$ then contains a stratum of the form (13.3.1). □

Corollary. *Let π be an irreducible smooth representation of G such that* $0 < \ell(\pi) \leqslant \ell(\chi\pi)$, *for every character χ of $F^\times$.* One and only one *of the following holds:*

(1) *π contains a split fundamental stratum.*
(2) *π contains a ramified simple stratum.*
(3) *π contains an unramified simple stratum.*

Proof. Since $\ell(\pi) > 0$, the representation π contains a fundamental stratum. By the theorem, this stratum is not essentially scalar. We can assume (13.1 Propositions) it is either ramified simple, unramified simple, or split. Proposition 13.2 shows that only one of these possibilities can occur. □

We will see in the next section that the Corollary partly reflects the initial partition of the irreducible representations of G into cuspidal and non-cuspidal ones.

13.4. We need more understanding of the structures underlying simple strata.

Definition. *An element $\alpha \in G \setminus Z$ is called* minimal over F *if the sub-algebra $E = F[\alpha]$ of A is a field and, setting $n = -\upsilon_E(\alpha)$, one of the following holds:*

(1) *E/F is totally ramified and n is odd;*
(2) *E/F is unramified and, for a prime element ϖ of F, the coset $\varpi^n\alpha+\mathfrak{p}_E$ generates the field extension $\boldsymbol{k}_E/\boldsymbol{k}$.*

The hypothesis $\alpha \notin Z$ implies that $[E{:}F] = 2$. We also have $f(E|F)\upsilon_E(\alpha) = \upsilon_F(\det\alpha)$, where $f(E|F)$ is the residue class degree $[\boldsymbol{k}_E : \boldsymbol{k}]$.

The definition does not really depend on α being an element of G: we could equally take a quadratic field extension E/F and apply it to elements α of $E \setminus F$.

Exercise. *Let E/F be a quadratic field extension, let $\alpha \in E^\times$, and write $n = -\upsilon_E(x)$. Show that α is minimal over F if and only if $\alpha+\mathfrak{p}_E^{1-n} \cap F = \emptyset$.*

Lemma. *Let $\alpha \in G$ be minimal over F; set $E = F[\alpha]$, $n = -\upsilon_E(\alpha)$, and choose a prime element ϖ of F. Define*

$$\alpha_0 = \begin{cases} \varpi^{(n+1)/2}\alpha & \text{if } E/F \text{ is ramified,} \\ \varpi^n\alpha & \text{if } E/F \text{ is unramified.} \end{cases} \tag{13.4.1}$$

We then have $\mathfrak{o}_E = \mathfrak{o}[\alpha_0]$.

Proof. In both cases, the ring $\mathfrak{o}[\alpha_0]$ contains a prime element of E and a $\boldsymbol{k}$-basis of $\mathfrak{o}_E/\varpi\mathfrak{o}_E$. □

There is a close connection between minimal elements and simple strata. In one direction:

Proposition. *Let $(\mathfrak{A}, n, \alpha)$ be a simple stratum in A. Then:*

(1) *α is minimal over F;*
(2) *$F[\alpha]^\times \subset \mathcal{K}_\mathfrak{A}$;*
(3) *$e(F[\alpha]|F) = e_\mathfrak{A}$;*
(4) *every $\alpha' \in \alpha+\mathfrak{P}^{1-n} = \alpha U^1_\mathfrak{A}$ is minimal over F, where $\mathfrak{P} = \mathrm{rad}\,\mathfrak{A}$.*

Proof. We fix a prime element ϖ of F. Suppose first that $(\mathfrak{A}, n, \alpha)$ is ramified. Thus $n = 2m+1$ is odd and the element $\alpha_0 = \varpi^{1+m}\alpha$ satisfies $\alpha_0\mathfrak{A} = \mathfrak{P}$ (13.1). Therefore $\upsilon_F(\det \alpha_0) = 1$ and $\upsilon_F(\mathrm{tr}\,\alpha_0) \geqslant 1$ (*cf.* (12.2.1)). The minimal polynomial of α_0 over F is thus an Eisenstein polynomial, so $E = F[\alpha] = F[\alpha_0]$ is a ramified quadratic field extension of F. Moreover, $\upsilon_E(\alpha_0) = 1$, $\upsilon_E(\alpha) = -n$, and n is odd. Thus α is minimal over F.

If $(\mathfrak{A}, n, \alpha)$ is unramified, we put $\alpha_0 = \varpi^n\alpha \in \mathfrak{A}$. The minimal polynomial $f(t)$ of α_0 over F remains irreducible on reduction modulo $\mathfrak{p}$. In particular, $f(t)$ is irreducible and $E = F[\alpha]$ is an unramified field extension of degree 2. Further, if $\tilde\alpha_0$ is the reduction of α_0 modulo $\mathfrak{p}_E$, we have $\boldsymbol{k}_E = \boldsymbol{k}[\tilde\alpha_0]$, as required.

In both cases, $\mathfrak{o}_E = \mathfrak{o}[\alpha_0] \subset \mathfrak{A}$. If $\mathcal{L}$ is the chain of $\mathfrak{A}$-lattices in $V = F^2$, every $L \in \mathcal{L}$ is an $\mathfrak{o}_E$-lattice. Moreover, $\mathcal{K}_\mathfrak{A}$ contains a prime element ϖ_E of E which is also a prime element of $\mathfrak{A}$: we take $\varpi_E = \alpha_0$ in the ramified case, $\varpi_E = \varpi$ in the unramified one. It follows that $\mathcal{L}$ is the chain of all $\mathfrak{o}_E$-lattices in V, and that $E^\times \subset \mathcal{K}_\mathfrak{A}$.

Finally, let $\alpha' \in \alpha+\mathfrak{P}^{1-n}$. The stratum $(\mathfrak{A}, n, \alpha')$ is equivalent to $(\mathfrak{A}, n, \alpha)$, hence simple, and the same argument applies. □

13.5. Thus simple strata give rise to minimal elements. The converse also holds:

Proposition. *Let $\alpha \in G$ be minimal over F. There exists a unique chain order $\mathfrak{A}$ in A such that $\alpha \in \mathcal{K}_\mathfrak{A}$. Moreover, $F[\alpha]^\times \subset \mathcal{K}_\mathfrak{A}$ and, if $n = -\upsilon_{F[\alpha]}(\alpha)$, the triple $(\mathfrak{A}, n, \alpha)$ is a simple stratum.*

Proof. Put $E = F[\alpha]$ and $n = -\upsilon_E(\alpha)$. Let $\mathfrak{A}$ be the unique chain order in A which is stable under conjugation by $E^\times$ (as in 12.4). In particular, $\alpha \in \mathcal{K}_\mathfrak{A}$.

Define α_0 as in (13.4.1). Let $\mathfrak{B}$ be a chain order with $\alpha \in \mathcal{K}_\mathfrak{B}$. We need to show that $\mathfrak{B} = \mathfrak{A}$. We have $\alpha_0 \in \mathcal{K}_\mathfrak{B}$ and $\upsilon_F(\det \alpha_0) \geqslant 0$. Therefore $\alpha_0 \in \mathfrak{B}$ and so, by 13.4 Lemma, $\mathfrak{o}_E \subset \mathfrak{B}$. Exactly as in the proof of 13.4 Proposition, we get $E^\times \subset \mathcal{K}_\mathfrak{B}$ and therefore $\mathfrak{B} = \mathfrak{A}$.

We now have $\alpha\mathfrak{A} = \mathfrak{P}^{-n}$, where $\mathfrak{P} = \mathrm{rad}\,\mathfrak{A}$. The final assertion is immediate, on comparing the definitions. □

Remark. Combining the last two propositions, we see that if $(\mathfrak{A}, n, \alpha)$ is a simple stratum and $\alpha' \in \alpha U^1_\mathfrak{A}$, then $F[\alpha']/F$ is a quadratic field extension with $e(F[\alpha']|F) = e(F[\alpha]|F)$. However, the fields $F[\alpha]$, $F[\alpha']$ need not be F-isomorphic.

14. Strata and the Principal Series

We start to exploit the theory of strata in the classification of the irreducible representations of G. We show that, once the obfuscatory effect of twisting (13.3) has been taken into account, one can distinguish between cuspidal and non-cuspidal representations of G using only fundamental strata. This, combined with a parallel result in level zero, is the content of the Exhaustion Theorem (14.5), the main result of the section.

The subgroups B, N, T, Z of G are as in 5.1.

14.1. We start with one of the cases of 13.3 Corollary. We use throughout the notation (12.1.2).

Proposition. *Let (π, V) be an irreducible smooth representation of G, and suppose that π contains a split fundamental stratum $(\mathfrak{M}, n, \alpha)$. One may choose $\alpha \in T$ and, in that case, the Jacquet module (π_N, V_N) contains the character $\psi_\alpha \mid U_{\mathfrak{M}}^n \cap T$.*

In particular, $V_N \neq 0$ and π is not cuspidal.

Proof. We are at liberty to replace the coset $\alpha+\mathfrak{P}_{\mathfrak{M}}^{1-n}$ by a $U_{\mathfrak{M}}$-conjugate. We can therefore take

$$\alpha = \varpi^{-n} \begin{pmatrix} a & 0 \\ 0 & b \end{pmatrix}, \tag{14.1.1}$$

where ϖ is a prime element of F and $a, b \in \mathfrak{o} \cap F^\times$, with $a \not\equiv b \pmod{\mathfrak{p}}$: in particular, $\alpha \in T$.

Let us write $\xi = \psi_\alpha \mid U_{\mathfrak{M}}^n$. It is enough to show that the space V^ξ has non-zero image in V_N.

Suppose, for a contradiction, that $V^\xi \subset V(N)$. Thus, for each $v \in V^\xi$, there is a compact open subgroup $N(v)$ of N such that

$$\int_{N(v)} \pi(u)v\, du = 0$$

(8.1). We write

$$N_j = \begin{pmatrix} 1 & \mathfrak{p}^j \\ 0 & 1 \end{pmatrix}.$$

The representation π is admissible (10.2 Corollary), so the space V^ξ is finite-dimensional. Therefore there exists $j \in \mathbb{Z}$ such that

$$\int_{N_j} \pi(u)v\, du = 0,$$

for all $v \in V^\xi$. We choose j maximal for this property, so there exists $v_1 \in V^\xi$ such that

$$\int_{N_{j+1}} \pi(u)v_1\,du \neq 0.$$

We set $t = \left(\begin{smallmatrix} \varpi & 0 \\ 0 & 1 \end{smallmatrix}\right)$. The element t then intertwines the character ξ, that is, the characters ξ, ξ^t, agree on the group

$$Y = U_{\mathfrak{M}}^n \cap t^{-1}U_{\mathfrak{M}}^n t = 1 + \begin{pmatrix} \mathfrak{p}^n & \mathfrak{p}^n \\ \mathfrak{p}^{n+1} & \mathfrak{p}^n \end{pmatrix}.$$

Lemma.

(1) *Any irreducible representation of $U_{\mathfrak{M}}^n$, containing $\xi \mid Y$, is of dimension one.*

(2) *Let ϕ be a character of $U_{\mathfrak{M}}^n$ such that $\phi \mid Y = \xi \mid Y$. There exists $x \in N_0$ such that $\phi^x = \xi$.*

Proof. The group Y contains $U_{\mathfrak{M}}^{n+1}$, and so (1) follows immediately. In part (2), we have $\phi = \psi_\delta$, where

$$\delta \equiv \varpi^{-n} \left(\begin{smallmatrix} a & x \\ 0 & b \end{smallmatrix}\right) \pmod{\mathfrak{p}^{1-n}\mathfrak{M}},$$

for some $x \in \mathfrak{o}$. For $y \in \mathfrak{o}$, we have

$$\begin{pmatrix} 1 & y \\ 0 & 1 \end{pmatrix}\begin{pmatrix} a & x \\ 0 & b \end{pmatrix}\begin{pmatrix} 1 & -y \\ 0 & 1 \end{pmatrix} = \begin{pmatrix} a & z \\ 0 & b \end{pmatrix},$$

where $z = x + (b-a)y$. We can surely choose $y \in \mathfrak{o}$ so that $z = 0$, as required. □

We consider the vector $v_2 = \pi(t^{-1})v_1$. By the definition of j,

$$\begin{aligned} \int_{N_j} \pi(u)v_2\,du &= \int_{N_j} \pi(ut^{-1})v_1\,du \\ &= \pi(t^{-1}) \int_{N_j} \pi(tut^{-1})v_1\,du \\ &= c\,\pi(t^{-1}) \int_{N_{j+1}} \pi(u)v_1\,du \neq 0, \end{aligned}$$

for some $c > 0$.

Let Φ be the set of characters ϕ of $U_{\mathfrak{M}}^n$ which agree with ξ on Y. By the lemma, we have $v_2 = \sum_{\phi\in\Phi} v_\phi$, for certain vectors $v_\phi \in V^\phi$. There exists $\phi \in \Phi$ such that

$$\int_{N_j} \pi(u)v_\phi\,du \neq 0.$$

By part (2) of the lemma, $\phi^x = \xi$, for some $x \in N_0$. Thus $v_3 = \pi(x^{-1})v_\phi$ lies in V^ξ and

$$\int_{N_j} \pi(u)v_3\,du \neq 0,$$

contrary to hypothesis. □

14.2. In the opposite direction, it is easy to spot a fundamental stratum in an induced representation $\mathrm{Ind}_B^G \chi$:

Proposition. *Let $\chi = \chi_1 \otimes \chi_2$ be a character of T and set $\Sigma = \mathrm{Ind}_B^G \chi$. Let n_i be the level of χ_i.*

(1) *If $n = \max(n_1, n_2) > 0$ and $\chi_1\chi_2^{-1} \mid U_F^n \neq 1$, then Σ contains a split fundamental stratum.*
(2) *If $n_1 = n_2 = n \neq 0$ and $\chi_1\chi_2^{-1} \mid U_F^n$ is trivial, then Σ contains an essentially scalar fundamental stratum.*
(3) *If $n_1 = n_2 = 0$, then Σ contains the trivial character of $U_{\mathfrak{I}}^1$.*

Proof. In case (1), we choose $a_i \in \mathfrak{p}^{-n}$ so that $\chi_i(1+x) = \psi(a_i x)$, $x \in \mathfrak{p}^n$. At least one of the a_i satisfies $v_F(a_i) = -n$, and $a_1 \not\equiv a_2 \pmod{\mathfrak{p}^{1-n}}$. Set

$$a = \begin{pmatrix} a_1 & 0 \\ 0 & a_2 \end{pmatrix}, \qquad N_n' = \begin{pmatrix} 1 & 0 \\ \mathfrak{p}^n & 1 \end{pmatrix}.$$

The triple $(\mathfrak{M}, n, a)$ is then a split fundamental stratum. Let $f \in \mathrm{Ind}_B^G \chi$ have support $BU_{\mathfrak{M}}^n = BN_n'$ and be fixed by N_n'. We then have $uf = \psi_a(u)f$, $u \in U_{\mathfrak{M}}^n$, so $(\mathfrak{M}, n, a)$ occurs in Σ. The other cases are similar. □

14.3. Let (π, V) be an irreducible smooth representation of G with $\ell(\pi) = 0$. We suppose that π contains the trivial character of $U_{\mathfrak{I}}^1$: this is the second case of 11.5 Theorem and the third of 14.2 Proposition. Write $I = U_{\mathfrak{I}}$. Since $I/U_{\mathfrak{I}}^1 \cong \boldsymbol{k}^\times \times \boldsymbol{k}^\times$, π contains a character ϕ of I trivial on $U_{\mathfrak{I}}^1$.

Proposition. *Let (π, V) be an irreducible smooth representation containing a character ϕ of I trivial on $U_{\mathfrak{I}}^1$. The canonical map $V \to V_N$ is injective on the isotypic space V^ϕ. In particular, (π, V) is not cuspidal.*

Proof. We use the notation

$$T^0 = \begin{pmatrix} U_F & 0 \\ 0 & U_F \end{pmatrix}, \quad N_j = \begin{pmatrix} 1 & \mathfrak{p}^j \\ 0 & 1 \end{pmatrix}, \quad N_j' = \begin{pmatrix} 1 & 0 \\ \mathfrak{p}^j & 1 \end{pmatrix},$$

for $j \in \mathbb{Z}$. We also put

$$t = \begin{pmatrix} \varpi & 0 \\ 0 & 1 \end{pmatrix},$$

for some prime element ϖ of F.

We fix a Haar measure μ on G and form the algebra $\mathcal{H}_\phi(G) = e_\phi * \mathcal{H}(G) * e_\phi$, in the notation of 4.4. This is the algebra of functions $h \in \mathcal{H}(G)$ such that $h(j_1 g j_2) = \phi(j_1 j_2)^{-1} h(g)$, for $g \in G$ and $j_1, j_2 \in I$. (Thus $\mathcal{H}_\phi(G) = \mathcal{H}(G, \phi^{-1})$, in the notation of 11.2.) We view the isotypic space V^ϕ as $\mathcal{H}_\phi(G)$-module, as in 4.4. The proof is based on the following:

Lemma. *Let $f \in \mathcal{H}_\phi(G)$ have support ItI, such that $f(t) = 1$. Then f is an invertible element of $\mathcal{H}_\phi(G)$.*

We defer consideration of this lemma to 14.4. Accepting it for the moment, we assume for a contradiction that there exists $v \in V^\phi$, $v \neq 0$, with zero image in V_N. Thus (8.1 Lemma (1)) there exists j such that

$$\int_{N_j} \pi(x) v \, dx = 0, \tag{14.3.1}$$

where dx is some Haar measure on N. The representation (π, V) is admissible (10.2 Corollary), so $\dim V^\phi$ is finite. We may therefore choose j maximal for the property that there exists $v \in V^\phi$, $v \neq 0$, satisfying (14.3.1).

We consider the element $w = \pi(t)v$. This has the properties

$$0 = \int_{N_{j+1}} \pi(x) w \, dx,$$

$$\pi(y) w = \phi(y) w, \qquad y \in N_1' T^0 N_1.$$

We now put

$$u = \pi(e_\phi) w = q^{-1} \sum_{z \in N_0/N_1} \pi(z) w.$$

Thus $u \in V^\phi$ and

$$\int_{N_{j+1}} \pi(x) u \, dx = 0.$$

However, $u = \mu(ItI)^{-1} \pi(f) v$ so, by the lemma, $u \neq 0$. We have contradicted the definition of j, and the proposition is proved. □

Remark. In the proposition, one can show that the canonical map $V^\phi \to V_N^{\phi|T^0}$ is an isomorphism.

14.4. We return to the lemma of 14.3. The restriction $\phi \mid T^0$ is the inflation of a character $\tilde{\phi} = \tilde{\phi}_1 \otimes \tilde{\phi}_2$ of $\boldsymbol{k}^\times \times \boldsymbol{k}^\times$.

We prove 14.3 Lemma in the case $\tilde{\phi}_1 \neq \tilde{\phi}_2$. Let $h \in \mathcal{H}_\phi(G)$ have support $It^{-1}I$, and $h(t^{-1}) = 1$. Consider the function $f * h$. This has support contained in $ItIt^{-1}I = IN_0'I$. Suppose $z \in N_0'$ lies in the support. Thus z intertwines the character ϕ (11.2 Lemma). However, $z \in U_{\mathfrak{M}}$, $\mathfrak{M} = \mathrm{M}_2(\mathfrak{o})$, and the image $\tilde{z}$ of z in $G_{\boldsymbol{k}} = \mathrm{GL}_2(\boldsymbol{k})$ therefore intertwines the character $\tilde{\phi}$ of the group

$B_{\boldsymbol{k}}$ of upper triangular matrices in $G_{\boldsymbol{k}}$. However, the character $\tilde{\phi}$ induces irreducibly to $G_{\boldsymbol{k}}$ (6.3 Proposition), so $\tilde{z} = 1$ and $z \in N_1'$. The support of $f * h$ is therefore contained in I and $f * h(1) = \mu(ItI) = h * f(1)$, by straightforward calculations. The lemma follows in this case.

We turn to the case $\tilde{\phi}_1 = \tilde{\phi}_2$. Let ϕ_1 be a character of $F^\times$ such that $\phi_1 \mid U_F$ is the inflation of $\tilde{\phi}_1$. The map $h \mapsto (\phi_1 \circ \det) \cdot h$ gives an algebra isomorphism $\mathcal{H}_\phi(G) \to \mathcal{H}(G, I)$, which preserves support of functions. This means we need only treat the case where $\tilde{\phi}$ is trivial. This requires a different technique, so we deal with it in 17.3 below.

In effect, we have reduced 14.3 Proposition to the case where (π, V) has an Iwahori-fixed vector. Irreducible representations (π, V) such that $V^I \neq 0$ have subtle and interesting features: we discuss them in more detail in §17. Here we note only one particularly significant instance, that of the *Steinberg representation* St_G introduced in 9.10:

Example. If $(\pi, V) = \mathrm{St}_G$, *then* $\dim V^I = 1$ *and* $\ell(\pi) = 0$.

For, if we set $(\Sigma, X) = \mathrm{Ind}_B^G 1_T$, then (7.3.3) implies $\dim X^I = 2$. The 1-dimensional G-subspace 1_G of X has only a one-dimensional space of I-fixed vectors, so $\dim V^I = 1$ also.

14.5. We can now characterize the irreducible cuspidal representations of G in terms of strata:

Exhaustion theorem. *Let* (π, V) *be an irreducible smooth representation of* G, *which satisfies* $\ell(\pi) \leqslant \ell(\chi\pi)$, *for every character* χ *of* $F^\times$. *The following are equivalent:*

(1) *The representation* π *is cuspidal.*
(2) *Either*
   (a) $\ell(\pi) = 0$ *and* π *contains a representation of* $U_{\mathfrak{M}} \cong \mathrm{GL}_2(\mathfrak{o})$ *inflated from an irreducible cuspidal representation of* $\mathrm{GL}_2(\boldsymbol{k})$, *or*
   (b) $\ell(\pi) > 0$ *and* π *contains a simple stratum.*

Proof. Suppose first that $\ell(\pi) = 0$. The result then follows from 11.5 Theorem and 14.3 Proposition.

We therefore assume $\ell(\pi) > 0$. If π does not contain a simple stratum, then it must contain a split fundamental stratum (13.3 Corollary). It is then not cuspidal (14.1 Proposition) and we have shown that (1) $\Rightarrow$ (2).

Conversely, let us assume that π is not cuspidal. We identity π with a G-subspace of a representation $\Sigma = \mathrm{Ind}_B^G \chi$, for some character $\chi = \chi_1 \otimes \chi_2$ of T (9.1 Proposition). Suppose first that Σ is irreducible. In particular, $\pi = \Sigma$. If some χ_i has level $\geqslant 1$, 14.2 Proposition says that Σ contains either a split or an essentially scalar fundamental stratum. The second possibility is excluded by hypothesis and 13.3 Theorem. Thus $\Sigma = \pi$ contains a split fundamental stratum and so cannot contain a simple stratum. If both χ_i have level zero, then $\Sigma = \pi$ contains the trivial character of $U_{\mathfrak{J}}^1$ *loc. cit.*, and so $\ell(\pi) = 0$, contrary to hypothesis.

We therefore assume that Σ is not irreducible. Thus π is either $\phi \circ \det$ or $\phi \cdot \mathrm{St}_G$, for some character ϕ of $F^\times$ (9.11). Each of these representations has normalized level l, where l is the level of ϕ. By the minimality hypothesis, $l = 0$ and π contains the trivial character of $U_{\mathfrak{J}}^1$. By 11.5 Theorem, π cannot satisfy condition (2).

This completes the proof of the theorem. □

15. Classification of Cuspidal Representations

In this section, we refine the Exhaustion Theorem (14.5) into a classification of the irreducible cuspidal representations of $G = \mathrm{GL}_2(F)$. The arguments rely on some quite remarkable intertwining properties of various characters defined by simple strata. While these are quite easy to state and use, their proofs require special techniques. We have therefore deferred them until the next section, working out their consequences in this one.

Throughout, ψ denotes a fixed character of F of level one.

15.1. In this paragraph, we fix a simple stratum $(\mathfrak{A}, n, \alpha)$ in A, with $n \geqslant 1$. The algebra $E = F[\alpha]$ is then a field such that $[E{:}F] = 2$, $E^\times \subset \mathcal{K}_{\mathfrak{A}}$, and α is minimal over F (13.4 Proposition). The function $\psi_\alpha = \psi_{A,\alpha}$ defines a character of $U_{\mathfrak{A}}^{[n/2]+1}$ which is trivial on $U_{\mathfrak{A}}^{n+1}$ (12.5).

Intertwining Theorem. *Let $(\mathfrak{A}, n, \alpha)$ be a simple stratum, and put $E = F[\alpha]$. Let $g \in G$. The following are equivalent:*

(1) *g intertwines the character ψ_α of $U_{\mathfrak{A}}^{[n/2]+1}$;*
(2) *g normalizes the character ψ_α of $U_{\mathfrak{A}}^{[n/2]+1}$;*
(3) *$g \in E^\times U_{\mathfrak{A}}^{[(n+1)/2]}$.*

Observe, in (3), that $E^\times \subset \mathcal{K}_{\mathfrak{A}}$ (13.4 Proposition), so $E^\times U_{\mathfrak{A}}^{[(n+1)/2]}$ is a subgroup of $\mathcal{K}_{\mathfrak{A}}$.

15.2. We will also need:

Conjugacy Theorem. *For $i = 1, 2$, let $(\mathfrak{A}, n, \alpha_i)$ be a simple stratum, $n \geqslant 1$. The characters ψ_{α_i} of $U_{\mathfrak{A}}^{[n/2]+1}$ intertwine in G if and only if they are conjugate by an element of $U_{\mathfrak{A}}$.*

We defer the proofs of these two theorems to §16. We remark that the second condition in the Conjugacy Theorem is equivalent to the cosets $\alpha_i + \mathfrak{P}^{-[n/2]} = \alpha_i U_{\mathfrak{A}}^{[(n+1)/2]}$ being $U_{\mathfrak{A}}$-conjugate: the argument is identical to 12.7. This, however, does not imply that the elements α_i are conjugate, nor even that the fields $F[\alpha_i]$ are isomorphic.

15.3. We reap the consequences for the representation theory of G. Let $(\mathfrak{A}, n, \alpha)$ be a simple stratum in A, with $n \geqslant 1$ and $E = F[\alpha]$, as in 15.1. We set

$$J_\alpha = E^\times U_{\mathfrak{A}}^{[(n+1)/2]}. \tag{15.3.1}$$

Thus $J_\alpha \subset \mathcal{K}_{\mathfrak{A}}$ is open in G; it contains and is compact modulo $Z \cong F^\times$.

Theorem. *With the preceding notation, let Λ be an irreducible representation of J_α which contains the character ψ_α of $U_{\mathfrak{A}}^{[n/2]+1}$. Then:*

(1) *The restriction of Λ to $U_{\mathfrak{A}}^{[n/2]+1}$ is a multiple of ψ_α.*
(2) *The representation*

$$\pi_\Lambda = \text{c-Ind}_{J_\alpha}^G \Lambda \tag{15.3.2}$$

is irreducible and cuspidal.

Proof. The restriction of Λ to the group $U_{\mathfrak{A}}^{[n/2]+1}$ is a direct sum of irreducible representations, any two of which are conjugate under J_α. Among these irreducible components is the character ψ_α, which is normalized by J_α (15.1), whence follows (1). Therefore, if $g \in G$ intertwines Λ, it must also intertwine ψ_α. Thus (15.1) $g \in J_\alpha$. Assertion (2) is now given by 11.4 Theorem. □

It will be convenient to have a special notation for this class of representations.

Definition. *Let $(\mathfrak{A}, n, \alpha)$, $n \geqslant 1$, be a simple stratum. Let $C(\psi_\alpha, \mathfrak{A})$ denote the set of equivalence classes of irreducible representations Λ of the group $J_\alpha = F[\alpha]^\times U_{\mathfrak{A}}^{[(n+1)/2]}$ such that $\Lambda \mid U_{\mathfrak{A}}^{[n/2]+1}$ is a multiple of ψ_α.*

15.4. We have a strong uniqueness property:

Theorem. *For $i = 1, 2$, let $(\mathfrak{A}_i, n_i, \alpha_i)$ be a simple stratum in A, $n_i \geqslant 1$, and let $\Lambda_i \in C(\psi_{\alpha_i}, \mathfrak{A}_i)$. Suppose that the representations*

$$\pi_{\Lambda_i} = \text{c-Ind}_{J_{\alpha_i}}^G \Lambda_i, \quad i = 1, 2,$$

are equivalent. Then $n_1 = n_2$ and there exists $g \in G$ such that

$$\mathfrak{A}_2 = g^{-1}\mathfrak{A}_1 g, \quad J_{\alpha_2} = g^{-1} J_{\alpha_1} g, \quad \Lambda_2 = \Lambda_1^g.$$

If $\mathfrak{A}_1 = \mathfrak{A}_2$, we may choose $g \in U_{\mathfrak{A}_1}$.

Proof. We identify $\pi_{\Lambda_1} = \pi_{\Lambda_2} = \pi$, say. The representation π contains each simple stratum $(\mathfrak{A}_i, n_i, \alpha_i)$. These strata are either both ramified or both unramified (13.3 Corollary). Both $\mathfrak{A}_i$ are conjugate to $\mathfrak{J}$ in the first case, to $\mathfrak{M}$ in the second (notation of (12.1.2)). In other words, we can assume $\mathfrak{A}_1 = \mathfrak{A}_2 = \mathfrak{A}$, say. Further, $n_i/e_{\mathfrak{A}} = \ell(\pi)$ (12.9 Theorem), so $n_1 = n_2 = n$, say. The characters ψ_{α_i} of $U_{\mathfrak{A}}^{[n/2]+1}$ intertwine in G (11.1), and so are $U_{\mathfrak{A}}$-conjugate (15.2), say $\psi_{\alpha_2} = \psi_{\alpha_1}^g$, $g \in U_{\mathfrak{A}}$. The G-normalizers J_{α_i} of these characters ψ_{α_i} are therefore conjugate under g.

Consider the representation $\Lambda_3 = \Lambda_1^g$ of J_{α_2}. The restriction of Λ_3 to $U_{\mathfrak{A}}^{[n/2]+1}$ is a multiple of ψ_{α_2}, and Λ_3 intertwines with Λ_2. If this intertwining is realized by an element h of G, then h also intertwines ψ_{α_2} and lies in J_{α_2}. It therefore fixes Λ_3, whence $\Lambda_3 \cong \Lambda_2$, as required. □

15.5. It will be convenient to introduce a new term:

Definition. *A* cuspidal type *in G is a triple $(\mathfrak{A}, J, \Lambda)$, where $\mathfrak{A}$ is a chain order in A, J is a subgroup of $\mathcal{K}_{\mathfrak{A}}$ and Λ is an irreducible smooth representation of J, of one of the following kinds:*

(1) *$\mathfrak{A} \cong \mathfrak{M}$, $J = ZU_{\mathfrak{A}}$, and $\Lambda \mid U_{\mathfrak{A}}$ is the inflation of an irreducible cuspidal representation of the group $U_{\mathfrak{A}}/U_{\mathfrak{A}}^1 \cong \mathrm{GL}_2(\boldsymbol{k})$;*
(2) *there is a simple stratum $(\mathfrak{A}, n, \alpha)$, $n \geqslant 1$, such that $J = J_\alpha$ and $\Lambda \in C(\psi_\alpha, \mathfrak{A})$;*
(3) *there is a triple $(\mathfrak{A}, J, \Lambda_0)$ satisfying (1) or (2), and a character χ of $F^\times$, such that $\Lambda \cong \Lambda_0 \otimes \chi \circ \det$.*

Note that the class of cuspidal types is defined independently of the choice of the character ψ, and is stable under conjugation by G.

There is an overlap between case (3) of the definition and the others. This will cause no problems but, for further comments, see 15.9 below.

The component $\mathfrak{A}$ in this notation is, strictly speaking, redundant, since $\mathfrak{A}$ is in all cases the unique chain order such that $\mathcal{K}_{\mathfrak{A}} \supset J$. It is more convenient, however, to retain it.

If $(\mathfrak{A}, J, \Lambda)$ is a cuspidal type in G, the representation $c\text{-}\mathrm{Ind}_J^G \Lambda$ is irreducible and cuspidal (15.3 Theorem).

We come to the main result.

Induction Theorem. *Let π be an irreducible cuspidal representation of $G = \mathrm{GL}_2(F)$. There exists a cuspidal type $(\mathfrak{A}, J, \Lambda)$ in G such that $\pi \cong c\text{-}\mathrm{Ind}_J^G \Lambda$. The representation π determines $(\mathfrak{A}, J, \Lambda)$ uniquely, up to G-conjugacy.*

Proof. Let π be an irreducible cuspidal representation of G. In case (3) of the definition above, we have

$$c\text{-}\mathrm{Ind}_J^G(\Lambda_0 \otimes \chi \circ \det) \cong \chi \cdot c\text{-}\mathrm{Ind}_J^G \Lambda_0,$$

so it is enough to treat the case where π satisfies $\ell(\pi) \leqslant \ell(\chi\pi)$, for all characters χ of $F^\times$.

If $\ell(\pi) = 0$, the result is given by the Exhaustion Theorem (14.5) and 11.5 Theorem. We assume $\ell(\pi) > 0$. By the Exhaustion Theorem, there is a simple stratum $(\mathfrak{A}, n, \alpha)$, $n \geqslant 1$, such that π contains the character ψ_α of $U_{\mathfrak{A}}^n$. The representation π therefore contains a character ξ of $U_{\mathfrak{A}}^{[n/2]+1}$ such that $\xi \mid U_{\mathfrak{A}}^n = \psi_\alpha$. The character ξ is of the form ψ_β, for some $\beta \equiv \alpha \pmod{\mathfrak{P}^{1-n}}$, where $\mathfrak{P} = \operatorname{rad}\mathfrak{A}$. By 13.4, 13.5 Propositions, the stratum $(\mathfrak{A}, n, \beta)$ is also simple. The representation π then contains some irreducible smooth representation Λ of J_β such that $\Lambda \mid U_{\mathfrak{A}}^{[n/2]+1}$ contains ψ_β. This restriction is therefore a multiple of ψ_β (15.3 Theorem), so the triple $(\mathfrak{A}, J_\beta, \Lambda)$ is a cuspidal type occurring in π. Since the representation $\pi_\Lambda = c\text{-}\mathrm{Ind}_{J_\beta}^G \Lambda$ is irreducible, we conclude that $\pi \cong \pi_\Lambda$, as desired.

If $(\mathfrak{A}', J', \Lambda')$ is another cuspidal type occurring in π, then $\ell(\pi) > 0$ (12.9) and 13.3 Corollary shows that Λ' is of the second kind in the definition. The uniqueness statement therefore follows from 15.4 Theorem. □

Consequently:

Corollary (Classification Theorem). *The map*

$$(\mathfrak{A}, J, \Lambda) \longmapsto \pi_\Lambda = c\text{-}\mathrm{Ind}_J^G \Lambda$$

induces a bijection between the set of conjugacy classes of cuspidal types in G and the set of equivalence classes of irreducible cuspidal representations of G.

Remark. Thus an irreducible cuspidal representation π of G contains a cuspidal type $(\mathfrak{A}, J, \Lambda)$. If $\ell(\pi) = 0$, this type is of the first kind in 15.5 Definition. If $0 < \ell(\pi) \leqslant \ell(\chi\pi)$, for all characters χ of $F^\times$, then it is of the second kind. Otherwise, it is of the third kind.

15.6. Corollary 15.5 reduces the study of cuspidal representations of G to that of cuspidal types in G. We therefore need to investigate the structure of cuspidal types. In the definition of these, it is only the second case of which there is anything to say. We therefore take a simple stratum $(\mathfrak{A}, n, \alpha)$, $n \geqslant 1$, $E = F[\alpha]$, and describe the representations $\Lambda \in C(\psi_\alpha, \mathfrak{A})$.

We need some intermediate groups:

$$H_\alpha^1 = U_E^1 U_{\mathfrak{A}}^{[n/2]+1}, \quad J_\alpha^1 = J_\alpha \cap U_{\mathfrak{A}}^1 = U_E^1 U_{\mathfrak{A}}^{[(n+1)/2]}. \tag{15.6.1}$$

Observe that $J_\alpha^1 = H_\alpha^1$ if and only if n is odd.

Proposition 1. *Suppose that n is odd.*

(1) *Every $\Lambda \in C(\psi_\alpha, \mathfrak{A})$ has dimension 1, and*
(2) *two characters $\Lambda_1, \Lambda_2 \in C(\psi_\alpha, \mathfrak{A})$ intertwine in G if and only if $\Lambda_1 = \Lambda_2$.*

Proof. We have $E^\times \cap U_{\mathfrak{A}}^{[n/2]+1} = U_E^{[n/2]+1}$. If ϕ is a character of $E^\times$ agreeing with ψ_α on $U_E^{[n/2]+1}$, the map

$$\Lambda : xu \longmapsto \phi(x)\psi_\alpha(u), \quad x \in E^\times,\ u \in U_{\mathfrak{A}}^{[n/2]+1},$$

defines a character of J_α (since ψ_α is invariant under conjugation by $E^\times$). Any irreducible representation of J_α containing ψ_α is of this form. This proves (1).

In (2), any $g \in G$ which intertwines Λ_1 with Λ_2 must intertwine ψ_α itself, and hence lie in J_α. It therefore fixes each Λ_i and the assertion follows. □

We turn to the case where n is even. In particular, E/F is unramified and $\mathfrak{A} \cong \mathfrak{M}$. In parallel to the first case, ψ_α admits extension to a character of the group H^1_α. We have to consider representations of the intermediate group J^1_α.

Lemma. *Suppose that n is even, and let θ be a character of H^1_α extending ψ_α. There is a unique irreducible representation η_θ of J^1_α such that $\eta_\theta \mid H^1_\alpha$ contains θ. Moreover,*

(1) $\dim \eta_\theta = q$, *and*
(2) $\eta_\theta \mid H^1_\alpha$ *is a multiple of* θ.

We give the proof later, in 16.4. Using the same notation, we prove:

Proposition 2.

(1) *An element $g \in G$ intertwines η_θ if and only if $g \in J_\alpha$.*
(2) *The representation η_θ admits extension to an irreducible representation of J_α, and any such extension lies in $C(\psi_\alpha, \mathfrak{A})$.*
(3) *A representation $\Lambda \in C(\psi_\alpha, \mathfrak{A})$ satisfies $\Lambda \mid J^1_\alpha \cong \eta_\theta$, for a uniquely determined character θ of H^1_α.*
(4) *Two representations $\Lambda_1, \Lambda_2 \in C(\psi_\alpha, \mathfrak{A})$ intertwine in G if and only if they are equivalent.*

Proof. The intertwining statements follow the standard course, so we do not repeat the details. In (2), the representation η_θ surely admits extension to a representation ρ of $F^\times J^1_\alpha$. Since η_θ is stable under conjugation by J_α, so is ρ. Since n is even, the field extension E/F is unramified, so $J_\alpha/F^\times J^1_\alpha$ is cyclic, of order $q+1$. Thus ρ extends to J_α. □

15.7. We note a useful consequence of these discussions:

Proposition. *Let $(\mathfrak{A}, J, \Lambda)$ be a cuspidal type in G. Set $J^0 = J \cap U_{\mathfrak{A}}$ and $\lambda = \Lambda \mid J^0$.*

(1) *The representation λ is irreducible.*
(2) *An element $g \in G$ intertwines λ if and only if $g \in J$.*
(3) *Let (π, V) be an irreducible smooth representation of G which contains λ. Then π is cuspidal and λ occurs in π with multiplicity one.*

Proof. The first assertion follows immediately from the constructions in (15.6). The second is parallel to earlier arguments. For the third, we consider the natural representation of J on the isotypic space V^λ. This is a direct sum of irreducible representations of J extending λ and agreeing with ω_π on Z. If we choose one such component Λ, we have $\pi \cong c\text{-Ind}\,\Lambda$. Any other component must intertwine with, hence be equivalent to, Λ. However, $\operatorname{Hom}_J(\Lambda, \pi) = \operatorname{End}_G(V)$ is one-dimensional. That is, Λ occurs in π with multiplicity one, and $V^\lambda = V^\Lambda$. □

15.8. In a cuspidal type $(\mathfrak{A}, J, \Lambda)$, the representation Λ has a singularly straightforward structure, but the group J may be difficult to specify without supplementary information. It is sometimes more convenient, therefore, to use a variant of the construction in which the group is standard but the representation may be less explicit.

Definition. *A* cuspidal inducing datum *in G is a pair $(\mathfrak{A}, \Xi)$, where $\mathfrak{A}$ is a chain order in A and Ξ is an irreducible smooth representation of $\mathcal{K}_\mathfrak{A}$ of the form $\Xi = \operatorname{Ind}_J^{\mathcal{K}_\mathfrak{A}} \Lambda$, for some cuspidal type $(\mathfrak{A}, J, \Lambda)$.*

It is straightforward to give a direct definition of cuspidal inducing datum, parallel to 15.5. By transitivity of induction, we have:

Proposition 1.

(1) *If $(\mathfrak{A}, \Xi)$ is a cuspidal inducing datum in G, the representation $\pi_\Xi = c\text{-Ind}_{\mathcal{K}_\mathfrak{A}}^G \Xi$ is an irreducible cuspidal representation of G.*
(2) *The map $(\mathfrak{A}, \Xi) \mapsto \pi_\Xi$ induces a bijection between the set of G-conjugacy classes of cuspidal inducing data in G and the set equivalence classes of irreducible cuspidal representations of G.*

If $(\mathfrak{A}, \Xi)$ is a cuspidal inducing datum, we use the notation:

$$\ell_\mathfrak{A}(\Xi) = \min\{n \geqslant 0 : U_\mathfrak{A}^{n+1} \subset \operatorname{Ker} \Xi\} = e_\mathfrak{A} \ell(\pi_\Xi). \qquad (15.8.1)$$

An analogue of 15.7 Proposition also holds here:

Proposition 2. *Let $(\mathfrak{A}, \Xi)$ be a cuspidal inducing datum in G. Set $\xi = \Xi \mid U_\mathfrak{A}$.*

(1) *The representation ξ is irreducible, and occurs in π_Ξ with multiplicity one.*
(2) *An element $g \in G$ intertwines ξ if and only if $g \in \mathcal{K}_\mathfrak{A}$.*

Proof. Let $(\mathfrak{A}, J, \Lambda)$ be a cuspidal type which induces Ξ. If we set $\lambda = \Lambda \mid J \cap U_\mathfrak{A}$, then $\xi = \operatorname{Ind}_{J\cap U_\mathfrak{A}}^{U_\mathfrak{A}} \lambda$. All assertions now follow from 15.7 Proposition. □

Exercise. *Let $\mathfrak{A}$ be a chain order in A and let Θ be an irreducible smooth representation of $\mathcal{K}_\mathfrak{A}$. Let n be the least integer $\geqslant 0$ such that $U_\mathfrak{A}^{n+1} \subset \operatorname{Ker}\Theta$. Suppose that $n \geqslant 1$ and that there is a simple stratum $(\mathfrak{A}, n, \alpha)$ such that $\Theta \mid U_\mathfrak{A}^n$ contains ψ_α. Show that $(\mathfrak{A}, \Theta)$ is a cuspidal inducing datum in G.*

15.9. Looking back to the definition (15.5) of cuspidal type, we see that there is a degree of overlap between the third kind and the other two. That between the first and third is insignificant: if $(\mathfrak{A}, J, \Lambda)$ is a cuspidal type of the first kind, then so is $(\mathfrak{A}, J, \chi\circ\det\otimes\Lambda)$ if and only if χ has level zero.

For the other case, let $(\mathfrak{A}, n, \alpha)$, $n \geqslant 1$ be a simple stratum and let χ be a character of $F^\times$. If χ is of level zero, the map

$$\Lambda \longmapsto \chi\Lambda = \Lambda\otimes\chi\circ\det$$

is a permutation of the set $C(\psi_\alpha, \mathfrak{A})$.

Suppose, on the other hand, that χ has level $l \geqslant 1$. If $l > n/e_{\mathfrak{A}}$, then $(\mathfrak{A}, J, \chi\circ\det\otimes\Lambda)$ is unequivocally of the third kind, so assume that $e_{\mathfrak{A}}l \leqslant n$. The character $\chi\circ\det \mid U^1_{\mathfrak{A}}$ thus has level $e_{\mathfrak{A}}l \leqslant n$. If we take $c \in \mathfrak{p}^{-l}$ such that $\chi(1+x) = \psi(cx)$ for $x \in \mathfrak{p}^{[l/2]+1}$, then $\chi\circ\det \mid U^{[n/2]+1}_{\mathfrak{A}} = \psi_c$. The triple $(\mathfrak{A}, n, \alpha+c)$ is a simple stratum. With this notation, we have:

Proposition. *The map $\Lambda \mapsto \chi\Lambda$ is a bijection $C(\psi_\alpha, \mathfrak{A}) \to C(\psi_{\alpha+c}, \mathfrak{A})$.*

16. Intertwining of Simple Strata

We prove the results quoted in §15 concerning intertwining properties of various characters defined by simple strata.

16.1. We start with the Intertwining Theorem of 15.1. Throughout, we write $\mathfrak{P} = \operatorname{rad}\mathfrak{A}$.

The implication (2) $\Rightarrow$ (1) is trivial. We show next that (3) $\Rightarrow$ (2). The group $E^\times U^{[(n+1)/2]}_{\mathfrak{A}}$ is contained in $\mathcal{K}_{\mathfrak{A}}$ and so normalizes $U^{[n/2]+1}_{\mathfrak{A}}$. If we take $u \in E^\times$, $y \in \mathfrak{P}^{[(n+1)/2]}$ and $x \in \mathfrak{P}^{[n/2]+1}$, we get

$$\begin{aligned}\psi_\alpha\big(u(1+y)(1+x)(1+y)^{-1}u^{-1}\big) &= \psi_A\big(\alpha u((1+y)(1+x)(1+y)^{-1}-1)u^{-1}\big)\\ &= \psi_A\big(u^{-1}\alpha u((1+y)(1+x)(1+y)^{-1}-1)\big)\\ &= \psi_A\big(\alpha((1+y)(1+x)(1+y)^{-1}-1)\big);\end{aligned}$$

We have $(1+y)(1+x)(1+y)^{-1} \equiv 1+x \pmod{\mathfrak{P}^{n+1}}$, so the last expression reduces to $\psi_\alpha(1+x)$, as required.

We next show that (1) $\Rightarrow$ (2). We proceed via the following very useful lemma:

Lemma. *Let $(\mathfrak{A}, n, \alpha)$, $(\mathfrak{A}, n, \beta)$ be simple strata with $n \geqslant 1$. Let $g \in G$ and suppose that g intertwines ψ_α with ψ_β on $U^n_{\mathfrak{A}}$: then $g \in \mathcal{K}_{\mathfrak{A}}$ and the characters $\psi_\alpha \mid U^n_{\mathfrak{A}}$, $\psi_\beta \mid U^n_{\mathfrak{A}}$ are conjugate under g.*

Proof. By hypothesis and 12.7 Proposition, we have

$$(\alpha+\mathfrak{P}^{1-n})\cap g^{-1}(\beta+\mathfrak{P}^{1-n})g \neq \emptyset.$$

Let $\gamma \in (\alpha+\mathfrak{P}^{1-n}) \cap g^{-1}(\beta+\mathfrak{P}^{1-n})g$. As $\gamma \in \alpha+\mathfrak{P}^{1-n} = \alpha U^1_{\mathfrak{A}}$, it is minimal over F and $F[\gamma]^\times \subset \mathcal{K}_{\mathfrak{A}}$ (13.4). If $\mathcal{L}$ is the chain of $\mathfrak{A}$-lattices in $V = F^2$, then $\mathcal{L}$ is the chain of all $\mathfrak{o}_{F[\gamma]}$-lattices in V. On the other hand, $g\gamma g^{-1} \in \beta U^1_{\mathfrak{A}}$ so $g^{-1}\mathcal{L}$ is the chain of all $\mathfrak{o}_{F[\gamma]}$-lattices in V. That is, $g^{-1}\mathcal{L} = \mathcal{L}$ and so $g \in \mathcal{K}_{\mathfrak{A}}$.

Finally, the cosets $\alpha U^1_{\mathfrak{A}}$ and $g^{-1}\beta U^1_{\mathfrak{A}} g = g^{-1}\beta g U^1_{\mathfrak{A}}$ intersect, and so they are equal. □

Returning to the main proof, if $g \in G$ intertwines ψ_α on $U_{\mathfrak{A}}^{[n/2]+1}$, it surely intertwines $\psi_\alpha \mid U^n_{\mathfrak{A}}$. By the lemma, it lies in $\mathcal{K}_{\mathfrak{A}}$. In particular, g normalizes the group $U_{\mathfrak{A}}^{[n/2]+1}$. Since it intertwines $\psi_\alpha \mid U_{\mathfrak{A}}^{[n/2]+1}$, it must actually normalize this character. Thus (1) ⇒ (2).

16.2. We have to show that (2) ⇒ (3). Let $g \in G$ normalize the character ψ_α of $U_{\mathfrak{A}}^{[n/2]+1}$; in particular, $g \in \mathcal{K}_{\mathfrak{A}}$ (16.1 Lemma). Moreover, conjugation by g fixes the coset $\alpha U_{\mathfrak{A}}^{[(n+1)/2]} = \alpha+\mathfrak{P}^{-[n/2]}$.

We have $\mathcal{K}_{\mathfrak{A}} = E^\times U_{\mathfrak{A}}$; we may as well, therefore, take $g \in U_{\mathfrak{A}}$. The intertwining condition then translates into $g^{-1}\alpha g \equiv \alpha \pmod{\mathfrak{P}^{-[n/2]}}$ or, equivalently,

$$\alpha g \alpha^{-1} \equiv g \pmod{\mathfrak{P}^{[(n+1)/2]}}. \tag{16.2.1}$$

The following lemma, applied with $k = [(n+1)/2]$, then completes the proof of the Intertwining Theorem.

Lemma. *Let $g \in \mathfrak{A}$, $k \in \mathbb{Z}$, $k \geqslant 1$; then $\alpha g\alpha^{-1} \equiv g \pmod{\mathfrak{P}^k}$ if and only if $g \in \mathfrak{o}_E+\mathfrak{P}^k$.*

Proof. The "if" implication is clear. We prove the converse by induction on k. We first assume $k \geqslant 2$ and, by inductive hypothesis, $g \in \mathfrak{o}_E+\mathfrak{P}^{k-1}$. We may as well take $g \in \mathfrak{P}^{k-1}$. If ϖ_E is a prime element of E, we can write $g = \varpi_E^{k-1} g_0$, where $g_0 \in \mathfrak{A}$ satisfies $\alpha g_0 \alpha^{-1} \equiv g_0 \pmod{\mathfrak{P}}$. We are thus reduced to the case $k = 1$.

We deal with this case by direct computation. Suppose first that E/F is unramified. Conjugating, we may take $\mathfrak{A} = \mathfrak{M}$. We have $\alpha = \varpi^{-n}\alpha_0$, where ϖ is a prime element of F and $\alpha_0 \in \mathfrak{M}$. By definition, the image $\tilde{\alpha}_0$ of α_0 in $\mathfrak{M}/\varpi\mathfrak{M} \cong \mathrm{M}_2(\boldsymbol{k})$ generates a quadratic field extension $\boldsymbol{l}$ of $\boldsymbol{k}$. Moreover, $\boldsymbol{l}$ is the image of $\mathfrak{o}_E$ in $\mathrm{M}_2(\boldsymbol{k})$. The centralizer of $\boldsymbol{l}$ in $\mathrm{M}_2(\boldsymbol{k})$ is $\boldsymbol{l}$ itself (5.3), and the result follows in this case.

Suppose, on the other hand, that E/F is totally ramified; we conjugate to achieve $\mathfrak{A} = \mathfrak{J}$. The element α is of the form $u\Pi^{-n}$, for some $u \in U_{\mathfrak{J}}$ and some prime element Π of $\mathfrak{J}$. (We recall also that n is odd in this case.) We have $\mathfrak{J}/\mathfrak{P}_{\mathfrak{J}} = \boldsymbol{k} \oplus \boldsymbol{k}$, and conjugation by a prime element interchanges the factors. The result again follows on observing that the image of $\mathfrak{o}_E$ in $\boldsymbol{k} \oplus \boldsymbol{k}$ consists of the elements (x, x). □

16.3. We now prove the Conjugacy Theorem of 15.2. We are given an element $g \in G$ which intertwines the characters ψ_{α_i} of $U_{\mathfrak{A}}^{[n/2]+1}$. An argument identical to that of 12.7 Proposition gives

$$(\alpha_1 + \mathfrak{P}^{-[n/2]}) \cap g^{-1}(\alpha_2 + \mathfrak{P}^{-[n/2]})g \neq \emptyset.$$

In particular,

$$(\alpha_1 + \mathfrak{P}^{1-n}) \cap g^{-1}(\alpha_2 + \mathfrak{P}^{1-n})g \neq \emptyset.$$

(16.1) Lemma yields $g \in \mathcal{K}_{\mathfrak{A}}$; since $\mathcal{K}_{\mathfrak{A}} = F[\alpha_2]^\times U_{\mathfrak{A}}$, we may as well take $g \in U_{\mathfrak{A}}$. The cosets

$$\alpha_1 + \mathfrak{P}^{-[n/2]} = \alpha_1 U_{\mathfrak{A}}^{[(n+1)/2]},$$
$$g^{-1}(\alpha_2 + \mathfrak{P}^{-[n/2]})g = g^{-1}\alpha_2 g\, U_{\mathfrak{A}}^{[(n+1)/2]},$$

intersect. They are therefore equal, as required. □

16.4. We have to prove 15.6 Lemma. The proof has two components, the first of which is an application of the theory of the preceding paragraphs.

To simplify the notation, we set $n = 2m$, so that $J_\alpha^1 = U_E^1 U_{\mathfrak{A}}^m$ and $H_\alpha^1 = U_E^1 U_{\mathfrak{A}}^{m+1}$. We define

$$\mathcal{V} = J_\alpha^1 / H_\alpha^1 \cong U_{\mathfrak{A}}^m / U_E^m U_{\mathfrak{A}}^{m+1} \cong \mathfrak{P}^m / \mathfrak{p}_E^m + \mathfrak{P}^{m+1}.$$

Thus $\mathcal{V}$ is a $\boldsymbol{k}$-vector space of dimension 2 and, in particular, an elementary abelian p-group, where p is the characteristic of $\boldsymbol{k}$.

For $x, y \in J_\alpha^1$, we consider the quantity $\theta[x, y]$, where $[x, y]$ denotes the commutator $xyx^{-1}y^{-1}$. Using standard commutator identities, the fact that θ is fixed under conjugation by J_α^1 implies $\theta[xx', y] = \theta[x, y]\theta[x', y]$ and symmetrically. Moreover, $\theta[x, y] = 1$ if either x or y lies in H_α^1. In all, we have a bi-additive pairing

$$h_\alpha : \mathcal{V} \times \mathcal{V} \longrightarrow \mathbb{C}^\times,$$

induced by $(x, y) \mapsto \theta[x, y]$. Further, h_α is alternating, in that $h_\alpha(x, x) = 1$ for all x.

Explicitly, if $x, y \in \mathfrak{P}^m$, we have

$$h_\alpha(1+x, 1+y) = \psi_\alpha[1+x, 1+y] = \psi_A\big(\alpha(xy-yx)\big).$$

The key point is:

Lemma 1. *The pairing h_α on $\mathcal{V}$ is non-degenerate. The group $J_\alpha^1/\mathrm{Ker}\,\theta$ has centre $H_\alpha^1/\mathrm{Ker}\,\theta$.*

Proof. This reduces to the following assertion: if $x \in \mathfrak{P}^m$ satisfies

$$\psi_A(\alpha(xy-yx)) = 1, \quad y \in \mathfrak{P}^m,$$

then $x \in \mathfrak{p}_E^m + \mathfrak{P}^{m+1}$. As $\psi_A(\alpha(xy-yx)) = \psi_A((\alpha x - x\alpha)y)$, this is equivalent to $\alpha x - x\alpha \in \mathfrak{P}^{1-m}$, or $\alpha x \alpha^{-1} \equiv x \pmod{\mathfrak{P}^{m+1}}$. Lemma 16.2 now implies $x \in H_\alpha^1$, as required.

If we view θ as a character of $H_\alpha^1/\operatorname{Ker}\theta$, it is faithful. If x lies in the centre of $J_\alpha^1/\operatorname{Ker}\theta$, then $h_\alpha(x, \mathcal{V}) = 1$, and $x \in H_\alpha^1/\operatorname{Ker}\theta$, as required for the second assertion. □

We recall a standard group-theoretic argument:

Lemma 2. *Let G be a finite group, with cyclic centre N, such that $V = G/N$ is an elementary abelian p-group. Let χ be a faithful character of N. The pairing $h_\chi : V \times V \to \mathbb{C}^\times$ induced by*

$$(x, y) \longmapsto \chi[x, y],$$

is nondegenerate. There is a unique irreducible representation ζ of G such that $\zeta \mid N$ contains χ. Moreover:

(1) *$\zeta \mid N$ is a multiple of χ;*
(2) *$\dim \zeta = |V|^{1/2}$;*
(3) *$\operatorname{Ind}_N^G \chi = \zeta^{|V|^{1/2}}$;*
(4) *if H is a subgroup of G, containing N, such that $(G{:}H) = |V|^{1/2}$, and such that h_χ is null on H/N, then $\zeta = \operatorname{Ind}_H^G \phi$, for any character ϕ of H such that $\phi \mid N = \chi$.*

Proof. As V is an elementary abelian p-group, the values $h_\chi(V, V)$ lie in the group $\mu_{\mathbb{C}}(p)$ of p-th roots of unity in $\mathbb{C}$. Composing with some group isomorphism $\mu_{\mathbb{C}}(p) \cong \mathbb{F}_p$, the form h_χ becomes an $\mathbb{F}_p$-bilinear form $V \times V \to \mathbb{F}_p$. It is alternating; it is nondegenerate because, if $x \in G$ satisfies $\chi[x, y] = 1$ for all $y \in G$, then $[x, y] = 1$ and x is central in G. In particular, $\dim_{\mathbb{F}_p} V = 2d$, say, is even. There is a vector subspace U of V, of dimension d, such that $h_\chi(u, u') = 0$, for all $u, u' \in U$. Since h_χ is nondegenerate,

(a) *U is its own orthogonal complement in V, and*
(b) *any linear map $U \to \mathbb{F}_p$ is of the form $u \mapsto h_\chi(u, v)$, for a uniquely determined element $v \in V/U$.*

Returning to the multiplicative context, let H be the inverse image of U in G. The triviality of h_χ on U then implies H is abelian. In particular, χ admits extension to a character χ' of H. By property (b), any two such extensions are G-conjugate, and the G-normalizer of χ' is precisely H itself. We deduce that $\eta = \operatorname{Ind}_H^G \chi'$ is irreducible. Its dimension is $p^d = |V|^{1/2}$. If η' is an irreducible representation of G containing χ, then $\eta' \mid H$ is a sum of characters, all extending χ. All such characters are G-conjugate, so $\eta' \mid H$ contains χ' and $\eta' \cong \eta$. All other assertions follow easily. $\square$

In the context of 15.6, we apply Lemma 2 with $G = J^1_\alpha/K$, where $K = \operatorname{Ker}\theta$. The subgroup $N = H^1_\alpha/K \cong \theta(H^1_\alpha)$ is certainly central in G. The nondegeneracy of h_α (Lemma 1) implies that N is the centre of G. Lemma 15.6 now follows directly from Lemma 2.

17. Representations with Iwahori-Fixed Vector

Let $I = U_{\mathfrak{I}}$ be the standard Iwahori subgroup of $G = \mathrm{GL}_2(F)$. The class of irreducible representations (π, V) of G for which $V^I \neq 0$ is particularly subtle and interesting. We study these in order to finish the proof of 14.3 Lemma, but we also need an analysis of the coefficients of the Steinberg representation for use in Chapter VI.

Throughout the section, we use the following notation:

$$w = \begin{pmatrix} 0 & 1 \\ 1 & 0 \end{pmatrix}, \quad T^0 = \begin{pmatrix} U_F & 0 \\ 0 & U_F \end{pmatrix}, \quad N_j = \begin{pmatrix} 1 & \mathfrak{p}^j \\ 0 & 1 \end{pmatrix}, \quad N'_j = \begin{pmatrix} 1 & 0 \\ \mathfrak{p}^j & 1 \end{pmatrix},$$

for $j \in \mathbb{Z}$. In particular, we have $I = N'_1 T^0 N_0$ (with the factors in any order).

17.1. We fix a prime element ϖ of F. Let $\mathbb{W}$ be the set (indeed group) of all matrices of the form

$$\begin{pmatrix} \varpi^a & 0 \\ 0 & \varpi^b \end{pmatrix} \quad \text{or} \quad \begin{pmatrix} 0 & \varpi^a \\ \varpi^b & 0 \end{pmatrix},$$

for $a, b \in \mathbb{Z}$.

Proposition. *The elements of $\mathbb{W}$ form a set of coset representatives for $I\backslash G/I$.*

Proof. We start from the Cartan decomposition

$$G = \bigcup_{a \leqslant b} K \begin{pmatrix} \varpi^a & 0 \\ 0 & \varpi^b \end{pmatrix} K,$$

where $K = \mathrm{GL}_2(\mathfrak{o})$ (7.2.2). We decompose each coset $K \left(\begin{smallmatrix} \varpi^a & 0 \\ 0 & \varpi^b \end{smallmatrix}\right) K$ into (I, I)-double cosets. The presence of a central factor makes no difference, so we may as well treat the coset $C_a = K \left(\begin{smallmatrix} \varpi^{-a} & 0 \\ 0 & 1 \end{smallmatrix}\right) K$, for an integer $a \geqslant 0$. If $a = 0$, the coset C_0 is just K and we know already that $K = I \cup IwI$ (7.3.2). We therefore assume that $a \geqslant 1$. We show that C_a is the disjoint union of the cosets

$$I \begin{pmatrix} \varpi^{-a} & 0 \\ 0 & 1 \end{pmatrix} I, \quad I \begin{pmatrix} 1 & 0 \\ 0 & \varpi^{-a} \end{pmatrix} I, \quad I \begin{pmatrix} 0 & \varpi^{-a} \\ 1 & 0 \end{pmatrix} I, \quad I \begin{pmatrix} 0 & 1 \\ \varpi^{-a} & 0 \end{pmatrix} I. \tag{17.1.1}$$

All of these cosets are certainly contained in C_a. Let μ be the Haar measure on G for which $\mu(I) = 1$. Thus $\mu(K) = q+1$ and

$$\mu(C_a) = (q+1)^2 q^{a-1}.$$

We have

$$\begin{aligned} \begin{pmatrix} \varpi^{-a} & 0 \\ 0 & 1 \end{pmatrix} I \begin{pmatrix} \varpi^{-a} & 0 \\ 0 & 1 \end{pmatrix}^{-1} &= N_{-a}\, T^0\, N'_{a+1}, \\ I \cap \begin{pmatrix} \varpi^{-a} & 0 \\ 0 & 1 \end{pmatrix} I \begin{pmatrix} \varpi^{-a} & 0 \\ 0 & 1 \end{pmatrix}^{-1} &= N_0\, T^0\, N'_{a+1}. \end{aligned}$$

This group has index q^a in I, so

$$\mu(I \left(\begin{smallmatrix} \varpi^{-a} & 0 \\ 0 & 1 \end{smallmatrix}\right) I) = q^a.$$

Similarly,

$$\begin{aligned} \mu(I \left(\begin{smallmatrix} 1 & 0 \\ 0 & \varpi^{-a} \end{smallmatrix}\right) I) &= q^a, \\ \mu(I \left(\begin{smallmatrix} 0 & \varpi^{-a} \\ 1 & 0 \end{smallmatrix}\right) I) &= q^{a-1}, \\ \mu(I \left(\begin{smallmatrix} 0 & 1 \\ \varpi^{-a} & 0 \end{smallmatrix}\right) I) &= q^{a+1}. \end{aligned}$$

Comparing measures, we see that the only possible equality in the list (17.1.1) is between the first two, given by the diagonal matrices. If they were the same, we'd have an equation

$$x \begin{pmatrix} \varpi^{-a} & 0 \\ 0 & 1 \end{pmatrix} = \begin{pmatrix} 1 & 0 \\ 0 & \varpi^{-a} \end{pmatrix} y,$$

with $x, y \in I$. Examining only the $(1,1)$-entry in these matrices shows this is impossible. The measures of the cosets (17.1.1) add up to that of C_a, while any coset IgI is open of positive measure. We deduce that C_a is the disjoint union of the four cosets (17.1.1), as required. □

We remark that the group $\mathbb{W}$ is an example of an *affine Weyl group*. It has a normal subgroup

$$\mathbb{W}_0 = \{x \in \mathbb{W} : \|\det x\| = 1\},$$

and it is the semi-direct product $\mathbb{W} = \langle \Pi \rangle \ltimes \mathbb{W}_0$, where $\Pi = \left(\begin{smallmatrix} 0 & 1 \\ \varpi & 0 \end{smallmatrix}\right)$ is the standard prime element of $\mathfrak{I}$.

17.2. Using the same Haar measure μ, we consider some special elements of the Hecke algebra $\mathcal{H}(G, I) = e_I * \mathcal{H}(G) * e_I$ of compactly supported functions on G which are invariant under translation by I on each side.

For $g \in G$, we let $[g] \in \mathcal{H}(G, I)$ denote the characteristic function of IgI. In particular, $[\Pi]$ is the characteristic function of $I\Pi I = I\Pi = \Pi I$.

Lemma 1. *For any $g \in G$, we have $[\Pi] * [g] = [\Pi g]$ and $[g] * [\Pi] = [g\Pi]$. In particular, $[\Pi]$ is invertible in $\mathcal{H}(G, I)$, with inverse $[\Pi^{-1}]$.*

Proof. Since Π normalizes I, this reduces to an easy calculation. □

Lemma 2. *The function $[w]$ satisfies the relation*

$$[w] * [w] + (1-q)[w] - qe_I = 0.$$

Proof. The support of $[w] * [w]$ is contained in $IwIwI \subset \mathrm{GL}_2(\mathfrak{o}) = I \cup IwI$, so $[w] * [w] = a[w] + be_I$, for some constants a, b. We first compute

$$[w] * [w](1) = \int_G [w](x)[w](x^{-1})\, d\mu(x) = \mu(IwI) = q,$$

since $IwI = \bigcup_{x \in N_0/N_1} xwI$. Next,

$$[w] * [w](w) = \int_G [w](x)[w](x^{-1}w)\, d\mu(x) = \sum_{y \in N_0/N_1} [w](wyw).$$

The term $y \in N_1$ contributes nothing, since $[w](w^2) = [w](1) = 0$. For $y \in N_0 \smallsetminus N_1$, we use the identity

$$\begin{pmatrix} 1 & 0 \\ a & 1 \end{pmatrix} = \begin{pmatrix} -a^{-1} & 1 \\ 0 & a \end{pmatrix} w \begin{pmatrix} 1 & a^{-1} \\ 0 & 1 \end{pmatrix}$$

to get $[w] * [w](w) = q-1$. Therefore $[w] * [w] = (q-1)[w] + qe_I$, as required. □

17.3. We now complete the proof of 14.3 Lemma by observing that, in our present notation, the characteristic function of ItI (as in 14.3) is $[t] = [w]*[\Pi]$, which is an invertible element of $\mathcal{H}(G, I)$. This proves the Lemma and the Proposition of 14.3, and fills the last gap in the proof of the Exhaustion Theorem (14.5).

17.4. We turn to a different aspect of the matter. Let (π, V) be an irreducible smooth representation of G. One says that π is *square-integrable* (modulo the centre Z of G) if $|\omega_\pi| = 1$ and, for a Haar measure $d\dot{g}$ on G/Z, we have:

$$\int_{G/Z} |\langle \check{v}, \pi(g)v\rangle|^2\, d\dot{g} < \infty, \tag{17.4.1}$$

for every $v \in V$, $\check{v} \in \check{V}$.

Exactly in parallel with 10.1 Proposition (2), (17.4.1) holds for all vectors $\check{v} \otimes v$ if it holds for one such vector $\check{v} \otimes v \neq 0$.

One says that an irreducible smooth representation π of G is *essentially square-integrable,* or is in the *discrete series* of G, if it is of the form $\phi\pi_0$, for a character ϕ of $F^\times$ and a square-integrable representation π_0 of G.

17.5. If π is cuspidal, then certainly it is essentially square-integrable; it is square-integrable if and only if $|\omega_\pi| = 1$. Thus it is the non-cuspidal representations which are of interest here:

Theorem.

(1) *The Steinberg representation of G is square-integrable.*
(2) *Let π be an irreducible smooth representation of G, and suppose that π is not cuspidal. It is square-integrable if and only if it is of the form $\pi \cong \phi \cdot \mathrm{St}_G$, for a character ϕ of $F^\times$ satisfying $|\phi| = 1$.*

17.6. To prove part (1) of 17.5 Theorem, we need to produce one non-trivial coefficient of St_G which is square-integrable modulo Z. We write $\mathrm{St}_G = (\pi, V)$, and identify it with the unique irreducible G-subspace of $(\Sigma, W) = \mathrm{Ind}_B^G\, \delta_B^{-1}$ (*cf.* (9.10.4). (9.10.5)). Thus V is the kernel of the G-homomorphism

$$\begin{aligned} W &\longrightarrow \mathbb{C}, \\ f &\longmapsto \int_{B\backslash G} f(g)\, d\bar{g}, \end{aligned}$$

where G acts trivially on $\mathbb{C}$ and $d\bar{g}$ is a semi-invariant measure on $B\backslash G$. Equivalently,

$$V = \{f \in W : \int_K f(k)\, dk = 0\},$$

where $K = \mathrm{GL}_2(\mathfrak{o})$ and dk is some Haar measure on K. As in 14.4, $\dim V^I = 1$, and so V^I is spanned by the function $\theta \in W$ defined by:

$$\begin{aligned} \theta(gj) &= \theta(g), \quad g \in G,\ j \in I \\ \theta(1) &= 1, \quad \theta(w) = -q^{-1}. \end{aligned}$$

We have $(\check{\Sigma}, \check{W}) \equiv \mathrm{Ind}_B^G\, 1_T$; we let $\tau \in \check{W}$ be the characteristic function of BI. The function

$$f(g) \equiv \langle \tau, \pi(g)\theta \rangle = \int_K \tau(k)\theta(kg)\, dk = \int_I \theta(jg)\, dj$$

is then a non-trivial coefficient of π, fixed on either side by I.

We henceforward work with the Haar measure μ on G for which $\mu(I) = 1$. The algebra $\mathcal{H}(G, I)$ acts on V^I in the natural way; we let it act on $\check{V}^I \otimes V^I$ via the second factor. We identify $\check{V}^I \otimes V^I$ with a subspace of $\mathcal{C}(\pi)$. This subspace has dimension 1, and is spanned by f. The action of $\mathcal{H}(G, I)$ is given by

$$\phi \cdot f(g) = f * \check{\phi}(g), \quad \phi \in \mathcal{H}(G, I),$$

and there exists $\alpha_\phi \in \mathbb{C}$ such that

$$\phi \cdot f = \alpha_\phi f, \quad \phi \in \mathcal{H}(G, I). \tag{17.6.1}$$

Writing $f(g) = \langle \tau, \pi(g)\theta \rangle$ as above, we have $\phi \cdot f(g) = \langle \tau, \pi(g)\pi(\phi)\theta \rangle$, so also

$$\pi(\phi)\,\theta = \alpha_\phi\,\theta, \quad \phi \in \mathcal{H}(G, I). \tag{17.6.2}$$

The map $\phi \mapsto \alpha_\phi$ is an algebra homomorphism $\mathcal{H}(G, I) \to \mathbb{C}$.

17.7. Take $\varPi$ as in 17.1.

Proposition. *Let $g \in G$ and, as before, let μ be the Haar measure on G such that $\mu(I) = 1$ and $[g]$ the characteristic function of IgI. We then have*

$$\alpha_{[g]} = [g] \cdot f(1) = \mu(IgI)\, f(g).$$

In particular,

$$\alpha_{[w]} = -1, \quad \alpha_{[\varPi]} = -1.$$

Proof. The first equality is a matter of definition. For the second,

$$[g] \cdot f(1) = \int_G f(x)\,[g](x)\, d\mu(x) = f(g)\,\mu(IgI),$$

as required.

The first of the special cases is given by the original definition of f. In the second, we get

$$\alpha_{[\varPi]} = f(\varPi) = \delta_B \left(\begin{smallmatrix} 1 & 0 \\ 0 & \varpi \end{smallmatrix} \right)^{-1} f(w) = -1,$$

as desired. □

The volume $\mu(IgI)$ depends only on the coset $ZIgI$ so, at least formally, we have

$$\int_{G/Z} |f(g)|^2\, d\dot{g} = \sum_{g \in ZI\backslash G/I} |\alpha_{[g]}|^2 \mu(IgI)^{-1}, \tag{17.7.1}$$

where $d\dot{g}$ is the Haar measure on G/Z for which IZ/Z has measure 1. To prove part (1) of 17.5 Theorem, we have to check that this series is convergent.

17.8. To do this, we use a systematic description of the algebra $\mathcal{H}(G, I)$. Since $\varPi^2 = \varpi 1_G \in Z$, a set of representatives for the space $ZI\backslash G/I$ is given by the matrices $\varPi^m x$, with $m \in \{0, 1\}$ and $x \in \mathbb{W}_0$.

Besides the standard permutation matrix $w \in \mathbb{W}_0$, we shall also need the element

$$w' = \varPi w \varPi^{-1} = \begin{pmatrix} 0 & \varpi^{-1} \\ \varpi & 0 \end{pmatrix}.$$

Lemma 1. *The set $S = \{w, w'\}$ generates the group $\mathbb{W}_0$.*

Proof. Simple calculations give

$$w'w = \begin{pmatrix} \varpi^{-1} & 0 \\ 0 & \varpi \end{pmatrix}, \quad ww'w = \begin{pmatrix} 0 & \varpi \\ \varpi^{-1} & 0 \end{pmatrix},$$

and so on:

$$\begin{pmatrix} \varpi^{-a} & 0 \\ 0 & \varpi^{a} \end{pmatrix} = (w'w)^a, \quad \begin{pmatrix} 0 & \varpi^{a} \\ \varpi^{-a} & 0 \end{pmatrix} = w(w'w)^a, \qquad a \geqslant 0. \qquad (17.8.1)$$

Conjugating by $\varPi$, we get

$$\begin{pmatrix} \varpi^{a} & 0 \\ 0 & \varpi^{-a} \end{pmatrix} = (ww')^a, \quad \begin{pmatrix} 0 & \varpi^{-(a+1)} \\ \varpi^{(a+1)} & 0 \end{pmatrix} = w'(ww')^a, \qquad a \geqslant 0. \quad (17.8.2)$$

This proves Lemma 1. □

Given $x \in \mathbb{W}_0$, there exist elements $w_1, w_2, \dots w_r \in S$ such that

$$x = w_1 w_2 \dots w_r.$$

The *length* of x, denoted $\ell(x)$, is the least integer $r \geqslant 0$ for which there is such an expression. There is one element of length 0 (namely 1), and two elements of length 1 (namely w and w'). The expressions (17.8.1), (17.8.2) are clearly minimal; it follows that $\mathbb{W}_0$ has exactly two elements of length a, for every integer $a \geqslant 1$.

Lemma 2. *Let μ be the Haar measure on G for which $\mu(I) = 1$. If $x \in \mathbb{W}_0$, let $[x]$ denote the characteristic function of IxI.*

(1) *We have $\mu(IgI) = \mu(I\varPi gI) = q^{\ell(g)}$, $g \in \mathbb{W}_0$.*
(2) *Let $g \in \mathbb{W}_0$, let $x \in S = \{w, w'\}$, and suppose that $\ell(xg) = \ell(g)+1$. Then $[xg] = [x] * [g]$.*
(3) *For $g \in \mathbb{W}_0$, we have $\alpha_{[g]} = (-1)^{\ell(g)}$.*

Proof. The list above, together with the calculations in 17.1, give part (1). Likewise (2), while (3) follows from 17.7 Lemma. □

17.9. We now complete the proof of 17.5 Theorem (1), starting from (17.7.1):

$$\int_{G/Z} |f(g)|^2 \, d\dot{g} = \sum_{g \in ZI\backslash G/I} |\alpha_{[g]}|^2 \mu(IgI)^{-1}$$
$$= 2 \sum_{g \in \mathbb{W}_0} q^{-\ell(g)}.$$

Given an integer $b \geqslant 1$, there are exactly 2 elements of $\mathbb{W}_0$ of length b, so this series converges. □

Remark. The same argument shows that, for the coefficient f above, the integral

$$\int_{G/Z} |f(g)|^{1+\epsilon} \, d\dot{g}$$

converges for any $\epsilon > 0$. It follows that it also converges for any $f \in \mathcal{C}(\pi)$.

17.10. We now prove 17.5 Theorem (2). The trivial (one-dimensional) representation of G is certainly not square-integrable, so it is enough to show:

Proposition. *Let χ be a character of T such that $(\pi, V) = \mathrm{Ind}_B^G \chi$ is irreducible. The representation π is then not essentially square-integrable.*

Proof. Twisting by an unramified character of $F^\times$, we can assume that $\omega_\pi = \chi \mid Z$ has absolute value 1. If π were essentially square-integrable it would be square-integrable. We have to show it is not.

If $r \geqslant 1$ is an integer, let $K_r = U_{\mathfrak{M}}^r = 1+\mathfrak{p}^r \mathrm{M}_2(\mathfrak{o})$. We choose r sufficiently large to ensure that χ is trivial on $K_r \cap T$. We define a function $\phi \in V$ by

$$\operatorname{supp}\phi = BK_r,$$
$$\phi(bk) = \chi(b), \quad b \in B,\ k \in K_r.$$

We have $(\check{\pi}, \check{V}) \cong \mathrm{Ind}_B^G \delta_B^{-1}\chi^{-1}$. We similarly take $\phi' \in \check{V}^{K_r}$ to have support BK_r and $\phi'(1) = 1$. We consider the coefficient

$$f : g \longmapsto \langle \phi', \pi(g)\phi\rangle = \int_{B\backslash G} \phi'(h)\phi(hg)\, d\bar{h}$$
$$= \int_K \phi'(k)\phi(kg)\, dk,$$

for a semi-invariant measure $d\bar{h}$ on $B\backslash G$ or a Haar measure dk on K. We show that

$$\int_{G/Z} |f(g)|^2\, d\dot{g} \tag{17.10.1}$$

diverges, where $d\dot{g}$ is a Haar measure on G/Z.

Let $\chi = \chi_1 \otimes \chi_2$, let ϖ be a prime element of F, and define $t \in T$ as follows:

$$t = \begin{cases} \begin{pmatrix} \varpi & 0 \\ 0 & 1 \end{pmatrix} & \text{if } |\chi_1(\varpi)| \geqslant q^{-1/2}, \\ \begin{pmatrix} \varpi^{-1} & 0 \\ 0 & 1 \end{pmatrix} & \text{if } |\chi_1(\varpi)| < q^{-1/2}. \end{cases}$$

We take $n > r$, and consider the contribution to the integral from the set $C_n = ZK_r t^n K_r$. There is a constant $c > 0$ such that $\dot{\mu}(C_n) = cq^n$ (where $d\dot{g} = d\dot{\mu}(g)$). As usual, we set

$$N_k' = \begin{pmatrix} 1 & 0 \\ \mathfrak{p}^k & 1 \end{pmatrix}, \quad k \in \mathbb{Z}.$$

In the first case, the support of the function $k \mapsto \phi'(k)\phi(kt^n)$, $k \in K$, contains $(K \cap B)N_0'$, so there is a constant $c_1 > 0$ such that

$$|f(g)| > c_1 q^{-n/2}, \quad g \in C_n.$$

In the second case, the support of this function contains $(K \cap B)N'_{n+r}$, and again

$$|f(g)| > c_2 q^{-n/2}, \quad g \in C_n,$$

for a constant $c_2 > 0$. In either case, therefore, there is a constant $c_3 > 0$ such that

$$\int_{C_n} |f(g)|^2 \, d\dot{g} > c_3, \quad n > r,$$

and the integral (17.10.1) therefore diverges. This completes the proof of 17.5 Theorem (2). □

Exercises.

(1) *Calculate the formal degree (see* (10a.2) *Remark) of the Steinberg representation.*

(2) *An irreducible representation* (π, V) *is called* spherical *if* $V^K \neq 0$, $K = \mathrm{GL}_2(\mathfrak{o})$. *Show that* π *is spherical if and only if it is a composition factor of* $\mathrm{Ind}_B^G(\chi_1 \otimes \chi_2)$ *with both* χ_i *unramified and* π *is not a twist of the Steinberg representation. Show that, if* π *is spherical, then* $\dim V^K = 1$.

(3) Alternative version: *Let* $K = \mathrm{GL}_2(\mathfrak{o})$*; show that the algebra* $\mathcal{H}(G, K)$ *is commutative. Deduce that any simple* $\mathcal{H}(G,K)$*-module has dimension one and, if* (π, V) *is spherical, then* $\dim V^K = 1$.

Further reading. The discussion of hereditary (or chain) orders and fundamental strata has a close analogue for $\mathrm{GL}_n(F)$: see [69] for the theory of hereditary orders and [11] for fundamental strata. The concept of fundamental stratum, and with it 12.9 Theorem, has been generalized to arbitrary connected reductive groups in [66].

This chapter follows exactly the course of the classification of the irreducible cuspidal representations of $\mathrm{GL}_n(F)$ [19], minus several levels of complexity. It is effectively Kutzko's original treatment [50], [51] ameliorated by hindsight. Carayol's account [21] of $\mathrm{GL}_p(F)$ (p prime) is in much the same spirit. It provides a more general, but easily accessible, example. The same sort of approach is effective in constructing cuspidal representations of much more general groups: see, for example, [73] or [80].

The importance of representations with Iwahori-fixed vector was first recognized in [7]. A survey of the general case is given in [24]. The significance of these ideas for investigating the structure of induction functors is discussed, for $\mathrm{GL}_n(F)$, in [19] and for general groups in [20].

5

Parametrization of Tame Cuspidals

The cuspidal types, and hence the cuspidal representations of G, constructed in §15 are visibly related to (multiplicative) characters of quadratic field extensions of F. When the characteristic p of the finite field $\boldsymbol{k} = \mathfrak{o}/\mathfrak{p}$ is odd, this relation can be made precise and informative: the connection is the subject of this chapter. When $p = 2$, only certain types are amenable to such a convenient description. However, it will be useful, and costs no extra effort, to carry this case along. We therefore impose no restriction on the prime number p.

18. Admissible Pairs

We start with some purely field-theoretic constructions. Throughout this section, ψ denotes a character of F of level one.

18.1. Let E/F be a finite field extension. We recall that E/F is *tamely ramified* if $p \nmid e(E|F)$. Equivalently, E/F is tamely ramified if and only if $\mathrm{Tr}_{E/F}(\mathfrak{o}_E) = \mathfrak{o}$.

In particular, every *quadratic* field extension E/F is tamely ramified if $p \neq 2$ while, in the case $p = 2$, a quadratic extension E/F is tamely ramified if and only if it is unramified. We gather some elementary properties:

Lemma. *Let E/F be a tamely ramified field extension, and set $e = e(E|F)$.*

(1) *For* $r \in \mathbb{Z}$, *we have* $\mathrm{Tr}_{E/F}(\mathfrak{p}_E^{1+r}) = \mathfrak{p}^{1+[r/e]} = \mathfrak{p}_E^{1+r} \cap F$.
(2) *For* $m \geqslant 1$, *the norm map* $\mathrm{N}_{E/F}$ *induces an isomorphism*

$$U_E^{em}/U_E^{em+1} \xrightarrow{\approx} U_F^m/U_F^{m+1}$$

satisfying

$$\mathrm{N}_{E/F}(1+x) \equiv 1 + \mathrm{Tr}_{E/F}(x) \pmod{\mathfrak{p}^{m+1}}, \quad x \in \mathfrak{p}_E^{em}.$$

(3) *The norm* $\mathrm{N}_{E/F}$ *induces a map* $U_E/U_E^1 \to U_F/U_F^1$ *which is surjective if* E/F *is unramified, and has kernel and cokernel of order* $\gcd(e, q-1)$ *if* E/F *is totally ramified.*

For E/F as in the lemma, we set $\psi_E = \psi \circ \mathrm{Tr}_{E/F}$.

Proposition.

(1) *The character* ψ_E *has level one.*
(2) *Let* χ *be a character of* $F^\times$ *of level* $m \geqslant 1$, *and let* $\chi_E = \chi \circ \mathrm{N}_{E/F}$. *The character* χ_E *has level em. If* $c \in \mathfrak{p}^{-m}$ *satisfies* $\chi(1+x) = \psi(cx)$, $x \in \mathfrak{p}^{[m/2]+1}$, *then*

$$\chi_E(1+y) = \psi_E(cy), \quad y \in \mathfrak{p}_E^{[em/2]+1}.$$

Proof. Part (1) and the first assertion of (2) follow from part (2) of the lemma. If $x \in \mathfrak{p}_E^{[em/2]+1}$, then

$$\mathrm{N}_{E/F}(1+x) \equiv 1 + \mathrm{Tr}_{E/F}(x) \pmod{\mathfrak{p}^{m+1}},$$

and the last assertion follows. □

18.2. We consider a pair $(E/F, \chi)$, where E/F is a *tamely ramified* quadratic field extension and χ is a character of $E^\times$.

Definition. *The pair* $(E/F, \chi)$ *is called* admissible *if*

(1) χ *does not factor through the norm map* $\mathrm{N}_{E/F} : E^\times \to F^\times$ *and,*
(2) *if* $\chi \mid U_E^1$ *does factor through* $\mathrm{N}_{E/F}$, *then* E/F *is unramified.*

Admissible pairs $(E/F, \chi)$, $(E'/F, \chi')$ are said to be *F-isomorphic* if there is an F-isomorphism $j : E \to E'$ such that $\chi = \chi' \circ j$. In the case $E = E'$, this amounts to $\chi' = \chi^\sigma$, for some $\sigma \in \mathrm{Gal}(E/F)$.

We write $\mathbb{P}_2(F)$ for the set of F-isomorphism classes of admissible pairs $(E/F, \chi)$.

If $(E/F, \chi)$ is an admissible pair, and if ϕ is a character of $F^\times$, the pair $(E/F, \chi \otimes \phi_E)$ is also admissible, where $\phi_E = \phi \circ \mathrm{N}_{E/F}$.

Let $(E/F, \chi)$ be an admissible pair, and let n be the level of χ. We say that $(E/F, \chi)$ is *minimal* if $\chi \mid U_E^n$ does not factor through $\mathrm{N}_{E/F}$. Clearly, any admissible pair $(E/F, \chi)$ is isomorphic to one of the form $(E/F, \chi' \otimes \phi_E)$, for a character ϕ of $F^\times$ and a minimal pair $(E/F, \chi')$.

Proposition. *Let E/F be a tamely ramified quadratic field extension, and let χ be a character of $E^\times$ of level $m \geqslant 1$. Let $\alpha \in \mathfrak{p}_E^{-m}$ satisfy $\chi(1+x) = \psi_E(\alpha x)$, $x \in \mathfrak{p}_E^m$. Then $(E/F, \chi)$ is a minimal (admissible) pair if and only if the element α is minimal over F.*

Proof. Proposition 18.1 shows that $\chi \mid U_E^m$ factors through $\mathrm{N}_{E/F}$ if and only if there exists $c \in F$ such that $c \equiv \alpha \pmod{\mathfrak{p}_E^{1-m}}$.

On the other hand, the definition of minimal element amounts to the following: if $\alpha \in E$ and $v_E(\alpha) = -m$, then α is *not* minimal over F if and only if there exists $c \in F$ such that $\alpha \equiv c \pmod{\mathfrak{p}_E^{1-m}}$ (*cf.* 13.4 Exercise). The result now follows. □

Exercise.

(1) *Let $(E_1/F, \chi_1)$, $(E_2/F, \chi_2)$ be admissible pairs. Choose an F-embedding $j_i : E_i \to A$, $i = 1, 2$. Show that the pairs $(E_i/F, \chi_i)$ are F-isomorphic if and only if there exists $g \in G$ such that $j_2(E_2) = gj_1(E_1)g^{-1}$ and, for $x \in E_1^\times$, $\chi_1(x) = \chi_2(j_2^{-1}(gj_1(x)g^{-1}))$.*
(2) *Let E_1/F, E_2/F be tamely ramified quadratic field extensions. Show, as consequences of* 18.1 Lemma, *that $\mathrm{N}_{E_i/F}(E_i^\times)$ is a subgroup of $F^\times$ of index* 2, *and that E_1 is F-isomorphic to E_2 if and only if $\mathrm{N}_{E_1/F}(E_1^\times) = \mathrm{N}_{E_2/F}(E_2^\times)$.*

Remark. Exercise (2) is a simple instance of *local class field theory*, for which see §29 below.

19. Construction of Representations

In this section, we attach to an admissible pair $(E/F, \chi)$ an irreducible cuspidal representation π_χ of $G = \mathrm{GL}_2(F)$.

19.1. We start with the special case of an admissible pair $(E/F, \chi)$ in which χ has *level zero.* Thus, by definition, E/F is unramified.

Lemma. *Let E/F be an unramified quadratic extension, let χ be a character of $E^\times$ of level zero, and let $\sigma \in$ $\mathrm{Gal}(E/F)$, $\sigma \neq 1$. The following are equivalent:*

(1) *the pair $(E/F, \chi)$ is admissible;*
(2) $\chi \neq \chi^\sigma$;
(3) $\chi \mid U_E \neq \chi^\sigma \mid U_E$.

Proof. Since E/F is unramified, we have $E^\times = F^\times U_E$, so surely (2) ⇔ (3). Since E/F is cyclic, the kernel of $\mathrm{N}_{E/F}$ consists of all elements x^σ/x, $x \in E^\times$.

Thus χ factors through $\mathrm{N}_{E/F}$ if and only if $\chi = \chi^\sigma$. The second condition in the definition of admissibility is empty in this case, so (1) $\Leftrightarrow$ (2). $\square$

Returning to the admissible pair $(E/F, \chi)$ of level zero, we write $\boldsymbol{k}_E = \mathfrak{o}_E/\mathfrak{p}_E$: thus $\boldsymbol{k}_E/\boldsymbol{k}$ is a quadratic field extension. We choose an F-embedding $E \to A$, and let $\mathfrak{A}$ be the unique chain order with $E^\times \subset \mathcal{K}_{\mathfrak{A}}$ (*cf.* 12.4). Conjugating by an element of G, we can take $\mathfrak{A} = \mathfrak{M} = \mathrm{M}_2(\mathfrak{o})$. This gives an embedding of $\mathfrak{o}_E$ in $\mathfrak{M}$ and hence a $\boldsymbol{k}$-embedding of $\boldsymbol{k}_E$ in $\mathrm{M}_2(\boldsymbol{k})$.

The character $\chi \mid U_E$ is the inflation of a character $\tilde{\chi}$ of $\boldsymbol{k}_E^\times$. Condition (3) in the lemma is equivalent to $\tilde{\chi}$ being a *regular* character of $\boldsymbol{k}_E^\times$ (*cf.* 6.4). As in 6.4, $\tilde{\chi}$ gives rise to an irreducible cuspidal representation $\tilde{\lambda}$ of $\mathrm{GL}_2(\boldsymbol{k})$ ($\tilde{\lambda} = \pi_{\tilde{\chi}}$ in the notation of §6). Let λ be the inflation of $\tilde{\lambda}$ to a representation of $U_{\mathfrak{M}} = \mathrm{GL}_2(\mathfrak{o})$. The restriction of λ to U_F is a multiple of $\chi \mid U_F$ (as follows from (6.4.1)); we therefore extend λ to a representation Λ of $\mathcal{K}_{\mathfrak{M}} = F^\times U_{\mathfrak{M}}$ by deeming that $\Lambda \mid F^\times$ be a multiple of χ.

The triple $(\mathfrak{M}, \mathcal{K}_{\mathfrak{M}}, \Lambda)$ is then a cuspidal type. We set

$$\pi_\chi = c\text{-}\mathrm{Ind}_{\mathcal{K}_{\mathfrak{M}}}^G \Lambda.$$

Thus π_χ is an irreducible cuspidal representation of G such that $\ell(\pi) = 0$. Isomorphic pairs $(E/F, \chi)$ (of level zero) give rise to conjugate pairs $(\boldsymbol{k}_E^\times, \tilde{\chi})$ in $\mathrm{GL}_2(\boldsymbol{k})$, so the equivalence class of π_χ depends only on the F-isomorphism class of $(E/F, \chi)$.

Write $\mathbb{P}_2(F)_0$ for the set of isomorphism classes of admissible pairs $(E/F, \chi)$ in which χ has level zero. Likewise, let $\mathcal{A}_2^0(F)_0$ denote the set of equivalence classes of irreducible cuspidal representations π of G such that $\ell(\pi) = 0$.

Proposition. *The map* $(E/F, \chi) \mapsto \pi_\chi$ *induces a bijection*

$$\mathbb{P}_2(F)_0 \xrightarrow{\approx} \mathcal{A}_2^0(F)_0. \tag{19.1.1}$$

Further, if $(E/F, \chi) \in \mathbb{P}_2(F)_0$*, then:*

(1) *if ϕ is a character of $F^\times$ of level zero, then* $\pi_{\chi\phi_E} = \phi\pi_\chi$;
(2) *if* $\pi = \pi_\chi$*, then* $\omega_\pi = \chi \mid F^\times$;
(3) *the pair* $(E/F, \check{\chi})$ *is admissible and* $\check{\pi}_\chi = \pi_{\check{\chi}}$.

Proof. The analysis of cuspidal representations of $\mathrm{GL}_2(\boldsymbol{k})$ in 6.4, together with the lemma, shows that any cuspidal type of level zero arises from an admissible pair $(E/F, \chi) \in \mathbb{P}_2(F)_0$. The Exhaustion Theorem (14.5) then implies that the map (19.1.1) is surjective.

To prove injectivity, suppose we have pairs $(E_i/F, \chi_i)$ such that the representations π_{χ_i} are equivalent. The extensions E_i/F are unramified and so F-isomorphic: we may as well take $E_1 = E_2 = E$, say. The central character relation gives $\chi_1 \mid F^\times = \chi_2 \mid F^\times$. Corollary 15.5 implies that the cuspidal representations $\pi_{\tilde{\chi}_i}$ of $\mathrm{GL}_2(\boldsymbol{k})$ are equivalent (notation of 6.4). The characters $\tilde{\chi}_i$ of $\boldsymbol{k}_E^\times$ are then Galois-conjugate (6.4 Theorem (2)). It follows that the characters $\chi_i \mid U_E$ are Galois-conjugate and then that the pairs $(E/F, \chi_i)$ are F-isomorphic. Thus (19.1.1) is injective and hence bijective.

Properties (1) and (2) are immediate from the construction, while (3) is given by the construction and 11.4 Remark 3. □

Exercise. *Find an example of a representation* $\pi \in \boldsymbol{A}_2^0(F)_0$ *and a* ramified *character* ϕ *of* $F^\times$ *such that* $\phi\pi \cong \pi$.

19.2. We now fix a character $\psi \in \widehat{F}$, of level one. Let $(E/F, \chi)$ be a *minimal* admissible pair such that χ has level $n \geqslant 1$. We set $\psi_E = \psi \circ \mathrm{Tr}_{E/F}$, $\psi_A = \psi \circ \mathrm{tr}_A$.

Next, we choose an element $\alpha \in \mathfrak{p}_E^{-n}$ such that $\chi(1{+}x) = \psi_E(\alpha x)$, $x \in \mathfrak{p}_E^{[n/2]+1}$. We choose an F-embedding of E in $A = \mathrm{M}_2(F)$ and we let $\mathfrak{A}$ be the unique chain order in A such that $E^\times \subset \mathcal{K}_\mathfrak{A}$ (12.4). Then $e_\mathfrak{A} = e(E|F)$ and the triple $(\mathfrak{A}, n, \alpha)$ is a simple stratum (18.2, 13.5).

Attached to the simple stratum $(\mathfrak{A}, n, \alpha)$, we have the subgroups J_α, J_α^1, H_α^1, as in §15. The next step is to define an irreducible representation $\Lambda \in C(\psi_\alpha, \mathfrak{A})$ (notation of 15.5).

19.3. Suppose in this paragraph that $n = 2m{+}1$ is *odd*. The desired representation Λ is then the character of $J_\alpha = E^\times U_\mathfrak{A}^{m+1}$ given by

$$\Lambda \mid U_\mathfrak{A}^{m+1} = \psi_\alpha, \quad \Lambda \mid E^\times = \chi. \tag{19.3.1}$$

Since $\mathrm{tr}_A \mid E = \mathrm{Tr}_{E/F}$ and $E \cap U_\mathfrak{A}^{m+1} = U_E^{m+1}$, these two conditions are consistent. The triple $(\mathfrak{A}, J_\alpha, \Lambda)$ is a cuspidal type in G, and so

$$\pi_\chi = c\text{-}\mathrm{Ind}_{J_\alpha}^G \Lambda$$

is an irreducible cuspidal representation of G containing the fundamental stratum $(\mathfrak{A}, n, \alpha)$. Thus

$$\ell(\pi_\chi) = n/e(E|F), \quad \omega_{\pi_\chi} = \chi \mid F^\times. \tag{19.3.2}$$

19.4. In this paragraph, we assume that $(E/F, \chi)$ is a minimal pair in which χ has *even* level $n = 2m > 0$. In particular, E/F is unramified. We define a character θ of $H_\alpha^1 = U_E^1 U_\mathfrak{A}^{m+1}$ by

$$\theta(ux) = \chi(u)\,\psi_\alpha(x), \quad x \in U_\mathfrak{A}^{m+1},\ u \in U_E^1. \tag{19.4.1}$$

These conditions are consistent, as before. We let $\eta = \eta_\theta$ be the unique irreducible representation of $J_\alpha^1 = U_E^1 U_\mathfrak{A}^m$ which contains θ (15.6 Lemma).

The cyclic group $\boldsymbol{\mu}_E$ of roots of unity, of order prime to p in E, acts on J^1_α by conjugation, fixing the representation η (15.6 Proposition 2). The subgroup $\boldsymbol{\mu}_F$ acts trivially.

Proposition. *There is a unique irreducible representation $\tilde\eta$ of $\boldsymbol{\mu}_E/\boldsymbol{\mu}_F \ltimes J^1_\alpha$ such that $\tilde\eta \mid J^1_\alpha \cong \eta$ and*

$$\operatorname{tr}\tilde\eta(\zeta u) = -\theta(u),$$

for $u \in H^1_\alpha$ and every $\zeta \in \boldsymbol{\mu}_E/\boldsymbol{\mu}_F$, $\zeta \neq 1$.

We prove this later, in §22. We need the following consequence:

Corollary. *There is a unique irreducible representation Λ of J_α such that*

(1) $\Lambda \mid J^1_\alpha \cong \eta$;
(2) $\Lambda \mid F^\times$ *is a multiple of* $\chi \mid F^\times$;
(3) *for every* $\zeta \in \boldsymbol{\mu}_E \smallsetminus \boldsymbol{\mu}_F$, *we have* $\operatorname{tr}\Lambda(\zeta) = -\chi(\zeta)$.

Proof. The stated conditions certainly determine Λ uniquely. We have to show it exists.

We identify $\boldsymbol{\mu}_E/\boldsymbol{\mu}_F$ with $E^\times/F^\times U^1_E$. We take the representation $\tilde\eta$ given by the proposition, and view it as a representation of $\boldsymbol{\mu}_E/\boldsymbol{\mu}_F \ltimes J^1_\alpha/\operatorname{Ker}\theta$. We inflate it to a representation ν of the group $E^\times \ltimes J^1_\alpha/\operatorname{Ker}\theta$ (in which $E^\times$ acts on $J^1_\alpha/\operatorname{Ker}\theta$ via the isomorphism $E^\times/U^1_E F^\times \cong \boldsymbol{\mu}_E/\boldsymbol{\mu}_F$). We next define a character $\tilde\chi$ of $E^\times \ltimes J^1_\alpha/\operatorname{Ker}\theta$ by deeming it to be trivial on $J^1_\alpha/\operatorname{Ker}\theta$ and to agree with χ on $E^\times$. We form the representation $\tilde\Lambda = \tilde\chi \otimes \nu$ of $E^\times \ltimes J^1_\alpha/\operatorname{Ker}\theta$.

There is a surjective group homomorphism $E^\times \ltimes J^1_\alpha/\operatorname{Ker}\theta \to J_\alpha/\operatorname{Ker}\theta$ given by $(x,j) \mapsto xj \pmod{\operatorname{Ker}\theta}$. The kernel of this map is the group of elements

$$(x^{-1}, x\operatorname{Ker}\theta), \quad x \in U^1_E.$$

The representation $\tilde\Lambda$ is trivial on this kernel, so $\tilde\Lambda$ is the inflation of an irreducible representation Λ_1 of $J_\alpha/\operatorname{Ker}\theta$. The inflation of Λ_1 to J_α is the representation Λ demanded by the Corollary. □

The representation Λ of the corollary lies in $C(\psi_\alpha, \mathfrak{A})$. We define:

$$\pi_\chi = c\text{-}\operatorname{Ind}_{J_\alpha}^G \Lambda. \tag{19.4.2}$$

Thus π_χ is an irreducible cuspidal representation of G satisfying

$$\ell(\pi_\chi) = n, \quad \omega_{\pi_\chi} = \chi \mid F^\times. \tag{19.4.3}$$

19.5. We have to check that the construction of π_χ is independent of choices:

Proposition. *Let $(E/F, \chi)$ be a minimal pair in which χ has positive level. The representation π_χ depends, up to equivalence, only on the isomorphism* *class of the pair $(E/F, \chi)$. In particular, it is independent of the choices of ψ, α and of the embedding $E \to A$.*

Moreover, if ϕ is a character of $F^\times$ such that $(E/F, \chi\phi_E)$ is also minimal, then $\pi_{\chi\phi_E} = \phi\pi_\chi$.

Proof. Changing the choices of ψ and of α (but fixing the embedding) changes neither the group J_α nor the representation Λ. Any two F-embeddings of E in A are G-conjugate, and so give rise to conjugate cuspidal types $(\mathfrak{A}, J_\alpha, \Lambda)$. Moreover, if we have minimal pairs $(E_i/F, \chi_i)$ which are F-isomorphic, and $j_i : E_i \to F$ is an F-embedding, the isomorphism $(E_1/F, \chi_1) \to (E_2/F, \chi_2)$ can be realized by a G-conjugation taking $j_1(E_1)$ to $j_2(E_2)$ which matches the characters (*cf.* 18.2 Exercise).

In the final assertion, the character $\chi\phi_E$ gives rise to the cuspidal type $(\mathfrak{A}, J_\alpha, \Lambda \otimes \phi \circ \det)$ (*cf.* 15.9). $\square$

19.6. Let $(E/F, \chi)$ be an admissible pair. As in 18.2, there is a character ϕ of $F^\times$ and a character χ' of $E^\times$ such that $(E/F, \chi')$ is minimal and $\chi = \chi'\phi_E$. We define

$$\pi_\chi = \phi\pi_{\chi'}. \tag{19.6.1}$$

The result is independent of the choice of decomposition $\chi = \chi'\phi_E$, by the final assertion of 19.5 Proposition. Immediately:

$$\ell(\pi_\chi) = n/e(E|F), \quad \omega_{\pi_\chi} = \chi \mid F^\times, \tag{19.6.2}$$

where n is the level of χ.

In all cases, the equivalence class of the representation π_χ depends only on the isomorphism class of the admissible pair $(E/F, \chi)$. Writing $\mathcal{A}_2^0(F)$ for the set of equivalence classes of irreducible cuspidal representations of $G = \mathrm{GL}_2(F)$, we have a map

$$\begin{aligned} \mathbb{P}_2(F) &\longrightarrow \mathcal{A}_2^0(F), \\ (E/F, \chi) &\longmapsto \pi_\chi, \end{aligned} \tag{19.6.3}$$

defined independently of all choices.

20. The Parametrization Theorem

From now on, we denote by $\mathcal{A}_2^0(F)$ the set of equivalence classes of irreducible cuspidal representations of $G = \mathrm{GL}_2(F)$.

20.1. Let π be an irreducible cuspidal representation of $G = \mathrm{GL}_2(F)$. We say that π is *unramified* if there exists an unramified character $\phi \neq 1$ of $F^\times$ such that $\phi\pi \cong \pi$.

We denote by $\mathcal{A}_2^{\mathrm{nr}}(F)$ the set of unramified classes in $\mathcal{A}_2^0(F)$. A representation $\pi \in \mathcal{A}_2^0(F) \smallsetminus \mathcal{A}_2^{\mathrm{nr}}(F)$ will be called *totally ramified.*

20.2. We come to the main result of the section:

Tame Parametrization Theorem. *The map* $(E/F,\chi) \mapsto \pi_\chi$ *of* (19.6.3) *induces a bijection*

$$\begin{aligned} \mathbb{P}_2(F) &\xrightarrow{\approx} \mathcal{A}_2^0(F) \quad \text{if } p \neq 2, \text{ or} \\ \mathbb{P}_2(F) &\xrightarrow{\approx} \mathcal{A}_2^{\mathrm{nr}}(F) \quad \text{if } p = 2. \end{aligned} \tag{20.2.1}$$

If $(E/F,\chi) \in \mathbb{P}_2(F)$, *then:*

(1) *if* χ *has level* $\ell(\chi)$, *then* $\ell(\pi_\chi) = \ell(\chi)/e(E|F)$;
(2) $\omega_{\pi_\chi} = \chi \mid F^\times$;
(3) *the pair* $(E/F,\check{\chi})$ *is admissible and* $\pi_{\check{\chi}} = \check{\pi}_\chi$;
(4) *if* ϕ *is a character of* $F^\times$, *then* $\pi_{\chi\phi_E} = \phi\pi_\chi$.

Properties (1), (2) and (4) have already been observed. For (3), write $\pi_\chi = c\text{-Ind}_J^G \Lambda$, for the cuspidal type $(\mathfrak{A}, J, \Lambda)$ constructed from $(E/F,\chi)$. The pair $(E/F,\check{\chi})$ gives rise to the type $(\mathfrak{A}, J, \Lambda^\vee)$, and so (3) follows from 11.4 Remark 3. We therefore have only to show that the map (20.2.1) is a bijection.

Remark. The properties (1)–(4) do not determine the bijection (20.2.1) uniquely. It is canonical, in that it does not depend on any of the choices made in its construction, but the lack of a simple, external characterization is a severe drawback. It does, however, serve to provide a usefully explicit approximation to the Langlands correspondence of Chapter VIII (which is specified by a short list of properties).

20.3. Let $\pi \in \mathcal{A}_2^0(F)$. As a prelude to proving the theorem, we demonstrate that useful information can be gleaned from the group of characters ϕ of $F^\times$ such that $\phi\pi \cong \pi$. Note first that, comparing central characters, the relation $\phi\pi \cong \pi$ implies $\phi^2 = 1$.

Lemma. *Let* π *be an irreducible cuspidal representation of* G, *containing a cuspidal inducing datum* $(\mathfrak{A}, \Xi)$. *The representation* π *is unramified if and only if* $\mathfrak{A} \cong \mathfrak{M}$.

Proof. Let ϕ be the unramified character of $F^\times$ of order 2. If $\mathfrak{A} \cong \mathfrak{M}$, then $\det \mathcal{K}_\mathfrak{A} = (F^\times)^2 U_F \subset \operatorname{Ker} \phi$; consequently $\Xi \otimes (\phi \circ \det) \cong \Xi$ and

$$\phi\pi = c\text{-Ind}_{\mathcal{K}_\mathfrak{A}}^G (\Xi \otimes (\phi \circ \det)) \cong c\text{-Ind}_{\mathcal{K}_\mathfrak{A}}^G (\Xi) = \pi.$$

Conversely, suppose that $\mathfrak{A} \cong \mathfrak{J}$ (notation of (12.1.2)). In this case, $\pi = c\text{-Ind}_J^G \Lambda$, for a character Λ of a group J on which $\phi \circ \det$ is not trivial. The relation $\phi\pi \cong \pi$ would imply that the characters Λ, $\Lambda \otimes (\phi \circ \det)$ intertwine in G, contrary to 15.6 Proposition 1. □

Proposition. *Suppose* $p \neq 2$, *and let* $\pi \in \mathcal{A}_2^0(F)$ *be totally ramified.*

(1) *There exists a unique character ϕ of $F^\times$, $\phi \neq 1$, such that $\phi\pi \cong \pi$. The character ϕ is ramified, of level* 0, *and of order* 2.
(2) *Let $(\mathfrak{A}, n, \alpha)$ be a simple stratum, with $n \geqslant 1$, and suppose that $\pi = \theta\pi_0$, for a character θ of $F^\times$ and a representation π_0 containing the character $\psi_\alpha \mid U_{\mathfrak{A}}^{[n/2]+1}$. The field $E = F[\alpha]$ then satisfies $\mathrm{N}_{E/F}(E^\times) = \mathrm{Ker}\,\phi$.*

Proof. Nothing is changed if we replace π by a twist, so we can assume $\ell(\pi) \leqslant \ell(\xi\pi)$, for all characters ξ of $F^\times$. The Exhaustion Theorem and the lemma together imply that there is a ramified simple stratum $(\mathfrak{A}, n, \alpha)$ such that π contains the character ψ_α of $U_{\mathfrak{A}}^{[n/2]+1}$. The integer $n = 2m+1$ is odd; putting $E = F[\alpha]$, we have $J_\alpha = E^\times U_{\mathfrak{A}}^{m+1}$, and the representation π contains a cuspidal type $(\mathfrak{A}, J_\alpha, \Lambda)$.

Since E/F is totally tamely ramified, $\det E^\times = \mathrm{N}_{E/F}(E^\times) \supset U_F^1$. On the other hand, $\det U_{\mathfrak{A}}^{m+1} \subset \det U_{\mathfrak{A}}^1 = U_F^1$. Therefore

$$\det J_\alpha = \mathrm{N}_{E/F}(E^\times), \tag{20.3.1}$$

which is a subgroup of $F^\times$ of index 2 (18.2 Exercise 2). If ϕ is the non-trivial character of $F^\times$ vanishing on $\mathrm{N}_{E/F}(E^\times)$, then $\Lambda \otimes \phi \circ \det = \Lambda$, whence $\pi = c\text{-Ind}\,\Lambda = c\text{-Ind}\,\Lambda \otimes \phi \circ \det = \phi\pi$. Certainly, ϕ is ramified, of level zero, and of order 2. This also proves part (2).

To prove uniqueness, let ξ be a character of $F^\times$ such that $\xi\pi \cong \pi$. Since $\xi^2 = 1$ and p is odd, the restriction $\xi \mid U_F^1$ is trivial and so ξ has level zero. The representation π thus contains the two characters Λ, $\Lambda \otimes \xi \circ \det$ of J_α. These extend ψ_α on $U_{\mathfrak{A}}^{m+1}$ and so are equal (15.6 Proposition 1). Thus ξ vanishes on $\mathrm{N}_{E/F}(E^\times)$, and ξ is therefore either trivial or equal to ϕ. □

21. Tame Intertwining Properties

We now prove the Tame Parametrization Theorem.

21.1. The first, and main, step in the proof is to show:

Proposition. *The map $(E/F, \chi) \mapsto \pi_\chi$ is injective on isomorphism classes of minimal pairs.*

Proof. Suppose we have minimal pairs $(E_i/F, \chi_i)$, $i = 1, 2$, such that $\pi_{\chi_1} \cong \pi_{\chi_2} = \pi$, say. If π is unramified, then so are both extensions E_i/F (20.3 Lemma). Otherwise, there is a non-trivial, ramified character ϕ of $F^\times$ such that $\phi\pi \cong \pi$, and E_i/F satisfies $\mathrm{N}_{E_i/F}(E_i^\times) = \mathrm{Ker}\,\phi$ (20.3 Proposition). In both cases therefore, $E_1 \cong E_2$ (*cf.* 18.2 Exercise (2)), so we can take $E_1 = E_2 = E$, say. Further, the level of χ_i is $n = e(E|F)\ell(\pi)$.

We choose an F-embedding of E in A, and let $\mathfrak{A}$ be the chain order normalized by (the image of) $E^\times$. We form the group $J = E^\times U_{\mathfrak{A}}^{[(n+1)/2]}$. The pair

$(E/F, \chi_i)$ gives a cuspidal type $(\mathfrak{A}, J, \Lambda_i)$. The restriction of Λ_i to $U_{\mathfrak{A}}^{[n/2]+1}$ is a multiple of ψ_{α_i}, for some $\alpha_i \in \mathfrak{p}_E^{-n}$. These two cuspidal types intertwine in G and, in particular, the characters ψ_{α_1}, ψ_{α_2} of $U_{\mathfrak{A}}^{[n/2]+1}$ are $U_{\mathfrak{A}}$-conjugate (15.2). We prove, under these hypotheses:

Lemma. *There exists* $u \in U_{\mathfrak{A}}$ *such that* $u\alpha_2u^{-1} \in E$ *and* $u\alpha_2u^{-1} \equiv \alpha_1 \pmod{\mathfrak{p}_E^{-[n/2]}}$.

Before proving this, we complete the main argument. The lemma shows that, after applying a suitable conjugation, we can assume that the α_i both lie in E and that the characters ψ_{α_i} of $U_{\mathfrak{A}}^{[n/2]+1}$ are the same. The representations Λ_i intertwine in G, and so are equivalent (15.6). If n is odd, the character χ_i is $\Lambda_i \mid E^\times$; if n is even, χ_i is given by the formula of 19.4 Corollary. In either case, we get $\chi_1 = \chi_2$, as required. □

21.2. We prove the lemma of 21.1. By hypothesis, the cosets $\alpha_i U_{\mathfrak{A}}^{[(n+1)/2]}$ are $U_{\mathfrak{A}}$-conjugate. In particular, the cosets $\alpha_i U_{\mathfrak{A}}^1$ are $U_{\mathfrak{A}}$-conjugate. Under these circumstances:

Lemma 1. *There exists* $\sigma \in \mathrm{Gal}(E/F)$ *such that* $\alpha_2 \equiv \alpha_1^\sigma \pmod{U_E^1}$.

Proof. Suppose first that E/F is unramified, and pick a prime element ϖ of F. Thus $\alpha_i \equiv \varpi^{-n}\zeta_i \pmod{U_E^1}$, for some root of unity $\zeta_i \in \boldsymbol{\mu}_E \smallsetminus \boldsymbol{\mu}_F$. The cosets $\zeta_i U_{\mathfrak{A}}^1$ are still $U_{\mathfrak{A}}$-conjugate. The images $\tilde{\zeta}_i$ of ζ_i in $\mathfrak{A}/\mathfrak{P} \cong \mathrm{M}_2(\boldsymbol{k})$ are therefore conjugate under $\mathrm{GL}_2(\boldsymbol{k})$. We have $\boldsymbol{k}[\tilde{\zeta}_1] = \boldsymbol{k}[\tilde{\zeta}_2] \cong \mathbb{F}_{q^2}$, so the $\tilde{\zeta}_i$ are Galois-conjugate. It follows that the ζ_i are Galois-conjugate.

Suppose now that E/F is totally ramified: thus $p \neq 2$ and $U_E = U_F U_E^1$. Thus there exists $y \in U_F$ such that $\alpha_2 \equiv y\alpha_1 \pmod{U_E^1}$. If the cosets $\alpha_i U_{\mathfrak{A}}^1$ are to be $U_{\mathfrak{A}}$-conjugate, the only possibilities (comparing norms) are $y \equiv \pm 1 \pmod{U_E^1}$. In this case, if σ is the non-trivial F-automorphism of E, we have $\alpha_1^\sigma \equiv -\alpha_1 \pmod{U_E^1}$, and the lemma again follows. □

The next point to observe is that, if we write $x \mapsto j(x)$ for our chosen embedding of E in A, and if $\sigma \in \mathrm{Gal}(E/F)$, the embedding $x \mapsto j(x^\sigma)$ likewise carries $E^\times$ into $\mathcal{K}_{\mathfrak{A}}$. It is therefore $U_{\mathfrak{A}}$-conjugate to j. We conclude:

Lemma 2. *There exists* $u \in U_{\mathfrak{A}}$ *such that* $uEu^{-1} = E$ *and* $u\alpha_2u^{-1} \equiv \alpha_1 \pmod{U_E^1}$.

We have therefore reduced to the case $\alpha_1 \equiv \alpha_2 \pmod{U_{\mathfrak{A}}^1}$ or, equivalently, $\alpha_1 \equiv \alpha_2 \pmod{U_E^1}$. It is now enough to prove:

Lemma 3. *Suppose that* $\alpha_1 \equiv \alpha_2 \pmod{U_E^1}$, *and that the cosets* $\alpha_1 U_{\mathfrak{A}}^{[(n+1)/2]}$, $\alpha_2 U_{\mathfrak{A}}^{[(n+1)/2]}$ *are* $U_{\mathfrak{A}}$*-conjugate. Then* $\alpha_1 \equiv \alpha_2 \pmod{U_E^{[(n+1)/2]}}$.

Proof. By hypothesis, there exists $u \in U_{\mathfrak{A}}$ such that $u\alpha_2 u^{-1} \equiv \alpha_1 \pmod{\mathfrak{P}^{1-n}}$, $\mathfrak{P} = \operatorname{rad}\mathfrak{A}$. By 16.2 Lemma, $u \in U_E U^1_{\mathfrak{A}}$, so we may as well take $u = 1+x \in U^1_{\mathfrak{A}}$. We now proceed inductively, using:

Lemma 4. *Let $m \geqslant 1$ be an integer. Suppose that $\alpha_1 \equiv \alpha_2 \pmod{\mathfrak{p}_E^{m-n}}$, and suppose there exists $x \in \mathfrak{P}^m$ such that $(1+x)\alpha_2(1+x)^{-1} \equiv \alpha_1 \pmod{\mathfrak{P}^{m+1-n}}$. Then $x \in \mathfrak{p}_E^m + \mathfrak{P}^{m+1}$ and $\alpha_1 \equiv \alpha_2 \pmod{\mathfrak{p}_E^{m+1-n}}$.*

Proof. We write $\alpha_2 = \alpha_1 + c$, where $c \in \mathfrak{p}_E^{m-n}$ and expand

$$(1+x)\alpha_2(1+x)^{-1} \equiv \alpha_1 + c - \alpha_1 x + x\alpha_1 \pmod{\mathfrak{P}^{m+1-n}}.$$

That is, $\alpha_1 x - x\alpha_1 \equiv c \pmod{\mathfrak{P}^{m+1-n}}$. Consider the character $y \mapsto \psi_E(cy)$, $y \in \mathfrak{p}_E^{n-m}$. We have

$$\psi_E(cy) = \psi_A(cy) = \psi_A\big((\alpha_1 x - x\alpha_1)y\big) = \psi_A\big((y\alpha_1 - \alpha_1 y)x\big) = 1,$$

since y commutes with α_1. This implies $c \in \mathfrak{p}_E^{m+1-n}$, since ψ_E has level one. That is, $\alpha_2 \equiv \alpha_1 \pmod{\mathfrak{p}_E^{m+1-n}}$. Also, $\alpha_1 x - x\alpha_1 \in \mathfrak{P}^{m+1-n}$, and 16.2 Lemma implies $x \in \mathfrak{p}_E^m + \mathfrak{P}^{m+1}$, as required. □

This proves Lemma 3, and also 21.1 Lemma. □

21.3. If $(E/F, \chi)$ is a minimal pair of positive level, the representation π_χ constructed above contains a simple stratum. It therefore satisfies $\ell(\pi_\chi) \leqslant \ell(\phi\pi_\chi)$, for any character ϕ of $F^\times$ (13.3). The converse also holds:

Proposition. *Let $\pi \in \mathcal{A}_2^0(F)$ satisfy $0 < \ell(\pi) \leqslant \ell(\phi\pi)$ for all characters ϕ of $F^\times$. Suppose, in the case $p = 2$, that π is unramified. There exists a minimal pair $(E/F, \chi) \in \mathbb{P}_2(F)$ such that $\pi \cong \pi_\chi$.*

Proof. The representation π contains a cuspidal type $(\mathfrak{A}, J, \Lambda)$ attached to a simple stratum $(\mathfrak{A}, n, \alpha)$. We have $J = J_\alpha$ and $\Lambda \mid U_{\mathfrak{A}}^{[n/2]+1}$ is a multiple of ψ_α. Setting $E = F[\alpha]$, we have $J_\alpha = E^\times U_{\mathfrak{A}}^{[(n+1)/2]}$.

If n is odd, we put $\chi = \Lambda \mid E^\times$ to get the desired minimal pair $(E/F, \chi)$. If n is even, we let θ be the unique character of H^1_α occurring in $\Lambda \mid H^1_\alpha$.

Lemma. *There exists a unique character χ of $E^\times$ such that*

$$\chi \mid U^1_E = \theta \mid U^1_E, \quad \chi \mid F^\times = \omega_\pi, \quad \chi(\zeta) = -\operatorname{tr}\Lambda(\zeta),$$

for every $\zeta \in \mu_E \smallsetminus \mu_F$. The pair $(E/F, \chi)$ is a minimal admissible pair.

Proof. Let χ' be a character of $E^\times$ agreeing with θ on U^1_E and ω_π on $F^\times$. The pair $(E/F, \chi')$ is then admissible and minimal. As in 19.4 Corollary, there is a unique representation $\Lambda' \in C(\psi_\alpha, \mathfrak{A})$ such that $\operatorname{tr}\Lambda'(\zeta) = -\chi'(\zeta)$, $\zeta \in \mu_E \smallsetminus \mu_F$. There is a unique character ϕ of the group $J_\alpha/F^\times J^1_\alpha \cong \mu_E/\mu_F$ such that $\Lambda \cong \phi \otimes \Lambda'$. We view ϕ as a character of $E^\times$ via the isomorphism $E^\times/F^\times U^1_E \cong \mu_E/\mu_F$; the desired character is then $\chi = \phi\chi'$. The remaining assertions are immediate. □

This completes the proof of the proposition. □

21.4. We now finish the proof of the Tame Parametrization Theorem (20.2). We take $\pi \in \mathcal{A}_2^0(F)$ ($p \neq 2$) or $\mathcal{A}_2^{\mathrm{nr}}(F)$ (if $p = 2$). We choose a character ϕ of $F^\times$ to minimize the level of $\rho = \phi^{-1}\pi$. Thus $\ell(\rho) \leqslant \ell(\theta\rho)$ for all characters θ of $F^\times$. By 21.3 or 19.1, there exists $(E/F, \xi) \in \mathbb{P}_2(F)$ such that $\rho = \pi_\xi$. Setting $\chi = \xi\phi_E$, we get $\pi = \pi_\chi$ and so (20.2.1) is surjective.

Suppose now we have pairs $(E_i/F, \chi_i) \in \mathbb{P}_2(F)$, $i = 1, 2$, with $\pi_{\chi_1} \cong \pi_{\chi_2}$. Twisting by a character of $F^\times$, we can assume that $(E_1/F, \chi_1)$ is minimal. The representation π_{χ_i} then has minimal level relative to twisting, so $(E_2/F, \chi_2)$ is minimal (19.6). Thus (21.1, 19.1) $(E_1/F, \chi_1)$ is isomorphic to $(E_2/F, \chi_2)$.

22. A Certain Group Extension

We have to prove 19.4 Proposition. This is part of a larger episode from the representation theory of finite groups, but we focus as closely as possible on the very special case to hand.

22.1. We fix a prime number p. We consider a finite p-group G, with cyclic centre Z, such that $V = G/Z$ is an elementary abelian p-group. Let θ be a faithful character of Z. The alternating pairing

$$h : V \times V \longrightarrow \mathbb{C}^\times,$$

induced by $(x, y) \mapsto \theta[x, y]$, is then nondegenerate. There exists a unique irreducible representation η of G such that $\eta \mid Z$ contains θ (*cf.* 16.4).

In addition, we are given a finite cyclic group A of automorphisms of G, of order relatively prime to p. We assume that A acts trivially on Z. It therefore fixes (the equivalence class of) the representation η. Since A is cyclic, the representation η admits extension to an irreducible representation of the semi-direct product $AG = A \ltimes G$.

Lemma. *Let η_0 be some irreducible representation of AG such that $\eta_0 \mid G \cong \eta$, and let χ range over the irreducible characters of $AG/G \cong A$. The representations $\chi \otimes \eta_0$ are then distinct and*

$$\mathrm{Ind}_G^{AG}\, \eta = \sum_\chi \chi \otimes \eta_0.$$

Proof. The determinant of the representation $\chi \otimes \eta_0$ is $\chi^{\dim \eta} \det \eta_0$. Since $\dim \eta$ is relatively prime to $|A|$, these determinants are distinct, whence so are the representations $\chi \otimes \eta_0$. They all occur in $\mathrm{Ind}_G^{AG}\, \eta$, and the second assertion follows on comparing dimensions. □

22.2. We are only concerned with a very special case, so we impose some correspondingly restrictive hypotheses:

Hypotheses.

(a) $|V| = q^2$, *where* q *is some power of* p;
(b) $|A| = q{+}1$;
(c) *if* $a \in A$, $a \neq 1$, *then* a *has only the trivial fixed point in* V.

Observe that condition (c) implies V to be *simple* as $\mathbb{F}_p[A]$-module: otherwise, V would have an A-subspace of cardinality $\leqslant q$ on which a generator of A would act as a cyclic automorphism with an orbit of $q{+}1$ elements.

Lemma. *Under these hypotheses, there is a unique irreducible representation* η_1 *of* AG *such that* $\eta_1 \mid G \cong \eta$ *and* $\operatorname{tr}\eta_1(a) = -1$, *for all* $a \in A$, $a \neq 1$.

Proof. Let η_i, $1 \leqslant i \leqslant q{+}1$, be the distinct extensions of η to AG. Consider the induced representation

$$\xi = \operatorname{Ind}_{AZ}^{AG} 1_A \otimes \theta,$$

where $1_A \otimes \theta$ denotes the character $(a, z) \mapsto \theta(z)$ of $AZ \cong A \times Z$. This takes the form

$$\xi = \sum_{i=1}^{q+1} m_i \eta_i,$$

for integers $m_i \geqslant 0$. We have $\xi \mid G \cong \operatorname{Ind}_Z^G \theta = q\eta$, so

$$\sum_{i=1}^{q+1} m_i = q. \qquad (22.2.1)$$

Consider also the restriction

$$\xi \mid AZ = \sum_{g \in AZ\backslash AG/AZ} \operatorname{Ind}_{g^{-1}AZg \cap AZ}^{AZ} (1_A \otimes \theta)^g.$$

The coset space $AZ\backslash AG/AZ$ is just the orbit space V/A. Hypothesis (c) implies $g^{-1}AZg \cap AZ = Z$ unless g represents the trivial orbit (of 1_G). This gives us

$$\xi \mid AZ = 1_A \otimes \theta + (q{-}1)(\operatorname{Reg}_A) \otimes \theta, \qquad (22.2.2)$$

where Reg_A denotes the regular representation of A. Thus

$$q = \langle 1_A \otimes \theta, \xi \rangle_{AZ} = \langle \xi, \xi \rangle_{AG} = \sum_i m_i^2.$$

Combining this with (22.2.1), we see that all the integers m_i are equal to 1, except for one which is zero. We re-number so that $m_1 = 0$, $m_i = 1$ for $i \geqslant 2$. Thus $\xi = \operatorname{Reg}_A \otimes \eta_1 - \eta_1$, or

$$\eta_1 = \operatorname{Ind}_G^{AG} \eta - \xi.$$

Let $a \in A$, $a \neq 1$. From (22.2.2), we get $\operatorname{tr}\xi(a) = 1$, so $\operatorname{tr}\eta_1(a) = -1$. Thus η_1 is the representation demanded by the lemma. □

22.3. It will be useful later to know more about the conjugacy class structure of the group AG. With the hypotheses of 22.2:

Lemma. *Let $a \in A$, $a \neq 1$.*

(1) *Let $g \in G$; the element ag is then G-conjugate to one of the form az, $z \in Z$.*
(2) *For $z \in Z$, the G-conjugacy class and the AG-conjugacy class of az are the same, and this class has q^2 elements.*
(3) *For $z_1, z_2 \in Z$, the elements az_1, az_2 are AG-conjugate if and only if $z_1 = z_2$.*

Proof. We first observe that the commutator map

$$\begin{aligned} V &\longrightarrow V, \\ v &\longmapsto ava^{-1}v^{-1} \end{aligned}$$

is bijective. Assertion (1) follows immediately. In (2), the group A centralizes az, so the AG-conjugacy class is the G-conjugacy class. As g ranges over $AG/AZ = G/Z$, the conjugates $gazg^{-1} = az(a^{-1}gag^{-1})$ are distinct, and there are q^2 of them. This proves (2). The commutator $a^{-1}gag^{-1}$ lies in Z if and only if $g \in Z$, so (3) holds. □

22.4. To prove 19.4 Proposition, we have to show that the situation of 19.4 satisfies the hypotheses of 22.2. The group G is $J^1_\alpha/\mathrm{Ker}\,\theta$; its centre $H^1_\alpha/\mathrm{Ker}\,\theta$, of which θ is a faithful character (16.4). Thus $V = J^1_\alpha/H^1_\alpha \cong \mathfrak{A}/\mathfrak{o}_E+\mathfrak{P}$, which has order q^2.

The group A is $\boldsymbol{\mu}_E/\boldsymbol{\mu}_F$, which has order $q+1$ and acts on $\mathfrak{A}/\mathfrak{o}_E+\mathfrak{P}$ by conjugation. We can identify $\mathfrak{A}/\mathfrak{P}$ with $\mathrm{M}_2(\boldsymbol{k})$, and the image of $\mathfrak{o}_E$ is the residue field $\boldsymbol{k}_E$. We choose $\sigma \in \mathrm{GL}_2(\boldsymbol{k})$ which normalizes $\boldsymbol{k}_E$ and acts on it (by conjugation) as the non-trivial element of $\mathrm{Gal}(\boldsymbol{k}_E/\boldsymbol{k})$. We then have $\mathrm{M}_2(\boldsymbol{k}) = \boldsymbol{k}_E \oplus \sigma\boldsymbol{k}_E$. Under the natural conjugation action of $\boldsymbol{\mu}_E \cong \boldsymbol{k}_E^\times$, a root of unity ζ fixes the factor $\boldsymbol{k}_E$ and acts on $\sigma\boldsymbol{k}_E$ as right multiplication by ζ^q/ζ. Thus all the hypotheses of 22.2 are satisfied. □

Further reading. The constructions of this chapter are more general than they appear. Fixing $n \geqslant 2$ and assuming, for simplicity, that $n \not\equiv 0 \pmod p$, one can use the same definition for admissible pairs of degree n. The idea originates with Howe [45]; that paper describes a method of constructing, from an admissible pair $(E/F, \chi)$ of degree n, an irreducible cuspidal representation π_χ of $\mathrm{GL}_n(F)$, and shows that the map $(E/F, \chi) \mapsto \pi_\chi$ is injective. Moy [65] subsequently proved, by an indirect method, that the map is surjective. A more direct proof, based on [19] and closer in spirit to the methods here, is given in [17]. More surprisingly, Yu [89] shows how to construct cuspidal representations of quite general groups in this way, under the hypothesis that the residual characteristic p is much larger than the rank of the group.

6

Functional Equation

In this chapter, we take an irreducible smooth representation π of $G = \mathrm{GL}_2(F)$ and attach to it a pair of invariants $L(\pi, s)$, $\varepsilon(\pi, s, \psi)$. Here, s is a complex variable and ψ is a non-trivial character of F. The *L-function* $L(\pi, s)$ is an elementary function of the form $f(q^{-s})^{-1}$, where $f(t)$ is a complex polynomial of degree at most two. The *local constant* $\varepsilon(\pi, s, \psi)$ is of the form cq^{-ms}, for a non-zero constant c and an integer m.

This theory generalizes a classical one for characters of $F^\times$. Indeed, for non-cuspidal representations of G, the L-function and local constant are expressed directly in terms of the corresponding objects for characters of $F^\times$. We therefore start the chapter with an account of the one-dimensional theory.

For cuspidal representations π of G, the picture is rather different. One invariably has $L(\pi, s) = 1$, and the local constant $\varepsilon(\pi, s, \psi)$ assumes greater importance. This generalizes the one-dimensional theory in a different way. If χ is a *ramified* character of $F^\times$, the local constant $\varepsilon(\chi, s, \psi)$ is given in terms of a classical, or abelian, local Gauss sum. In the two-dimensional case, the local constant $\varepsilon(\pi, s, \psi)$ of a cuspidal representation π can be expressed in terms of a "non-abelian Gauss sum" attached to a cuspidal inducing datum occurring in π: the parallel is very close. These Gauss sums provide us with a very effective computational tool.

The main consequence in the present chapter is the Converse Theorem. This asserts that an irreducible smooth representation π of G is determined, up to equivalence, by the pair of functions $\chi \mapsto L(\chi\pi, s)$, $\chi \mapsto \varepsilon(\chi\pi, s, \psi)$, where χ ranges over the characters of $F^\times$. There is, in fact, an absolute dichotomy: if π is not cuspidal, it is determined by the L-functions $L(\chi\pi, s)$ while, if π is cuspidal, the L-functions give no information whatsoever, and the representation π is determined by the local constants $\varepsilon(\chi\pi, s, \psi)$.

23. Functional Equation for GL(1)

We start with a detailed review of the theory of the local functional equation for characters of $F^\times = \mathrm{GL}_1(F)$. As well as providing an illustrative introduction to the corresponding theory for $\mathrm{GL}_2(F)$, many of the results and calculations have direct relevance for later parts of the chapter.

23.1. Recall that, if $a \in F$ and $\psi \in \widehat{F}$, then $a\psi$ denotes the character $x \mapsto \psi(ax)$. Further, if $\psi \neq 1$, the map $a \mapsto a\psi$ gives an isomorphism $F \cong \widehat{F}$ (1.7).

We fix $\psi \in \widehat{F}$, $\psi \neq 1$, and a Haar measure μ on F. For $\Phi \in C_c^\infty(F)$, we define the *Fourier transform* $\hat{\Phi}$ of Φ (relative to μ and ψ) by

$$\hat{\Phi}(x) = \int_F \Phi(y)\psi(xy)\, d\mu(y), \quad x \in F.$$

The integrand is locally constant and the integral reduces to a finite sum.

Proposition.

(1) *For $\Phi \in C_c^\infty(F)$, the function $\hat{\Phi}$ lies in $C_c^\infty(F)$.*

(2) *There is a positive real number $c = c(\psi, \mu)$ such that*

$$\hat{\hat{\Phi}}(x) = c\Phi(-x), \quad \Phi \in C_c^\infty(F),\ x \in F. \tag{23.1.1}$$

(3) *For given ψ, there is a unique Haar measure μ_ψ for which $c(\psi, \mu_\psi) = 1$. This measure satisfies $\mu_\psi(\mathfrak{o}) = q^{l/2}$, where l is the level of ψ.*

(4) *For $a \in F^\times$, we have $\mu_{a\psi} = \|a\|^{\frac{1}{2}} \mu_\psi$.*

Proof. Let l be the level of ψ, and let Φ_j be the characteristic function of $\mathfrak{p}^j$. For $a \in F$, the character $a\psi \mid \mathfrak{p}^j$ is trivial if and only if $a \in \mathfrak{p}^{l-j}$. The support of $\hat{\Phi}_j$ is therefore $\mathfrak{p}^{l-j}$ and

$$\hat{\Phi}_j(x) = \mu(\mathfrak{o})\, q^{-j}, \quad x \in \mathfrak{p}^{l-j}.$$

Assertions (1) and (2) therefore hold for all functions Φ_j, with $c = \mu(\mathfrak{o})^2 q^{-l}$.

Let $\Phi \in C_c^\infty(F)$, $a \in F$, and let Ψ denote the function $x \mapsto \Phi(x-a)$. We then have

$$\hat{\Psi}(x) = \int_F \Phi(y-a)\psi(xy)\, d\mu(y) = a\psi(x)\hat{\Phi}(x).$$

The function $a\psi$ is locally constant, so $\hat{\Psi}$ lies in $C_c^\infty(F)$ provided $\hat{\Phi}$ does. Calculating the Fourier transform again, we get

$$\hat{\hat{\Psi}}(x) = \hat{\hat{\Phi}}(a+x).$$

If parts (1) and (2) of the Lemma hold for Φ, they also therefore hold for Ψ with the same value of c. However, they do hold for the functions Φ_j, and the functions $x \mapsto \Phi_j(x-a)$, $a \in F$, $j \in \mathbb{Z}$, span $C_c^\infty(F)$ (7.4). The first two parts of the proposition therefore hold for all $\Phi \in C_c^\infty(F)$, with $c = \mu(\mathfrak{o})^2 q^{-l}$, as above.

For $b > 0$, we have $c(\psi, b\mu) = b^2 c(\psi, \mu)$; put another way, to achieve $c(\psi, \mu) = 1$, we must have $\mu(\mathfrak{o})^2 q^{-l} = 1$, so part (3) follows. Part (4) is an immediate consequence. □

The measure μ_ψ is called *the self-dual Haar measure on F, relative to ψ*. Using μ_ψ to compute the Fourier transform:

$$\hat{\Phi}(x) = \int_F \Phi(y)\, \psi(xy)\, d\mu_\psi(y),$$

(23.1.1) gives the *Fourier inversion formula*

$$\hat{\hat{\Phi}}(x) = \Phi(-x), \quad \Phi \in C_c^\infty(F),\ x \in F. \tag{23.1.2}$$

23.2. Let μ^* be a Haar measure on $F^\times$. Let χ be a character of $F^\times$, let $\Phi \in C_c^\infty(F)$, and choose a prime element ϖ of F.

For $m \in \mathbb{Z}$, let $\mathfrak{e}_m$ denote the characteristic function of the set $\varpi^m U_F = \mathfrak{p}^m \smallsetminus \mathfrak{p}^{m+1}$. For $\Phi \in C_c^\infty(F)$, the function $\Phi_m = \mathfrak{e}_m \Phi$ lies in $C_c^\infty(F^\times) \subset C_c^\infty(F)$. It is identically zero for $m \ll 0$. We may therefore set

$$z_m = z_m(\Phi, \chi) = \int_{\varpi^m U_F} \Phi(x)\chi(x)\, d\mu^*(x), \quad m \in \mathbb{Z},$$

and define a formal Laurent series $Z(\Phi, \chi, X) \in \mathbb{C}((X))$ by

$$Z(\Phi, \chi, X) = \sum_{m \in \mathbb{Z}} z_m X^m.$$

Clearly, $\Phi \mapsto Z(\Phi, \chi, X)$ is a linear map $C_c^\infty(F) \to \mathbb{C}((X))$. We denote its image

$$\mathcal{Z}(\chi) = \mathcal{Z}(\chi, X) = \{Z(\Phi, \chi, X) : \Phi \in C_c^\infty(F)\}.$$

For $a \in F^\times$ and $\Phi \in C_c^\infty(F)$, we denote by $a\Phi$ the function $x \mapsto \Phi(a^{-1}x)$. We have

$$Z(a\Phi, \chi, X) = \chi(a)\, X^{\upsilon_F(a)}\, Z(\Phi, \chi, X). \tag{23.2.1}$$

Thus $\mathcal{Z}(\chi)$ is a module over the ring $\mathbb{C}[X, X^{-1}]$ of Laurent polynomials in X.

This ring has useful properties: particularly, it is a principal ideal domain and its unit group consists of the monomials aX^b, $a \in \mathbb{C}^\times$, $b \in \mathbb{Z}$.

Proposition. *Let χ be a character of $F^\times$; then*

$$\mathcal{Z}(\chi, X) = P_\chi(X)^{-1}\,\mathbb{C}[X, X^{-1}],$$

where

$$P_\chi(X) = \begin{cases} 1-\chi(\varpi)X & \text{if } \chi \text{ is unramified,} \\ 1 & \text{otherwise.} \end{cases}$$

Proof. Suppose first that $\Phi(0) = 0$. Thus $\Phi \mid F^\times$ lies in $C_c^\infty(F^\times)$ and $Z(\Phi, \chi, X)$ has only finitely many non-zero coefficients. That is, $Z(\Phi, \chi, X) \in \mathbb{C}[X, X^{-1}]$. Further, if Φ is the characteristic function of a sufficiently small neighbourhood of 1, then $Z(\Phi, \chi, X)$ is a positive constant. Thus $1 \in \mathcal{Z}(\chi)$ and

$$\{Z(\Phi, \chi, X) : \Phi \in C_c^\infty(F^\times)\} = \mathbb{C}[X, X^{-1}].$$

Let Φ_0 be the characteristic function of $\mathfrak{o}$; we have

$$Z(\Phi_0, \chi, X) = \sum_{m \geqslant 0} \chi(\varpi^m)\, X^m \int_{U_F} \chi(x)\, d\mu^*(x). \tag{23.2.2}$$

The inner integral is $\mu^*(U_F)$ if χ is unramified, and is zero otherwise. That is,

$$\mu^*(U_F)^{-1}\, Z(\Phi_0, \chi, X) = \begin{cases} (1-\chi(\varpi)X)^{-1} & \text{if } \chi \text{ is unramified,} \\ 0 & \text{otherwise.} \end{cases}$$

Since $C_c^\infty(F)$ is spanned by Φ_0 and $C_c^\infty(F^\times)$, the result follows. □

23.3. We take a character $\psi \in \widehat{F}$, $\psi \neq 1$. For $\Phi \in C_c^\infty(F)$, let $\hat{\Phi}$ denote the Fourier transform of Φ, calculated using the Haar measure on F which is self-dual relative to ψ.

Theorem. *Let χ be a character of $F^\times$. There is a unique rational function $c(\chi, \psi, X) \in \mathbb{C}(X)$ such that*

$$Z(\hat{\Phi}, \check{\chi}, \tfrac{1}{qX}) = c(\chi, \psi, X)\, Z(\Phi, \chi, X),$$

for all $\Phi \in C_c^\infty(F)$.

Proof. Consider the space Λ of linear maps $\lambda : C_c^\infty(F) \to \mathbb{C}(X)$ satisfying

$$\lambda(a\Phi) = \chi(a)\, X^{v_F(a)}\, \lambda(\Phi), \quad \Phi \in C_c^\infty(F),\ a \in F^\times. \tag{23.3.1}$$

Surely Λ is a $\mathbb{C}(X)$-vector space. It contains the map

$$\lambda_0 : \Phi \longmapsto Z(\Phi, \chi, X),$$

which is non-zero (23.2 Proposition). For $\Phi \in C_c^\infty(F)$ and $a \in F^\times$, a simple calculation yields

$$\widehat{a\Phi} = \|a\|\, a^{-1}\hat{\Phi}.$$

It follows that the map

$$\lambda_1 : \Phi \longmapsto Z(\hat{\Phi}, \check{\chi}, 1/qX)$$

also lies in Λ. The theorem is thus a consequence of:

Lemma. *The space* Λ *has dimension one over* $\mathbb{C}(X)$.

Proof. Choose $n \geqslant 0$ such that $U_F^n \subset \operatorname{Ker}\chi$. Let Φ_k denote the characteristic function of U_F^k, $k \geqslant 1$. We consider the map

$$\begin{aligned} \Lambda &\longrightarrow \mathbb{C}(X), \\ \lambda &\longmapsto \lambda(\Phi_n). \end{aligned}$$

We show this is injective, and the lemma will follow.

Suppose that $\lambda(\Phi_n) = 0$. The defining property (23.3.1) gives $\lambda(a\Phi_k) = \lambda(\Phi_k)$, for $k \geqslant n$, $a \in U_F^n$ and $\lambda \in \Lambda$, so

$$\lambda(\Phi_k) = q^{n-k}\lambda(\Phi_n) = 0, \quad k \geqslant n.$$

Any $\Phi \in C_c^\infty(F^\times)$ is a finite linear combination of $F^\times$-translates of functions Φ_k, $k \geqslant n$. So, if $\lambda(\Phi_n) = 0$ then $\lambda(\Phi) = 0$ for all $\Phi \in C_c^\infty(F^\times)$. For $\Phi \in C_c^\infty(F)$, therefore, the value $\lambda(\Phi)$ depends only on $\Phi(0)$. Thus $\lambda(a\Phi) = \lambda(\Phi)$, for all $a \in F^\times$, and (23.3.1) implies that $\lambda(\Phi) = 0$, whence $\lambda = 0$. □

This completes the proof of the theorem. □

We introduce a more traditional notation. We let s be a complex variable, and set

$$\begin{aligned} \zeta(\Phi, \chi, s) &= Z(\Phi, \chi, q^{-s}), \\ L(\chi, s) &= P_\chi(q^{-s})^{-1}, \\ \gamma(\chi, s, \psi) &= c(\chi, \psi, q^{-s}). \end{aligned} \tag{23.3.2}$$

In particular,

$$\zeta(\Phi, \chi, s) = \int_{F^\times} \Phi(x)\chi(x)\|x\|^s \, d\mu^*(x) \tag{23.3.3}$$

in the following sense. The calculations in the proof of 23.2 Proposition (especially (23.2.2)) show that this integral converges, absolutely and uniformly in vertical strips, in some half-plane $\operatorname{Re} s > s_0$. It there represents a rational function in q^{-s} and so admits analytic continuation to a meromorphic function on the whole s-plane.

The two languages are equivalent, and the relation between them is transparent. We use whichever seems more convenient at the time.

23.4. In the classical language,

$$L(\chi, s) = \begin{cases} (1-\chi(\varpi)q^{-s})^{-1} & \text{if } \chi \text{ is unramified,} \\ 1 & \text{otherwise.} \end{cases} \tag{23.4.1}$$

The L-function $L(\chi, s)$ thus says nothing about χ when χ is ramified. If, however, χ is unramified, then it is determined completely by its L-function.

Corollary 1. *Let χ_1, χ_2 be unramified characters of $F^\times$. The following are equivalent:*

(1) *the meromorphic functions $L(\chi_1, s)$, $L(\chi_2, s)$ have a pole in common;*
(2) *the meromorphic functions $L(\chi_1, s)$, $L(\chi_2, s)$ have the same sets of poles;*
(3) $\chi_1 = \chi_2$.

The structure of the function $\gamma(\chi, s, \psi)$ is of particular importance. We define a rational function $\varepsilon(\chi, s, \psi) \in \mathbb{C}(q^{-s})$ by

$$\varepsilon(\chi, s, \psi) = \gamma(\chi, s, \psi)\,\frac{L(\chi, s)}{L(\check{\chi}, 1-s)}\,.$$

Corollary 2. *The function $\varepsilon(\chi, s, \psi)$ satisfies the functional equation*

$$\varepsilon(\chi, s, \psi)\,\varepsilon(\check{\chi}, 1-s, \psi) = \chi(-1). \tag{23.4.2}$$

It is of the form

$$\varepsilon(\chi, s, \psi) = q^{(\frac{1}{2}-s)n(\chi,\psi)}\,\varepsilon(\chi, \tfrac{1}{2}, \psi),$$

for some $n(\chi, \psi) \in \mathbb{Z}$.

Proof. The functional equation in 23.3 Theorem reads

$$\zeta(\hat{\Phi}, \check{\chi}, 1-s) = \gamma(\chi, s, \psi)\,\zeta(\Phi, \chi, s). \tag{23.4.3}$$

Applying this twice, we get

$$\zeta(\hat{\hat{\Phi}}, \chi, s) = \gamma(\check{\chi}, 1-s, \psi)\,\gamma(\chi, s, \psi)\,\zeta(\Phi, \chi, s).$$

Fourier inversion (23.1.2) gives $\zeta(\hat{\hat{\Phi}}, \chi, s) = \chi(-1)\zeta(\Phi, \chi, s)$, whence

$$\gamma(\check{\chi}, 1-s, \psi)\,\gamma(\chi, s, \psi) = \varepsilon(\check{\chi}, 1-s, \psi)\,\varepsilon(\chi, s, \psi) = \chi(-1),$$

and the relation (23.4.2) follows.

We can re-write (23.4.3) in the form

$$\frac{\zeta(\hat{\Phi}, \check{\chi}, 1-s)}{L(\check{\chi}, 1-s)} = \varepsilon(\chi, s, \psi)\,\frac{\zeta(\Phi, \chi, s)}{L(\chi, s)}\,, \tag{23.4.4}$$

the fraction on either side lying in $\mathbb{C}[q^s, q^{-s}]$. As in 23.2 Proposition, we may choose Φ such that $\zeta(\Phi, \chi, s) = L(\chi, s)$. We deduce that $\varepsilon(\chi, s, \psi) \in \mathbb{C}[q^s, q^{-s}]$. Likewise, we have $\varepsilon(\check{\chi}, 1-s, \psi) \in \mathbb{C}[q^s, q^{-s}]$. The functional equation (23.4.2) implies that $\varepsilon(\chi, s, \psi)$ is an invertible element of the ring $\mathbb{C}[q^s, q^{-s}]$, and hence equal to aq^{ms}, for some $a \in \mathbb{C}^\times$, $m \in \mathbb{Z}$. This can surely be written in the required form. □

Remark. The relation (23.4.4) is generally referred to as "Tate's (local) functional equation". The function $\varepsilon(\chi, s, \psi)$ is the *Tate local constant* of χ (relative to ψ).

23.5. It will be essential to have a table of values for the local constant $\varepsilon(\chi, s, \psi)$. The dependence on ψ is transparent (*cf.* 23.1 Proposition (4)):

Lemma 1. *Let $a \in F^\times$; we then have*

$$\varepsilon(\chi, s, a\psi) = \chi(a)\, \|a\|^{s-\frac{1}{2}}\, \varepsilon(\chi, s, \psi).$$

We therefore only calculate $\varepsilon(\chi, s, \psi)$ for characters ψ *of level one.* The self-dual Haar measure for ψ then satisfies

$$\int_{\mathfrak{o}} dx = q^{1/2}. \tag{23.5.1}$$

For the purposes of calculation, it will be convenient to choose the Haar measure d^*x on $F^\times$ such that

$$\int_{U_F} d^*x = 1.$$

Proposition. *Suppose that χ is unramified and that ψ has level one. If ϖ is a prime element of F, then*

$$\varepsilon(\chi, s, \psi) = q^{s-\frac{1}{2}}\chi(\varpi)^{-1}.$$

Proof. If Φ denotes the characteristic function of $\mathfrak{o}$, we get $\zeta(\Phi, \chi, s) = L(\chi, s)$, as in the proof of 23.2 Proposition. On the other side, $\hat{\Phi}(x) = q^{1/2}\Phi(\varpi^{-1}x)$, so

$$\begin{aligned}\zeta(\hat{\Phi}, \chi^{-1}, s) &= q^{1/2} \int_{F^\times} \Phi(\varpi^{-1}x)\, \chi(x)^{-1} \|x\|^s\, d^*x \\ &= q^{1/2-s}\chi(\varpi)^{-1} \int_{F^\times} \Phi(x)\, \chi(x)^{-1} \|x\|^s\, d^*x \\ &= q^{1/2-s}\chi(\varpi)^{-1} L(\chi^{-1}, s).\end{aligned}$$

It follows that

$$\varepsilon(\chi, s, \psi) = \frac{\zeta(\hat{\Phi}, \chi^{-1}, 1-s)}{L(\chi^{-1}, 1-s)} = q^{s-\frac{1}{2}}\chi(\varpi)^{-1},$$

as required. □

Theorem. *Suppose that χ has level $n \geqslant 0$ and is not unramified. Let $\psi \in \widehat{F}$ have level one. Then*

$$\varepsilon(\chi, s, \psi) = q^{n(\frac{1}{2}-s)} \sum_{x \in U_F/U_F^{n+1}} \chi(\alpha x)^{-1}\psi(\alpha x)/q^{(n+1)/2},$$

for any $\alpha \in F^\times$ such that $v_F(\alpha) = -n$.

Proof. Let Φ be the characteristic function of U_F^{n+1}, so that

$$\zeta(\Phi, \chi, s) = \mu^*(U_F^{n+1}).$$

The Fourier transform $\hat{\Phi}$ has support $\mathfrak{p}^{-n}$ and

$$\hat{\Phi}(y) = q^{\frac{1}{2}-n-1}\psi(y), \quad y \in \mathfrak{p}^{-n}.$$

Thus

$$\begin{aligned} \zeta(\hat{\Phi}, \chi^{-1}, s) &= q^{\frac{1}{2}-n-1} \int_{\mathfrak{p}^{-n}\smallsetminus\{0\}} \psi(y)\chi(y)^{-1}\|y\|^s\, d^*y \\ &= q^{\frac{1}{2}-n-1} \sum_{m \geqslant -n} z_m q^{-ms}, \end{aligned}$$

where

$$z_m = \int_{\mathfrak{p}^m \smallsetminus \mathfrak{p}^{m+1}} \psi(y)\chi(y)^{-1}\, d^*y.$$

We know (23.4 Corollary 2) that $\zeta(\hat{\Phi}, \chi^{-1}, s)$ must reduce to a monomial in q^s, so only one of the coefficients z_m is non-zero. The first one is

$$z_{-n} = \int_{U_F} \psi(\alpha u)\chi^{-1}(\alpha u)\, d^*u,$$

for any $\alpha \in F$ with $v_F(\alpha) = -n$. The integrand is constant on cosets of U_F^{n+1} and so the integral reduces to

$$\int_{U_F} \psi(\alpha u)\chi^{-1}(\alpha u)\, d^*u = \mu^*(U_F^{n+1}) \sum_{x \in U_F/U_F^{n+1}} \chi(\alpha x)^{-1}\psi(\alpha x).$$

The result will follow, therefore, when we show:

Lemma 2. *The coefficients z_m, $m \geqslant 1-n$, are all zero.*

Proof. We take $\beta \in F$, with $v_F(\beta) = m > -n$, and consider the coefficient

$$z_m = \int_{U_F} \psi(\beta u)\chi(\beta u)^{-1}\, d^*u. \tag{23.5.2}$$

Taking $v \in U_F^n$, we have

$$\int_{U_F} \psi(\beta u)\chi(\beta u)^{-1}\, d^*u = \int_{U_F} \psi(\beta uv)\chi(\beta uv)^{-1}\, d^*u.$$

Setting $v = 1+y$, the first factor reduces to $\psi(\beta u)\psi(\beta uy) = \psi(\beta u)$, since $\beta uy \in \mathfrak{p}^{m+n} \subset \mathfrak{p} \subset \operatorname{Ker}\psi$. Thus

$$\int_{U_F} \psi(\beta u)\chi(\beta u)^{-1}\, d^*u = \chi(v)^{-1} \int_{U_F} \psi(\beta u)\chi(\beta u)^{-1}\, d^*u.$$

We may certainly choose v so that $\chi(v) \neq 1$, so the integral (23.5.2) vanishes, as required. □

This completes the proof of the theorem. □

Remark. A variation on the same technique shows that

$$\int_{U_F} \psi(\beta u)\chi(\beta u)^{-1}\, d^*u = 0$$

for any $\beta \in F^\times$ such that $v_F(\beta) \neq -n$.

Exercise. *Let χ be a ramified character of $F^\times$ and $\psi \in \widehat{F}$, $\psi \neq 1$. Let $d^*x = \|x\|^{-1}dx$, where dx is the self-dual Haar measure on F relative to ψ. Let $c \in F^\times$. Show that*

$$\int_{U_F} \chi(cx)^{-1}\psi(cx)\, d^*x = \begin{cases} \varepsilon(\chi, 1, \psi) & \text{if } v_F(c) = -n(\chi,\psi), \\ 0 & \text{otherwise.} \end{cases}$$

23.6. The trigonometric sum appearing in 23.5 Theorem is of particular interest. Formally, let χ be a ramified character of $F^\times$ of level $n \geqslant 0$, and let $\psi \in \widehat{F}$ have level one. Let $c \in F$ satisfy $v_F(c) = -n$. The sum

$$\tau(\chi,\psi) = \sum_{x \in U_F/U_F^{n+1}} \check{\chi}(cx)\psi(cx) \tag{23.6.1}$$

is called the *Gauss sum* of χ (relative to ψ). The definition (23.6.1) is independent of the choice of coset representatives x, and of the element c. In these terms, 23.5 Theorem reads:

$$\varepsilon(\chi, s, \psi) = q^{n(\frac{1}{2}-s)}\, \tau(\chi,\psi)/q^{(n+1)/2}. \tag{23.6.2}$$

The functional equation for the local constant (23.4.2) yields the classical relation

$$\tau(\chi,\psi)\,\tau(\check{\chi},\psi) = \chi(-1)\, q^{n+1}. \tag{23.6.3}$$

One can usefully simplify the defining expression (23.6.1) in most cases:

Proposition. *Suppose that χ has level $n \geqslant 1$. Let $c \in F$ satisfy*

$$\chi(1+x) = \psi(cx), \quad x \in \mathfrak{p}^{[n/2]+1}.$$

Then

$$\tau(\chi,\psi) = q^{[(n+1)/2]} \sum_y \check{\chi}(cy)\psi(cy), \tag{23.6.4}$$

where y ranges over $U_F^{[(n+1)/2]}/U_F^{[n/2]+1}$.

Proof. In the defining sum (23.6.1), write $x = y(1+z)$, with $y \in U_F/U_F^{[n/2]+1}$ and $z \in \mathfrak{p}^{[n/2]+1}/\mathfrak{p}^{n+1}$. The typical term is then

$$\check{\chi}(cy(1+z))\psi(cy(1+z)) = \check{\chi}(cy)\psi(cy)\,\psi(c(y-1)z).$$

Summing:

$$\tau(\chi,\psi) = \sum_y \check{\chi}(cy)\psi(cy) \sum_z \psi(c(y-1)z).$$

The map $z \mapsto \psi(c(y-1)z)$ is a character of $\mathfrak{p}^{[n/2]+1}/\mathfrak{p}^{n+1}$, since $c(y-1) \in \mathfrak{p}^{-n}$. It is the trivial character if and only if $y \equiv 1 \pmod{\mathfrak{p}^{n-[n/2]}}$, that is, if and only if $y \in U_F^{[(n+1)/2]}$. The inner sum therefore vanishes unless $y \in U_F^{[(n+1)/2]}$, in which case it takes the value $(\mathfrak{p}^{[n/2]+1} : \mathfrak{p}^{n+1}) = q^{[(n+1)/2]}$, as required. □

Remark. In an equally valid convention, what we have called $\tau(\chi,\psi)$ would be denoted $\tau(\check{\chi},\psi)$. Our version is slightly more convenient for present purposes.

23.7. The Gauss sum $\tau(\chi,\psi)$, in the case where χ is of level zero, relates to a particularly well-known species of classical Gauss sum.

For the moment, let $\Bbbk$ be a finite field; let θ be a character of $\Bbbk^\times$ and η a *non-trivial* character of $\Bbbk$. We can form the classical Gauss sum

$$\mathfrak{g}_{\Bbbk}(\theta,\eta) = \sum_{x\in\Bbbk^\times} \check{\theta}(x)\eta(x).$$

In the case where χ is a ramified character of $F^\times$ of level zero, the restriction $\chi \mid U_F$ is the inflation of a character $\tilde{\chi}$ of $\boldsymbol{k}^\times$. Likewise, as ψ is of level one, the restriction $\psi \mid \mathfrak{o}$ is the inflation of a non-trivial character $\tilde{\psi}$ of $\boldsymbol{k}$. Immediately:

$$\tau(\chi,\psi) = \mathfrak{g}_{\boldsymbol{k}}(\tilde{\chi},\tilde{\psi}). \tag{23.7.1}$$

(Observe that, when χ is *unramified*, this does not give the correct formula for the local constant.)

23.8. The formula of 23.6 Proposition has an important application:

Stability theorem. *Let θ, χ be characters of $F^\times$, of level $l \geqslant 0$, $n \geqslant 1$ respectively. Suppose that $2l < n$. Let $\psi \in \widehat{F}$, $\psi \neq 1$, and let $c \in F$ satisfy $\chi(1+x) = \psi(cx)$, $x \in \mathfrak{p}^{[n/2]+1}$. Then*

$$\varepsilon(\theta\chi, s, \psi) = \theta(c)^{-1}\varepsilon(\chi, s, \psi).$$

Proof. By 23.5 Lemma 1, it is enough to treat the case where ψ has level one. The character $\theta\chi$ has level n, and agrees with χ on $U_F^{[(n+1)/2]}$. Applying 23.6 Proposition, we get

$$\tau(\theta\chi, \psi) = q^{[(n+1)/2]} \sum_y \check{\theta}\check{\chi}(cy)\psi(cy),$$

where the sum is taken over $u \in U_F^{[(n+1)/2]}/U_F^{[n/2]+1}$. Since θ is trivial on $U_F^{[(n+1)/2]}$, this reduces to $\theta(c)^{-1}\tau(\chi, \psi)$, as required. □

24. Functional Equation for GL(2)

We turn to the functional equation of Godement and Jacquet, which generalizes the ideas of §23 to representations of $G = \mathrm{GL}_2(F)$. As in the classical case, there are two steps. The first leads to an L-function, and the second is a functional equation giving rise to a local constant. In this section, we state the results in general, but prove them only for cuspidal representations. We calculate the local constants of cuspidal representations in §25. We deal with the non-cuspidal representations separately, in §26.

We use the description of cuspidal representations given in §15. One can proceed quite well without this level of detail but there is no point neglecting it, since it will be needed for the computation of local constants. We state the results here in a classical language of analytic functions. The method of proof we adopt for cuspidal representations is, however, algebraic and developed from that of §23.

24.1. Let $\mathfrak{M} = \mathrm{M}_2(\mathfrak{o}) \subset \mathrm{M}_2(F) = A$. In parallel to the one-dimensional case, the space $C_c^\infty(A)$ is spanned by the characteristic functions of the sets $a+\mathfrak{p}^j\mathfrak{M}$, $a \in A$, $j \in \mathbb{Z}$.

We fix $\psi \in \widehat{F}$, $\psi \neq 1$, and we set $\psi_A = \psi \circ \mathrm{tr}_A$. We define the Fourier transform $\hat{\Phi}$ of $\Phi \in C_c^\infty(A)$ by

$$\hat{\Phi}(x) = \int_A \Phi(y)\psi_A(xy)\, d\mu(y), \quad \Phi \in C_c^\infty(A),$$

relative to a Haar measure μ on A. Then $\hat{\Phi} \in C_c^\infty(A)$ and, exactly as in 23.1, there is a unique Haar measure μ_ψ^A for which the Fourier inversion formula

$$\hat{\hat{\Phi}}(x) = \Phi(-x), \quad \Phi \in C_c^\infty(A),$$

holds. This is the *self dual Haar measure on A, relative to* ψ. As in 23.1, we get the relations

$$\begin{aligned} \mu_\psi^A(\mathfrak{M}) &= q^{2l}, \\ \mu_{a\psi}^A &= \|a\|^2 \mu_\psi^A, \end{aligned} \tag{24.1.1}$$

where $a \in F^\times$ and l is the level of ψ.

24.2. Let (π, V) be an irreducible smooth representation of G. Let $\mathcal{C}(\pi)$ be the space of coefficients of π (10.1). As in the proof of 10.1 Proposition, the space $\mathcal{C}(\pi)$ carries a smooth representation of $G \times G$ and the canonical map $\check{V} \otimes V \to \mathcal{C}(\pi)$ is a $(G \times G)$-isomorphism. In particular, $\mathcal{C}(\pi)$ is irreducible over $G \times G$.

We consider integrals of the form

$$\zeta(\Phi, f, s) = \int_G \Phi(x)\, f(x)\, \|\det x\|^s \, d\mu^*(x), \tag{24.2.1}$$

for $\Phi \in C_c^\infty(A)$ and $f \in \mathcal{C}(\pi)$, where s is a complex variable and μ^* is a Haar measure on G. We usually abbreviate $d^*x = d\mu^*(x)$.

Theorem 1. *Let (π, V) be an irreducible smooth representation of G.*

(1) *There exists $s_0 \in \mathbb{R}$ such that the integral (24.2.1) converges, absolutely and uniformly in vertical strips in the region $\operatorname{Re} s > s_0$, for all Φ and f. The integral represents a rational function in q^{-s}.*
(2) *Define*

$$\mathcal{Z}(\pi) = \mathcal{Z}(\pi, X) = \{\zeta(\Phi, f, s+\tfrac{1}{2}) : \Phi \in C_c^\infty(A), f \in \mathcal{C}(\pi)\}.$$

There is a unique polynomial $P_\pi(X) \in \mathbb{C}[X]$, satisfying $P_\pi(0) = 1$, and

$$\mathcal{Z}(\pi) = P_\pi(q^{-s})^{-1}\, \mathbb{C}[q^s, q^{-s}].$$

One sets

$$L(\pi, s) = P_\pi(q^{-s})^{-1}. \tag{24.2.2}$$

This definition is, we note, independent of the choice of Haar measure μ^*.

Next, we choose $\psi \in \widehat{F}$, $\psi \neq 1$. For $\Phi \in C_c^\infty(A)$, we denote by $\hat{\Phi}$ the Fourier transform of Φ using the Haar measure on A which is self-dual with respect to ψ.

If (π, V) is an irreducible smooth representation of G and $f \in \mathcal{C}(\pi)$, the function $\check{f} : g \mapsto f(g^{-1})$ lies in $\mathcal{C}(\check{\pi})$. Indeed, the map $f \mapsto \check{f}$ gives a linear isomorphism $\mathcal{C}(\pi) \cong \mathcal{C}(\check{\pi})$.

Theorem 2. *Let (π, V) be an irreducible smooth representation of G. There is a unique rational function $\gamma(\pi, s, \psi) \in \mathbb{C}(q^{-s})$ such that*

$$\zeta(\hat{\Phi}, \check{f}, \tfrac{3}{2}-s) = \gamma(\pi, s, \psi)\, \zeta(\Phi, f, \tfrac{1}{2}+s), \tag{24.2.3}$$

for all $\Phi \in C_c^\infty(A)$, $f \in \mathcal{C}(\pi)$.

Before starting the proofs of these theorems, we deduce:

Corollary. *Define*

$$\varepsilon(\pi, s, \psi) = \gamma(\pi, s, \psi)\, \frac{L(\pi, s)}{L(\check{\pi}, 1-s)}\,.$$

The function $\varepsilon(\pi, s, \psi)$ *satisfies the functional equation*

$$\varepsilon(\pi, s, \psi)\, \varepsilon(\check{\pi}, 1-s, \psi) = \omega_\pi(-1). \tag{24.2.4}$$

Moreover, there exist $a \in \mathbb{C}^\times$ *and* $b \in \mathbb{Z}$ *such that* $\varepsilon(\pi, s, \psi) = aq^{bs}$.

Proof. The equation (24.2.3), applied to $(\hat{\Phi}, \check{f}, 1-s)$ in place of (Φ, f, s), reads

$$\zeta(\hat{\hat{\Phi}}, f, \tfrac{1}{2}+s) = \gamma(\check{\pi}, 1-s, \psi)\, \zeta(\hat{\Phi}, \check{f}, \tfrac{3}{2}-s),$$

so

$$\zeta(\hat{\hat{\Phi}}, f, \tfrac{1}{2}+s) = \gamma(\check{\pi}, 1-s, \psi)\, \gamma(\pi, s, \psi)\, \zeta(\Phi, f, \tfrac{1}{2}+s).$$

The Fourier inversion formula (24.1) gives $\zeta(\hat{\hat{\Phi}}, f, s) = \omega_\pi(-1)\zeta(\Phi, f, s)$, whence follows the relation (24.2.4).

By definition, we can find $\Phi_i \in C_c^\infty(A)$, $f_i \in \mathcal{C}(\pi)$, $1 \leqslant i \leqslant r$, such that

$$\sum_{i=1}^{r} \zeta(\Phi_i, f_i, s+\tfrac{1}{2}) = L(\pi, s).$$

The equation (24.2.3) gives us

$$L(\check{\pi}, 1-s)^{-1} \sum_{i=1}^{r} \zeta(\hat{\Phi}_i, \check{f}_i, \tfrac{3}{2}-s) = \varepsilon(\pi, s, \psi).$$

By definition, the left hand side lies in $\mathbb{C}[q^s, q^{-s}]$, so $\varepsilon(\pi, s, \psi) \in \mathbb{C}[q^s, q^{-s}]$. Likewise, $\varepsilon(\check{\pi}, 1-s, \psi)$ lies in $\mathbb{C}[q^s, q^{-s}]$. The relation (24.2.4) now implies that $\varepsilon(\pi, s, \psi)$ is an invertible element of the ring $\mathbb{C}[q^s, q^{-s}]$, and hence of the desired form. □

24.3. The relation (24.2.3) is called the "Godement-Jacquet functional equation" for π. The quantity $\varepsilon(\pi, s, \psi)$ is the *Godement-Jacquet local constant* of π. We may write

$$\varepsilon(\pi, s, \psi) = q^{n(\pi,\psi)(\frac{1}{2}-s)}\, \varepsilon(\pi, \tfrac{1}{2}, \psi), \tag{24.3.1}$$

for an integer $n(\pi, \psi)$.

The choice of Haar measure μ^* on G has no effect on the definitions of the L-function $L(\pi, s)$ and $\gamma(\pi, s, \psi)$. The choice of Haar measure μ on A, used for computing Fourier transforms, is dictated by the character ψ. Changing ψ has no effect on the L-function, but both $\gamma(\pi, s, \psi)$, $\varepsilon(\pi, s, \psi)$ do genuinely depend on ψ. Following the definitions through, one finds:

Proposition. *Let (π, V) be an irreducible smooth representation of G. Let $\psi \in \widehat{F}$, $\psi \neq 1$, and let $a \in F^\times$. Then:*

$$\varepsilon(\pi, s, a\psi) = \omega_\pi(a)\, \|a\|^{2s-1}\, \varepsilon(\pi, s, \psi),$$
$$\gamma(\pi, s, a\psi) = \omega_\pi(a)\, \|a\|^{2s-1}\, \gamma(\pi, s, \psi).$$

24.4. We make some general remarks concerning the integrals (24.2.1), and describe them in terms of power series, much as in §23.

The space $C_c^\infty(A)$ carries a smooth representation of the locally profinite group $G \times G$ by

$$(g,h)\Phi : x \longmapsto \Phi(g^{-1}xh), \quad g, h \in G,\ \Phi \in C_c^\infty(A),\ x \in A.$$

This extends to an action of $\mathcal{H}(G \times G)$ on $C_c^\infty(A)$, as in 4.2. Identifying the algebra $\mathcal{H}(G \times G)$ with $\mathcal{H}(G) \otimes \mathcal{H}(G)$ (*cf.* 3.2), the action is given by

$$(\phi_1 \otimes \phi_2) : \Phi \longmapsto \phi_1 * \Phi * \check{\phi}_2, \quad \Phi \in C_c^\infty(A),\ \phi_i \in \mathcal{H}(G),$$

where $\check{\phi}_2$ is the function $g \mapsto \phi_2(g^{-1})$. Explicitly,

$$\phi_1 * \Phi * \check{\phi}_2(x) = \iint_{G\times G} \phi_1(g)\, \Phi(g^{-1}xh)\, \phi_2(h)\, d^*g\, d^*h.$$

Similarly for $\mathcal{C}(\pi)$.

These actions interact predictably with the Fourier transform. A simple calculation yields:

$$((g,h)\Phi)\hat{} = \|\det gh^{-1}\|^2\, (h,g)\hat{\Phi}, \quad g, h \in G,\ \Phi \in C_c^\infty(A). \qquad (24.4.1)$$

For an integer m, let

$$G_m = \{x \in G : v_F(\det x) = m\},$$

and let $\mathfrak{e}_m$ denote the characteristic function of G_m in A. Define

$$\Phi_m = \mathfrak{e}_m \Phi, \quad \Phi \in C_c^\infty(A). \qquad (24.4.2)$$

Lemma 1. *Let $\Phi \in C_c^\infty(A)$ and $m \in \mathbb{Z}$. The function Φ_m lies in $C_c^\infty(G)$ for all m, and is identically zero for $m \ll 0$.*

Proof. The function Φ_m is locally constant on G, so we have to show that its support is compact.

Let $K_0 = \mathrm{GL}_2(\mathfrak{o})$ and let ϖ be a prime element of F. Consider the set of pairs of integers (a, b), such that $a \leqslant b$ and

$$\mathrm{supp}(\Phi) \cap K_0 \begin{pmatrix} \varpi^a & 0 \\ 0 & \varpi^b \end{pmatrix} K_0 \neq \emptyset.$$

The integers a, b are individually bounded below by this condition. If we further impose the condition $a+b = m$, there are only finitely many such pairs (a, b). Thus $\operatorname{supp} \Phi_m$ meets only finitely many K_0-double cosets contained in G_m, as required. □

Let (π, V) be an irreducible smooth representation of G. For $m \in \mathbb{Z}$, $\Phi \in C_c^\infty(A)$, and $f \in \mathcal{C}(\pi)$, we define

$$z_m(\Phi, f) = \int_G \Phi_m(x) f(x)\, d^*x = \int_{G_m} \Phi(x) f(x)\, d^*x. \tag{24.4.3}$$

The integrand here lies in $C_c^\infty(G)$, and is zero for $m \ll 0$ (Lemma 1), so we can define a formal Laurent series $Z(\Phi, f, X) \in \mathbb{C}((X))$ by

$$Z(\Phi, f, X) = \sum_{m \in \mathbb{Z}} z_m(\Phi, f) X^m.$$

We put

$$\mathcal{Z}(\pi) = \{Z(\Phi, f, q^{-1/2}X) : \Phi \in C_c^\infty(A), f \in \mathcal{C}(\pi)\}.$$

Lemma 2.

(1) *The space* $\mathcal{Z}(\pi)$ *is a* $\mathbb{C}[X, X^{-1}]$*-module, containing* $\mathbb{C}[X, X^{-1}]$.
(2) *If* $f_0 \in \mathcal{C}(\pi)$, $f_0 \neq 0$, *the set of series*

$$\{Z(\Phi, f_0, q^{-1/2}X) : \Phi \in C_c^\infty(A)\},$$

generates $\mathcal{Z}(\pi)$ *as* $\mathbb{C}[X, X^{-1}]$*-module.*

Proof. For $\Phi \in C_c^\infty(A)$, $f \in \mathcal{C}(\pi)$ and $g, h \in G$, we have

$$Z((g,h)\Phi, (g,h)f, X) = X^{\upsilon_F(\det gh^{-1})} Z(\Phi, f, X). \tag{24.4.4}$$

It follows that $\mathcal{Z}(\pi)$ is a $\mathbb{C}[X, X^{-1}]$-module. Suppose $f(1) \neq 0$ and let K be a compact open subgroup of G such that $K \times K$ fixes f. If Φ is the characteristic function of K, then $Z(\Phi, f, X)$ is a positive constant. We deduce that $\mathcal{Z}(\pi)$ contains $\mathbb{C}[X, X^{-1}]$.

As noted in 24.2, the space $\mathcal{C}(\pi)$ is irreducible over $G \times G$. Thus every $f \in \mathcal{C}(\pi)$ is a finite linear combination of functions $(g, h)f_0$, for various $g, h \in G$. Assertion (2) now follows from (24.4.4). □

24.5. In this paragraph, we prove 24.2 Theorem 1 under the assumption that π is *cuspidal.* More precisely:

Proposition. *Let* (π, V) *be an irreducible cuspidal representation of* G*. Then*

$$\mathcal{Z}(\pi) = \mathbb{C}[X, X^{-1}].$$

Proof. As in 15.8, there is a cuspidal inducing datum $(\mathfrak{A}, \Xi)$ which occurs in π. We write $\xi = \Xi \mid U_{\mathfrak{A}}$. The representation ξ is irreducible and $V^{\xi} = V^{\Xi}$ (15.8 Proposition 2). We let $e_{\xi} \in \mathcal{H}(G)$ be the idempotent attached to ξ, as in 4.4.

We choose elements $v \in V^{\xi}$ and $\check{v} \in \check{V}^{\check{\xi}}$ so that the coefficient

$$f_0 = \gamma_{\check{v}\otimes v} : g \longmapsto \langle \check{v}, \pi(g)v \rangle$$

satisfies $f_0(1) = 1$. We have $\check{e}_{\xi} * f_0 * \check{e}_{\xi} = f_0$ (and $\check{e}_{\xi} = e_{\check{\xi}}$), so f_0 has support contained in $\mathcal{K}_{\mathfrak{A}}$ (15.8 Proposition 2, 11.1 Proposition 2). Also, since every g in the support of e_{ξ} satisfies $\upsilon_F(\det g) = 0$, a simple calculation yields

$$Z(\Phi, f_0, X) = Z(e_{\xi} * \Phi * e_{\xi}, f_0, X), \quad \Phi \in C_c^{\infty}(A). \tag{24.5.1}$$

Lemma. *Take $f_0 = \gamma_{\check{v}\otimes v} \in \mathcal{C}(\pi)$ as above.*

(1) *Let Φ be the characteristic function of $\mathfrak{A}$; then $Z(\Phi, f_0, X) = 0$.*
(2) *Let $\Phi \in C_c^{\infty}(A)$ and suppose that $e_{\xi} * \Phi * e_{\xi} = \Phi$. Then $\Phi \in C_c^{\infty}(G)$ and the support of Φ is contained in $\mathcal{K}_{\mathfrak{A}}$.*
(3) *We have $Z(\Phi, f_0, X) \in \mathbb{C}[X, X^{-1}]$, for all $\Phi \in C_c^{\infty}(A)$.*

Proof. In (1), the function Φ is invariant under translation, on both sides, by $U_{\mathfrak{A}}$. It follows that $e_{\xi} * \Phi * e_{\xi} = 0$, whence, by (24.5.1), $Z(\Phi, f_0, X) = 0$.

In (2), the relation $e_{\xi} * \Phi * e_{\xi} = \Phi$ implies, via 11.1 Proposition 2, that $\operatorname{supp}\Phi \cap G \subset \mathcal{K}_{\mathfrak{A}}$. We next show that $\operatorname{supp}\Phi \subset G$. Suppose, on the contrary, that $\operatorname{supp}\Phi$ contains a singular matrix x. It therefore contains an open neighbourhood $\mathcal{N}$ of x. The open set $\mathcal{N}$ contains a diagonalizable matrix x' with eigenvalues $a_1, a_2 \in F^{\times}$, such that $\|a_1\| \neq \|a_2\|$. (The matrix x is G-conjugate to a triangular matrix, and the assertion is obvious for triangular matrices.) Thus $x' \in G \smallsetminus \mathcal{K}_{\mathfrak{A}}$ but $x' \in \operatorname{supp}\Phi \cap G$. This is impossible, so $\operatorname{supp}\Phi \subset G$, as asserted.

It follows that $\operatorname{supp}\Phi$ is contained in a union of cosets $\Pi^r U_{\mathfrak{A}}$, $r \in \mathbb{Z}$, where Π is a prime element of $\mathfrak{A}$. Consider the set of integers r such that $\operatorname{supp}\Phi \cap \Pi^r U_{\mathfrak{A}}$ is non-empty. This set is bounded below, since $\operatorname{supp}\Phi$ is compact. Since $\Phi(0) = 0$, it is also bounded above.

This proves (2), and (3) follows from (24.5.1). □

The proposition follows from part (3) of the lemma and 24.4 Lemma 2. □

For integrals of the form $\zeta(\Phi, f_0, s)$, the convergence statement of 24.2 Theorem 1(1) is implied by the fact that $\operatorname{supp} f_0 \subset \mathcal{K}_{\mathfrak{A}}$. The general case follows from (24.4.4). Replacing X by q^{-s}, all assertions of 24.2 Theorem 1 have been proved in this case. Moreover:

Corollary. *Let (π, V) be an irreducible cuspidal representation of G. The L-function of π is then trivial:*

$$L(\pi, s) = 1.$$

24.6. We now prove 24.2 Theorem 2 in the case where π is cuspidal, starting with:

Proposition. *Let (π, V) be an irreducible cuspidal representation of G. There is a unique rational function $c(\pi, X, \psi) \in \mathbb{C}(X)$ such that*

$$Z(\hat{\Phi}, \check{f}, \tfrac{1}{q^2X}) = c(\pi, X, \psi)\, Z(\Phi, f, X), \tag{24.6.1}$$

for all $\Phi \in C_c^\infty(A)$ and all $f \in \mathcal{C}(\pi)$.

Proof. We define an action of the group $G \times G$ on the field $\mathbb{C}(X)$ of rational functions by

$$(g, h) f(X) = X^{\upsilon_F(\det gh^{-1})}\, f(X).$$

We consider the space

$$\mathcal{L} = \mathrm{Hom}_{G\times G}\big(C_c^\infty(A) \otimes_{\mathbb{C}} \mathcal{C}(\pi), \mathbb{C}(X)\big).$$

This is a vector space over the field $\mathbb{C}(X)$, with the key property:

Lemma. *The space $\mathcal{L}$ has $\mathbb{C}(X)$-dimension one.*

We give the proof in the next paragraph, after completing the proof of the proposition.

Consider the pairings $C_c^\infty(A) \times \mathcal{C}(\pi) \to \mathbb{C}(X)$ given by

$$\begin{aligned} \lambda : (\Phi, f) &\longmapsto Z(\Phi, f, X), \\ \lambda' : (\Phi, f) &\longmapsto Z(\hat{\Phi}, \check{f}, 1/q^2X). \end{aligned}$$

The first of these belongs to $\mathcal{L}$ by (24.4.4). Combining (24.4.4) with (24.4.1), we see that $\lambda' \in \mathcal{L}$ also. The existence of the rational function $c(\pi, X, \psi)$ now follows from the lemma. □

In (24.6.1), we put $X = q^{-s}$ to obtain

$$\zeta(\hat{\Phi}, \check{f}, 2-s) = c(\pi, q^{-s}, \psi)\, \zeta(\Phi, f, s).$$

Setting

$$c(\pi, q^{-\frac{1}{2}-s}, \psi) = \gamma(\pi, s, \psi), \tag{24.6.2}$$

Theorem 2 of 24.2 follows in this cuspidal case. Here we have

$$\gamma(\pi, s, \psi) = \varepsilon(\pi, s, \psi) = c(\pi, q^{-\frac{1}{2}-s}, \psi), \tag{24.6.3}$$

by 24.5 Corollary. Moreover, $c(\pi, X, \psi)$ is a *monomial* in X.

24.7. We prove the lemma of 24.6. Let

$$\mathfrak{h} : C_c^\infty(A) \otimes \mathcal{C}(\pi) \longrightarrow \mathbb{C}(X)$$

be a $G \times G$-equivariant pairing. Let $f_0 \in \mathcal{C}(\pi)$ be the coefficient introduced in 24.5. Since f_0 spans $\mathcal{C}(\pi)$ as $G \times G$-space, the pairing $\mathfrak{h}$ is determined by its restriction to the subspace $C_c^\infty(A) \otimes f_0$. Since every g in the support of the idempotent e_ξ has $v_F(\det g) = 0$, the property $\check{e}_\xi * f_0 * \check{e}_\xi = f_0$ implies

$$\mathfrak{h}(\Phi \otimes f_0) = \mathfrak{h}(e_\xi * \Phi * e_\xi \otimes f_0),$$

for every $\Phi \in C_c^\infty(A)$. It follows (24.5 Lemma) that $\mathfrak{h}$ is determined by its restriction to $C_c^\infty(G) \otimes \mathcal{C}(\pi)$. Next,

$$\mathrm{Hom}_{G\times G}(C_c^\infty(G) \otimes \mathcal{C}(\pi), \mathbb{C}(X)) \cong \mathrm{Hom}_{G\times G}(\mathcal{C}(\pi), C_c^\infty(G)^* \otimes \mathbb{C}(X)),$$

where $C_c^\infty(G)^*$ denotes the abstract linear dual of $C_c^\infty(G)$. The image of any $G \times G$-map $\mathcal{C}(\pi) \to C_c^\infty(G)^* \otimes \mathbb{C}(X)$ lies in the subspace of smooth vectors. Since $\mathbb{C}(X)$ is smooth over $G \times G$, this subspace is $C_c^\infty(G)^\vee \otimes \mathbb{C}(X)$, so

$$\mathrm{Hom}_{G\times G}(C_c^\infty(G) \otimes \mathcal{C}(\pi), \mathbb{C}(X)) \cong \mathrm{Hom}_{G\times G}(\mathcal{C}(\pi), C_c^\infty(G)^\vee \otimes \mathbb{C}(X)).$$

We describe the $G \times G$-space $C_c^\infty(G)^\vee$. Let H denote the diagonal subgroup $\{(g,g) : g \in G\}$ of $G\times G$. For $\phi \in C_c^\infty(G)$, we define a function $f_\phi : G\times G \to \mathbb{C}$ by $f_\phi(g_1, g_2) = \phi(g_1^{-1} g_2)$. The map $\phi \mapsto f_\phi$ is then a $G \times G$-isomorphism $C_c^\infty(G) \to c\text{-}\mathrm{Ind}_H^{G\times G}(1_H)$. Thus

$$C_c^\infty(G)^\vee \cong \mathrm{Ind}_H^{G\times G}(1_H).$$

There is a canonical $G \times G$-injection

$$\mathrm{Ind}_H^{G\times G}(1_H) \otimes \mathbb{C}(X) \longrightarrow \mathrm{Ind}_H^{G\times G}(\mathbb{C}(X))$$

given by mapping the tensor $f \otimes \phi(X)$ to the function $y \mapsto f(y)\phi(X)$, $y \in G \times G$. This induces

$$\mathrm{Hom}_{G\times G}(\mathcal{C}(\pi), C_c^\infty(G)^\vee \otimes \mathbb{C}(X)) \hookrightarrow \mathrm{Hom}_H(\mathcal{C}(\pi), \mathbb{C}(X)).$$

Since H acts trivially on $\mathbb{C}(X)$, this is the same as $\mathrm{Hom}_G(\mathcal{C}(\pi), \mathbb{C}(X))$, where G acts on $\mathcal{C}(\pi) \cong \check{V} \otimes V$ in the natural (diagonal) way, and trivially on $\mathbb{C}(X)$. We have

$$\begin{aligned}\mathrm{Hom}_G(\mathcal{C}(\pi), \mathbb{C}(X)) &\cong \mathrm{Hom}_G(\check{V} \otimes V, \mathbb{C}) \otimes \mathbb{C}(X) \\ &\cong \mathrm{End}_G(V) \otimes \mathbb{C}(X) \cong \mathbb{C}(X),\end{aligned}$$

and the result follows. □

24.8. It is sometimes useful to assemble the series $Z(\Phi, f, X)$, or the integrals $\zeta(\Phi, f, s)$, into a more compact form: we will need this in the next section.

Let (π, V) be a irreducible smooth representation of G. We shall assume π is *cuspidal,* although this is not strictly necessary. For $m \in \mathbb{Z}$ and $\Phi \in C_c^\infty(A)$, define Φ_m as in (24.4.2). Then (24.4 Lemma 1) $\Phi_m \in \mathcal{H}(G)$ and we can form

$$z_m(\Phi, \pi) = \int_G \Phi_m(x)\pi(x)\, d^*x = \pi(\Phi_m).$$

If K is a compact open subgroup of G, such that $K \times K$ fixes Φ, this operator annihilates $V(K)$ (2.3) and is, in effect, an endomorphism of the finite-dimensional space V^K. We define

$$Z(\Phi, \pi, X) = \sum_{m\in\mathbb{Z}} z_m(\Phi, \pi)X^m = \sum_{m\in\mathbb{Z}} \pi(\Phi_m)X^m.$$

This is a formal Laurent series with coefficients in $\mathrm{End}_{\mathbb{C}}(V)$: indeed, if $K \times K$ fixes Φ, the coefficients lie in $\mathrm{End}_{\mathbb{C}}(V^K)$. If we take $v \in V$, $\check{v} \in \check{V}$ and form the coefficient function $f = \gamma_{\check{v}\otimes v} \in \mathcal{C}(\pi)$, then

$$Z(\Phi, f, X) = \langle \check{v}, Z(\Phi, \pi, X)v\rangle.$$

Choosing a basis of V^K and the dual basis of $\check{V}^K$, one may think of $Z(\Phi, \pi, X)$ as a finite matrix whose entries are various series $Z(\Phi, f, X)$.

We can also form a zeta-function

$$\zeta(\Phi, \pi, s) = Z(\Phi, \pi, q^{-s}).$$

This can be expressed as an integral,

$$\zeta(\Phi, \pi, s) = \int_G \Phi(x)\pi(x)\|\det x\|^s\, d^*x,$$

convergent in a half-plane.

As an operator on $\check{V}$, $\zeta(\Phi, \check{\pi}, s)$ has a transpose $\check{\zeta}(\Phi, \check{\pi}, s) \in \mathrm{End}_{\mathbb{C}}(V)$, defined in 2.10 by:

$$\langle \zeta(\Phi, \check{\pi}, s)\check{v}, v\rangle = \langle \check{v}, \check{\zeta}(\Phi, \check{\pi}, s)v\rangle.$$

In these terms, the functional equation (24.2.3) reads:

$$\check{\zeta}(\hat{\Phi}, \check{\pi}, \tfrac{3}{2}-s) = \varepsilon(\pi, s, \psi)\, \zeta(\Phi, \pi, s+\tfrac{1}{2}). \tag{24.8.1}$$

25. Cuspidal Local Constants

We give an explicit formula for the local constant $\varepsilon(\pi, s, \psi)$, when π is an irreducible cuspidal representation of $G = \mathrm{GL}_2(F)$. Throughout this section, ψ denotes a character of F *of level one.* We put $\psi_A = \psi \circ \mathrm{tr}_A$.

The discussion centres on a *non-abelian Gauss sum*, generalizing the abelian Gauss sum appearing in (23.6.1). Indeed, the whole discussion closely parallels that in 23.5, and leads to a Stability Theorem generalizing 23.8.

25.1. Let $(\mathfrak{A}, \Xi)$ be a cuspidal inducing datum in G (15.8), say $\Xi : \mathcal{K}_{\mathfrak{A}} \to \mathrm{Aut}_{\mathbb{C}}(W)$. Let $n = \ell_{\mathfrak{A}}(\Xi)$ (15.8.1).

We define an element $\mathcal{T}(\Xi, \psi)$ of $\mathrm{End}_{\mathbb{C}}(\check{W})$ by

$$\mathcal{T}(\Xi, \psi) = \sum_{x \in U_{\mathfrak{A}}/U_{\mathfrak{A}}^{n+1}} \check{\Xi}(cx)\, \psi_A(cx), \tag{25.1.1}$$

where $c \in \mathcal{K}_{\mathfrak{A}}$ satisfies $c\mathfrak{A} = \mathfrak{P}^{-n}$, $\mathfrak{P} = \mathrm{rad}\,\mathfrak{A}$.

Lemma.

(1) *The definition* (25.1.1) *is independent of the choices of coset representatives* $x \in U_{\mathfrak{A}}/U_{\mathfrak{A}}^{n+1}$ *and of* $c \in \mathcal{K}_{\mathfrak{A}}$ *such that* $c\mathfrak{A} = \mathfrak{P}^{-n}$.
(2) *There exists* $\tau(\Xi, \psi) \in \mathbb{C}$ *such that*

$$\mathcal{T}(\Xi, \psi) = \tau(\Xi, \psi)\, 1_{\check{W}}. \tag{25.1.2}$$

Proof. For $y \in U_{\mathfrak{A}}^{n+1}$, we have $\check{\Xi}(cxy) = \check{\Xi}(cx)$, by definition. Also, $\psi_A(cxy) = \psi_A(cx)\psi_A(cx(y-1))$. Since $cx \in \mathfrak{P}^{-n}$, $(y-1) \in \mathfrak{P}^{n+1}$, we have $\psi_A(cx(y-1)) = 1$ (12.5). This proves the first assertion in (1) and the second follows immediately.

For $g \in \mathcal{K}_{\mathfrak{A}}$, we have

$$\check{\Xi}(g)\, \mathcal{T}(\Xi, \psi)\, \check{\Xi}(g)^{-1} = \sum_{x \in U_{\mathfrak{A}}/U_{\mathfrak{A}}^{n+1}} \check{\Xi}(gcg^{-1}\, gxg^{-1})\, \psi_A(cx).$$

However, $\psi_A(cx) = \psi_A(gcxg^{-1})$; the set $\{gxg^{-1} : x \in U_{\mathfrak{A}}/U_{\mathfrak{A}}^{n+1}\}$ is a set of coset representatives for $U_{\mathfrak{A}}/U_{\mathfrak{A}}^{n+1}$, and $gcg^{-1} \in \mathcal{K}_{\mathfrak{A}}$ generates $\mathfrak{P}^{-n}$. It follows from (1) that

$$\check{\Xi}(g)\, \mathcal{T}(\Xi, \psi)\, \check{\Xi}(g)^{-1} = \mathcal{T}(\Xi, \psi),$$

whence $\mathcal{T}(\Xi, \psi)$ is a scalar operator, as required for (2). □

The complex number $\tau(\Xi, \psi)$ is the *non-abelian Gauss sum* of Ξ. One can equally express it in the form

$$\tau(\Xi, \psi) = \frac{1}{\dim \Xi} \sum_{x \in U_{\mathfrak{A}}/U_{\mathfrak{A}}^{n+1}} \mathrm{tr}(\check{\Xi}(cx))\, \psi_A(cx), \tag{25.1.3}$$

for the same c as before.

25.2. We use the Gauss sum to calculate the local constant $\varepsilon(\pi, s, \psi)$ of an irreducible cuspidal representation π of G. We define $\ell(\pi)$ as in 12.6.

Theorem. *Let (π, V) be an irreducible cuspidal representation of G, and let ψ be a character of F of level one. Let $(\mathfrak{A}, \Xi)$ be a cuspidal inducing datum which occurs in π. Let $n = \ell_{\mathfrak{A}}(\Xi) = e_{\mathfrak{A}}\ell(\pi)$. Writing $\mathfrak{P} = \operatorname{rad}\mathfrak{A}$, we have*

$$\begin{aligned}\varepsilon(\pi, s, \psi) &= (\mathfrak{P}^{-n} : \mathfrak{A})^{(\frac{1}{2}-s)/2}\, \frac{\tau(\Xi, \psi)}{(\mathfrak{A} : \mathfrak{P}^{n+1})^{\frac{1}{2}}} \\ &= q^{2\ell(\pi)(\frac{1}{2}-s)}\, \frac{\tau(\Xi, \psi)}{(\mathfrak{A} : \mathfrak{P}^{n+1})^{\frac{1}{2}}}\,.\end{aligned} \tag{25.2.1}$$

Proof. Fix a Haar measure μ^* on G and write $d^*g = d\mu^*(g)$. In this argument, it will be more convenient to use the operator-valued integrals $\zeta(\Phi, \pi, s)$ of 24.8.

We temporarily write $K = U_{\mathfrak{A}}^{n+1}$, $\mathcal{K} = \mathcal{K}_{\mathfrak{A}}$. We choose $\Phi = e_K$; this gives

$$\zeta(\Phi, \pi, s) = \pi(e_K).$$

Let $\xi = \Xi \mid U_{\mathfrak{A}}$: we know (15.8 Proposition 2) that ξ is irreducible and $V^{\Xi} = V^{\xi}$. Consequently, if $v \in V^{\xi}$, then

$$\pi(e_\xi)\,\pi(g)\,\pi(e_\xi)\,v = \begin{cases} \Xi(g)\,v & \text{if } g \in \mathcal{K}, \\ 0 & \text{otherwise,}\end{cases}$$

while $\pi(e_\xi)\,\pi(g)\,\pi(e_\xi)$ annihilates the K-complement of V^{ξ} in V. A direct calculation yields

$$\pi(e_\xi)\,\zeta(\Phi, \pi, s)\,\pi(e_\xi) = \zeta(e_\xi * \Phi * e_\xi, \pi, s) = \pi(e_\xi).$$

We abbreviate $\Psi = e_\xi * \Phi * e_\xi = e_\xi$. We then have $\hat{\Psi} = \check{e}_\xi * \hat{\Phi} * \check{e}_\xi$; since $\check{e}_\xi = e_{\check{\xi}}$, the support of $\hat{\Psi}$ is contained in $\mathcal{K}$ (24.5 Lemma). The transpose of the functional equation (24.8.1) gives

$$\begin{aligned}\varepsilon(\pi, s{-}\tfrac{1}{2}, \psi)\,\check{\pi}(\check{e}_\xi) &= \zeta(\hat{\Psi}, \check{\pi}, 2{-}s) \\ &= \int_{\mathcal{K}} \hat{\Psi}(x)\,\check{\pi}(x)\,\|\det x\|^{2-s}\,d^*x \\ &= \int_{\mathcal{K}} \check{e}_\xi * \hat{\Psi} * \check{e}_\xi(x)\,\check{\pi}(x)\,\|\det x\|^{2-s}\,d^*x \\ &= \int_{\mathcal{K}} \hat{\Psi}(x)\,\check{\pi}(\check{e}_\xi)\check{\pi}(x)\check{\pi}(\check{e}_\xi)\,\|\det x\|^{2-s}\,d^*x.\end{aligned}$$

As an operator on $\check{V}^{\check{\xi}}$, this reduces to

$$\begin{aligned}\varepsilon(\pi, s{-}\tfrac{1}{2}, \psi)\,\check{\pi}(\check{e}_\xi) &= \int_{\mathcal{K}} \hat{\Psi}(x)\,\check{\Xi}(x)\,\|\det x\|^{2-s}\,d^*x \\ &= \int_{\mathcal{K}} \hat{\Phi}(x)\,\check{\Xi}(x)\,\|\det x\|^{2-s}\,d^*x.\end{aligned}$$

We calculate $\hat{\Phi}$. The self-dual Haar measure μ on A satisfies $\mu(\mathfrak{A}) = (\mathfrak{A} : \mathfrak{P})^{1/2}$. The support of $\hat{\Phi}$ is $\mathfrak{P}^{-n}$ and

$$\hat{\Phi}(x) = \mu^*(K)^{-1}\mu(\mathfrak{P}^{n+1})\,\psi_A(x), \quad x \in \mathfrak{P}^{-n}.$$

So:

$$\int_K \hat{\Phi}(x)\,\check{\Xi}(x)\,\|\det x\|^{2-s}\,d^*x$$
$$= \frac{\mu(\mathfrak{P}^{n+1})}{\mu^*(K)} \int_{\mathfrak{P}^{-n}\cap K} \psi_A(x)\,\check{\Xi}(x)\,\|\det x\|^{2-s}\,d^*x.$$

The range of integration is effectively $\bigcup_{r\geqslant -n} \Pi^r U_{\mathfrak{A}}$, where Π is a prime element of $\mathfrak{A}$. We know that $\varepsilon(\pi, s, \psi)$ is a monomial in q^{-s}, so only one of the "shells" $\Pi^r U_{\mathfrak{A}}$ contributes.

Lemma. *We have*

$$\int_{\Pi^r U_{\mathfrak{A}}} \psi_A(x)\,\check{\Xi}(x)\,\|\det x\|^{2-s}\,d^*x = 0$$

for all $r \geqslant 1-n$.

Proof. As an operator on $\check{V}$, the integral in question stabilizes $\check{V}^{\xi}$ and annihilates all other $U_{\mathfrak{A}}$-isotypic subspaces of $\check{V}$.

We abbreviate $c = \Pi^r$, for some $r \geqslant 1-n$. The factor $\|\det x\|^{2-s}$ in the integral is a non-zero constant, so we have to consider

$$\int_{cU_{\mathfrak{A}}} \psi_A(x)\,\check{\Xi}(x)\,d^*x = \int_{cU_{\mathfrak{A}}} \psi_A(xu)\,\check{\Xi}(xu)\,d^*x,$$

for any $u \in U_{\mathfrak{A}}$. Let $u \in U_{\mathfrak{A}}^n$. Writing $u = 1+y$, we have $\psi_A(xu) = \psi_A(x)$, since $xy \in \mathfrak{P} \subset \operatorname{Ker}\psi_A$. Thus

$$\int_{cU_{\mathfrak{A}}} \psi_A(x)\,\check{\Xi}(x)\,d^*x = \int_{cU_{\mathfrak{A}}} \psi_A(x)\,\check{\Xi}(x)\,d^*x \cdot \check{\Xi}(u).$$

The operator

$$\int_{cU_{\mathfrak{A}}} \psi_A(x)\,\check{\Xi}(x)\,d^*x$$

therefore annihilates the subspace $\check{V}^{\xi}(U_{\mathfrak{A}}^n)$ (notation of 2.3). Since $\check{V}^{\xi}$ has no $U_{\mathfrak{A}}^n$-fixed points, this operator is zero, as required. □

Returning to the theorem, the shell $r = -n$ must give a non-zero integral, and the result follows. □

The proof gives the relation

$$\varepsilon(\pi, s, \psi) \dim \Xi = \frac{\mu(\mathfrak{P}^{n+1})}{\mu^*(U_{\mathfrak{A}}^{n+1})} \int_{\Pi^{-n} U_{\mathfrak{A}}} \operatorname{tr} \check{\Xi}(x)\, \psi_A(x) \,\| \det x \|^{\frac{3}{2}-s}\, d^*x. \tag{25.2.2}$$

As a further consequence, we get the following functional equation for the Gauss sum,

$$\tau(\Xi, \psi)\, \tau(\check{\Xi}, \psi) = \omega_{\Xi}(-1)\, (\mathfrak{A} : \mathfrak{P}^{n+1}),$$

where ω_{Ξ} is the central character of Ξ.

Remark. The formula (25.2.1) says that, when π is cuspidal and ψ has level one, the integer $n(\pi, \psi)$ of (24.3.1) satisfies

$$n(\pi, \psi) = 2\ell(\pi). \tag{25.2.3}$$

25.3. For an argument in §27, we need to generalize this calculation. Keeping ψ of level one as above, one defines $\mathcal{T}(\Xi, a\psi)$, $a \in F^{\times}$, by the same formula (25.1.1), except that the element $c \in \mathcal{K}_{\mathfrak{A}}$ has to satisfy $c\mathfrak{A} = a^{-1}\mathfrak{P}^{-n}$. Clearly, its eigenvalue satisfies

$$\tau(\Xi, a\psi) = \omega_{\Xi}(a)\, \tau(\Xi, \psi),$$

and so, comparing with 24.3,

$$\varepsilon(\pi, s, a\psi) = (a^{-1}\mathfrak{P}^{-n} : \mathfrak{A})^{(\frac{1}{2}-s)/2}\, \frac{\tau(\Xi, a\psi)}{(\mathfrak{A} : \mathfrak{P}^{n+1})^{\frac{1}{2}}}\,. \tag{25.3.1}$$

In this more general context, the analogue of the 25.2 Lemma says:

$$\int_{\Pi^r U_{\mathfrak{A}}} \check{\Xi}(x)\, a\psi_A(x) \| \det x \|^{2-s}\, d^*x = 0, \quad r \geqslant 1-n-e_{\mathfrak{A}} v_F(a). \tag{25.3.2}$$

25.4. When the cuspidal inducing datum $(\mathfrak{A}, \Xi)$ is of level zero, the Gauss sum $\tau(\Xi, \psi)$ can be given a more familiar form by using the classical Gauss sum $\mathfrak{g}_{\Bbbk}(\theta, \eta)$ of 23.7.

As Ξ has level zero, we can take $\mathfrak{A} = \mathrm{M}_2(\mathfrak{o})$; the representation $\xi = \Xi \mid U_{\mathfrak{A}}$ is the inflation of an irreducible cuspidal representation $\tilde{\xi}$ of $U_{\mathfrak{A}}/U_{\mathfrak{A}}^1 = \mathrm{GL}_2(\boldsymbol{k})$. We use the description of such representations given in §6. Let $\boldsymbol{l}/\boldsymbol{k}$ be the quadratic field extension of $\boldsymbol{k}$. There is a regular character θ of $\boldsymbol{l}^{\times}$ such that $\tilde{\xi} \cong \pi_{\theta}$, in the notation of that section. One can calculate $\tau(\Xi, \psi)$ explicitly, using (25.1.3) and the character table (6.4.1). The character $\psi \mid \mathfrak{o}$ is the inflation of a character $\tilde{\psi}$ of $\boldsymbol{k}$; we set $\tilde{\psi}_{\boldsymbol{l}} = \tilde{\psi} \circ \mathrm{Tr}_{\boldsymbol{l}/\boldsymbol{k}}$. In the definition of $\mathcal{T}(\Xi, \psi)$, we take $c = 1$. An entertaining calculation gives:

$$\tau(\Xi, \psi) = -q\, \mathfrak{g}_{\boldsymbol{l}}(\theta, \tilde{\psi}_{\boldsymbol{l}}). \tag{25.4.1}$$

25.5. It will be helpful to translate the formalism of Gauss sums of cuspidal inducing data back into the language of cuspidal types.

Lemma. *Let $(\mathfrak{A}, \Xi)$ be a cuspidal inducing datum, and set $n = \ell_{\mathfrak{A}}(\Xi)$, $\mathfrak{P} = \operatorname{rad}\mathfrak{A}$. Let $(\mathfrak{A}, J, \Lambda)$ be a cuspidal type which induces Ξ. The Gauss sum $\tau(\Xi, \psi)$ is then the unique eigenvalue of the scalar operator*

$$\sum_{x \in J \cap U_{\mathfrak{A}}/U_{\mathfrak{A}}^{n+1}} \Lambda^\vee(cx)\psi_A(cx), \tag{25.5.1}$$

for any $c \in J$ such that $c\mathfrak{A} = \mathfrak{P}^{-n}$.

Proof. For notational convenience, we work with $\check{\Xi}$ in place of Ξ. Thus $\tau(\check{\Xi}, \psi)$ is the unique eigenvalue of the scalar operator

$$\sum_{x \in U_{\mathfrak{A}}/U_{\mathfrak{A}}^{n+1}} \Xi(cx)\psi_A(cx),$$

for any $c \in \mathcal{K}_{\mathfrak{A}}$ generating $\mathfrak{P}^{-n}$.

Let Ξ act on the space W; the isotypic space W^Λ is then irreducible over J and J-equivalent to Λ. Let e_Λ denote the J-projection $W \to W^\Lambda$. For $g \in \mathcal{K}_{\mathfrak{A}}$, we have

$$e_\Lambda\, \Xi(g)\, e_\Lambda = \begin{cases} \Lambda(g) & \text{if } g \in J, \\ 0 & \text{otherwise.} \end{cases}$$

Choosing $c \in J$, we get the result immediately. □

We can simplify further:

Proposition. *Choose $c \in J$ such that $\Lambda \mid U_{\mathfrak{A}}^{[n/2]+1}$ is a multiple of ψ_c. The Gauss sum $\tau(\Xi, \psi)$ is the unique eigenvalue of the scalar operator*

$$(\mathfrak{A} : \mathfrak{P}^{[(n+1)/2]}) \sum_y \Lambda^\vee(cy)\psi_A(cy), \tag{25.5.2}$$

where y ranges over $U_{\mathfrak{A}}^{[(n+1)/2]}/U_{\mathfrak{A}}^{[n/2]+1}$.

Proof. In the sum (25.5.1), we write $x = y(1+z)$, $y \in J^0/U_{\mathfrak{A}}^{[(n+1)/2]}$, $z \in \mathfrak{P}^{[n/2]+1}/\mathfrak{P}^{n+1}$. The sum over z vanishes unless $z \mapsto \psi(-cz + cyz)$ is the trivial character of $\mathfrak{P}^{[n/2]+1}/\mathfrak{P}^{n+1}$, that is, unless $y \in U_{\mathfrak{A}}^{[(n+1)/2]}$. In that case, it takes the value $(\mathfrak{P}^{[n/2]+1} : \mathfrak{P}^{n+1}) = (\mathfrak{A} : \mathfrak{P}^{[(n+1)/2]})$, as required. □

We will rely heavily on this result for comparisons of local constants. We therefore re-formulate it explicitly in these terms.

Corollary. *Let $(\mathfrak{A}, J, \Lambda)$ be a cuspidal type in G. Let n be the least integer $\geqslant 0$ such that $U_{\mathfrak{A}}^{n+1} \subset \operatorname{Ker} \Lambda$. Choose $c \in J$ such that $\Lambda \mid U_{\mathfrak{A}}^{[n/2]+1}$ is a multiple of ψ_c. If $\pi = c\text{-Ind}_J^G \Lambda$, then*

$$\varepsilon(\pi, \tfrac{1}{2}, \psi) = q^a \sum_x \operatorname{tr} \Lambda^\vee(cx)\psi_A(cx),$$

where x ranges over $U_{\mathfrak{A}}^{[(n+1)/2]}/U_{\mathfrak{A}}^{[n/2]+1}$ and

$$q^a \dim \Lambda = \begin{cases} 1 & \text{if } n \text{ is odd,} \\ (\mathfrak{A} : \mathfrak{P})^{-1/2} & \text{if } n \text{ is even.} \end{cases}$$

25.6. We have to prove an analogue of the Stability Theorem (23.8). To this end, we introduce another species of trigonometric sum.

Let $\mathfrak{A}$ be a chain order with radical $\mathfrak{P}$ and $e = e_{\mathfrak{A}}$. Let χ be a character of $F^\times$ of level $l \geqslant 1$. We view $\chi \circ \det$ as character of $\mathcal{K}_{\mathfrak{A}}$, when it has level el. We can form a Gauss sum

$$\tau_{\mathfrak{A}}(\chi, \psi) = \sum_{x \in U_{\mathfrak{A}}/U_{\mathfrak{A}}^{el+1}} \check{\chi}(\det cx)\psi_A(cx),$$

for any $c \in F^\times$ such that $c\mathfrak{o} = \mathfrak{p}^{-l}$. An easier version of the proof of 25.5 Proposition yields:

Lemma. *Suppose that $c \in F$ satisfies $\chi(1+x) = \psi(cx)$, $x \in \mathfrak{p}^{[l/2]+1}$. Then $\chi \circ \det \mid U_{\mathfrak{A}}^{[el/2]+1} = \psi_c$ and*

$$\tau_{\mathfrak{A}}(\chi, \psi) = (\mathfrak{A} : \mathfrak{P}^{[(el+1)/2]}) \sum_y \check{\chi}(\det cy)\psi_A(cy),$$

where y ranges over $U_{\mathfrak{A}}^{[(el+1)/2]}/U_{\mathfrak{A}}^{[el/2]+1}$.

A direct calculation now yields:

$$\frac{\tau_{\mathfrak{A}}(\chi, \psi)}{(\mathfrak{A} : \mathfrak{P}^{el+1})^{1/2}} = \frac{\tau(\chi, \psi)^2}{q^{l+1}}, \tag{25.6.1}$$

where $\tau(\chi, \psi)$ is defined in (23.6.1). We can define, purely formally at the moment,

$$\varepsilon(\chi \circ \det, s, \psi) = (\mathfrak{P}^{-el} : \mathfrak{A})^{(\frac{1}{2}-s)/2}\, \tau_{\mathfrak{A}}(\chi, \psi)/(\mathfrak{A} : \mathfrak{P}^{el+1})^{1/2}, \tag{25.6.2}$$

for some $\mathfrak{A}$ and ψ of level one. We note that this is independent of the choice of $\mathfrak{A}$, by (25.6.1).

We use 24.3 to formally extend the definition (25.6.2) to arbitrary characters of F by setting

$$\varepsilon(\chi \circ \det, s, a\psi) = \chi(a)^2 \, \|a\|^{2s-1} \, \varepsilon(\chi \circ \det, s, \psi), \quad a \in F^\times. \tag{25.6.3}$$

Remark. Theorem 26.1 will show that the definitions (25.6.2), (25.6.3) are correct: see particularly the propositions of 26.6, 26.7.

25.7. We prove:

Stability Theorem. *Let π be an irreducible cuspidal representation of G and let χ be a character of $F^\times$, of level m, such that $m > 2\ell(\pi)$. Let $\mu \in \widehat{F}$, $\mu \neq 1$. Let $c \in F^\times$ satisfy $\chi(1+x) = \mu(cx)$, $x \in \mathfrak{p}^{[m/2]+1}$. Then:*

$$\varepsilon(\chi\pi, s, \mu) = \omega_\pi(c)^{-1}\,\varepsilon(\chi \circ \det, s, \mu).$$

Proof. By (25.6.3), 24.3, it is enough to treat the case where μ has level one.

The representation π contains a cuspidal type $(\mathfrak{A}, J, \Lambda)$, so $\chi\pi$ contains the type $(\mathfrak{A}, J, \chi\Lambda)$, where $\chi\Lambda : g \mapsto \chi(\det g)\Lambda(g)$. Put $e = e_{\mathfrak{A}}$. As representation of J, Λ has level $e\ell(\pi)$ and $\chi\Lambda$ has level em. The hypothesis on levels implies Λ is trivial on the group $U_{\mathfrak{A}}^{[(em+1)/2]}$, and the result follows straightaway, from 25.5 Corollary. □

26. Functional Equation for Non-Cuspidal Representations

In this section, we prove Theorems 1 and 2 of 24.2 for non-cuspidal irreducible representations of G: we use the classical language of zeta-integrals and analytic functions, since it is marginally easier here. The subgroups B, N, T, Z of G are as in 5.1.

26.1. We also compute the associated L-functions and local constants:

Theorem. *Let $\chi = \chi_1 \otimes \chi_2$ be a character of the group T, and let π be a G-composition factor of $\iota_B^G\,\chi$. For any $\psi \in \widehat{F}$, $\psi \neq 1$, we have*

$$\begin{aligned} L(\pi, s) &= L(\chi_1, s)\,L(\chi_2, s), \\ \varepsilon(\pi, s, \psi) &= \varepsilon(\chi_1, s, \psi)\,\varepsilon(\chi_2, s, \psi), \end{aligned} \tag{26.1.1}$$

except when $\pi \cong \phi \cdot \mathrm{St}_G$, for an unramified character ϕ of $F^\times$. In this exceptional case, we have

$$L(\pi, s) = L(\phi, s+\tfrac{1}{2}), \quad \varepsilon(\pi, s, \psi) = -\varepsilon(\phi, s, \psi). \tag{26.1.2}$$

The L-functions and local constants of characters of $F^\times$ have been worked out in §23.

26.2. Let $\chi = \chi_1 \otimes \chi_2$ be a character of T. We set $(\pi, V) = \iota_B^G\,\chi = \mathrm{Ind}_B^G\,\chi \otimes \delta_B^{-1/2}$. While π is not necessarily irreducible, we can still consider the space $\mathcal{C}(\pi)$ of coefficients of π. If (σ, W) is a G-subspace of (π, V), we can identify $\mathcal{C}(\sigma)$ with a $G \times G$-subspace of $\mathcal{C}(\pi)$: it is the linear span of coefficients $\gamma_{\check{x}\otimes w}$, $w \in W$, $\check{x} \in \check{V}$. Similarly, if $Y = V/W$ and τ denotes the natural representation of G on Y, then $\mathcal{C}(\tau)$ is the span of the coefficients $\gamma_{\check{x}\otimes v}$, $v \in V$ and $\check{x} \in \check{V}$ such that $\langle \check{x}, W\rangle = 0$.

We choose a Haar measure μ_G on G and abbreviate $d\mu_G(g) = dg$. We form zeta-integrals

$$\zeta(\Phi, f, s) = \int_G \Phi(g) f(g) \,\| \det g\|^s \, dg, \quad \Phi \in C_c^\infty(A),\ f \in \mathcal{C}(\pi),$$

as before, and we write

$$\mathcal{Z}(\pi) = \mathcal{Z}(\pi, q^{-s}) = \{\zeta(\Phi, f, s+\tfrac{1}{2}) : \Phi \in C_c^\infty(A),\ f \in \mathcal{C}(\pi)\}.$$

We first prove an analogue of 24.2 Theorem 1 for the representation π.

Proposition. *Let $\chi = \chi_1 \otimes \chi_2$ be a character of T, and put $(\pi, V) = \iota_B^G \chi$.*

(1) *There exists $s_0 \in \mathbb{R}$, depending only on χ, such that $\zeta(\Phi, f, s)$ converges, absolutely and uniformly in vertical strips, in the region* $\operatorname{Re} s > s_0$.
(2) *The integral $\zeta(\Phi, f, s)$ represents a rational function of q^{-s} and*

$$\mathcal{Z}(\pi, q^{-s}) = \mathcal{Z}(\chi_1, q^{-s})\mathcal{Z}(\chi_2, q^{-s}).$$

Proof. We transfer the zeta-integrals on G to the corresponding objects on $T \cong F^\times \times F^\times$.

Let D be the algebra of diagonal matrices in A; thus $T = D^\times$ and the space $C_c^\infty(D)$ is canonically isomorphic to $C_c^\infty(F) \otimes C_c^\infty(F)$ (3.2).

We choose a Haar measure dn on the group N of upper triangular unipotent matrices in G.

Lemma. *Let $\Phi \in C_c^\infty(A)$. There exists a unique function $\Phi_T \in C_c^\infty(D)$ such that*

$$\Phi_T(t) = \|t_1\| \int_N \Phi(tn)\, dn, \quad t = \left(\begin{smallmatrix} t_1 & 0 \\ 0 & t_2 \end{smallmatrix}\right) \in T.$$

The map $\Phi \mapsto \Phi_T$ is a linear surjection $C_c^\infty(A) \to C_c^\infty(D)$.

Proof. The integral defining Φ_T converges absolutely to a function on T, and the map $\Phi \mapsto \Phi_T$ is certainly linear. The space $C_c^\infty(A)$ is spanned by functions of the form

$$\Phi = (\phi_{ij}) : (a_{ij}) \longmapsto \prod_{i,j} \phi_{ij}(a_{ij}), \tag{26.2.1}$$

with "coordinates" $\phi_{ij} \in C_c^\infty(F)$. For such a function Φ, we have (using the obvious isomorphism $N \cong F$)

$$\Phi_T(t) = \phi_{11}(t_1)\, \phi_{22}(t_2)\, \phi_{21}(0) \int_F \phi_{12}(n)\, dn,$$

Thus $\Phi_T \in C_c^\infty(D)$, and so the same applies to arbitrary $\Phi \in C_c^\infty(A)$. Surjectivity is now clear. □

We next describe the coefficients of π. Let $\theta \in V$, $\tau \in \check{V}$; thus θ, τ are functions $G \to \mathbb{C}$ satisfying

$$\theta(ntg) = \delta_B(t)^{-1/2}\chi(t)\,\theta(g), \quad \tau(ntg) = \delta_B(t)^{-1/2}\chi(t)^{-1}\tau(g),$$

for $n \in N$, $t \in T$, $g \in G$. The pair (τ, θ) gives a coefficient

$$f(g) = \langle \tau, \pi(g)\theta \rangle = \int_{B\backslash G} \tau(x)\,\theta(xg)\,d\dot{x},$$

for a positive semi-invariant measure $d\dot{x}$ on $B\backslash G$. If we put $K = \mathrm{GL}_2(\mathfrak{o})$, there is a Haar measure dk on K such that

$$f(g) = \int_K \tau(k)\,\theta(kg)\,dk$$

(*cf.* 7.6 Corollary). We expand formally:

$$\begin{aligned}\zeta(\Phi, f, s) &= \int_G \int_K \Phi(g)\,\tau(k)\,\theta(kg)\,\|\det g\|^s\,dk\,dg \\ &= \int_K \int_G \Phi(k^{-1}g)\,\tau(k)\,\theta(g)\,\|\det g\|^s\,dg\,dk \\ &= \int_K \int_K \int_B \Phi(k^{-1}bk')\,\tau(k)\,\theta(bk')\,\|\det b\|^s\,db\,dk'\,dk,\end{aligned}$$

for a left Haar measure db on B (7.6.2).

There exists an open subgroup K_1 of K which fixes θ and τ under right translation and satisfies

$$\Phi(k^{-1}gk') = \Phi(g), \quad g \in G,\ k, k' \in K_1.$$

The last expression for $\zeta(\Phi, f, s)$ reduces to

$$\zeta(\Phi, f, s) = \mu(K_1)^2 \sum_{i,j} \int_B \Phi^{ij}(b)\,\theta(bk_i)\,\tau(k_j)\,\|\det b\|^s\,db, \tag{26.2.2}$$

where k_i, k_j range independently over K/K_1 and $\Phi^{ij} : x \mapsto \Phi(k_i^{-1}xk_j)$.

We write $db = dt\,dn$ for Haar measures dt on T, dn on N. A typical term in the sum in (26.2.2) then reduces to

$$\begin{aligned}&\theta(k_i)\,\tau(k_j) \int_T \int_N \Phi^{ij}(tn)\,\chi(t)\,\delta_B(t)^{-1/2}\,\|\det b\|^s\,dtdn \\ &\qquad\qquad = \theta(k_i)\,\tau(k_j) \int_T \Phi_T^{ij}(t)\,\chi(t)\,\|\det t\|^{s-\frac{1}{2}}\,dt.\end{aligned}$$

The right hand side has the required convergence properties (23.3, concluding remarks). This argument also shows that $\zeta(\Phi, f, s+\frac{1}{2}) \in \mathcal{Z}(\chi_1)\mathcal{Z}(\chi_2)$, and so $\mathcal{Z}(\pi) \subset \mathcal{Z}(\chi_1)\mathcal{Z}(\chi_2)$.

To prove the opposite containment, we take $\phi_i \in C_c^\infty(F)$, $i = 1, 2$, such that $\zeta(\phi_i, \chi_i, s) = L(\chi_i, s)$. We can assume that

$$\phi_i(uxv) = \chi_i(uv)^{-1}\phi_i(x), \quad u, v \in U_F.$$

We choose an integer $r \geqslant 0$, at least as large as the level of each χ_i, and form the group $H = (K \cap B)N'_{r+1}$, where

$$N'_j = \begin{pmatrix} 1 & 0 \\ \mathfrak{p}^j & 1 \end{pmatrix}.$$

We then have $H = (H \cap N)(H \cap T)(H \cap N')$, and the factors may be taken in any order. We define a character $\tilde{\chi}$ of H to be trivial on the unipotent factors $H \cap N$, $H \cap N'$, and to agree with χ on $K \cap T$.

We choose $\Phi \in C_c^\infty(A)$ such that $\Phi_T = \phi_1 \otimes \phi_2$; we may pick Φ to satisfy

$$\Phi(h_1 x h_2) = \tilde{\chi}(h_1 h_2)^{-1}\Phi(x), \quad x \in A,\ h_i \in H.$$

We take θ, τ to have support BH and to be fixed by $H \cap N'$, normalized so that $\theta(1) = \tau(1) = 1$.

We return to the expression

$$\zeta(\Phi, f, s) = \int_K \int_K \int_B \Phi(k^{-1}bk')\, \|\det b\|^s \tau(k)\, \theta(bk')\, db\, dk\, dk'.$$

On following through the procedure above, this reduces to

$$\zeta(\Phi, f, s) = c \int_B \Phi(b)\, \theta(b)\, \|\det b\|^s\, db = c\, L(\chi_1, s-\tfrac{1}{2})\, L(\chi_2, s-\tfrac{1}{2}),$$

for a constant $c > 0$. Therefore $\mathcal{Z}(\pi) = \mathcal{Z}(\chi_1)\, \mathcal{Z}(\chi_2)$, as desired. □

26.3. We now deal with the functional equation for the (possibly reducible) representation $\pi = \iota_B^G \chi$. We have to prove:

Proposition. *Let $\chi = \chi_1 \otimes \chi_2$ be a character of T and set $(\pi, V) = \iota_B^G \chi$. Let $\psi \in \widehat{F}$, $\psi \neq 1$. There is a unique rational function $\gamma(\pi, s, \psi) \in \mathbb{C}(q^{-s})$ such that*

$$\zeta(\hat{\Phi}, \check{f}, \tfrac{3}{2}-s) = \gamma(\pi, s, \psi)\, \zeta(\Phi, f, s+\tfrac{1}{2}),$$

for all $\Phi \in C_c^\infty(A)$ and all $f \in \mathcal{C}(\pi)$. Moreover,

$$\gamma(\pi, s, \psi) = \gamma(\chi_1, s, \psi)\, \gamma(\chi_2, s, \psi). \tag{26.3.1}$$

Proof. We first need to work out the relation between the map $\Phi \mapsto \Phi_T$ and the various Fourier transforms. Let μ_F be the self-dual Haar measure on F with respect to the character ψ. If we identify $A = \mathrm{M}_2(F)$ with $F \oplus F \oplus F \oplus F$ in the obvious way, then $\mu_F \otimes \mu_F \otimes \mu_F \otimes \mu_F$ is the self-dual Haar measure on A relative to the character $\psi_A = \psi \circ \mathrm{tr}_A$. The restriction $\psi_D = \psi_A \mid D$ is a non-trivial character of D, and the self-dual measure on D, relative to ψ_D, is $\mu_F \otimes \mu_F$. We use these measures to compute Fourier transform in $C_c^\infty(A)$, $C_c^\infty(D)$. We can view μ_F as a Haar measure on N via the obvious isomorphism $N \cong F$, and we define the map $\Phi \mapsto \Phi_T$ of 26.2 using this measure.

Lemma. *For $\Phi \in C_c^\infty(A)$, we have $(\hat{\Phi})_T = \widehat{\Phi_T}$.*

Proof. It is enough to treat a function $\Phi = (\phi_{ij})$ as in 26.2. We can compute the Fourier transform term by term:

$$\hat{\Phi} = \begin{pmatrix} \hat{\phi}_{11} & \hat{\phi}_{21} \\ \hat{\phi}_{12} & \hat{\phi}_{22} \end{pmatrix}.$$

As in Lemma 26.2,

$$\Phi_T(\delta) = \phi_{11}(\delta_1)\,\phi_{22}(\delta_2)\,\phi_{21}(0) \int_F \phi_{12}(x)\,d\mu_F(x), \quad \delta = \left(\begin{smallmatrix} \delta_1 & 0 \\ 0 & \delta_2 \end{smallmatrix}\right) \in D.$$

Likewise,

$$\begin{aligned} \hat{\Phi}_T(\delta) &= \hat{\phi}_{11}(\delta_1)\,\hat{\phi}_{22}(\delta_2)\,\hat{\phi}_{12}(0) \int_F \hat{\phi}_{21}(x)\,d\mu_F(x) \\ &= \hat{\phi}_{11}(\delta_1)\,\hat{\phi}_{22}(\delta_2)\,\phi_{21}(0) \int_F \phi_{12}(x)\,d\mu_F(x) \\ &= \widehat{\Phi_T}(\delta), \end{aligned}$$

as required. □

We now take f as in 26.2:

$$f(g) = \int_{B\backslash G} \tau(x)\,\theta(xg)\,d\dot{x},$$

so that

$$\check{f}(g) = f(g^{-1}) = \int_K \tau(kg)\,\theta(k)\,dk.$$

Using the same procedure as before, we get

$$\begin{aligned} \zeta(\hat{\Phi}, \check{f}, s) &= \int_G \int_K \hat{\Phi}(g)\,\tau(kg)\,\theta(k)\,\|\det g\|^s\,dk dg \\ &= \int_B \int_K \int_K \hat{\Phi}(k^{-1}bk')\,\tau(bk')\,\theta(k)\,\|\det b\|^s\,dk dk' db. \end{aligned}$$

Choosing the same open subgroup K_1 and coset representatives k_i as before, this reduces to

$$\zeta(\hat{\Phi}, \check{f}, s) = \mu(K_1) \sum_{i,j} \theta(k_i)\, \tau(k_j) \int_T \hat{\Phi}_T^{ji}(t)\, \chi(t)^{-1}\, \| \det t \|^{s-\frac{1}{2}}\, dt,$$

where $\hat{\Phi}^{ji} : x \mapsto \hat{\Phi}(k_j^{-1} x k_i)$. We certainly have $\hat{\Phi}^{ji} = \widehat{\Phi^{ij}}$, so the lemma gives $\hat{\Phi}_T^{ji} = \big(\Phi_T^{ij}\big)\widehat{\ }$, and we deduce that

$$\zeta(\hat{\Phi}, \check{f}, \tfrac{3}{2}-s) = \gamma(\chi_1, s, \psi)\, \gamma(\chi_2, s, \psi)\, \zeta(\Phi, f, s+\tfrac{1}{2}),$$

as required. □

26.4. This completes the proofs of 24.2 Theorems 1 and 2 for *irreducible* representations of the form $\iota_B^G \chi$. We can also write down the L-functions and local constants:

Proposition. *Let $\chi = \chi_1 \otimes \chi_2$ be a character of T such that $\pi = \iota_B^G \chi$ is irreducible. Then:*

$$\begin{aligned} L(\pi, s) &= L(\chi_1, s)\, L(\chi_2, s), \\ \varepsilon(\pi, s, \psi) &= \varepsilon(\chi_1, s, \psi)\, \varepsilon(\chi_2, s, \psi), \end{aligned} \tag{26.4.1}$$

for any $\psi \in \widehat{F}$, $\psi \neq 1$.

Proof. The L-function relation reflects the equality $\mathcal{Z}(\pi) = \mathcal{Z}(\chi_1)\mathcal{Z}(\chi_2)$ (26.2) and the ε-relation comes from the corresponding relation between the γ's (26.3.1). □

26.5. Let χ be a character of T, set $\Sigma = \iota_B^G \chi$, and let π be a G-composition factor of Σ. We prove the theorems of 24.2 for π.

First, we have $\mathcal{C}(\pi) \subset \mathcal{C}(\Sigma)$, so the convergence statements hold for π and we have $\mathcal{Z}(\pi) \subset \mathcal{Z}(\Sigma)$.

On the other hand, if $f(1) \neq 0$ and Φ is the characteristic function of a sufficiently small compact open subgroup of G, then $\zeta(\Phi, f, s)$ is a non-zero constant. Therefore

$$\mathbb{C}\big[q^{-s}, q^{s}\big] \subset \mathcal{Z}(\pi) \subset \mathcal{Z}(\Sigma), \tag{26.5.1}$$

and, in the obvious notation (*cf.* 24.2), $P_\pi(t)$ divides $P_\Sigma(t)$.

Likewise, for $f \in \mathcal{C}(\pi)$, we have $\check{f} \in \mathcal{C}(\check{\pi}) \subset \mathcal{C}(\check{\Sigma})$, whence

$$\zeta(\hat{\Phi}, \check{f}, \tfrac{3}{2}-s) = \gamma(\Sigma, s, \psi)\, \zeta(\Phi, f, s+\tfrac{1}{2}),$$

so (24.2.3) holds with $\gamma(\pi, s, \psi) = \gamma(\Sigma, s, \psi)$. The expression

$$\gamma(\pi, s, \psi) = \varepsilon(\pi, s, \psi) L(\pi, s) / L(\check{\pi}, 1-s)$$

holds, for a monomial $\varepsilon(\pi, s, \psi)$, as in 24.2 Corollary. This completes the proofs of Theorems 1 and 2 of 24.2. □

We display one of the conclusions of the preceding arguments:

Corollary. *Let $\chi = \chi_1 \otimes \chi_2$ be a character of T and let π be a G-composition factor of $\Sigma = \iota_B^G \chi$. We have*

$$\gamma(\pi, s, \psi) = \gamma(\iota_B^G \chi, s, \psi) = \gamma(\chi_1, s, \psi)\gamma(\chi_2, s, \psi),$$

and $P_\pi(t)$ divides $P_\Sigma(t) = P_{\chi_1}(t)P_{\chi_2}(t)$.

26.6. We compute the L-function $L(\pi, s)$ in the cases where π is a composition factor of a *reducible* representation $\Sigma = \iota_B^G \chi$. Thus $\chi = \phi\delta_B^{\pm 1/2}$, for a character ϕ of $F^\times$ (9.11). One case is completely straightforward:

Proposition. *Let π be a composition factor of a representation $\iota_B^G\, \phi\delta_B^{\pm 1/2}$, and suppose that the character ϕ of $F^\times$ is not unramified. Then*

$$L(\pi, s) = 1, \quad \varepsilon(\pi, s, \psi) = \varepsilon(\phi, s-\tfrac{1}{2}, \psi)\, \varepsilon(\phi, s+\tfrac{1}{2}, \psi).$$

Proof. Writing $\Sigma = \iota_B^G\, \phi\delta_B^{\pm 1/2}$, we have

$$\mathbb{C}[q^{-s}, q^s] \subset \mathcal{Z}(\pi) \subset \mathcal{Z}(\Sigma) = \mathbb{C}[q^{-s}, q^s],$$

so $L(\pi, s) = 1$. By 26.3 Proposition, we have

$$\begin{aligned} \varepsilon(\pi, s, \psi) = \gamma(\pi, s, \psi) &= \gamma(\phi, s+\tfrac{1}{2}, \psi)\, \gamma(\phi, s-\tfrac{1}{2}, \psi) \\ &= \varepsilon(\phi, s+\tfrac{1}{2}, \psi)\, \varepsilon(\phi, s-\tfrac{1}{2}, \psi), \end{aligned}$$

as claimed. □

26.7. It remains to treat the composition factors of representations $\iota_B^G\, \phi\delta_B^{\pm 1/2}$, where ϕ is unramified. These factors are the one dimensional representation $\phi \circ \det$ of G and the twist $\phi \cdot \mathrm{St}_G$ of the Steinberg representation. The first of them can be dealt with by a direct computation:

Proposition. *Let ϕ be an unramified character of $F^\times$ and put $\pi = \phi \circ \det$. Then:*

$$\begin{aligned} L(\pi, s) &= L(\phi, s-\tfrac{1}{2})\, L(\phi, s+\tfrac{1}{2}), \\ \varepsilon(\pi, s, \psi) &= \varepsilon(\phi, s-\tfrac{1}{2}, \psi)\, \varepsilon(\phi, s+\tfrac{1}{2}, \psi). \end{aligned}$$

In particular, if ψ has level one, then

$$\varepsilon(\phi \circ \det, s, \psi) = \phi(\varpi)^{-2}\, q^{2s-1},$$

for any prime element ϖ of F.

Proof. Introducing a translation on s, we can assume ϕ is trivial. The constant function $g \mapsto 1$ is a coefficient of 1_G. Let Φ denote the characteristic function of $\mathrm{M}_2(\mathfrak{o})$. Using the Cartan decomposition (7.2.2), one finds that

$$\int_G \Phi(g) \| \det g\|^s \, d\mu_G(g) = \mu_G(K)\, (1-q^{-s})^{-1}(1-q^{1-s})^{-1}.$$

The polynomial $P_\pi(t)$ therefore has degree at least 2, and the first assertion follows from 26.5 Corollary. Likewise the second. □

26.8. We take $\pi = \mathrm{St}_G$ and let Σ denote one of the representations $\iota_B^G \, \delta_B^{\pm 1/2}$. Writing 1_F for the trivial character of $F^\times$ and $\zeta_F(s) = (1-q^{-s})^{-1}$, we get

$$\begin{aligned}\gamma(\Sigma, s, \psi) &= \gamma(1_F, s+\tfrac{1}{2}, \psi)\,\gamma(1_F, s-\tfrac{1}{2}, \psi)\\ &= q^{2s-1}\,\zeta_F(\tfrac{1}{2}-s)\zeta_F(\tfrac{3}{2}-s)/\zeta_F(s-\tfrac{1}{2})\zeta_F(s+\tfrac{1}{2})\\ &= -q^{s-\frac{1}{2}}\,\zeta_F(\tfrac{3}{2}-s)/\zeta_F(s+\tfrac{1}{2}) = \gamma(\pi, s, \psi).\end{aligned}$$

Since $\pi \cong \check{\pi}$ (9.10.5), we have $\gamma(\pi, s, \psi) = \varepsilon(\pi, s, \psi)L(\pi, s)/L(\pi, 1-s)$, so there are only the possibilities $L(\pi, s) = \zeta_F(s-\frac{1}{2})\zeta_F(s+\frac{1}{2})$ or $L(\pi, s) = \zeta_F(s+\frac{1}{2})$. A similar remark applies to $\phi \cdot \mathrm{St}_G$, where ϕ is unramified. We prove:

Proposition. *Let ϕ be an unramified character of $F^\times$ and let $\pi = \phi \cdot \mathrm{St}_G$; then*

$$L(\pi, s) = L(\phi, s+\tfrac{1}{2}) \quad \textit{and} \quad \varepsilon(\pi, s, \psi) = -\varepsilon(\phi, s, \psi).$$

Proof. The formula for ε follows from that for the L-function and the relation $\gamma(\pi, s, \psi) = \gamma(\Sigma, s, \psi)$. To determine the L-function, we again need only treat the case where ϕ is trivial. We take $f \in \mathcal{C}(\pi)$ and consider the zeta-integral

$$\zeta(\Phi, f, s) = \int_G \Phi(g) f(g)\, \|\det g\|^s \, dg. \tag{26.8.1}$$

The key point is:

Lemma. *The integral* (26.8.1) *converges absolutely for* $\mathrm{Re}\, s > \frac{1}{2}$ *and any* $\Phi \in C_c^\infty(A)$.

Proof. We first choose Haar measures $d^\times x$ on $Z = F^\times$ and $d\dot{g}$ on G/Z so that $dg = d^\times x \, d\dot{g}$. Since we are only concerned with absolute convergence, we may as well take Φ to be the characteristic function of $\mathrm{M}_2(\mathfrak{o})$ and s real. We consider the integral

$$\int_{G/Z} |f(g)| \int_{F^\times} \Phi(gx)\, \|\det gx\|^s \, d^\times x \, d\dot{g}.$$

The inner integral, call it $\mathcal{J}_s(g)$, depends only on $ZKgK$, where $K = \mathrm{GL}_2(\mathfrak{o})$, so we may take

$$g = \begin{pmatrix} \varpi^a & 0 \\ 0 & 1 \end{pmatrix},$$

for an integer $a \geqslant 0$ and a prime element ϖ of F. For this element g, we get

$$\mathcal{J}_s(g) = \int_{F^\times} \Phi(gx)\, \|\det gx\|^s \, d^\times x = c_1\, q^{-as}(1-q^{-2s})^{-1},$$

where c_1 is a positive constant depending on choice of measure. For $a \neq 0$, the coset KgK has measure $q^{a-1}(q+1)\mu(K)$. The function $\mathcal{J}_s(g)$ is therefore square-integrable over G/Z, provided $\mathrm{Re}\, s > \frac{1}{2}$. The coefficient $f(g)$ is square-integrable (17.5 Theorem), so the lemma is proved. □

It follows that the function $\zeta(\Phi, f, s+\frac{1}{2})$ is holomorphic in the region $\operatorname{Re} s > 0$. The same therefore applies to $L(\pi, s)$ and the proposition is proved. □

We have also finished the proof of 26.1 Theorem. □

Exercise. *Let $\mathfrak{A}$ be a chain order in A and let Ξ be an irreducible smooth representation of $\mathcal{K}_{\mathfrak{A}}$ of level $n \geqslant 0$. One can define the Gauss sum $\tau(\Xi, \psi)$ exactly as in* 25.1. *The representation Ξ is called* nondegenerate *if $n = 0$, or, if $n \geqslant 1$, $\Xi \mid U_{\mathfrak{A}}^n$ contains a character ψ_c with $c \in \mathcal{K}_{\mathfrak{A}}$.*

(1) *Show that $\tau(\Xi, \psi) \neq 0$ if and only if Ξ is nondegenerate.*
(2) *Let π be an irreducible smooth representation of G such that*
  (a) *π contains a nondegenerate representation Ξ of $\mathcal{K}_{\mathfrak{A}}$;*
  (b) *$L(\pi, s) = 1$.*

Show that $\varepsilon(\pi, s, \psi)$ is given by the formula (25.2.1).

Exercise/Remark. *Show that an irreducible representation π of V is spherical* (Exercises, §17) *if and only if the polynomial $P_\pi(X)$ has degree* 2.

27. Converse Theorem

We prove a fundamental theorem, on which everything depends. It is the analogue, for representations of $\mathrm{GL}_2(F)$, of the famous theorem of Hecke on classical modular forms which can claim to be the starting point of the entire theory.

27.1. We recall that, if χ is a character of $F^\times$ and π is an irreducible smooth representation of G, then $\chi\pi$ denotes the representation $g \mapsto \chi(\det g)\pi(g)$. The result to be proved is:

Converse Theorem. *Let $\psi \in \widehat{F}$, $\psi \neq 1$. Let π_1, π_2 be irreducible smooth representations of $G = \mathrm{GL}_2(F)$. Suppose that*

$$L(\chi\pi_1, s) = L(\chi\pi_2, s) \quad \text{and} \quad \varepsilon(\chi\pi_1, s, \psi) = \varepsilon(\chi\pi_2, s, \psi),$$

for all characters χ of $F^\times$. We then have $\pi_1 \cong \pi_2$.

27.2. We can separate the proof into two parts, using the following:

Proposition. *An irreducible smooth representation π of G is cuspidal if and only if $L(\phi\pi, s) = 1$, for all characters ϕ of $F^\times$.*

Proof. If π is cuspidal, then so is $\phi\pi$ and $L(\phi\pi, s) = 1$ (Corollary 24.5). Conversely, suppose that π is not cuspidal, and hence a composition factor of $\iota_B^G\,\chi$, for some character $\chi = \chi_1 \otimes \chi_2$ of T. The representation $\phi\pi$ is then a composition factor of $\iota_B^G\,\phi\chi$, and $\phi\chi = \phi\chi_1 \otimes \phi\chi_2$. We choose $\phi = \chi_2^{-1}$, and 26.1 Theorem then gives $L(\phi\pi, s) \neq 1$. □

27.3. In the theorem, therefore, if one of the π_i is in the principal series then so is the other. To prove the theorem for such representations, we show that an irreducible, principal series representation π of G is determined by the function $\phi \mapsto L(\phi\pi, s)$ of characters ϕ of $F^\times$. We do this using the list of L-functions in 26.1 Theorem.

We can assume to start with that π has been chosen, within the set of twists $\phi\pi$, so that $L(\pi, s) \neq 1$. If $L(\pi, s)$ has degree 2, it is of the form $L(\chi_1, s)L(\chi_2, s)$, for unramified characters χ_i of $F^\times$. The set $\{\chi_1, \chi_2\}$ is determined by the function $L(\chi_1, s)L(\chi_2, s)$ (23.4 Corollary 1). If $\chi_1\chi_2^{-1}$ is not one of the characters $x \mapsto \|x\|^{\pm 1}$, the representation $\boldsymbol{\iota}_B^G(\chi_1 \otimes \chi_2)$ is irreducible, hence equivalent to π, and it is the only irreducible representation of G with L-function $L(\chi_1, s)L(\chi_2, s)$. If, on the other hand, $\chi_1\chi_2^{-1}$ is one of the characters $x \mapsto \|x\|^{\pm 1}$, then π is the one-dimensional composition factor of the reducible representation $\boldsymbol{\iota}_B^G(\chi_1 \otimes \chi_2)$.

We therefore assume that the non-trivial function $L(\pi, s)$ is of degree 1, i.e., $L(\pi, s) = L(\theta, s)$, for some unramified character θ of $F^\times$. There are only two cases giving rise to such L-functions. In the first, π is an irreducible representation $\boldsymbol{\iota}_B^G(\theta' \otimes \theta)$, where θ' is a ramified character of $F^\times$. In the second, $\pi = \theta' \cdot \mathrm{St}_G$, where θ' is the character $x \mapsto \theta(x)\|x\|^{-1/2}$.

We distinguish these cases as follows. In the first, there is a *ramified* character ϕ of $F^\times$ such that $L(\phi\pi, s) \neq 1$: for example, we could take $\phi^{-1} = \theta'$. In the second, by contrast, we have $L(\phi\pi, s) = 1$ for all ramified characters ϕ.

In the first case, we recover the character θ' as follows. We choose a ramified character ϕ such that $L(\phi\pi, s) \neq 1$. Thus $L(\phi\pi, s) = L(\theta'', s)$, for a uniquely determined unramified character θ'' of $F^\times$. We then have $\theta' = \phi^{-1}\theta''$.

This completes the proof of 27.1 Theorem for non-cuspidal representations.

27.4. We turn to cuspidal representations. We start with a minor simplification.

Lemma. *Let π_1, π_2 be irreducible cuspidal representations of G, and suppose there exists $\nu \in \widehat{F}$, $\nu \neq 1$, such that $\varepsilon(\chi\pi_1, s, \nu) = \varepsilon(\chi\pi_2, s, \nu)$, for all characters χ of $F^\times$. We then have $\omega_{\pi_1} = \omega_{\pi_2}$ and $\varepsilon(\chi\pi_1, s, \mu) = \varepsilon(\chi\pi_2, s, \mu)$ for all χ and all $\mu \in \widehat{F}$, $\mu \neq 1$.*

Proof. Let χ be a character of $F^\times$ of level $m(\chi)$. Assuming $m(\chi) \geqslant 1$, there exists $\delta(\chi) \in F^\times$ such that $\chi(1+x) = \nu(\delta(\chi)x)$, $x \in \mathfrak{p}^{[m(\chi)/2]+1}$. If $\|\delta(\chi)\|$ is sufficiently large, we have (25.7 Theorem)

$$\varepsilon(\chi\pi_i, \tfrac{1}{2}, \nu) = \omega_{\pi_i}^{-1}(\delta(\chi))\, \varepsilon(\chi \circ \det, \tfrac{1}{2}, \nu).$$

The relation $\varepsilon(\chi\pi_1, \frac{1}{2}, \nu) = \varepsilon(\chi\pi_2, \frac{1}{2}, \nu)$ implies that the characters ω_{π_i} agree on all elements of $F^\times$ of sufficiently large valuation. Therefore $\omega_{\pi_1} = \omega_{\pi_2}$. The final assertion follows from 24.3 Proposition. □

27.5. We need to make some preparations. We fix a chain order $\mathfrak{A}$, with radical $\mathfrak{P}$, and an integer $n \geqslant 0$ relatively prime to $e = e_{\mathfrak{A}}$. We abbreviate $\mathcal{K} = \mathcal{K}_{\mathfrak{A}}$, and put

$$\begin{aligned} \mathcal{K}(n) &= \{g \in \mathcal{K} : g\mathfrak{A} = \mathfrak{P}^{-n}\}, \\ F(n) &= \{x \in F^{\times} : \upsilon_F(x) = -n\}. \end{aligned}$$

We record some elementary identities:

Lemma.

(1) *If* $x \in \mathcal{K}(n)$, *then* $\det x \in F(2n/e)$ *and* $\upsilon_F(\operatorname{tr}_A x) \geqslant -[n/e]$.
(2) *We have* $\det U_{\mathfrak{A}}^{n+1} = U_F^{[n/e]+1}$ *and* $\operatorname{tr}_A(\mathfrak{P}) = \mathfrak{p}$.

Proof. This follows easily from the explicit forms (12.1.2). □

The group $\mathcal{K}$ acts by conjugation on the coset space $\mathcal{K}(n)/U_{\mathfrak{A}}^{n+1}$: we denote the orbit space by $\operatorname{Ad}\mathcal{K}\backslash\mathcal{K}(n)/U_{\mathfrak{A}}^{n+1}$. We have a map

$$\begin{aligned} \Psi : \operatorname{Ad}\mathcal{K}\backslash\mathcal{K}(n)/U_{\mathfrak{A}}^{n+1} &\longrightarrow F(2n/e)/U_F^{[n/e]+1} \times \mathfrak{p}^{-[n/e]}/\mathfrak{p}, \\ xU_{\mathfrak{A}}^{n+1} &\longmapsto \big(\det x, \operatorname{tr}_A x\big). \end{aligned}$$

Proposition.

(1) *Let* $x, y \in \mathcal{K}(n)$, *and suppose that* x *is minimal over* F. *If* $\Psi(yU_{\mathfrak{A}}^{n+1}) = \Psi(xU_{\mathfrak{A}}^{n+1})$, *then* y *is minimal over* F.
(2) *Let* $x \in \mathcal{K}(n)$ *be minimal over* F. *The number of cosets* $yU_{\mathfrak{A}}^{n+1}$ *contained in the* $\operatorname{Ad}\mathcal{K}$*-orbit of* $xU_{\mathfrak{A}}^{n+1}$ *depends only on* n/e.
(3) *The map* Ψ *is injective on orbits of minimal elements; it is bijective in the case* $e = 2$.

Proof. In the case $e = 2$, all elements of $\mathcal{K}(n)$ are minimal over F. If $e = 1$, the minimality of $x \in \mathcal{K}(n)$ is expressed in terms of the characteristic polynomial of $\varpi^n x$ modulo $\mathfrak{p}$, where ϖ is a prime element of F. Assertion (1) follows.

In (2), the number of cosets in question is the index in $\mathcal{K}/U_{\mathfrak{A}}^{n+1}$ of the centralizer of $xU_{\mathfrak{A}}^{n+1}$. Lemma 16.2 shows that this centralizer is $F[x]^{\times}U_{\mathfrak{A}}^{n+1}$, and the assertion follows readily.

In (3), we take $x \in \mathcal{K}(n)$, minimal over F. We set $d = \det x$, $t = \operatorname{tr}_A x$. We can assume that $\mathfrak{A}$ is $\mathfrak{M}$ or $\mathfrak{I}$. Using rational canonical form (5.3), x is G-conjugate to a matrix

$$x' = \begin{pmatrix} 0 & -d \\ 1 & t \end{pmatrix}.$$

We conjugate x' by a diagonal matrix $\left(\begin{smallmatrix} \varpi^a & 0 \\ 0 & 1 \end{smallmatrix}\right)$, to get a matrix

$$x'' = \begin{pmatrix} 0 & -\varpi^a d \\ \varpi^{-a} & t \end{pmatrix}.$$

Taking $a = [n/e]$, we get $x'' \in \mathcal{K}$. Since x is minimal over F, any element of G conjugating x to x'' lies in $\mathcal{K}$. That is, x is $\mathcal{K}$-conjugate to x''. Likewise, if $y \in \mathcal{K}(n)$ is minimal, it is $\mathcal{K}$-conjugate to the matrix

$$y'' = \begin{pmatrix} 0 & -\varpi^a d' \\ \varpi^{-a} & t' \end{pmatrix}, \quad a = [n/e],$$

where $d' = \det y$, $t' = \mathrm{tr}_A\, y$. The hypothesis $\Psi(yU_{\mathfrak{A}}^{n+1}) = \Psi(xU_{\mathfrak{A}}^{n+1})$ is then equivalent to $y'' \equiv x'' \pmod{U_{\mathfrak{A}}^{n+1}}$, as required for (3).

In the case $e = 2$, we have already remarked that $\mathcal{K}(n)$ consists solely of minimal elements. Surjectivity of Ψ in this case follows easily: one simply writes down a matrix of the form $\left(\begin{smallmatrix} 0 & * \\ * & * \end{smallmatrix}\right)$, lying in $\mathcal{K}(n)$, with the desired determinant and trace. □

27.6. Let π be an irreducible cuspidal representation of G satisfying $\ell(\pi) \leqslant \ell(\chi\pi)$, for all characters χ of $F^\times$. The representation π contains a cuspidal inducing datum $(\mathfrak{A}, \Xi)$, unique up to conjugacy. Setting $n = \ell_{\mathfrak{A}}(\Xi)$ (15.8.1), we have $\gcd(n, e_{\mathfrak{A}}) = 1$ and $\ell(\pi) = n/e_{\mathfrak{A}}$. In particular, the rational number $\ell(\pi)$ determines the conjugacy class of the chain order $\mathfrak{A}$. It determines rather more:

Lemma. *Let π be an irreducible cuspidal representation of G such that $\ell(\pi) \leqslant \ell(\chi\pi)$, for all characters χ of $F^\times$. Let $(\mathfrak{A}, \Xi)$ be a cuspidal inducing datum contained in π, and set $n = \ell_{\mathfrak{A}}(\Xi)$. We then have:*

$$\dim \Xi = \begin{cases} (q-1)q^n & \text{if } \ell(\pi) = n \in \mathbb{Z}, \\ (q-1)q^{(n-1)/2} & \text{if } \ell(\pi) = n/2 \notin \mathbb{Z}. \end{cases}$$

Proof. Suppose first that $\ell(\pi) \notin \mathbb{Z}$. Thus n is odd and we may take $\mathfrak{A} = \mathfrak{J}$. There is a simple stratum $(\mathfrak{J}, n, \alpha)$, and a cuspidal type $(\mathfrak{J}, J_\alpha, \Lambda)$ which induces Ξ. We have $J_\alpha = F[\alpha]^\times U_{\mathfrak{J}}^{[n/2]+1}$ and $\dim \Lambda = 1$ (15.6). The field extension $F[\alpha]/F$ is totally ramified, and $\dim \Xi = (\mathcal{K}_{\mathfrak{J}} : J_\alpha) = (q-1)q^{(n-1)/2}$, as required.

If $\ell(\pi) = n \in \mathbb{Z}$ and n is odd, a similar calculation applies, except that the field extension $F[\alpha]/F$ is unramified.

Consider next the case $\ell(\pi) = 0$. We may take $\mathfrak{A} = \mathfrak{M}$. The representation $\Xi \mid U_{\mathfrak{M}}$ is the inflation of an irreducible cuspidal representation of $\mathrm{GL}_2(\boldsymbol{k})$, hence of dimension $q-1$ (6.4).

Finally, we have the case where $\ell(\pi) = n \in \mathbb{Z}$, $n = 2m > 0$. We can take $\mathfrak{A} = \mathfrak{M}$; the representation Ξ is induced from a cuspidal type $(\mathfrak{M}, J_\alpha, \Lambda)$ based on a simple stratum $(\mathfrak{M}, n, \alpha)$. We have $J_\alpha = F[\alpha]^\times U_{\mathfrak{M}}^m$, the field extension $F[\alpha]/F$ is unramified, and $\dim \Lambda = q$ (15.6). The result follows. □

27.7. If $r \geqslant 0$ is an integer, let Γ_r denote the group of characters of $F^\times$ which are trivial on U_F^r.

Theorem. *Let π be an irreducible cuspidal representation of G such that $\ell(\pi) \leqslant \ell(\chi\pi)$, for all characters χ of $F^\times$. Let $(\mathfrak{A}, \Xi)$ be a cuspidal inducing datum occurring in π and write $n = \ell_{\mathfrak{A}}(\Xi)$, $e = e_{\mathfrak{A}}$ and $r = [n/e]$.*

Let ψ be a character of F of level 1. There is a constant $k > 0$, depending only on $\ell(\pi) = n/e$, such that

$$\operatorname{tr} \check{\Xi}(x) = k \sum_{\substack{\chi \in \Gamma_{r+1}/\Gamma_0, \\ c \in U_F/U_F^{r+1}}} \varepsilon(\chi\pi, \tfrac{1}{2}, c\psi)\, \chi(\det x)\, c\psi(-\operatorname{tr}_A x), \tag{27.7.1}$$

for every $x \in \mathcal{K}(n)$ which is minimal over F.

Proof. We choose Haar measure on G so that $U_{\mathfrak{A}}$ has measure 1. For $\chi \in \Gamma_{r+1}$ and $\mu \in \widehat{F}$ of level one, (25.2.2) gives a constant $k_1 > 0$ such that

$$\int_{\mathcal{K}(n)} \operatorname{tr} \check{\Xi}(x)\, \chi(\det x)^{-1} \mu(\operatorname{tr}_A x)\, dx = k_1\, \varepsilon(\chi\pi, \tfrac{1}{2}, \mu).$$

By 27.6 Lemma, the constant k_1 depends only on $\ell(\pi)$.

If, on the other hand, $\mu \in \widehat{F}$ has level $\leqslant 0$, (25.3.2) gives

$$\int_{\mathcal{K}(n)} \operatorname{tr} \check{\Xi}(x)\, \chi(\det x)^{-1} \mu(\operatorname{tr}_A x)\, dx = 0.$$

For $a \in F(2n/e)/U_F^{r+1}$ and $b \in \mathfrak{p}^{-r}/\mathfrak{p}$, we define

$$\mathfrak{e}(a, b) = \sum_x \operatorname{tr} \check{\Xi}(x),$$

where x ranges over the elements of $\mathcal{K}(n)/U_{\mathfrak{A}}^{n+1}$ such that $\Psi(x) = (a, b)$. By 27.5 Proposition, there is a constant $k_2 > 0$, depending only on $\ell(\pi)$, such that

$$\operatorname{tr} \check{\Xi}(x) = k_2\, \mathfrak{e}(\det x, \operatorname{tr}_A x)$$

for all *minimal* $x \in \mathcal{K}(n)$.

We define the Fourier transform

$$\Upsilon(\chi, \mu) = \sum_{\substack{a \in F(2n/e)/U_F^{r+1}, \\ b \in \mathfrak{p}^{-r}/\mathfrak{p}}} \mathfrak{e}(a, b)\, \chi(a)^{-1}\, \mu(b),$$

for $\chi \in \Gamma_{r+1}$ and $\mu \in (\mathfrak{p}^{-r}/\mathfrak{p})\widehat{\ }$. There is then a constant $k_3 > 0$, depending only on $\ell(\pi)$, such that

$$\Upsilon(\chi, \mu) = k_3 \int_{\mathcal{K}(n)} \operatorname{tr} \check{\Xi}(x)\, \chi(\det x)^{-1} \mu(\operatorname{tr}_A x)\, dx.$$

Therefore

$$\Upsilon(\chi,\mu) = \begin{cases} k_4\,\varepsilon(\chi\pi,\frac{1}{2},\mu) & \text{if } \mu \text{ has level one,} \\ 0 \quad \text{otherwise,} \end{cases}$$

where again $k_4 > 0$ depends only on $\ell(\pi)$. We take the Fourier transform of Υ on the finite abelian group $\Gamma_{r+1}/\Gamma_0 \times (\mathfrak{p}^{-r}/\mathfrak{p})\widehat{\ }$. To get the result, it remains only to observe that the elements of $(\mathfrak{p}^{-r}/\mathfrak{p})\widehat{\ }$ of level one are the characters $c\psi$, for $c \in U_F/U_F^{r+1}$. □

27.8. We return to the Converse Theorem. We are given irreducible cuspidal representations π_1, π_2 of G, satisfying

$$\varepsilon(\chi\pi_1, s, \psi) = \varepsilon(\chi\pi_2, s, \psi), \tag{27.8.1}$$

for all characters χ of $F^\times$. We may assume that ψ has level one (27.4).

We may also assume that $\ell(\pi_1) \leqslant \ell(\chi\pi_1)$ for all χ. (25.2.1) gives

$$\varepsilon(\chi\pi_i, s, \psi) = q^{2\ell(\chi\pi_i)(\frac{1}{2}-s)}\,\varepsilon(\chi\pi_i, \tfrac{1}{2}, \psi).$$

We deduce that $\ell(\pi_1) = \ell(\pi_2) \leqslant \ell(\chi\pi_2)$ for all χ. Thus there is a chain order $\mathfrak{A}$ and a cuspidal inducing datum $(\mathfrak{A}, \Xi_i)$ such that π_i contains Ξ_i, $i = 1, 2$. We set $n = \ell_{\mathfrak{A}}(\Xi_i)$, so that $\ell(\pi_i) = n/e_{\mathfrak{A}}$. Theorem 27.7 gives

$$\operatorname{tr}\check{\Xi}_1(x) = \operatorname{tr}\check{\Xi}_2(x), \tag{27.8.2}$$

for all $x \in \mathcal{K}_{\mathfrak{A}}(n)$ which are minimal over F. We recall (27.4) that

$$\omega_{\pi_1} = \omega_{\pi_2}. \tag{27.8.3}$$

Taking account of (27.8.2) and (27.8.3), the Converse Theorem for cuspidal representations will follow from:

Proposition. *Let $\mathfrak{A}$ be a chain order in A and, for $i = 1, 2$, let $(\mathfrak{A}, \Xi_i)$ be a cuspidal inducing datum such that $\ell_{\mathfrak{A}}(\chi\Xi_i) \geqslant \ell_{\mathfrak{A}}(\Xi_i)$ for all characters χ of $F^\times$. Suppose further that*

(a) *$\ell_{\mathfrak{A}}(\Xi_1) = \ell_{\mathfrak{A}}(\Xi_2) = n$, say;*
(b) *the restrictions $\Xi_i \mid F^\times$ are multiples of the same character ω;*
(c) *$\operatorname{tr}\Xi_1(x) = \operatorname{tr}\Xi_2(x)$, for all F-minimal elements $x \in \mathcal{K}_{\mathfrak{A}}(n)$.*

We then have $\Xi_1 \cong \Xi_2$.

Proof. We have to divide into a small number of cases. Suppose first that $n = 0$. We may then take $\mathfrak{A} = \mathfrak{M} = \mathrm{M}_2(\mathfrak{o})$. The representation $\Xi_i \mid U_{\mathfrak{M}}$ is the inflation of an irreducible cuspidal representation $\tilde{\xi}_i$ of $\mathrm{GL}_2(\boldsymbol{k})$. Using the notation of (6.4), we have $\tilde{\xi}_i = \pi_{\theta_i}$, where θ_i is a regular character of $\boldsymbol{l}^\times$ and

$\boldsymbol{l}/\boldsymbol{k}$ is the quadratic field extension, viewed as a sub-algebra of $\mathrm{M}_2(\boldsymbol{k})$. Using the character table (6.4.1), we get

$$\operatorname{tr} \pi_{\theta_i}(\zeta) = -\big(\theta_i(\zeta)+\theta_i^q(\zeta)\big),$$

for every $\zeta \in \boldsymbol{l} \smallsetminus \boldsymbol{k}$. However, an element $x \in \mathcal{K}_{\mathfrak{M}}(n) = \mathrm{GL}_2(\mathfrak{o})$ is F-minimal if and only if its image in $\mathrm{GL}_2(\boldsymbol{k})$ is conjugate to an element of $\boldsymbol{l} \smallsetminus \boldsymbol{k}$. We conclude that $\theta_1 = \theta_2$ or θ_2^q and hence that $\Xi_1 \mid U_{\mathfrak{M}} \cong \Xi_2 \mid U_{\mathfrak{M}}$. Since $\mathcal{K}_{\mathfrak{M}} = F^\times U_{\mathfrak{M}}$, we deduce that $\Xi_1 \cong \Xi_2$, as required.

We therefore assume $n \geqslant 1$, and use the machinery of §15. There is a cuspidal type $(\mathfrak{A}, J, \Lambda_1)$ which induces Ξ_1. The hypothesis $\ell_{\mathfrak{A}}(\Xi_1) \leqslant \ell_{\mathfrak{A}}(\chi\Xi_1)$ implies that there is a simple stratum $(\mathfrak{A}, n, \alpha)$ such that $J = J_\alpha$ and $\Lambda_1 \in C(\psi_\alpha, \mathfrak{A})$ (all notation as in 15.3, 15.5). Write $E = F[\alpha]$, $H_\alpha^1 = U_E^1 U_{\mathfrak{A}}^{[n/2]+1}$. The restriction $\Lambda_1 \mid H_\alpha^1$ is then a multiple of a character θ. We take a Haar measure μ_G on G; the character of the representation Λ_1 is then given by

$$\operatorname{tr} \Lambda_1(j) = \mu_G(H_\alpha^1)^{-1} \int_{H_\alpha^1} \operatorname{tr} \Xi_1(ju)\, \theta(u)^{-1}\, d\mu_G(u), \quad j \in J_\alpha. \tag{27.8.4}$$

Lemma. *There exists $j \in J_\alpha \cap \mathcal{K}_{\mathfrak{A}}(n)$ which is minimal over F and such that* $\operatorname{tr} \Lambda_1(j) \neq 0$.

Proof. If n is odd, then $\dim \Lambda_1 = 1$ (15.6 Proposition 1) and $\operatorname{tr} \Lambda_1(j) \neq 0$ for all $j \in J_\alpha$.

Suppose therefore that n is even. Thus E/F is unramified. According to the Tame Parametrization Theorem (20.2), the cuspidal type Λ_1 is constructed from an admissible pair following the procedure of 19.4: such types certainly have the desired property. □

For an element j as in the Lemma, we have

$$\int_{H_\alpha^1} \operatorname{tr} \Xi_2(ju)\, \theta(u)^{-1}\, d\mu_G(u) = \int_{H_\alpha^1} \operatorname{tr} \Xi_1(ju)\, \theta(u)^{-1}\, d\mu_G(u) \neq 0.$$

We conclude that Ξ_2 contains the character θ of H_α^1 and hence a cuspidal type $(\mathfrak{A}, J_\alpha, \Lambda_2)$, with $\Lambda_2 \mid J_\alpha^1 \cong \Lambda_1 \mid J_\alpha^1$ (15.6). Since the central characters agree, the irreducible representations $\Lambda_i \mid F^\times J_\alpha^1$ are the same. However, the set $J_\alpha \smallsetminus F^\times J_\alpha^1$ consists of minimal elements: in all cases, it is the set of minimal elements in $F^\times \mathcal{K}_{\mathfrak{A}}(n)$. The character formula (27.8.4), and its analogue for Λ_2, implies that the characters $\operatorname{tr} \Lambda_i$ also agree on $J_\alpha \smallsetminus F^\times J_\alpha^1$. This implies $\Lambda_1 \cong \Lambda_2$ and $\Xi_1 \cong \Xi_2$, as required. □

This finishes the proof of the Converse Theorem (27.1). We give a different proof, for cuspidal representations, in §37 below: see, in particular, 37.3 Remark.

Further reading. The material of §23 is based on Tate's thesis [81], although the approach we take here is more representation-theoretic and suggested by ideas of Weil [86]. The global context is also treated in [81] in an illuminating manner. §24 is based on [35] and [46], but we have taken a more algebraic approach, as in §23. The arguments of [35] are valid for $\mathrm{GL}_n(F)$ (and more). The same applies to our modified version. The description of cuspidal local constants in terms of Gauss sums is likewise valid for $\mathrm{GL}_n(F)$ and has its origins in [12], [11].

7

Representations of Weil Groups

We now transfer attention to the arithmetic of the field F, expressed via the finite-dimensional, smooth representations of its Weil group.

In order to define the Weil group $\mathcal{W}_F$ of F, we start by recalling some basic results from Galois theory. We discuss the category of representations of $\mathcal{W}_F$ in which we shall be interested. We then recall, without proofs, the elements of local class field theory.

The core of the chapter concerns the theory of L-functions and local constants attached to finite-dimensional, semisimple, smooth representations of $\mathcal{W}_F$. This theory generalizes that of §23, but in a direction different from the main thrust of Chapter VI. Local class field theory sets up a canonical isomorphism between the groups of characters of $\mathcal{W}_F$ and of $F^\times$, with the result that $L(\rho, s)$, $\varepsilon(\rho, s, \psi)$ are defined for representations ρ of $\mathcal{W}_F$ of dimension one. The objective is to extend to arbitrary finite dimension.

For the L-function, this is quite straightforward, albeit guided by global experience. The local constant is, in this situation, specified by a short list of formal properties concerning its interaction with the process of induction of representations of Weil groups (of finite extension fields of F). The proof that such an object exists is quite elaborate. The only practical way known for dealing with it involves transferring the problem to *global* fields, where some strong properties of L-functions can be invoked. We cannot assemble here the necessary apparatus from algebraic number theory; our account is brief but, for a reader familiar with such matters, it is complete.

There is another aspect. Our aim is to relate the irreducible smooth representations of $G = \mathrm{GL}_2(F)$ to the 2-dimensional representations of $\mathcal{W}_F$: we give the precise statement in the next chapter. It turns out that the standard concept of representation of $\mathcal{W}_F$ is inadequate to this rôle, reflecting the reducibility of certain principal series representations of G. One has to expand the framework by introducing the "Deligne representations" of $\mathcal{W}_F$: these are ordinary representations with a further element of structure.

This approach gives an imperfect idea of the significance of Deligne representations, as they arise quite naturally in a broader context. In many areas of arithmetic or algebraic geometry, one is led to consider continuous representations of $\mathcal{W}_F$ not over $\mathbb{C}$, but over the algebraic closure $\overline{\mathbb{Q}}_\ell$ of the field $\mathbb{Q}_\ell$ of ℓ-adic numbers, $\ell \neq p$. These so-called ℓ-adic representations of $\mathcal{W}_F$ are parametrized by Deligne representations over $\mathbb{C}$. We have included the argument, as it gives a more satisfactory reason for the introduction of the Deligne representations. It also provides a broader context in which to consider the Langlands correspondence and raise further issues in the next chapter.

28. Weil Groups and Representations

We start by recalling some features of the Galois theory of F. In the first five paragraphs, the proofs are all standard or reasonably straightforward, so we have omitted them. Throughout, p means the characteristic of the residue class field $\boldsymbol{k} = \mathfrak{o}/\mathfrak{p}$.

28.1. We choose a separable algebraic closure $\overline{F}$ of F. We set $\Omega_F = \mathrm{Gal}(\overline{F}/F)$, and view it as a profinite group with its natural (Krull) topology. Thus

$$\Omega_F = \varprojlim \mathrm{Gal}(E/F),$$

where E/F ranges over finite Galois extensions with $E \subset \overline{F}$.

If K/F is a finite field extension, with $K \subset \overline{F}$, we denote by Ω_K the open subgroup $\mathrm{Gal}(\overline{F}/K)$ of Ω_F.

If $\widetilde{F}/F$ is some other separable algebraic closure of F, an F-isomorphism $\widetilde{F} \cong \overline{F}$ induces an isomorphism $\mathrm{Gal}(\widetilde{F}/F) \cong \Omega_F$. This does depend on the choice of isomorphism $\widetilde{F} \cong \overline{F}$, but only up to inner automorphism. In particular, it induces a *canonical* bijection between the sets of equivalence classes of irreducible smooth representations of the two groups $\mathrm{Gal}(\widetilde{F}/F)$, Ω_F.

28.2. For the time being, we consider only extension fields E/F with $E \subset \overline{F}$. With this convention, the field F admits a unique unramified extension F_m/F

of degree m, for each $m \geqslant 1$. Let F_∞ denote the composite of all these fields. Thus F_∞/F is the unique maximal unramified extension of F.

The extension F_m/F is Galois and $\mathrm{Gal}(F_m/F)$ is cyclic. An F-automorphism of F_m is determined by its action on the residue field $\boldsymbol{k}_{F_m} \cong \mathbb{F}_{q^m}$. In particular, there is a unique element ϕ_m of $\mathrm{Gal}(F_m/F)$ which acts on $\boldsymbol{k}_{F_m}$ as $x \mapsto x^q$. We set $\Phi_m = \phi_m^{-1}$. The map $\Phi_m \mapsto 1$ gives a canonical isomorphism $\mathrm{Gal}(F_m/F) \to \mathbb{Z}/m\mathbb{Z}$. Taking the limit over m, we get a canonical isomorphism of topological groups

$$\mathrm{Gal}(F_\infty/F) \cong \varprojlim{}_{m\geqslant 1}\, \mathbb{Z}/m\mathbb{Z}, \tag{28.2.1}$$

and a unique element $\Phi_F \in \mathrm{Gal}(F_\infty/F)$ which acts on F_m as Φ_m, for all m. The element Φ_F is the *geometric Frobenius substitution* on F_∞. (Its inverse ϕ_F is the *arithmetic* Frobenius substitution.) An element of Ω_F is called a *geometric Frobenius element* (over F) if its image in $\mathrm{Gal}(F_\infty/F)$ is Φ_F.

It is conventional to denote the projective limit in (28.2.1) by $\widehat{\mathbb{Z}}$. The Chinese Remainder Theorem gives a canonical isomorphism of topological groups

$$\widehat{\mathbb{Z}} \cong \prod_{\ell} \mathbb{Z}_\ell,$$

where ℓ ranges over all prime numbers and $\mathbb{Z}_\ell$ is the (additive) group of ℓ-adic integers.

We set $\mathcal{I}_F = \mathrm{Gal}(\overline{F}/F_\infty)$: this is called the *inertia group* of F. We have an exact sequence of topological groups

$$1 \to \mathcal{I}_F \longrightarrow \Omega_F \longrightarrow \widehat{\mathbb{Z}} \to 0.$$

28.3. For each integer $n \geqslant 1$, $p \nmid n$, the field F_∞ has a unique extension E_n/F_∞ of degree n. If ϖ is a prime element of F, the extension E_n is generated over F_∞ by an n-th root of ϖ.

The extension E_n/F_∞ is cyclic, and the Galois group $\mathrm{Gal}(E_n/F_\infty)$ admits a canonical description. Choose $\alpha \in E_n$ such that $\alpha^n = \varpi$. The map

$$\begin{aligned} \mathrm{Gal}(E_n/F_\infty) &\longrightarrow \boldsymbol{\mu}_n, \\ \sigma &\longmapsto \sigma(\alpha)/\alpha, \end{aligned} \tag{28.3.1}$$

gives a canonical isomorphism of $\mathrm{Gal}(E_n/F_\infty)$ with the group $\boldsymbol{\mu}_n$ of n-th roots of unity in F_∞.

The composite E_∞ of all extensions E_n/F_∞ is the maximal tamely ramified extension of F. Taking the limit of the isomorphisms (28.3.1), we get topological isomorphisms

$$\mathrm{Gal}(E_\infty/F_\infty) \cong \varprojlim \boldsymbol{\mu}_n \cong \prod_{\ell \neq p} \mathbb{Z}_\ell. \tag{28.3.2}$$

(In the limit, n ranges over all positive integers not divisible by p; in the product, ℓ ranges over all prime numbers other than p.)

Let $\mathcal{P}_F = \mathrm{Gal}(\overline{F}/E_\infty)$; one calls $\mathcal{P}_F$ the *wild inertia group* of F. Any finite extension of E_∞ has degree a power of p, so the profinite group $\mathcal{P}_F$ is a pro p-group. It is the unique pro p-Sylow subgroup[1] of $\mathcal{I}_F$.

The isomorphism $\mathcal{I}_F/\mathcal{P}_F \cong \prod_{\ell\neq p}\mathbb{Z}_\ell$ (28.3.2) is uniquely determined up to multiplication by an element of $\prod\mathbb{Z}_\ell^\times$.

The group $\mathrm{Gal}(F_\infty/F)$ acts on $\mathrm{Gal}(E_\infty/F_\infty)$ by conjugation. An isomorphism (28.3.2) transfers this action to one on the group $\prod_{\ell\neq p}\mathbb{Z}_\ell$. Tracking through from (28.3.1), we get:

Proposition. *If $t_0 : \mathcal{I}_F/\mathcal{P}_F \to \prod_{\ell\neq p}\mathbb{Z}_\ell$ is a topological isomorphism, then*

$$t_0(\Phi_F\sigma\Phi_F^{-1}) = q^{-1}t_0(\sigma), \quad \sigma\in\mathcal{I}_F/\mathcal{P}_F. \tag{28.3.3}$$

28.4. Let ${}_a\mathcal{W}_F$ denote the inverse image in Ω_F of the cyclic subgroup $\langle\Phi_F\rangle$ of $\mathrm{Gal}(F_\infty/F)$ generated by Φ_F. Thus ${}_a\mathcal{W}_F$ is the dense subgroup of Ω_F generated by the Frobenius elements. It is normal in Ω_F and it fits into an exact sequence (of abstract groups)

$$1\to\mathcal{I}_F\longrightarrow {}_a\mathcal{W}_F\longrightarrow\mathbb{Z}\to 0.$$

Definition. *The* Weil group $\mathcal{W}_F$ *of F (relative to $\overline{F}/F$) is the topological group, with underlying abstract group ${}_a\mathcal{W}_F$, so that*

(1) *$\mathcal{I}_F$ is an open subgroup of $\mathcal{W}_F$, and*
(2) *the topology on $\mathcal{I}_F$, as subspace of $\mathcal{W}_F$, coincides with its natural topology as* $\mathrm{Gal}(\overline{F}/F_\infty)\subset\Omega_F$.

Thus $\mathcal{W}_F$ is locally profinite, and the identity map $\iota_F : \mathcal{W}_F \to {}_a\mathcal{W}_F\subset\Omega_F$ is a continuous injection.

The definition of $\mathcal{W}_F$ does depend on the choice of $\overline{F}/F$, but only up to inner automorphism of Ω_F.

We write $v_F : \mathcal{W}_F\to\mathbb{Z}$ for the canonical map taking a geometric Frobenius element to 1 and

$$\|x\| = q^{-v_F(x)}, \quad x\in\mathcal{W}_F.$$

28.5. Let E/F be a finite extension, $E\subset\overline{F}$. The canonical inclusion $\Omega_E\to\Omega_F$ induces a bijection ${}_a\mathcal{W}_E\to{}_a\mathcal{W}_F\cap\Omega_E$. This same map induces a homeomorphism of $\mathcal{W}_E$ with an open subgroup of $\mathcal{W}_F$ of finite index. More formally:

[1] For the notions of "order" and "Sylow subgroup", applied to a profinite group, see [75].

Proposition.

(1) *Let E/F be a finite extension, $E \subset \overline{F}$.*

(a) *The group $\mathcal{W}_F$ has a unique subgroup $\mathcal{W}_F^E$ such that*

$$\iota_F(\mathcal{W}_F^E) = {}_a\mathcal{W}_F \cap \Omega_E.$$

(b) *The subgroup $\mathcal{W}_F^E$ is open and of finite index in $\mathcal{W}_F$; it is normal in $\mathcal{W}_F$ if and only if E/F is Galois.*

(c) *The canonical map $\mathcal{W}_F^E\backslash\mathcal{W}_F \to \Omega_E\backslash\Omega_F$ is a bijection.*

(d) *The canonical map $\iota_E : \mathcal{W}_E \to \Omega_E$ induces a topological isomorphism $\mathcal{W}_E \cong \mathcal{W}_F^E$.*

(2) *The map $E/F \mapsto \mathcal{W}_F^E$ is a bijection between the set of finite extensions E of F inside $\overline{F}$ and the set of open subgroups of $\mathcal{W}_F$ of finite index.*

We henceforward identify $\mathcal{W}_E$ with the subgroup $\mathcal{W}_F^E$ of $\mathcal{W}_F$.

28.6. We consider basic aspects of the representation theory of the locally profinite group $\mathcal{W}_F$. The group Ω_F is profinite, so a smooth representation of Ω_F is semisimple (*cf.* 2.2). On the other hand, $\mathcal{W}_F$ admits the discrete group $\mathbb{Z}$ as a quotient and it therefore has smooth representations which are not semisimple. As we will see, the *irreducible* representations of $\mathcal{W}_F$ are quite closely related to those of Ω_F, but the reducible ones exhibit a greater variety.

We shall work with representations of $\mathcal{W}_F$ over $\mathbb{C}$. However, everything we say remains valid if $\mathbb{C}$ is replaced by any algebraically closed field of characteristic zero which is uncountable: the key point is that the proof of Schur's Lemma (2.6) is valid for such fields.

Lemma 1. *Let (ρ, V) be an irreducible smooth representation of $\mathcal{W}_F$. Then ρ has finite dimension.*

Proof. Let $v \in V$, $v \neq 0$. Since V is irreducible, the $\mathcal{W}_F$-translates of v span V. However, v is fixed by an open subgroup $\mathcal{J}$ of $\mathcal{I}_F$. This is of the form $\mathcal{J} = \mathcal{I}_F \cap \Omega_E$, for some finite extension E/F. We can assume E/F is Galois, and hence that $\mathcal{J}$ is normal in $\mathcal{W}_F$. It follows that $\mathcal{J} \subset \operatorname{Ker}\rho$.

Let $\Phi \in \mathcal{W}_F$ be a Frobenius element. It acts by conjugation on the finite group $\mathcal{I}_F/\mathcal{J}$, so some power Φ^d, $d \geqslant 1$, acts trivially. Thus $\rho(\Phi^d)$ commutes with $\rho(\mathcal{W}_F)$ and, since ρ is irreducible, $\rho(\Phi^d)$ must be a scalar (2.6 Schur's Lemma). It follows that the translates $\rho(x)v$, $x \in \mathcal{W}_F$, span a finite-dimensional space U. This space U is stable under $\rho(\mathcal{W}_F)$, so $U = V$ and $\dim V$ is finite. □

Remark. Let ρ be a finite-dimensional smooth representation of $\mathcal{W}_F$. The restriction of ρ to the profinite group $\mathcal{I}_F$ is thus semisimple, and the group $\rho(\mathcal{I}_F)$ is finite. All complications, therefore, arise from the Frobenius elements of $\mathcal{W}_F$.

Let ρ be an irreducible smooth representation of Ω_F. Thus $\rho \circ \iota_F$ is a smooth representation of $\mathcal{W}_F$ which we think of as the "restriction" of ρ to $\mathcal{W}_F$.

Lemma 2.

(1) *Let ρ be an irreducible smooth representation of Ω_F. The representation $\rho \circ \iota_F$ of $\mathcal{W}_F$ is then irreducible.*
(2) *If ρ_1, ρ_2 are irreducible smooth representations of Ω_F, then $\rho_1 \cong \rho_2$ if and only if $\rho_1 \circ \iota_F \cong \rho_2 \circ \iota_F$.*

Proof. Since $\operatorname{Ker} \rho$ is an open subgroup of Ω_F, we have $\Omega_F = \mathcal{W}_F \cdot \operatorname{Ker} \rho$ and $\rho(\mathcal{W}_F) = \rho(\Omega_F)$. Part (1) follows and, in (2), any $\mathcal{W}_F$-isomorphism $\rho_1 \to \rho_2$ is an Ω_F-isomorphism. □

A character χ of Ω_F or of $\mathcal{W}_F$ is called *unramified* if it is trivial on $\mathcal{I}_F$. If ρ is a smooth, finite-dimensional representation of either group, the map $x \mapsto \det(\rho(x))$ is a character, which we denote $\det \rho$.

Proposition. *Let τ be an irreducible smooth representation of $\mathcal{W}_F$. The following are equivalent:*

(1) *the group $\tau(\mathcal{W}_F)$ is finite;*
(2) *$\tau \cong \rho \circ \iota_F$, for some irreducible smooth representation ρ of Ω_F;*
(3) *the character $\det \tau$ has finite order.*

For any irreducible smooth representation τ of $\mathcal{W}_F$, there is an unramified character χ of $\mathcal{W}_F$ such that $\chi \otimes \tau$ satisfies (1)–(3).

Proof. Let $\Phi \in \mathcal{W}_F$ be a Frobenius element. Assuming (1) holds, the operator $\tau(\Phi)$ has finite order d say. For $a \in \widehat{\mathbb{Z}}$, there is an integer $\bar{a}$ such that $\bar{a} \equiv a \pmod{d\widehat{\mathbb{Z}}}$, and $\bar{a}$ is uniquely determined modulo $d\mathbb{Z}$. As in (28.2.1), an element $\omega \in \Omega_F$ can be written uniquely in the form $\omega = \Phi^a \sigma$, for some $a \in \widehat{\mathbb{Z}}$ and $\sigma \in \mathcal{I}_F$. We define a smooth representation ρ of Ω_F by setting $\rho(\Phi^a \sigma) = \tau(\Phi)^{\bar{a}}\, \tau(\sigma)$. The representation ρ is irreducible and it satisfies $\rho \circ \iota_F = \tau$, so (1) $\Rightarrow$ (2).

Surely (2) $\Rightarrow$ (3), so we assume that $\det \tau$ has finite order. We choose a Frobenius $\Phi \in \mathcal{W}_F$. There is an integer $k \geqslant 1$ such that $\tau(\Phi)^k$ commutes with the finite group $\tau(\mathcal{I}_F)$ and hence with the whole of $\tau(\mathcal{W}_F)$. Thus $\tau(\Phi)^k$ is a scalar. If $\det \tau$ has finite order, then so does the scalar $\tau(\Phi)^k$. Therefore $\tau(\mathcal{W}_F)$ is finite, as required for (1).

In any case, there is an integer $k \geqslant 1$ such that $\tau(\Phi^k)$ is a scalar c, say. If χ is an unramified character of $\mathcal{W}_F$ such that $\chi(\Phi)^k = c$, then the representation $\chi^{-1} \otimes \tau$ satisfies condition (1). □

28.7. If E/F is a finite, separable field extension inside $\overline{F}$, we write $\mathrm{Ind}_{E/F}$, rather than $\mathrm{Ind}_{\mathcal{W}_E}^{\mathcal{W}_F}$, for the functor of smooth induction from $\mathcal{W}_E$ to $\mathcal{W}_F$. Also, if ρ is a smooth representation of $\mathcal{W}_F$, we put $\rho_E = \mathrm{Res}_{E/F}\,\rho = \rho \mid \mathcal{W}_E$.

We consider the issue of *semisimplicity*, for representations of $\mathcal{W}_F$. Translating directly from 2.7 Lemma, we have:

Lemma. *Let E/F be a finite separable field extension, $E \subset \overline{F}$.*

(1) *Let ρ be a smooth representation of $\mathcal{W}_F$; then ρ is semisimple if and only if ρ_E is semisimple.*
(2) *Let τ be a smooth representation of $\mathcal{W}_E$; then τ is semisimple if and only if $\mathrm{Ind}_{E/F}\,\tau$ is semisimple.*

For each integer $n \geqslant 1$, we denote by $\mathcal{G}_n^{\mathrm{ss}}(F)$ the set of isomorphism classes of semisimple smooth representations of $\mathcal{W}_F$ of dimension n. We denote by $\mathcal{G}_n^0(F)$ the set of isomorphism classes of *irreducible* smooth representations of $\mathcal{W}_F$ of dimension n.

If E/F is a finite extension, $E \subset \overline{F}$, the lemma shows that we have induction and restriction maps

$$\begin{aligned} \mathrm{Ind}_{E/F} &: \mathcal{G}_n^{\mathrm{ss}}(E) \longrightarrow \mathcal{G}_{nd}^{\mathrm{ss}}(F), \\ \mathrm{Res}_{E/F} &: \mathcal{G}_n^{\mathrm{ss}}(F) \longrightarrow \mathcal{G}_n^{\mathrm{ss}}(E), \end{aligned} \tag{28.7.1}$$

where $d = [E{:}F]$.

Remark. Many choices were made in the definition of the Weil group, but $\mathcal{W}_F$ is, colloquially speaking, well-defined up to inner automorphism of Ω_F. The sets $\mathcal{G}_n^0(F)$, $\mathcal{G}_n^{\mathrm{ss}}(F)$ are thus well-defined and independent of all choices. Similarly for the maps (28.7.1). Put formally, suppose we have two separable field extensions E/F, E'/F' of finite degree d and an isomorphism $\alpha : E \to E'$ such that $\alpha(F) = F'$. We get a well-defined bijection $\mathcal{G}_n^{\mathrm{ss}}(F) \to \mathcal{G}_n^{\mathrm{ss}}(F')$, and similarly for the larger fields. Moreover, the diagram

$$\begin{array}{ccc} \mathcal{G}_n^{\mathrm{ss}}(E) & \longrightarrow & \mathcal{G}_n^{\mathrm{ss}}(E') \\ {\scriptstyle \mathrm{Ind}_{E/F}}\big\downarrow & & \big\downarrow{\scriptstyle \mathrm{Ind}_{E'/F'}} \\ \mathcal{G}_{nd}^{\mathrm{ss}}(F) & \longrightarrow & \mathcal{G}_{nd}^{\mathrm{ss}}(F') \end{array}$$

commutes. Similarly for restriction. In particular, the maps (28.7.1) are defined for any finite separable extension E/F, independent of any choice of F-embedding of E in $\overline{F}$.

In practice, it is often important to recognize semisimplicity of representations in terms of individual Frobenius elements.

Proposition. *Let (ρ, V) be a smooth representation of $\mathcal{W}_F$ of finite dimension, and let $\Phi \in \mathcal{W}_F$ be a Frobenius element. The following are equivalent:*

(1) *the representation* ρ *is semisimple;*
(2) *the automorphism* $\rho(\Phi) \in \mathrm{Aut}_{\mathbb{C}}(V)$ *is semisimple;*
(3) *the automorphism* $\rho(\Psi) \in \mathrm{Aut}_{\mathbb{C}}(V)$ *is semisimple, for every element* Ψ *of* $\mathcal{W}_F$.

Proof. Certainly (3) $\Rightarrow$ (2).

Suppose that (2) holds. The group $\rho(\mathcal{I}_F)$ is finite, so there exists an integer $d \geqslant 1$ such that $\rho(\Phi)^d$ commutes with it. The automorphism $\rho(\Phi)$ is semisimple, so $\rho(\Phi)^d$ is semisimple. The restriction of ρ to the open subgroup $\langle \Phi^d, \mathcal{I}_F \rangle$ is semisimple. The lemma implies that ρ is semisimple, so (2) $\Rightarrow$ (1).

Suppose that (1) holds, and let $\Psi \in \mathcal{W}_F$. If Ψ is an element of the profinite group $\mathcal{I}_F$, then $\rho(\Psi)$ is semisimple, so we assume $\Psi \notin \mathcal{I}_F$. Thus there is a finite extension E/F such that $\Psi \in \mathcal{W}_E$ and $\mathcal{W}_E$ is the group $\langle \Psi, \mathcal{I}_E \rangle$ generated by Ψ and $\mathcal{I}_E$. The representation $\rho \mid \mathcal{W}_E$ is semisimple, by the lemma, so we may as well assume $E = F$. We can further assume that ρ is irreducible. By 28.6 Proposition, $\rho(\Psi) = c\phi$, where $c \in \mathbb{C}^\times$ and ϕ has finite order. Thus $\rho(\Psi)$ is semisimple and we have shown (1) $\Rightarrow$ (3). □

29. Local Class Field Theory

We recall the broad outlines of local class field theory. For our purposes, an axiomatic account will suffice. We use it to develop a theory of L-functions and local constants attached to representations of the Weil group $\mathcal{W}_F$.

29.1. Let $\mathcal{W}_F^{\mathrm{der}}$ denote the closure of the commutator subgroup of $\mathcal{W}_F$, and write $\mathcal{W}_F^{\mathrm{ab}} = \mathcal{W}_F/\mathcal{W}_F^{\mathrm{der}}$. Thus $\mathcal{W}_F^{\mathrm{ab}}$ is a locally profinite abelian group (defined independently of the choices made in the definition of $\mathcal{W}_F$).

Local class field theory.

There is a canonical continuous group homomorphism

$$\boldsymbol{a}_F : \mathcal{W}_F \longrightarrow F^\times$$

with the following properties.

(1) *The map* $\boldsymbol{a}_F$ *induces a topological isomorphism* $\mathcal{W}_F^{\mathrm{ab}} \cong F^\times$.
(2) *An element* $x \in \mathcal{W}_F$ *is a geometric Frobenius if and only if* $\boldsymbol{a}_F(x)$ *is a prime element of* F.
(3) *We have* $\boldsymbol{a}_F(\mathcal{I}_F) = U_F$ *and* $\boldsymbol{a}_F(\mathcal{P}_F) = U_F^1$.
(4) *If* E/F *is a finite separable extension, the diagram*

$$\begin{array}{ccc} \mathcal{W}_E & \xrightarrow{\boldsymbol{a}_E} & E^\times \\ \downarrow & & \downarrow \mathrm{N}_{E/F} \\ \mathcal{W}_F & \xrightarrow{\boldsymbol{a}_F} & F^\times \end{array}$$

commutes.

(5) *Let $\alpha : F \to F'$ be an isomorphism of fields. The map α induces an isomorphism $\alpha : \mathcal{W}_F^{\mathrm{ab}} \to \mathcal{W}_{F'}^{\mathrm{ab}}$, and the diagram*

$$\begin{array}{ccc} \mathcal{W}_F^{\mathrm{ab}} & \xrightarrow{\alpha} & \mathcal{W}_{F'}^{\mathrm{ab}} \\ {\scriptstyle \boldsymbol{a}_F}\downarrow & & \downarrow{\scriptstyle \boldsymbol{a}_{F'}} \\ F^\times & \xrightarrow[\alpha]{} & F'^\times \end{array}$$

commutes.

The map $\boldsymbol{a}_F$ is the *Artin Reciprocity Map.*

Part (4) implies the more standard statements of local class field theory:

Consequences.

(1) *The map $E/F \mapsto \mathrm{N}_{E/F}(E^\times)$ gives a bijection between the set of finite abelian extensions E/F, $E \subset \overline{F}$, and the set of open subgroups of $F^\times$ of finite index.*

(2) *The map $\boldsymbol{a}_F$ induces an isomorphism $\mathrm{Gal}(E/F) \cong F^\times/\mathrm{N}_{E/F}(E^\times)$, for any finite abelian extension E/F.*

(3) *Let E/F be a finite separable extension. If E_1/F is the maximal abelian sub-extension of E/F, then $\mathrm{N}_{E/F}(E^\times) = \mathrm{N}_{E_1/F}(E_1^\times)$.*

There is a further property to be added to the list. For the moment, let G be some group and H a subgroup of G of finite index. Let $D(G)$ be the commutator subgroup of G. We choose a section $t : G/H \to G$ of the quotient map $G \to G/H$. For $x \in G/H$ and $g \in G$, define an element $h_{x,g} \in H$ by $gt(x) = t(gx)h_{x,g}$. We define a map

$$\begin{aligned} \mathrm{ver}_{G/H} : \ G/D(G) &\longrightarrow H/D(H), \\ g\,D(G) &\longmapsto \prod_{x \in G/H} h_{x,g}\,D(H). \end{aligned}$$

The map $\mathrm{ver}_{G/H}$, called the "transfer" or "Verlagerung", is a homomorphism of abelian groups $G/D(G) \to H/D(H)$. If G is locally compact, we write G^{der} for the closure of $D(G)$ in G and $G^{\mathrm{ab}} = G/G^{\mathrm{der}}$. If H is a closed subgroup of G of finite index, the transfer then induces a continuous homomorphism $G^{\mathrm{ab}} \to H^{\mathrm{ab}}$, which we continue to denote $\mathrm{ver}_{G/H}$.

In particular, if E/F is a finite separable field extension, the transfer gives a map $\mathrm{ver}_{E/F} : \mathcal{W}_F^{\mathrm{ab}} \to \mathcal{W}_E^{\mathrm{ab}}$.

Transfer theorem.

(6) *If E/F is a finite separable field extension, the diagram*

$$\begin{array}{ccc} \mathcal{W}_E^{\mathrm{ab}} & \xrightarrow{\boldsymbol{a}_E} & E^\times \\ {\scriptstyle \mathrm{ver}_{E/F}}\uparrow & & \uparrow \\ \mathcal{W}_F^{\mathrm{ab}} & \xrightarrow[\boldsymbol{a}_F]{} & F^\times \end{array}$$

commutes, where the map $F^\times \to E^\times$ is inclusion.

29.2. In particular, the Artin Reciprocity Map $\boldsymbol{a}_F$ gives an isomorphism $\chi \mapsto \chi \circ \boldsymbol{a}_F$ of the group of characters of $F^\times$ with that of $\mathcal{W}_F$. Because of property (3), unramified characters of $F^\times$ correspond with unramified characters of $\mathcal{W}_F$, and tamely ramified characters of $F^\times$ correspond with characters of $\mathcal{W}_F$ trivial on $\mathcal{P}_F$.

We henceforward identify the two sorts of character and omit $\boldsymbol{a}_F$ from the notation whenever possible. In particular, if $\rho \in \mathfrak{G}_n^{\mathrm{ss}}(F)$, we often view $\det \rho$ as a character of $F^\times$.

If K/F is a separable extension of finite degree d and 1_K denotes the trivial character of $\mathcal{W}_K$, we can form the regular representation

$$R_{K/F} = \mathrm{Ind}_{K/F}\, 1_K \in \mathfrak{G}_d^{\mathrm{ss}}(F), \tag{29.2.1}$$

and its determinant

$$\varkappa_{K/F} = \det R_{K/F}. \tag{29.2.2}$$

Thus $\varkappa_{K/F}$ is a character of $F^\times$ satisfying $\varkappa_{K/F}^2 = 1$. It arises in the following context:

Proposition. *Let K/F be a separable extension of degree d, let $\rho \in \mathfrak{G}_m^{\mathrm{ss}}(K)$ and put $\sigma = \mathrm{Ind}_{K/F}\,\rho \in \mathfrak{G}_{md}^{\mathrm{ss}}(F)$. Then*

$$\det \sigma = \varkappa_{K/F}^m \otimes \det \rho \mid F^\times.$$

Proof. This follows from the Transfer Theorem in 29.1 and the following general fact.

Lemma. *Let G be a group, let H be a subgroup of G of finite index, and ρ a finite dimensional representation of H. Set $\sigma = \mathrm{Ind}_H^G\,\rho$. Then*

$$\det \sigma = \det(R_{G/H})^{\dim \rho} \otimes (\det \rho \circ \mathrm{ver}_{G/H}),$$

where $R_{G/H} = \mathrm{Ind}_H^G\, 1_H$.

29.3. If χ is a character of $F^\times$, the quantities $L(\chi, s)$, $\varepsilon(\chi, s, \psi)$ have been defined in §23. We translate these definitions to characters $\chi \circ \boldsymbol{a}_F$ of $\mathcal{W}_F$ by setting

$$\begin{aligned} L(\chi \circ \boldsymbol{a}_F, s) &= L(\chi, s), \\ \varepsilon(\chi \circ \boldsymbol{a}_F, s, \psi) &= \varepsilon(\chi, s, \psi). \end{aligned} \tag{29.3.1}$$

We have to extend the definition (29.3.1) to finite-dimensional, semisimple, smooth representations of $\mathcal{W}_F$.

The L-function is straightforward. First, let $\sigma \in \mathfrak{G}_n^0(F)$. The case $n = 1$ is given by (29.3.1); otherwise, we set

$$L(\sigma, s) = 1, \quad \sigma \in \mathfrak{G}_n^0(F),\ n \geqslant 2. \tag{29.3.2}$$

We then use the rule

$$L(\sigma_1 \oplus \sigma_2, s) = L(\sigma_1, s)L(\sigma_2, s)$$

to extend the definition to $\sigma \in \mathcal{G}_n^{\mathrm{ss}}(F)$.

Remark. There is a more uniform version of the definition of $L(\sigma, s)$. Let (σ, V) be a finite-dimensional, semisimple smooth representation of $\mathcal{W}_F$. The space $V^{\mathcal{I}_F}$ of $\mathcal{I}_F$-fixed vectors in V then carries a natural representation $\sigma_{\mathcal{I}}$ of $\mathcal{W}_F$. If Φ is a geometric Frobenius in $\mathcal{W}_F$, we have

$$L(\sigma, s) = \det(1 - \sigma_{\mathcal{I}}(\Phi)q^{-s})^{-1}. \tag{29.3.3}$$

The formula (29.3.3) is the standard definition of the *Artin L-function.*

Exercise. Let K/F be a finite separable extension, let $\rho \in \mathcal{G}_n^{\mathrm{ss}}(K)$, and put $\sigma = \mathrm{Ind}_{K/F}\,\rho$. Show that $L(\sigma, s) = L(\rho, s)$.

29.4. The definition of the local constant is more involved. If E/F is a finite separable extension, we again write 1_E for the trivial character of $\mathcal{W}_E$. For $\psi \in \widehat{F}$, we set $\psi_E = \psi \circ \mathrm{Tr}_{E/F} \in \widehat{E}$. We also write $\mathcal{G}^{\mathrm{ss}}(F) = \bigcup_{n\geqslant 1} \mathcal{G}_n^{\mathrm{ss}}(F)$.

Theorem. *Let $\psi \in \widehat{F}$, $\psi \neq 1$, and let E/F range over finite extensions inside $\overline{F}$. There is a unique family of functions*

$$\begin{aligned} \mathcal{G}^{\mathrm{ss}}(E) &\longrightarrow \mathbb{C}[q^s, q^{-s}]^\times, \\ \rho &\longmapsto \varepsilon(\rho, s, \psi_E), \end{aligned}$$

with the following properties:

(1) *If χ is a character of $E^\times$, then*

$$\varepsilon(\chi \circ \boldsymbol{a}_E, s, \psi_E) = \varepsilon(\chi, s, \psi_E).$$

(2) *If $\rho_1, \rho_2 \in \mathcal{G}^{\mathrm{ss}}(E)$, then*

$$\varepsilon(\rho_1 \oplus \rho_2, s, \psi_E) = \varepsilon(\rho_1, s, \psi_E)\,\varepsilon(\rho_2, s, \psi_E).$$

(3) *If $\rho \in \mathcal{G}_n^{\mathrm{ss}}(E)$ and $E \supset K \supset F$, then*

$$\frac{\varepsilon(\mathrm{Ind}_{E/K}\,\rho, s, \psi_K)}{\varepsilon(\rho, s, \psi_E)} = \frac{\varepsilon(R_{E/K}, s, \psi_K)^n}{\varepsilon(1_E, s, \psi_E)^n}. \tag{29.4.1}$$

Remark. The uniqueness properties imply that the function ε is natural with respect to isomorphisms of fields and so it is defined independently of choices made in the construction of the Weil groups. For the same reason, the restriction to subfields of $\overline{F}$ is unnecessary.

The quantity $\varepsilon(\rho, s, \psi)$, $\rho \in \mathcal{G}^{ss}(F)$, is called the *Langlands-Deligne local constant* of ρ, relative to the character $\psi \in \widehat{F}$ and the complex variable s. We list some of its useful properties.

Proposition. *Let $\psi \in \widehat{F}$, $\psi \neq 1$, and let $\rho \in \mathcal{G}^{ss}(F)$.*

(1) *There is an integer $n(\rho, \psi)$ such that*

$$\varepsilon(\rho, s, \psi) = q^{n(\rho,\psi)(\frac{1}{2}-s)}\, \varepsilon(\rho, \tfrac{1}{2}, \psi).$$

(2) *Let $a \in F^\times$. Then:*

$$\begin{aligned} \varepsilon(\rho, s, a\psi) &= \det \rho(a)\, \|a\|^{\dim(\rho)(s-\frac{1}{2})}\, \varepsilon(\rho, s, \psi), \\ n(\rho, a\psi) &= n(\rho, \psi) + \upsilon_F(a) \dim \rho. \end{aligned}$$

In particular, $n(\rho, \psi)$ depends only on ρ and the level of ψ.

(3) *The local constants satisfy the functional equation*

$$\varepsilon(\rho, s, \psi)\, \varepsilon(\check{\rho}, 1-s, \psi) = \det \rho(-1).$$

(4) *There is an integer n_ρ such that, if χ is a character of $F^\times$ of level $k \geqslant n_\rho$, then*

$$\varepsilon(\chi \otimes \rho, s, \psi) = \det \rho(c(\chi))^{-1} \varepsilon(\chi, s, \psi)^{\dim \rho},$$

for any $c(\chi) \in F^\times$ such that $\chi(1+x) = \psi(c(\chi)x)$, $x \in \mathfrak{p}^{[k/2]+1}$.

We prove these assertions in the next section.

30. Existence of the Local Constant

We prove the Theorem and Proposition of 29.4. The proof has several steps, one of which makes essential use of machinery from the theory of global fields. We have separated out that part and put it at the end of the section. The ideas in it intervene nowhere else.

30.1. We start with some abstract machinery.

Let G be a profinite group, and let K_0G be the *Grothendieck group* of G. Thus K_0G is the free abelian group on symbols $[\rho]$, where ρ ranges over the set of isomorphism classes of irreducible smooth representations of G. We regard the set of isomorphism classes of finite-dimensional smooth representations of G as contained in K_0G: if ρ is such a representation, we write it as a sum of irreducible representations $\rho = \rho_1 \oplus \rho_2 \oplus \cdots \oplus \rho_r$, to obtain an element

$$[\rho] = \sum_{i=1}^{r} [\rho_i] \in K_0G.$$

We invariably abuse notation and write $[\rho] = \rho$.

There is a dimension map $K_0G \to \mathbb{Z}$, defined in the obvious way. We put

$$\widetilde{K}_0G = \bigcup_H K_0H,$$

where H ranges over the open subgroups of G. We denote the elements of $\widetilde{K}_0G$ as pairs (H, ρ), where H is an open subgroup of G and $\rho \in K_0H$.

The group K_0G contains the set $\Gamma(G) = \mathrm{Hom}(G, \mathbb{C}^\times)$. We write

$$\widetilde{\Gamma}(G) = \bigcup_H \Gamma(H) \subset \widetilde{K}_0G,$$

where H ranges over the open subgroups of G.

Suppose we have another profinite group $\overline{G}$ and a continuous surjective homomorphism $\overline{G} \to G$. Using inflation of representations, we can regard $\widetilde{K}_0G$ as a subset of $\widetilde{K}_0\overline{G}$.

Remark. Write $G = \varprojlim G_i$, where G_i ranges over the finite quotients of G. The set $\widetilde{K}_0G$ is then the union of its subsets $\widetilde{K}_0G_i$. This enables us to reduce most issues concerning $\widetilde{K}_0G$ to the case of a *finite* group G.

Definitions. *Let $\mathbb{A}$ be an abelian group and G a profinite group.*

(1) *An* induction constant *on G (with values in $\mathbb{A}$) is a function*

$$\mathcal{F} : \widetilde{K}_0G \longrightarrow \mathbb{A}$$

with the following properties:

(a) *for each open subgroup H of G, the map $\mathcal{F} \mid K_0H$ is a group homomorphism;*

(b) *if $H \subset J$ are open subgroups of G and $(H, \rho) \in K_0H$ has dimension zero, then*

$$\mathcal{F}(J, \mathrm{Ind}_H^J \rho) = \mathcal{F}(H, \rho).$$

(2) *A* division *on G (with values in $\mathbb{A}$) is a function $\mathcal{D} : \widetilde{\Gamma}(G) \to \mathbb{A}$.*

An induction constant $\mathcal{F}$ on G defines a division $\partial\mathcal{F}$ on G by restriction: we call $\partial\mathcal{F}$ the *boundary* of $\mathcal{F}$.

We shall be more interested in the converse relation: a division $\mathcal{D}$ on G will be called *pre-inductive* on G if it is of the form $\mathcal{D} = \partial\mathcal{F}$, for some induction constant $\mathcal{F}$ on G.

The starting point is a simple uniqueness property:

Lemma 1. *Let $\mathcal{F}$ be an induction constant on G. The division $\partial\mathcal{F}$ determines $\mathcal{F}$ uniquely, and $\mathcal{F}$ takes its values in the abelian group generated by the values of $\partial\mathcal{F}$.*

Proof. It is enough to treat the case where G is *finite.* We recall[2]:

Brauer induction theorem. *Let G be a finite group and ρ an irreducible representation of G of dimension $m \geqslant 1$. There then exist elements $(H_i, \chi_i) \in \widetilde{\Gamma}(G)$, $1 \leqslant i \leqslant r$, such that*

$$[\rho] - m[1_G] = \sum_{1\leqslant i\leqslant r} \operatorname{Ind}_{H_i}^G ([\chi_i] - [1_{H_i}]),$$

where 1_H denotes the trivial character of a group H.

In the lemma, let ρ be an irreducible representation of $H \subset G$ of dimension m. We use the Brauer induction theorem for H to write

$$[\rho] - m[1_H] = \sum_{1\leqslant i\leqslant r} \operatorname{Ind}_{H_i}^H ([\chi_i] - [1_{H_i}]), \tag{30.1.1}$$

for various $(H_i, \chi_i) \in \widetilde{\Gamma}(H)$. Writing $\mathbb{A}$ multiplicatively, this yields

$$\mathcal{F}(H, \rho) = \partial\mathcal{F}(H, 1_H)^m \prod_{1\leqslant i\leqslant r} \partial\mathcal{F}(H_i, \chi_i)\, \partial\mathcal{F}(H_i, 1_{H_i})^{-1}.$$

The value $\mathcal{F}(H, \rho)$ is thereby expressed in terms of values of $\partial\mathcal{F}$. □

Remark. Let $\mathcal{F}$ be an induction constant on G and let $(H, \rho) \in \widetilde{K}_0G$. Let 1_H be the trivial character of H and set $R_{G/H} = \operatorname{Ind}_H^G 1_H$. We then have

$$\mathcal{F}([\rho] - m[1_H]) = \mathcal{F}([\operatorname{Ind}_H^G \rho] - m[R_{G/H}]),$$

where $m = \dim \rho$ (*cf.* (29.4.1)).

The uniqueness property of Lemma 1 enables us to reduce to the case of finite groups:

Lemma 2. *Let G be a profinite group and let $\mathcal{D}$ be a division on G. Suppose there is a family $\mathcal{H}$ of open normal subgroups H of G such that*

(a) *the canonical map*

$$G \longrightarrow \varprojlim_{H\in\mathcal{H}} G/H$$

is an isomorphism, and

(b) *the restriction $\mathcal{D}_{G/H}$ of $\mathcal{D}$ to $\widetilde{\Gamma}(G/H)$ is pre-inductive on G/H, for all $H \in \mathcal{H}$.*

The division $\mathcal{D}$ is then pre-inductive on G. If $\mathcal{F}$ is the induction constant on G with boundary $\mathcal{D}$, then $\mathcal{D}_{G/H}$ is the boundary of $\mathcal{F} \mid \widetilde{K}_0G/H$.

[2] For this version of the well-known result, see [26].

30.2. We will use this apparatus in the following context. We take a finite Galois extension L/F of local fields and set $G = \mathrm{Gal}(L/F)$. Via local class field theory, we think of $\widetilde{\Gamma}(G)$ as the set of pairs (E,χ), where E ranges over the fields between L and F, and χ over the characters of $E^\times$ which are null on $\mathrm{N}_{L/E}(L^\times)$.

The core result to be proved is:

Theorem. *Let L/F be a finite Galois extension, and set $G = \mathrm{Gal}(L/F)$. There exists $\psi \in \widehat{F}$, $\psi \neq 1$, such that the division on G given by*

$$\mathcal{D}_\psi^{L/F} : (E,\chi) \longmapsto \varepsilon(\chi, s, \psi_E), \quad (E,\chi) \in \widetilde{\Gamma}(G),$$

is pre-inductive on G.

The proof is deferred to the end of the section. In preparation for it, however, we record a technical local result.

Lemma. *Let L/F be a finite Galois extension, and put $G = \mathrm{Gal}(L/F)$. There exists an integer $n_{L/F} \geqslant 1$ such that, if α is a character of $F^\times$ of level $\geqslant n_{L/F}$ and $(E,\chi) \in \widetilde{\Gamma}(G)$, the level of $\alpha \circ \mathrm{N}_{E/F}$ is greater than twice the level of χ.*

Proof. For $F \subset E \subset L$, let $\ell(E) \geqslant 1$ be an integer such that $U_E^{\ell(E)} \subset \mathrm{N}_{L/E}(L^\times)$. If $(E,\chi) \in \widetilde{\Gamma}(G)$, the character χ then has level $< \ell(E)$. We set $m(E) = 2\ell(E){+}1$. We choose an integer $n(E)$ such that $U_F^{n(E)} \subset \mathrm{N}_{E/F}(U_E^{m(E)})$. The integer $n_{L/F} = \max_E n(E)$, as E/F ranges over subextensions of L/F, then has the required property. □

30.3. For the time being, we assume 30.2 Theorem and deduce the other assertions of 29.4.

We first remove the restriction on ψ in 30.2 Theorem. By 30.1 Lemma 2, the division $(E,\chi) \mapsto \varepsilon(\chi, s, \psi_E)$ is pre-inductive on Ω_F, being (by definition) the boundary of the induction constant $(\Omega_E, \rho) \mapsto \varepsilon(\rho, s, \psi_E)$.

Take $a \in F^\times$, and consider the function

$$(E,\rho) \longmapsto \det\rho(a)\|a\|_E^{(s-\frac{1}{2})\dim\rho}, \quad (E,\rho) \in \widetilde{K}_0\Omega_F.$$

This is an induction constant on Ω_F, whence so is the function

$$(E,\rho) \longmapsto \det\rho(a)\|a\|_E^{(s-\frac{1}{2})\dim\rho}\,\varepsilon(\rho, s, \psi_E).$$

The boundary of this induction constant is

$$(E,\chi) \longmapsto \chi(a)\|a\|_E^{s-\frac{1}{2}}\varepsilon(\chi, s, \psi_E) = \varepsilon(\chi, s, a\psi_E).$$

This division is therefore pre-inductive, and the boundary of (by definition) the induction constant $(E,\rho) \mapsto \varepsilon(\rho, s, a\psi_E)$. Thus 30.2 Theorem holds for all

$\psi \in \widehat{F}$, $\psi \neq 1$, and this finishes the proof of 29.4 Theorem for representations of Galois groups.

We next prove parts (1) and (2) of 29.4 Proposition for representations of Galois groups. Part (1) is an immediate consequence of the corresponding property of local constants of characters and 30.1 Lemma 1. The first identity in (2) has just been proved, and the second follows immediately.

30.4. We have to extend the arguments to cover representations of Weil groups. We fix a prime element ϖ of F. Let ϕ be an unramified character of $F^\times$. We write $\phi(\varpi) = q^{-s(\phi)}$, for some $s(\phi) \in \mathbb{C}$. If E/F is a finite extension and ϖ_E is a prime of E, we also have $\phi_E(\varpi_E) = q_E^{-s(\phi)}$. Thus, if χ is a character of $E^\times$, we have

$$\varepsilon(\chi\phi_E, s, \psi_E) = \varepsilon(\chi, s{+}s(\phi), \psi_E).$$

Proposition. *Let $(\Omega_E, \rho) \in \widetilde{K}_0\Omega_F$, and let ϕ be an unramified character of $F^\times$ of finite order. Then $\varepsilon(\phi_E \otimes \rho, s, \psi_E) = \varepsilon(\rho, s{+}s(\phi), \psi_E)$.*

Proof. The maps $(\Omega_E, \rho) \mapsto \varepsilon(\phi_E \otimes \rho, s, \psi_E)$, $(\Omega_E, \rho) \mapsto \varepsilon(\rho, s{+}s(\phi), \psi_E)$ are induction constants, with the same boundary. They are therefore the same. □

Let E/F be a finite extension, let 1_E be the trivial character of $\mathcal{W}_E$ and set $R_{E/F} = \operatorname{Ind}_{E/F} 1_E$. Define

$$\lambda_{E/F}(s, \psi) = \frac{\varepsilon(R_{E/F}, s, \psi)}{\varepsilon(1_E, s, \psi_E)}. \tag{30.4.1}$$

Corollary. *The function $\lambda_{E/F}(s, \psi)$ is constant in s.*

Proof. Let ϕ be an unramified character of $F^\times$ of finite order. We have $\phi \otimes R_{E/F} \cong \operatorname{Ind}_{E/F} \phi_E$, so

$$\begin{aligned}\lambda_{E/F}(s, \psi) &= \frac{\varepsilon(\operatorname{Ind}_{E/F} \phi_E, s, \psi)}{\varepsilon(\phi_E, s, \psi_E)}\\ &= \frac{\varepsilon(R_{E/F}, s{+}s(\phi), \psi)}{\varepsilon(1_E, s{+}s(\phi), \psi_E)}\\ &= \lambda_{E/F}(s{+}s(\phi), \psi).\end{aligned}$$

Thus $\lambda_{E/F}(s{+}s(\phi), \psi) = \lambda_{E/F}(s, \psi)$, for all unramified characters ϕ of finite order. That is, $\lambda_{E/F}(s{+}\zeta, \psi) = \lambda_{E/F}(s, \psi)$ for all roots of unity $\zeta \in \mathbb{C}$. Since $\lambda_{E/F}(s, \psi)$ is a constant times a power of $q^{\frac{1}{2}-s}$, the result follows. □

We therefore write $\lambda_{E/F}(s, \psi) = \lambda_{E/F}(\psi)$: it is called the *Langlands constant.*

Remark. In terms of the Langlands constant, the defining relation (29.4.1) reads

$$\frac{\varepsilon(\mathrm{Ind}_{E/K}\,\rho, s, \psi_K)}{\varepsilon(\rho, s, \psi_E)} = \lambda_{E/K}(\psi_K)^n, \quad \rho \in \mathcal{G}_n^{ss}(E). \tag{30.4.2}$$

The functional equation 29.4 Proposition (3) reads

$$\lambda_{E/F}(\psi)^2 = \varkappa_{E/F}(-1), \tag{30.4.3}$$

so $\lambda_{E/F}(\psi)$ is a 4-th root of unity.

30.5. Let ρ be an irreducible smooth representation of $\mathcal{W}_E$. There is an unramified character ϕ of $\mathcal{W}_F$ such that $\phi_E \otimes \rho$ factors through a representation ρ_0 of Ω_E. Because of 30.4 Proposition, we can unambiguously define

$$\varepsilon(\rho, s, \psi_E) = \varepsilon(\rho_0, s{-}s(\phi), \psi_E). \tag{30.5.1}$$

The inductive properties required by 29.4 Theorem follow from 30.4 Corollary. The uniqueness property is proved by applying the Brauer induction theorem to ρ_0, as in 30.1.

30.6. We now prove 29.4 Proposition. We have already proved parts (1) and (2) for representations of Galois groups. The general case of (1) follows directly from the definition (30.5.1). The first assertion in (2) follows from the Galois case and the definition, and there is nothing to do in the second.

In parts (3) and (4), it is enough to consider representations of Galois groups. The functions

$$(\Omega_E, \tau) \longmapsto \begin{cases} \varepsilon(\tau, s, \psi_E)\,\varepsilon(\check\tau, 1{-}s, \psi_E), \\ \det \tau(-1), \end{cases}$$

on $K_0\Omega_F$ are both induction constants with boundary

$$(E, \phi) \longmapsto \varepsilon(\phi, s, \psi_E)\,\varepsilon(\check\phi, 1{-}s, \psi_E) = \phi(-1)$$

(23.4.2). Thus they are the same, as required for (3).

In part (4), it is enough to consider an irreducible representation ρ of Ω_F, of dimension m, say. We write

$$[\rho] - m[1_F] = \sum_{i=1}^{r} \mathrm{Ind}_{E_i/F}\big([\phi_i] - [1_{E_i}]\big), \tag{30.6.1}$$

for characters ϕ_i of $E_i^\times$, for various finite extensions E_i/F. We then have

$$\frac{\varepsilon(\chi \otimes \rho, s, \psi)}{\varepsilon(\chi, s, \psi)^m} = \prod_{i=1}^{r} \frac{\varepsilon(\chi_{E_i}\phi_i, s, \psi_{E_i})}{\varepsilon(\chi_{E_i}, s, \psi_{E_i})}.$$

If χ has sufficiently large level, the right hand side reduces to $\prod_{i=1}^{r} \phi_i(c(\chi))^{-1}$, giving (*cf.* 23.8)

$$\varepsilon(\chi \otimes \rho, s, \psi) = \varepsilon(\chi, s, \psi)^m \prod_i \phi_i(c(\chi))^{-1}.$$

The determinant of the defining relation (30.6.1) and 29.2 Lemma give

$$\prod_i \phi_i(c(\chi)) = \det \rho(c(\chi)),$$

as required. □

30.7. We turn to the proof of 30.2 Theorem.

We first recall, with extreme brevity, some matters concerning global fields. Let K be a global field; thus K is a finite extension of either the rational field $\mathbb{Q}$ or a field $\mathbb{F}_q(t)$ of rational functions, in one variable, over a finite field. We fix a separable algebraic closure $\overline{K}/K$ and write $\Omega_K = \mathrm{Gal}(\overline{K}/K)$.

Let $\boldsymbol{A}_K$ denote the adèle ring of K, and $\boldsymbol{I}_K$ its idèle group. Thus $K^\times$ embeds naturally in $\boldsymbol{I}_K$ as a discrete subgroup; we denote by $\boldsymbol{C}_K = \boldsymbol{I}_K/K^\times$ the idèle class group of K.

Let $P(K)$ denote the set of places of K. For $v \in P(K)$, we denote by K_v the completion of K at v. There is a canonical embedding of $K_v^\times$ in $\boldsymbol{C}_K$, for each $v \in P(K)$. Thus, if χ is a continuous homomorphism $\boldsymbol{C}_K \to \mathbb{C}^\times$, the restriction of χ to the image of $K_v^\times$ gives a character χ_v of $K_v^\times$.

If v is non-Archimedean, we can form the L-function $L(\chi_v, s)$ as in §23. If v is Archimedean, $L(\chi_v, s)$ is given by an explicit formula involving elementary functions which need not detain us. The product

$$L(\chi, s) = \prod_{v \in P(K)} L(\chi_v, s) \tag{30.7.1}$$

converges in a half-plane $\mathrm{Re}\, s > s_0$. It admits analytic continuation to a meromorphic function on the whole s-plane, where it satisfies a functional equation

$$L(\chi, s) = \varepsilon(\chi, s)\, L(\chi^{-1}, 1-s). \tag{30.7.2}$$

The factor $\varepsilon(\chi, s)$ is an exponential function, having neither poles nor zeros.

There are canonical embeddings of K and of K_v in $\boldsymbol{A}_K$, for each $v \in P(K)$. In particular, the image of K is a discrete subgroup of $\boldsymbol{A}_K$. Let ψ be a non-trivial unitary character of $\boldsymbol{A}_K/K$. We can restrict to get a character ψ_v of K_v for each v. The character ψ_v is invariably non-trivial. The ε-factor in (30.7.2) then satisfies

$$\varepsilon(\chi, s) = \prod_{v \in P(K)} \varepsilon(\chi_v, s, \psi_v). \tag{30.7.3}$$

In this expression, the factor $\varepsilon(\chi_v, s, \psi_v)$ (for v non-Archimedean) is that discussed in §23. (The definition of $\varepsilon(\chi_v, s, \psi_v)$, for Archimedean v, need not detain us.)

30.8. For each $v \in P(K)$, choose a separable algebraic closure $\overline{K}_v/K_v$ and write $\Omega_v = \mathrm{Gal}(\overline{K}_v/K_v)$. There is then an embedding of Ω_v in Ω_K as a closed subgroup, determined up to inner automorphism of Ω_K. If ρ is a finite-dimensional, smooth representation of Ω_K, then $\rho_v = \rho \mid \Omega_v$ is a smooth representation of Ω_v. Using (29.3.3) (which applies equally when v is Archimedean), we can define $L(\rho_v, s)$, and put

$$L(\rho, s) = \prod_{v \in P(K)} L(\rho_v, s). \tag{30.8.1}$$

This is the *Artin L-function* of the representation ρ. The basic facts here are:

Lemma.

(1) *Let L/K be a finite extension, $L \subset \overline{K}$, and let τ be a smooth representation of $\Omega_L = \mathrm{Gal}(\overline{K}/L)$ of finite dimension. Then*

$$L(\mathrm{Ind}_{L/K}\,\tau, s) = L(\tau, s).$$

(2) *Let ρ be a finite-dimensional smooth representation of Ω_K. The product* (30.8.1) *converges in some half-plane and admits analytic continuation to a meromorphic function on the whole s-plane. It satisfies a functional equation*

$$L(\rho, s) = \varepsilon(\rho, s)\, L(\check{\rho}, 1-s), \tag{30.8.2}$$

for a uniquely determined exponential function $\varepsilon(\rho, s)$ having neither zeros nor poles.

(3) *In the situation of* (1) *we have*

$$\varepsilon(\mathrm{Ind}_{L/K}\,\tau, s) = \varepsilon(\tau, s). \tag{30.8.3}$$

Sketch proof. Part (1) is given by an elementary argument, based on the definitions (30.8.1), (29.3.3). Part (2) then follows from the Brauer Induction Theorem (as in 30.1) and the case recalled in 30.7. Part (3) is an immediate consequence. □

30.9. We return to the situation of 30.2, with a finite Galois extension L/F of non-Archimedean local fields and $G = \mathrm{Gal}(L/F)$. We make a connection with the global situation:

Lemma 1. *There is a finite Galois extension $\boldsymbol{L}/\boldsymbol{F}$ of global fields, and a non-Archimedean place v_0 of $\boldsymbol{F}$, with the following properties:*

(1) *There is a unique place* u_0 *of* $\boldsymbol{L}$ *lying over* v_0 *and a topological isomorphism* $\boldsymbol{L}_{u_0} \cong L$ *which induces a topological isomorphism* $\boldsymbol{F}_{v_0} \cong F$.
(2) *The field* $\boldsymbol{F}$ *has no embedding in* $\mathbb{R}$.

Proof. In the case where F has characteristic zero, we use the Weak Approximation Theorem in $\mathbb{Q}$ to find a negative $x \in \mathbb{Q}^\times$ which is a square in F. We choose a global field $\boldsymbol{E}$ which is dense in L and contains a square root of x in F. In the case where F has positive characteristic, we simply choose the global field $\boldsymbol{E}$ to be dense in L.

In either case, we let $\boldsymbol{L}$ be the composite of the fields $\boldsymbol{E}^\sigma$, $\sigma \in \mathrm{Gal}(L/F)$, and set $\boldsymbol{F} = \boldsymbol{L} \cap F$. The extension $\boldsymbol{L}/\boldsymbol{F}$ is then Galois, and restriction induces an isomorphism $\mathrm{Gal}(L/F) \to \mathrm{Gal}(\boldsymbol{L}/\boldsymbol{F})$. Moreover, when F has characteristic zero, $\boldsymbol{F}$ contains the imaginary quadratic field $\mathbb{Q}(\sqrt{x})$, with the result that $\boldsymbol{L}$ satisfies (2).

The inclusion $\boldsymbol{F} \to F$ defines a place v_0 of $\boldsymbol{F}$ and an embedding $\boldsymbol{F}_{v_0} \to F$. Likewise, the inclusion $\boldsymbol{L} \to L$ defines a place u_0 of $\boldsymbol{L}$ lying over v_0 and, since $\boldsymbol{L}$ is dense in L, an isomorphism $\boldsymbol{L}_{u_0} \cong L$. The place u_0 is stable under $\mathrm{Gal}(\boldsymbol{L}/\boldsymbol{F}) = \mathrm{Gal}(L/F)$, hence is the only place of $\boldsymbol{L}$ lying over v_0. The extensions $\boldsymbol{L}/\boldsymbol{F}$, L/F have the same degree, so $\boldsymbol{F}_{v_0} = F$, as required. □

In particular, we have a canonical isomorphism $G \cong \mathrm{Gal}(\boldsymbol{L}/\boldsymbol{F})$. So, for any intermediate field E, $F \subset E \subset L$, there is a unique field $\boldsymbol{E}$, $\boldsymbol{F} \subset \boldsymbol{E} \subset \boldsymbol{L}$, with closure E in $L = \boldsymbol{L}_{u_0}$.

For each non-Archimedean place v of $\boldsymbol{F}$, let $\mathfrak{o}_v$ denote the discrete valuation ring in $\boldsymbol{F}_v$ and $\mathfrak{p}_v$ its maximal ideal: we use similar notations for extension fields.

For each non-Archimedean place $v \neq v_0$ of $\boldsymbol{F}$, we choose an integer $n_v \geqslant 0$ as follows. (30.9.1) For each $v \in P(\boldsymbol{F})$, let u be a place of $\boldsymbol{L}$ lying over v.

(1) *If* v *is unramified in* $\boldsymbol{L}$, *set* $n_v = 0$, *and*
(2) *choose* $n_v \geqslant n_{\boldsymbol{L}_u/\boldsymbol{F}_v}$ *if* v *is ramified in* $\boldsymbol{L}$.

(In (2), we use the notation of 30.2 Lemma.)

We need one more deep result[3]:

Lemma 2. *There exists a character* $\boldsymbol{\alpha}$ *of* $\boldsymbol{C_F}$, *of finite order, such that*

(1) $\boldsymbol{\alpha}_{v_0} = 1$, *and*
(2) *the level* l_v *of* $\boldsymbol{\alpha}_v$ *satisfies* $l_v \geqslant n_v$, *for every non-Archimedean* $v \neq v_0$.

For an intermediate field $\boldsymbol{F} \subset \boldsymbol{E} \subset \boldsymbol{L}$, we set $\boldsymbol{\alpha_E} = \boldsymbol{\alpha} \circ \mathrm{N}_{\boldsymbol{E}/\boldsymbol{F}}$.

We fix a non-trivial character Ψ of $\boldsymbol{A_F}/\boldsymbol{F}$ and, for $\boldsymbol{E}$ as before, set $\Psi_{\boldsymbol{E}} = \Psi \circ \mathrm{Tr}_{\boldsymbol{E}/\boldsymbol{F}}$. We put $\psi = \Psi_{v_0}$.

[3] This is a simple instance of the Grunwald-Wang Theorem, for which see [2].

For each non-Archimedean place v of $\boldsymbol{F}$, we choose $c_v \in \boldsymbol{F}_v$ such that $\boldsymbol{\alpha}_v(1+x) = \Psi_v(c_v x)$, $x \in \mathfrak{p}_v^{[l_v/2]+1}$, with the understanding that $c_v = 1$ if $l_v = 0$. Thus $c_v = 1$ for almost all non-Archimedean v. If v is Archimedean, we put $c_v = 1$. Let $\boldsymbol{c}$ denote the idèle (c_v).

Let us adjust our notation slightly, so that $G = \mathrm{Gal}(L/F)$, $\boldsymbol{G} = \mathrm{Gal}(\boldsymbol{L}/\boldsymbol{F})$. Under the canonical isomorphism $G \cong \boldsymbol{G}$, we have a bijection $\widetilde{\Gamma}(G) \cong \widetilde{\Gamma}(\boldsymbol{G})$, which we denote $(E,\chi) \mapsto (\boldsymbol{E},\boldsymbol{\chi})$. In this notation, $\boldsymbol{\chi}$ is a character of $\boldsymbol{C_E}$ trivial on norms $\mathrm{N}_{\boldsymbol{L}/\boldsymbol{E}}(\boldsymbol{C_L})$.

Let $(E,\chi) \in \widetilde{\Gamma}(G)$. Let w range over the non-Archimedean places of $\boldsymbol{E}$, with w_0 being the unique place of $\boldsymbol{E}$ lying over v_0. We obtain:

$$\varepsilon(\boldsymbol{\chi}_w \boldsymbol{\alpha}_{\boldsymbol{E},w}, s, \Psi_{\boldsymbol{E},w}) = \begin{cases} \boldsymbol{\chi}_w(c_v)^{-1}\varepsilon(\boldsymbol{\alpha}_{\boldsymbol{E},w}, s, \Psi_{\boldsymbol{E},w}), & w \neq w_0, \\ \varepsilon(\chi, s, \psi_E), \quad w = w_0. \end{cases}$$

In the first identity, v denotes the place of $\boldsymbol{F}$ lying under w. When v is unramified in $\boldsymbol{L}$, the character χ_w is also unramified and the identity follows immediately from the formulas in 23.5. Otherwise, the first identity follows from the Stability Theorem 23.8 and 30.2 Lemma. The statement is trivial at the Archimedean places, since $\boldsymbol{\chi}_w$ and $\boldsymbol{\alpha}_v$ are then both trivial. The second identity comes from the definition of $\boldsymbol{\alpha}$.

Now we expand using the product formula (30.7.3)

$$\begin{aligned} \boldsymbol{\varepsilon}(\boldsymbol{\chi}\boldsymbol{\alpha}_{\boldsymbol{E}}, s) &= \prod_{w \in P(\boldsymbol{E})} \varepsilon(\boldsymbol{\chi}_w \boldsymbol{\alpha}_{\boldsymbol{E}_w}, s, \Psi_{\boldsymbol{E},w}) \\ &= \varepsilon(\chi, s, \psi_E)\, \boldsymbol{\chi}(\boldsymbol{c})^{-1}\, \boldsymbol{a}(\boldsymbol{E}), \end{aligned}$$

where

$$\boldsymbol{a}(\boldsymbol{E}) = \prod_{w \neq w_0} \varepsilon(\boldsymbol{\alpha}_{\boldsymbol{E}_w}, s, \Psi_{\boldsymbol{E},w}).$$

The division $(\boldsymbol{E},\boldsymbol{\chi}) \mapsto \boldsymbol{\varepsilon}(\boldsymbol{\chi}\boldsymbol{\alpha}_{\boldsymbol{E}}, s)$ on $\boldsymbol{G}$ is pre-inductive (30.8.3). Also, the division $(\boldsymbol{E},\boldsymbol{\chi}) \mapsto \boldsymbol{\chi}(\boldsymbol{c})^{-1}$ is pre-inductive: it is $\partial\mathcal{F}$, where $\mathcal{F}(\boldsymbol{E},\rho) = \det\rho(\boldsymbol{c})^{-1}$. As the notation indicates, the quantity $\boldsymbol{a}(\boldsymbol{E})$ depends only on the field $\boldsymbol{E}$. The map $(\boldsymbol{E},\boldsymbol{\chi}) \mapsto \boldsymbol{a}(\boldsymbol{E})$ is a pre-inductive division, being the boundary of the induction constant $(\mathrm{Gal}(\boldsymbol{L}/\boldsymbol{E}),\rho) \mapsto \boldsymbol{a}(\boldsymbol{E})^{\dim\rho}$.

The division $(E,\chi) \mapsto \varepsilon(\chi, s, \psi_E)$ on G is therefore pre-inductive: it is the boundary of an induction constant which we denote

$$(E,\rho) \longmapsto \varepsilon(\rho, s, \psi_E).$$

Because of 30.1 Lemma 2, this proves 30.2 Theorem, for the particular character $\Psi_{v_0} = \psi \in \widehat{F}$. □

31. Deligne Representations

We introduce a more highly-structured family of representations of $\mathcal{W}_F$. In this section, we take a pragmatic and *ad hoc* approach. A more convincing motivation is given in §32.

31.1. A *Deligne representation* of $\mathcal{W}_F$ is a triple $(\rho, V, \mathfrak{n})$, in which (ρ, V) is a finite-dimensional, smooth representation of $\mathcal{W}_F$ and $\mathfrak{n} \in \mathrm{End}_{\mathbb{C}}(V)$ is *nilpotent* satisfying

$$\rho(x)\,\mathfrak{n}\,\rho(x)^{-1} = \|x\|\,\mathfrak{n}, \quad x \in \mathcal{W}_F.$$

One defines isomorphism between such triples in the obvious way.

A Deligne representation $(\rho, V, \mathfrak{n})$ is called *semisimple* if the smooth representation (ρ, V) of $\mathcal{W}_F$ is semisimple (*cf.* 28.7 Proposition). We shall write $\mathcal{G}_n(F)$ for the set of equivalence classes of n-dimensional, semisimple, Deligne representations of $\mathcal{W}_F$.

We can identify $\mathcal{G}_n^{\mathrm{ss}}(F)$ with the set of triples $(\rho, V, 0) \in \mathcal{G}_n(F)$, so

$$\mathcal{G}_n^0(F) \subset \mathcal{G}_n^{\mathrm{ss}}(F) \subset \mathcal{G}_n(F).$$

Example. Let $V = \mathbb{C}^n$, $n \geqslant 1$. Let $\mathfrak{n} \in \mathrm{M}_n(\mathbb{C})$ be the standard Jordan block of rank $n{-}1$: thus, if $\{v_0, v_1, \dots, v_{n-1}\}$ is the standard basis of V, we have $\mathfrak{n}v_{n-1} = 0$ and $\mathfrak{n}v_i = v_{i+1}$, $0 \leqslant i < n{-}1$. We define a smooth representation ρ_0 of $\mathcal{W}_F$ by $\rho_0(x)v_i = \|x\|^i v_i$, $0 \leqslant i \leqslant n{-}1$, $x \in \mathcal{W}_F$. We set

$$\rho(x) = \|x\|^{(1-n)/2}\rho_0(x).$$

The triple $(\rho, \mathbb{C}^n, \mathfrak{n})$ is then a semisimple Deligne representation of $\mathcal{W}_F$, denoted $\mathrm{Sp}(n)$.

31.2. The standard constructions of representation theory have analogues for Deligne representations, although they are not always the obvious ones. For example, if $(\rho, V, \mathfrak{n})$ is a Deligne representation, we can form the contragredient $(\check{\rho}, \check{V})$ of (ρ, V) and the transpose $\check{\mathfrak{n}} \in \mathrm{End}_{\mathbb{C}}(\check{V})$ of $\mathfrak{n}$. However, one sets

$$(\rho, V, \mathfrak{n})^\vee = (\check{\rho}, \check{V}, -\check{\mathfrak{n}}). \tag{31.2.1}$$

Likewise, the definition of tensor product is not quite obvious:

$$(\rho_1, V_1, \mathfrak{n}_1) \otimes (\rho_2, V_2, \mathfrak{n}_2) = (\rho_1 \otimes \rho_2, V_1 \otimes V_2, \mathrm{Id}_1 \otimes \mathfrak{n}_2 + \mathfrak{n}_1 \otimes \mathrm{Id}_2), \tag{31.2.2}$$

where Id_j is the identity endomorphism of V_j. The definition of direct sum

$$(\rho_1, V_1, \mathfrak{n}_1) \oplus (\rho_2, V_2, \mathfrak{n}_2) = (\rho_1 \oplus \rho_2, V_1 \oplus V_2, \mathfrak{n}_1 \oplus \mathfrak{n}_2)$$

is, however, the obvious one.

The reasons for these forms of standard definitions emerge with 32.7 Theorem below.

Exercise. *Say that a semisimple Deligne representation of* $\mathcal{W}_F$ *is* indecomposable *if it cannot be written as a direct sum of two non-zero Deligne representations. Show that the indecomposable, semisimple Deligne representations of* $\mathcal{W}_F$ *are those of the form* $\rho \otimes \mathrm{Sp}(n)$, *for* $n \geqslant 1$ *and* $\rho \in \mathcal{G}_n^0(F)$.

31.3. We extend the machinery of L-functions and local constants to the class of semisimple Deligne representations as follows. Given a triple $(\rho, V, \mathfrak{n}) \in \mathcal{G}_n(F)$, for some n, the space $V_\mathfrak{n} = \mathrm{Ker}\,\mathfrak{n}$ carries a semisimple representation $\rho_\mathfrak{n}$ of $\mathcal{W}_F$. We set

$$L((\rho, V, \mathfrak{n}), s) = L(\rho_\mathfrak{n}, s). \tag{31.3.1}$$

We form the contragredient $(\check\rho, \check V, -\check{\mathfrak{n}})$ of $(\rho, V, \mathfrak{n})$ as above, and put

$$\varepsilon((\rho, V, \mathfrak{n}), s, \psi) = \varepsilon(\rho, s, \psi)\, \frac{L(\check\rho, 1{-}s)}{L(\rho, s)}\, \frac{L(\rho_\mathfrak{n}, s)}{L(\check\rho_{\check{\mathfrak{n}}}, 1{-}s)}\,. \tag{31.3.2}$$

Example. Consider the case $(\rho, V, \mathfrak{n}) = \mathrm{Sp}(2)$. Setting $\zeta_F(s) = (1{-}q^{-s})^{-1}$, we have

$$L(\rho, s) = \zeta_F(s{-}\tfrac12)\,\zeta_F(s{+}\tfrac12) = L(\check\rho, s),$$

while

$$L(\rho_\mathfrak{n}, s) = \zeta_F(s{+}\tfrac12) = L(\check\rho_{\check{\mathfrak{n}}}, s).$$

Taking ψ of level one, we get $\varepsilon(\rho, s, \psi) = q^{2s-1}$ (23.5 Proposition), and hence

$$\begin{aligned}\varepsilon(\mathrm{Sp}(2), s, \psi) &= q^{2s-1}\, \frac{\zeta_F(\tfrac12{-}s)\zeta_F(\tfrac32{-}s)}{\zeta_F(s{-}\tfrac12)\zeta_F(s{+}\tfrac12)}\, \frac{\zeta_F(s{+}\tfrac12)}{\zeta_F(\tfrac32{-}s)} \\ &= q^{2s-1}\, \frac{(1{-}q^{\frac12-s})}{(1{-}q^{s-\frac12})} = -q^{s-\frac12}.\end{aligned}$$

32. Relation with ℓ-adic Representations

In this section, we give some motivation for the introduction of Deligne representations in §31.

Let ℓ be a prime number. Most of the representations of Ω_F and $\mathcal{W}_F$ which arise "in nature" are actually *ℓ-adic representations:* they act continuously on vector spaces over extensions of the field $\mathbb{Q}_\ell$. The primary examples of this phenomenon are provided by étale cohomology groups.

We now explain how such representations can be classified in terms of complex Deligne representations. We assume throughout that $\ell \neq p$, the residual characteristic of F. The case $\ell = p$ is more complicated and we say nothing of it.

32.1. Let G be a locally profinite group, and let C be a field of characteristic zero. A representation $\pi : G \to \mathrm{Aut}_C(V)$ of G on a C-vector space V is defined to be *smooth* if every $v \in V$ has open stabilizer in G, just as in the standard case $C = \mathbb{C}$. Smooth representations of G on C-vector spaces form a category $\mathrm{Rep}_C(G)$.

Let C' be another field of characteristic zero, and $\iota : C \to C'$ a field embedding. If (π, V) is a smooth representation of G over C, we get a representation $\pi_{C'}$ of G on $C' \otimes_C V$ by

$$\pi_{C'}(g) : 1 \otimes v \longmapsto 1 \otimes \pi(g)v.$$

This representation is smooth, and we have a functor

$$\begin{aligned} \mathrm{Rep}_C(G) &\longrightarrow \mathrm{Rep}_{C'}(G), \\ (\pi, V) &\longmapsto (\pi_{C'}, C' \otimes_C V). \end{aligned}$$

If ι is an isomorphism, this functor is an equivalence of categories.

Example. We consider the case where C is an algebraically closed field of characteristic zero, equipped with an isomorphism $\mathbb{C} \cong C$. (We have in mind the case where C is an algebraic closure $\overline{\mathbb{Q}}_\ell$ of the field $\mathbb{Q}_\ell$ of ℓ-adic numbers, for a prime $\ell \neq p$.) We take $G = \mathrm{GL}_2(F)$.

We have an equivalence of categories $\mathrm{Rep}(G) \cong \mathrm{Rep}_C(G)$, and hence a bijection between the sets of equivalence classes of irreducible smooth representations of G over $\mathbb{C}$ and over C. As before, let B be the group of upper triangular matrices in G. We also have an equivalence $\mathrm{Rep}(B) \cong \mathrm{Rep}_C(B)$, and a commutative diagram

$$\begin{array}{ccc} \mathrm{Rep}(B) & \longrightarrow & \mathrm{Rep}_C(B) \\ \mathrm{Ind}_B^G \downarrow & & \downarrow \mathrm{Ind}_B^G \\ \mathrm{Rep}(G) & \longrightarrow & \mathrm{Rep}_C(G) \end{array}$$

Note that we cannot use normalized induction ι_B^G because, for $x \in F^\times$, the quantity $\|x\|^{1/2}$ is a power of $q^{1/2}$ and has no clear meaning in C.

So, if we re-write the classification theorem 9.11 in terms of Ind rather than ι, it makes sense and remains true for representations over C. Similarly, the equivalences of categories induced by the isomorphism $C \cong \mathbb{C}$ are compatible with the compact induction functor, so the classification theorem for cuspidal representations (15.5) holds for representations over C. Informally speaking, the classification of irreducible smooth representations of G is essentially independent of the coefficient field.

In particular, there is no need to attempt to reproduce or replace the proofs of the classification theorems by working directly over the field C. Observe, however, that concepts like unitarity and square-integrability make no sense at all for representations over C.

32.2. Similarly, the notion of a Deligne representation of $\mathcal{W}_F$ makes sense over any field C of characteristic zero: a Deligne representation of $\mathcal{W}_F$ on a finite-dimensional C-vector space V is a triple $\sigma = (\rho, V, \mathfrak{n})$, where (ρ, V) is a smooth representation of $\mathcal{W}_F$ on V and $\mathfrak{n} \in \mathrm{End}_C(V)$ is nilpotent, satisfying

$$\rho(x)\,\mathfrak{n}\,\rho(x)^{-1} = \|x\|\,\mathfrak{n}, \quad x \in \mathcal{W}_F.$$

(Observe that $\|x\|$, as defined in 28.4, is a power of q and so has an unambiguous meaning as an element of C.)

A homomorphism $(\rho_1, V_1, \mathfrak{n}_1) \to (\rho_2, V_2, \mathfrak{n}_2)$ of Deligne representations over C is a C-linear map $f : V_1 \to V_2$ such that $f \circ \rho_1(g) = \rho_2(g) \circ f$, for all $g \in \mathcal{W}_F$, and $f \circ \mathfrak{n}_1 = \mathfrak{n}_2 \circ f$. The class of Deligne representations of $\mathcal{W}_F$ over C forms a category, which we denote $\text{D-Rep}_C(\mathcal{W}_F)$.

A field embedding $\iota : C \to C'$ gives rise to a functor

$$\begin{aligned} \text{D-Rep}_C(\mathcal{W}_F) &\longrightarrow \text{D-Rep}_{C'}(\mathcal{W}_F), \\ (\rho, V, \mathfrak{n}) &\longmapsto (\rho_{C'}, C' \otimes V, 1 \otimes \mathfrak{n}). \end{aligned}$$

If $C \to C'$ is an isomorphism, this is an equivalence of categories.

32.3. We shall make frequent use of an elementary device. Let C be a field of characteristic zero and V a finite-dimensional C-vector space. If $\mathfrak{n} \in \mathrm{End}_C(V)$ is *nilpotent,* the series

$$\exp \mathfrak{n} = 1 + \sum_{j \geqslant 1} \frac{\mathfrak{n}^j}{j!}$$

has only finitely many terms, and represents a *unipotent* element of $\mathrm{Aut}_C(V)$.

In the opposite direction, if $\mathfrak{u} = 1+\mathfrak{n} \in \mathrm{Aut}_C(V)$ is unipotent, we get a nilpotent element

$$\log \mathfrak{u} = \sum_{j \geqslant 1} (-1)^{j-1} \frac{\mathfrak{n}^j}{j}$$

of $\mathrm{End}_C(V)$. We have the standard relations $\log(\exp \mathfrak{n}) = \mathfrak{n}$, $\exp(\log \mathfrak{u}) = \mathfrak{u}$.

32.4. Let ℓ be a prime number. We take as coefficient field an algebraic closure $\overline{\mathbb{Q}}_\ell$ of $\mathbb{Q}_\ell$. In particular, $\overline{\mathbb{Q}}_\ell$ is a topological $\mathbb{Q}_\ell$-algebra, the topology being defined by a valuation $\overline{\mathbb{Q}}_\ell \to \mathbb{Q} \cup \{\infty\}$. This defines a metric on $\overline{\mathbb{Q}}_\ell$, but $\overline{\mathbb{Q}}_\ell$ is *not* complete with respect to this metric.

Let V be a $\overline{\mathbb{Q}}_\ell$-vector space of finite dimension d. The choice of a basis identifies V with $\overline{\mathbb{Q}}_\ell^d$ and $\mathrm{Aut}_{\overline{\mathbb{Q}}_\ell}(V)$ with $\mathrm{GL}_d(\overline{\mathbb{Q}}_\ell)$. Thus $\mathrm{Aut}_{\overline{\mathbb{Q}}_\ell}(V)$ inherits a topology, and it is independent of the choice of basis.

If G is a locally profinite group, a representation (π, V) of G on V is said to be continuous if the implied map $\pi : G \to \mathrm{Aut}_{\overline{\mathbb{Q}}_\ell}(V)$ is continuous, relative to this topology on $\mathrm{Aut}_{\overline{\mathbb{Q}}_\ell}(V)$. Observe that a smooth representation of G on V is automatically continuous, but the converse statement does not apply.

32.5. Henceforth, we assume $\ell \neq p$. We take $G = \mathcal{W}_F$.

We recall (28.3.1) that there is a continuous surjection $t : \mathcal{I}_F \to \mathbb{Z}_\ell$. Its kernel fits into an exact sequence

$$1 \to \mathcal{P}_F \longrightarrow \operatorname{Ker} t \longrightarrow \prod_{\substack{m \text{ prime,} \\ m \neq \ell, p}} \mathbb{Z}_m \to 0,$$

and so has pro-order prime to ℓ. Such a map t is therefore unique up to multiplication by an element of $\mathbb{Z}_\ell^\times$, and so it satisfies the analogue of (28.3.3):

$$t(gxg^{-1}) = \|g\|\, t(x), \quad x \in \mathcal{I}_F,\ g \in \mathcal{W}_F. \tag{32.5.1}$$

Moreover, the kernel of t contains no open subgroup of $\mathcal{I}_F$.

The following result is the key to analyzing ℓ-adic representations of $\mathcal{W}_F$.

Theorem. *Let (σ, V) be a finite-dimensional, continuous representation of $\mathcal{W}_F$ over $\overline{\mathbb{Q}}_\ell$, $\ell \neq p$. There is a unique nilpotent endomorphism $\mathfrak{n}_\sigma \in \operatorname{End}_{\overline{\mathbb{Q}}_\ell}(V)$ such that*

$$\sigma(x) = \exp(t(x)\mathfrak{n}_\sigma),$$

for all elements x of some open subgroup of $\mathcal{I}_F$.

Proof. The uniqueness of $\mathfrak{n}_\sigma$ is straightforward. For, suppose there is a nilpotent $\mathfrak{n} \in \operatorname{End}_{\overline{\mathbb{Q}}_\ell}(V)$ and an open subgroup H of $\mathcal{I}_F$ such that

$$\sigma(y) = \exp(t(y)\mathfrak{n}), \quad y \in H.$$

The character $t \mid H$ is non-zero and, for any $y \in H$ with $t(y) \neq 0$, the operator $\sigma(y) \in \operatorname{Aut}_{\overline{\mathbb{Q}}_\ell}(V)$ is unipotent. We get $\mathfrak{n} = t(y)^{-1} \log \sigma(y)$.

The issue is therefore the existence of $\mathfrak{n}_\sigma$. Let $\mathfrak{O}_\ell$ denote the integral closure of $\mathbb{Z}_\ell$ in $\overline{\mathbb{Q}}_\ell$. Fixing a basis of V allows us to identify V with $\overline{\mathbb{Q}}_\ell^d$, $d = \dim_{\overline{\mathbb{Q}}_\ell} V$, and $\operatorname{Aut}_{\overline{\mathbb{Q}}_\ell}(V)$ with $\mathrm{GL}_d(\overline{\mathbb{Q}}_\ell)$. For an integer $m \geqslant 1$, let K_m denote the subgroup $1 + \ell^m \mathrm{M}_d(\mathfrak{O}_\ell)$ of $\mathrm{GL}_d(\overline{\mathbb{Q}}_\ell)$. Then K_m is an open subgroup of $\mathrm{GL}_d(\overline{\mathbb{Q}}_\ell)$ normalizing K_{m+1}. The quotient K_m/K_{m+1} is a discrete abelian group of exponent ℓ.

We view σ as a continuous homomorphism $\mathcal{W}_F \to \mathrm{GL}_d(\overline{\mathbb{Q}}_\ell)$. Let J denote the set of $g \in \operatorname{Ker} t$ such that $\sigma(g) \in K_2$. Thus J is an open subgroup of $\operatorname{Ker} t$ and $\sigma(J) \subset K_2$. The image of $\sigma(J)$ in K_2/K_3 is an abelian group of exponent ℓ; since $\operatorname{Ker} t$ has pro-order relatively prime to ℓ, this implies $\sigma(J) \subset K_3$. Inductively, we deduce that $\sigma(J) \subset K_m$ for all $m \geqslant 2$, so $\sigma(J)$ is trivial.

Since J is an open subgroup of $\operatorname{Ker} t$, there is an open subgroup H_0 of $\mathcal{I}_F$ such that $H_0 \cap \operatorname{Ker} t \subset J$. Shrinking H_0 if necessary, we can assume $\sigma(H_0) \subset K_2$. There is an open, normal subgroup H of $\mathcal{W}_F$, of finite index in $\mathcal{W}_F$, such that $H_0 \supset H \cap \mathcal{I}_F$.

The restriction of σ to $H \cap \mathcal{I}_F$ therefore factors through a continuous homomorphism $\phi : t(H \cap \mathcal{I}_F) \to K_2$, that is, $\sigma(h) = \phi(t(h))$, $h \in H \cap \mathcal{I}_F$.

Combining this with (32.5.1), we have

$$\sigma(\Phi h \Phi^{-1})^q = \sigma(h), \quad h \in H \cap \mathcal{I}_F, \tag{32.5.2}$$

for every Frobenius element $\Phi \in \mathcal{W}_F$. Let $h \in H \cap \mathcal{I}_F$, and let $v \in V$ be an eigenvector for $\sigma(h)$, with eigenvalue α, say. Thus $\sigma(\Phi)v$ is an eigenvector for $\sigma(\Phi h \Phi^{-1})$ with eigenvalue α. By (32.5.2), α^q is also an eigenvalue for $\sigma(h)$. The set of eigenvalues for $\sigma(h)$ is therefore invariant under the map $\alpha \mapsto \alpha^q$. As $\sigma(h)$ is invertible, this implies that the eigenvalues of $\sigma(h)$ are all roots of unity in $\overline{\mathbb{Q}}_\ell$. Since $\sigma(h) \in K_2$, the element $(\sigma(h)-1_V)/\ell^2$ is integral (over $\mathbb{Z}_\ell$).

Lemma.

(1) *If $\alpha \in \overline{\mathbb{Q}}_\ell$ is a root of unity such that $(\alpha-1)/\ell^2$ is integral, then $\alpha = 1$.*
(2) *For $h \in H \cap \mathcal{I}_F$, the matrix $\sigma(h) \in \mathrm{GL}_d(\overline{\mathbb{Q}}_\ell)$ is unipotent.*

Proof. Suppose first that α has order ℓ^r, $r \geqslant 1$. The extension $\mathbb{Q}_\ell(\alpha)/\mathbb{Q}_\ell$ is then totally ramified and $\alpha-1$ is prime in $\mathbb{Q}_\ell(\alpha)$. Thus $(\alpha-1)/\ell$ is not integral except in the case $\ell = 2$ and $r = 1$. In all cases, therefore, $(\alpha-1)/\ell^2$ is not integral.

Suppose next that $\alpha \neq 1$, and that the order of α is divisible by a prime number other than ℓ. We can write $\alpha = \beta\gamma$, for roots of unity $\beta \neq 1$ and γ, of order prime to ℓ and a power of ℓ respectively. We have $\alpha-1 = \beta(\gamma-1)+(\beta-1)$. The quantity $\beta-1$ is a unit in $\mathbb{Q}_\ell(\alpha)$ and $(\gamma-1)$ is not. So, $\alpha-1$ is a unit and $(\alpha-1)/\ell^2$ cannot be integral.

This proves (1), and (2) follows immediately. □

We now choose $h_0 \in H \cap \mathcal{I}_F$ such that $t(h_0) \neq 0$ and set $\mathfrak{n} = t(h_0)^{-1} \log \sigma(h_0)$. This element $\mathfrak{n}$ is nilpotent and $\sigma(h_0) = \phi(t(h_0)) = \exp(t(h_0)\mathfrak{n})$.

Now put $A = \mathbb{Z}_\ell t(h_0) \subset \mathbb{Z}_\ell$; we have two continuous homomorphisms $A \to \mathrm{GL}_d(\overline{\mathbb{Q}}_\ell)$, namely $x \mapsto \phi(x)$ and $x \mapsto \exp(x\mathfrak{n})$. They coincide on h_0, hence on $\mathbb{Z}h_0$ and also on the closure A of $\mathbb{Z}h_0$. Putting $H' = t^{-1}(A)$ we have $\sigma(y) = \exp(t(y)\mathfrak{n})$, for $y \in H'$. Finally, H' is open in $\mathcal{I}_F$, since A is open in $\mathbb{Z}_\ell$. □.

Remark. In the context of the theorem, the representation (σ, V) is smooth if and only if the nilpotent element $\mathfrak{n}_\sigma$ is zero. In particular, if $\dim \sigma = 1$, then σ is smooth.

32.6. We draw some conclusions. Let (σ, V) be a continuous, finite-dimensional representation of $\mathcal{W}_F$ over $\overline{\mathbb{Q}}_\ell$. The uniqueness of $\mathfrak{n}_\sigma$ and (32.5.1) together imply

$$\sigma(g)\, \mathfrak{n}_\sigma\, \sigma(g)^{-1} = \|g\|\, \mathfrak{n}_\sigma, \quad g \in \mathcal{W}_F. \tag{32.6.1}$$

In particular, $\mathfrak{n}_\sigma$ commutes with $\sigma(\mathcal{I}_F)$.

Fixing a Frobenius element $\Phi \in \mathcal{W}_F$, we put

$$\sigma_\Phi(\Phi^a x) = \sigma(\Phi^a x)\exp(-t(x)\mathfrak{n}_\sigma), \quad a \in \mathbb{Z},\ x \in \mathcal{I}_F. \tag{32.6.2}$$

The map $\sigma_\Phi : \mathcal{W}_F \to \mathrm{Aut}_{\overline{\mathbb{Q}}_\ell}(V)$ is a homomorphism which, by the theorem, is trivial on some open subgroup of $\mathcal{I}_F$. It therefore provides a smooth representation of $\mathcal{W}_F$. By (32.6.1), the triple $(\sigma_\Phi, V, \mathfrak{n}_\sigma)$ is a Deligne representation of $\mathcal{W}_F$ on V.

Theorem. *Let* $\mathrm{Rep}^f_{\overline{\mathbb{Q}}_\ell}(\mathcal{W}_F)$ *be the category of finite-dimensional, continuous representations of* $\mathcal{W}_F$ *over* $\overline{\mathbb{Q}}_\ell$. *Let* $\Phi \in \mathcal{W}_F$ *be a Frobenius element, and* $t : \mathcal{I}_F \to \mathbb{Z}_\ell$ *a continuous surjection. The assignment* $(\sigma, V) \mapsto (\sigma_\Phi, V, \mathfrak{n}_\sigma)$ *of* (32.6.2) *is functorial, and induces an equivalence of categories*

$$\mathrm{Rep}^f_{\overline{\mathbb{Q}}_\ell}(\mathcal{W}_F) \longrightarrow \text{D-Rep}_{\overline{\mathbb{Q}}_\ell}(\mathcal{W}_F). \tag{32.6.3}$$

The isomorphism class of the Deligne representation $(\sigma_\Phi, V, \mathfrak{n}_\sigma)$ *depends only on the isomorphism class of* (σ, V), *and not on the choices of* Φ *and* t.

Proof. We first check that the association $(\sigma, V) \mapsto (\sigma_\Phi, V, \mathfrak{n}_\sigma)$ is given by a functor.

If (ρ, U) is a continuous representation of $\mathcal{W}_F$ on a finite-dimensional $\overline{\mathbb{Q}}_\ell$-vector space U, and if $\phi : V \to U$ is a $\mathcal{W}_F$-homomorphism, the uniqueness property gives

$$\mathfrak{n}_\rho \circ \phi = \phi \circ \mathfrak{n}_\sigma.$$

Thus ϕ induces a $\mathcal{W}_F$-homomorphism $(\sigma_\Phi, V, \mathfrak{n}_\sigma) \to (\rho_\Phi, U, \mathfrak{n}_\rho)$ of Deligne representations.

In the opposite direction, let $(\tau, V, \mathfrak{n})$ be a Deligne representation of $\mathcal{W}_F$ over $\overline{\mathbb{Q}}_\ell$. The formula

$$\tau^\Phi(\Phi^a x) = \tau(\Phi^a x)\exp(t(x)\mathfrak{n}), \quad a \in \mathbb{Z},\ x \in \mathcal{I}_F,$$

defines a representation of $\mathcal{W}_F$ on V. It is continuous because τ is smooth and the homomorphism

$$\begin{aligned} \mathcal{I}_F &\longrightarrow \mathrm{Aut}_C(V), \\ x &\longmapsto \exp(t(x)\mathfrak{n}), \end{aligned}$$

is continuous. Moreover, $\mathfrak{n} = \mathfrak{n}_{(\tau^\Phi)_\Phi}$.

If $(\tau_1, V_1, \mathfrak{n}_1)$ is another Deligne representation and $\phi : (\tau, V, \mathfrak{n}) \to (\tau_1, V_1, \mathfrak{n}_1)$ is a homomorphism, then the map $\phi : V \to V_1$ gives a homomorphism $(\tau^\Phi, V) \to (\tau_1^\Phi, V_1)$. Thus $(\tau, V, \mathfrak{n}) \mapsto (\tau^\Phi, V)$ is a functor, inverse to the first.

Now let $\Phi' = \Phi x$, $x \in \mathcal{I}_F$, be some other Frobenius. We define an automorphism $A \in \mathrm{Aut}_{\overline{\mathbb{Q}}_\ell}(V)$ by

$$A = \exp(\lambda t(x)\mathfrak{n}_\sigma),$$

where $\lambda = (q-1)^{-1}$. For $y \in \mathcal{I}_F$, we have

$$A \circ \sigma_\Phi(y) \circ A^{-1} = \sigma(y) \exp\big((\lambda t(x) - t(y) - \lambda t(x))\mathfrak{n}_\sigma\big) = \sigma_{\Phi'}(y),$$

and

$$\begin{aligned} A \circ \sigma_\Phi(\Phi) \circ A^{-1} &= \exp(\lambda t(x)\mathfrak{n}_\sigma)\, \sigma(\Phi) \exp(-\lambda t(x)\mathfrak{n}_\sigma) \\ &= \sigma(\Phi) \exp(\lambda t(x) q \mathfrak{n}_\sigma) \exp(-\lambda t(x)\mathfrak{n}_\sigma) \\ &= \sigma(\Phi) \exp(\lambda(q-1)t(x)\mathfrak{n}_\sigma) \\ &= \sigma(\Phi) \exp(t(x)\mathfrak{n}_\sigma). \end{aligned}$$

On the other hand,

$$\sigma_{\Phi'}(\Phi) = \sigma_{\Phi'}(\Phi' x^{-1}) = \sigma(\Phi)\ \exp(t(x)\mathfrak{n}_\sigma) = A \circ \sigma_\Phi(\Phi) \circ A^{-1}.$$

Thus $A \circ \sigma_\Phi(g) \circ A^{-1} = \sigma_{\Phi'}(g)$, for all $g \in \mathcal{W}_F$.

This shows that, up to isomorphism, $(\sigma_\Phi, V, \mathfrak{n}_\sigma)$ is independent of Φ.

We next have to show that the isomorphism class of $(\sigma_\Phi, V, \mathfrak{n}_\sigma)$ is unchanged on replacing the character t by αt, for some $\alpha \in \mathbb{Z}_\ell^\times$. To do this, it is enough to produce an automorphism B of V, which commutes with $\sigma(\mathcal{W}_F)$, such that $B\mathfrak{n}_\sigma B^{-1} = \alpha\mathfrak{n}_\sigma$.

Choose a positive integer a such that $\sigma(\Phi^a)$ is central in $\sigma(\mathcal{W}_F)$. Let V_λ be the "generalized λ-eigenspace" of $\sigma(\Phi^a)$, that is, $V_\lambda = \mathrm{Ker}\,(\sigma(\Phi^a) - \lambda 1_V)^{\dim V}$. The relation $\Phi^a\, \mathfrak{n}_\sigma = \|\Phi\|^a\, \mathfrak{n}_\sigma\, \Phi^a$ implies $\mathfrak{n}_\sigma\, V_{\lambda q^a} \subset V_\lambda$. Each V_λ is a $\sigma(\mathcal{W}_F)$-subspace of V. We define B by

$$Bv = \mu_\lambda v, \quad v \in V_\lambda,$$

for a family of elements $\mu_\lambda \in \mathbb{Z}_\ell^\times$ satisfying $\alpha\mu_{\lambda q^a} = \mu_\lambda$. This completes the proof. □

Remark. The definitions in 31.2 are chosen to correspond, via the equivalence of the theorem, to the standard constructions in the category $\mathrm{Rep}^f_{\overline{\mathbb{Q}}_\ell}(\mathcal{W}_F)$.

The theorem gives a *canonical* bijection between the set of isomorphism classes of finite-dimensional continuous representations of $\mathcal{W}_F$ over $\overline{\mathbb{Q}}_\ell$ and the set of isomorphism classes of Deligne representations of $\mathcal{W}_F$ over $\overline{\mathbb{Q}}_\ell$. As in 32.1, the choice of an isomorphism $\overline{\mathbb{Q}}_\ell \cong \mathbb{C}$ gives further bijection of these sets with the set of isomorphism classes of Deligne representations of $\mathcal{W}_F$ over $\mathbb{C}$.

32.7. It is necessary to have a more refined version. As in 31.1, a Deligne representation $(\rho, V, \mathfrak{n})$ over $\overline{\mathbb{Q}}_\ell$ is called *semisimple* if the underlying smooth representation (ρ, V) is semisimple, in the sense of 28.7.

A continuous, finite-dimensional representation (σ, V) of $\mathcal{W}_F$ is defined to be *Φ-semisimple* if the associated Deligne representation $(\sigma_\Phi, V, \mathfrak{n}_\sigma)$ is semisimple.

Proposition. *Let (σ, V) be a finite-dimensional continuous representation of $\mathcal{W}_F$ over $\overline{\mathbb{Q}}_\ell$. The following conditions are equivalent:*

(1) *(σ, V) is Φ-semisimple;*
(2) *there is a Frobenius element $\Psi \in \mathcal{W}_F$ such that $\sigma(\Psi)$ is semisimple;*
(3) *$\sigma(g)$ is semisimple, for every $g \in \mathcal{W}_F \smallsetminus \mathcal{I}_F$.*

Proof. We form the Deligne representation $(\sigma_\Psi, V, \mathfrak{n}_\sigma)$, as in 32.6. As $\sigma_\Psi(\Psi) = \sigma(\Psi)$, the equivalence of (1) and (2) follows from 28.7 Proposition.

Surely (3) implies (2). Conversely, take $g \in \mathcal{W}_F \smallsetminus \mathcal{I}_F$. If g is of the form Ψx, $x \in \mathcal{W}_F$, the operators $\sigma(g)$, $\sigma(\Psi)$ are conjugate, as in the proof of 32.6 Theorem. Thus $\sigma(g)$ is semisimple. In general, $g = \Psi^a x$, for some $x \in \mathcal{I}_F$ and some integer $a \neq 0$. Restricting to $\mathcal{W}_E$, where E/F is unramified of degree $|a|$, the same argument shows that $\sigma(g)$ is semisimple, as required. □

As a consequence, we get:

Theorem. *Let ℓ be a prime number, $\ell \neq p$, and let $n \geqslant 1$ be an integer. The sets of isomorphism classes of the following objects are in canonical bijection:*

(1) *n-dimensional, Φ-semisimple, continuous representations of $\mathcal{W}_F$ over $\overline{\mathbb{Q}}_\ell$;*
(2) *n-dimensional, semisimple, Deligne representations of $\mathcal{W}_F$ over $\overline{\mathbb{Q}}_\ell$.*

The choice of an isomorphism $\overline{\mathbb{Q}}_\ell \cong \mathbb{C}$ induces a bijection between these sets and the set of isomorphism classes of n-dimensional, semisimple, Deligne representations of $\mathcal{W}_F$ over $\mathbb{C}$.

Further reading The concept of Weil group originates in [85]. A formal summary, incorporating the more subtle global version, is given in [83]. Local class field theory can be found in many texts, of which [74], [76] have been particularly influential. More recent and extremely compact, is [68]. Older works tend to define the Artin Reciprocity isomorphism as the inverse of the one used now: it takes *arithmetic* Frobenius elements of the Weil group to prime elements of F.

Langlands' original proof of the existence of the local constant used only local methods, but was long and only ever published informally [56]. Deligne

[27] discovered a brief global argument. The version we give here is adapted from Tate's Durham lectures [82] on a variant of Deligne's proof. Various attempts, based on "canonical" versions of the Brauer induction theorem, have been made to provide a more manageable local construction, but they are not useful in our context.

The algebraic number theory of §30 is standard fare: [54] is probably closest to what we use, particularly as it covers the Artin L-function (for Galois representations). Most of the key facts are also in [68]. The careful summary in [61] may be found helpful. The first part (30.7) is a summary of the global part of Tate's thesis [81].

The ℓ-adic context is summarized in [83]. The key result 32.5 is due to Grothendieck and is outlined in an appendix to [78]. The paper [78] discusses a significant area in which ℓ-adic representations of Weil groups arise naturally.

8

The Langlands Correspondence

We have to hand all of the concepts necessary for the statement of the central result. It concerns the existence of a canonical bijection, at the level of isomorphism classes, between the two-dimensional, semisimple Deligne representations of the Weil group $\mathcal{W}_F$ and the irreducible smooth representations of $G = \mathrm{GL}_2(F)$. This bijection, the *Langlands correspondence,* is specified in terms of relations between the L-functions and local constants of the two kinds of representation.

It will be clear from the L-function relations that the correspondence must match the *non-irreducible* representations of $\mathcal{W}_F$ with the *principal series* representations of G. Given the classification theory in Chapter III and the calculations in Chapter VI, it is straightforward to write down the correspondence for such representations: this is the first step below. The interest is in the matching of the *irreducible* representations of $\mathcal{W}_F$ with the irreducible *cuspidal* representations of G.

When the residual characteristic p of the base field F is *odd,* we get the result very quickly by using the Tame Parametrization Theorem of Chapter V. All that is required is a little Galois theory and some elementary Gauss sum manipulations. The case $p = 2$ is considerably more involved and interesting: it requires several chapters to itself.

Before going on to that, we look back at §32, particularly 32.7 Theorem. This shows that the Langlands correspondence implies a bijection between classes of irreducible smooth representations of G and classes of the appropriate two-dimensional ℓ*-adic* representations of $\mathcal{W}_F$. However, it relies on choosing an isomorphism $\mathbb{C} \cong \overline{\mathbb{Q}}_\ell$: the rationality properties of the Langlands correspondence, relative to automorphisms of the coefficient field $\mathbb{C}$, therefore become relevant. As it happens, they are not quite what is required. We show how to modify the original (complex) correspondence to get a canonical ℓ-adic correspondence. (We observe that §34, §35 are independent of each other, and could be read in either order.)

33. The Langlands Correspondence

We recall that $\mathcal{G}_2(F)$ denotes the set of equivalence classes of 2-dimensional, semisimple, Deligne representations of the Weil group $\mathcal{W}_F$. Analogously, let $\mathcal{A}_2(F)$ denote the set of equivalence classes of irreducible smooth representations of $G = \mathrm{GL}_2(F)$.

33.1. We arrive at the formal statement of the motivating result. As before, if χ is a character of $F^\times$, we write simply χ for the corresponding character $\chi \circ \boldsymbol{a}_F$ of $\mathcal{W}_F$.

Langlands Correspondence. *Let $\psi \in \widehat{F}$, $\psi \neq 1$. There is a unique map*

$$\pi : \mathcal{G}_2(F) \longrightarrow \mathcal{A}_2(F)$$

such that

$$\begin{aligned} L(\chi\pi(\rho), s) &= L(\chi \otimes \rho, s), \\ \varepsilon(\chi\pi(\rho), s, \psi) &= \varepsilon(\chi \otimes \rho, s, \psi), \end{aligned} \tag{33.1.1}$$

for all $\rho \in \mathcal{G}_2(F)$ and all characters χ of $F^\times$.

The map π is a bijection, and (33.1.1) *holds for all $\psi \in \widehat{F}$, $\psi \neq 1$.*

The map π is the *Langlands correspondence* for $\mathrm{GL}_2(F)$. We use the notation π_F when we need to specify the base field F.

33.2. We make some preliminary remarks. First, there is at most one map π satisfying (33.1.1), because of the Converse Theorem (27.1).

We have a decomposition

$$\mathcal{G}_2(F) = \mathcal{G}_2^1(F) \cup \mathcal{G}_2^0(F),$$

where $\mathcal{G}_2^0(F)$ is the set of equivalence classes of *irreducible* smooth representations of $\mathcal{W}_F$ of dimension two, and $\mathcal{G}_2^1(F)$ consists of those classes of semisimple Deligne representations $(\rho, V, \mathfrak{n}) \in \mathcal{G}_2(F)$ for which the representation ρ of $\mathcal{W}_F$ is reducible. Likewise,

$$\mathcal{A}_2(F) = \mathcal{A}_2^1(F) \cup \mathcal{A}_2^0(F),$$

where $\mathcal{A}_2^0(F)$ (resp. $\mathcal{A}_2^1(F)$) consists of the classes of representations $\pi \in \mathcal{A}_2(F)$ which are cuspidal (resp. non-cuspidal).

Proposition.

(1) *Let* $\pi \in \mathcal{A}_2(F)$; *then* $\pi \in \mathcal{A}_2^0(F)$ *if and only if* $L(\chi\pi, s) = 1$ *for all characters* χ *of* $F^\times$.
(2) *Let* $\rho \in \mathcal{G}_2(F)$; *then* $\rho \in \mathcal{G}_2^0(F)$ *if and only if* $L(\chi \otimes \rho, s) = 1$ *for all characters* χ *of* $F^\times$.

Proof. The first assertion is 27.2 Proposition. The second follows immediately from the definitions in 29.3 and 31.3. □

33.3. Any map $\boldsymbol{\pi}$ satisfying (33.1.1) must therefore take $\mathcal{G}_2^1(F)$ to $\mathcal{A}_2^1(F)$ and $\mathcal{G}_2^0(F)$ to $\mathcal{A}_2^0(F)$. We deal with the first half of the problem.

Theorem. *There is a unique map*

$$\boldsymbol{\pi}^1 : \mathcal{G}_2^1(F) \longrightarrow \mathcal{A}_2^1(F)$$

with the property

$$L(\chi\boldsymbol{\pi}^1(\rho), s) = L(\chi \otimes \rho, s),$$

for all $\rho \in \mathcal{G}_2^1(F)$ *and all characters* χ *of* $F^\times$. *The map* $\boldsymbol{\pi}^1$ *is bijective, and it satisfies*

$$\boldsymbol{\pi}^1(\chi \otimes \rho) = \chi\boldsymbol{\pi}^1(\rho),$$
$$\varepsilon(\boldsymbol{\pi}^1(\rho), s, \psi) = \varepsilon(\rho, s, \psi),$$

for all ρ, χ, *and all* $\psi \in \widehat{F}$, $\psi \neq 1$.

Proof. We simply exhibit the map $\boldsymbol{\pi}^1$: the required properties are easily verified from the definitions in 29.3, 31.3, and the list of L-functions and local constants in 26.1 Theorem.

So, let $(\rho, V, \mathfrak{n}) \in \mathcal{G}_2^1(F)$. Since ρ is semisimple, it is of the form $\chi_1 \oplus \chi_2$, for characters χ_i of $F^\times$. We form the character $\chi = \chi_1 \otimes \chi_2$ of the group T of diagonal matrices in G, and consider the induced representation $\pi = \iota_B^G\, \chi$ of G. If π is irreducible, the criterion in 9.11 Lemma implies that $\mathfrak{n} = 0$. We set $\boldsymbol{\pi}^1(\rho) = \pi = \iota_B^G\, \chi$.

This leaves the case where $\iota_B^G\, \chi$ is not irreducible. Thus there is a character ϕ of $F^\times$ such that $\chi_1(x) = \phi(x)\|x\|^{-1/2}$ and $\chi_2(x) = \phi(x)\|x\|^{1/2}$. There are two elements $(\rho, V, \mathfrak{n})$ of $\mathcal{G}_2^1(F)$ in which ρ has $\mathcal{W}_F$-composition factors χ_1, χ_2; they are distinguished by whether or not $\mathfrak{n}$ is zero. We put

$$\boldsymbol{\pi}^1(\rho, V, 0) = \phi \circ \det$$

in the one case and (*cf.* 31.3 Example)

$$\boldsymbol{\pi}^1(\rho, V, \mathfrak{n}) = \phi \cdot \mathrm{St}_G, \quad \mathfrak{n} \neq 0$$

in the other. □

33.4. The heart of the matter is therefore:

Theorem. *Let $\psi \in \widehat{F}$, $\psi \neq 1$. There is a unique map*

$$\boldsymbol{\pi} : \mathcal{G}_2^0(F) \longrightarrow \mathcal{A}_2^0(F)$$

with the property

$$\varepsilon(\chi \otimes \rho, s, \psi) = \varepsilon(\chi\boldsymbol{\pi}(\rho), s, \psi), \tag{33.4.1}$$

for all $\rho \in \mathcal{G}_2^0(F)$, all characters χ of $F^\times$.

Moreover, the map $\boldsymbol{\pi}$ is a bijection, and (33.4.1) *holds for all $\psi \in \widehat{F}$, $\psi \neq 1$.*

Again, the uniqueness of the map follows from 27.1, 27.2. We can make some helpful simplifications.

Proposition. *Let $\boldsymbol{\pi}$ be a map satisfying* (33.4.1) *relative to some $\psi \in \widehat{F}$, $\psi \neq 1$. Then:*

(1) *If $\rho \in \mathcal{G}_2^0(F)$ and $\pi = \boldsymbol{\pi}(\rho)$, then $\omega_\pi = \det \rho$.*
(2) *The map $\boldsymbol{\pi}$ satisfies* (33.4.1) *for all $\psi \in \widehat{F}$, $\psi \neq 1$.*

Proof. The first assertion follows from comparing the stability theorem 25.7 with its analogue 29.4 Proposition (4). The second then follows from 24.3 Proposition and 29.4 Proposition (2). □

Exercise. *Let $\boldsymbol{\pi}$ be a map satisfying the requirements of the theorem.*

(a) *Show that $\boldsymbol{\pi}(\chi \otimes \rho) = \chi\boldsymbol{\pi}(\rho)$, for $\rho \in \mathcal{G}_2^0(F)$ and any character χ of $F^\times$.*
(b) *If $\rho \in \mathcal{G}_2^0(F)$ and $\pi = \boldsymbol{\pi}(\rho)$, show that $\boldsymbol{\pi}(\check{\rho}) = \check{\pi}$.*

34. The Tame Correspondence

In this section, we prove the theorem of 33.4 in almost all cases, using an explicit construction based on that of Chapter V. Throughout, p denotes the characteristic of the finite field $\boldsymbol{k} = \mathfrak{o}/\mathfrak{p}$. We allow the case $p = 2$, although the results are complete only when $p \neq 2$.

34.1. We define a subset $\mathcal{G}_2^{\mathrm{nr}}(F)$ of $\mathcal{G}_2^0(F)$ as follows: the class $\rho \in \mathcal{G}_2^0(F)$ lies in $\mathcal{G}_2^{\mathrm{nr}}(F)$ if there is an unramified character $\chi \neq 1$ of $\mathcal{W}_F$ such that $\chi \otimes \rho \cong \rho$. If $\rho \in \mathcal{G}_2^0(F) \smallsetminus \mathcal{G}_2^{\mathrm{nr}}(F)$, we say that ρ is *totally ramified.*

Let $\mathbb{P}_2(F)$ denote the set of isomorphism classes of admissible pairs, as in 18.2. If $(E/F, \xi) \in \mathbb{P}_2(F)$, we can view ξ as a character of $\mathcal{W}_E$ and form the induced representation $\rho_\xi = \mathrm{Ind}_{E/F}\,\xi$ of $\mathcal{W}_F$.

Theorem. *If $(E/F, \xi)$ is an admissible pair, the representation ρ_ξ of $\mathcal{W}_F$ is irreducible. The map $(E/F, \xi) \mapsto \rho_\xi$ induces a bijection*

$$\begin{aligned} \mathbb{P}_2(F) &\xrightarrow{\approx} \mathcal{G}_2^0(F) \quad \text{if } p \neq 2, \text{ or} \\ \mathbb{P}_2(F) &\xrightarrow{\approx} \mathcal{G}_2^{\mathrm{nr}}(F) \quad \text{if } p = 2. \end{aligned} \tag{34.1.1}$$

Proof. Let $(E/F, \xi) \in \mathbb{P}_2(F)$ and $\sigma \in \mathrm{Gal}(E/F)$, $\sigma \neq 1$. Since ξ does not factor through $\mathrm{N}_{E/F}$, the characters ξ, ξ^σ of $E^\times$ are distinct. The Artin map $\mathcal{W}_E^{\mathrm{ab}} \cong E^\times$ is $\mathrm{Gal}(E/F)$-equivariant (29.1 property (5)) so the characters ξ, ξ^σ of $\mathcal{W}_E$ are distinct and the representation $\rho_\xi = \mathrm{Ind}_{E/F}\, \xi$ is consequently irreducible. The equivalence class of ρ_ξ surely only depends on the isomorphism class of $(E/F, \xi)$.

Lemma. *Let $(E/F, \xi)$ be an admissible pair, set $\rho_\xi = \mathrm{Ind}_{E/F}\, \xi$. Let $\varkappa = \varkappa_{E/F}$ be the non-trivial character of $F^\times$ which is null on $\mathrm{N}_{E/F}(E^\times)$. We then have $\rho_\xi \cong \varkappa \otimes \rho_\xi$. In particular, $\rho_\xi \in \mathcal{G}_2^{\mathrm{nr}}(F)$ if and only if E/F is unramified.*

Proof. We have $\varkappa \otimes \rho \cong \mathrm{Ind}_{E/F}(\varkappa_E \otimes \xi)$, where $\varkappa_E = \varkappa \circ \mathrm{N}_{E/F} = 1$. This gives one implication in the final assertion. Conversely, suppose $\rho_\xi \in \mathcal{G}_2^{\mathrm{nr}}(F)$. Let χ be the unramified quadratic character of $F^\times$. If σ generates $\mathrm{Gal}(E/F)$, then $\xi^\sigma/\xi = \chi_E$. This implies that $\xi \mid U_E^1$ factors through $\mathrm{N}_{E/F}$, whence, by the definition of admissible pair, the extension E/F is unramified. □

We now show that the map (34.1.1) is injective. Let $(E_i/F, \xi_i) \in \mathbb{P}_2(F)$, $i = 1, 2$, and suppose that $\rho_{\xi_1} \cong \rho_{\xi_2}$. If $E_1/F \cong E_2/F$, we may as well take $E_1 = E_2 = E$. The restriction $\rho_{\xi_1} \mid \mathcal{W}_E$ is $\xi_1 \oplus \xi_1^\sigma$, where $\sigma \in \mathrm{Gal}(E/F)$, $\sigma \neq 1$. It follows that $\xi_2 = \xi_1$ or ξ_1^σ, and $(E/F, \xi_2) \cong (E/F, \xi_1)$ in either case.

We therefore assume $E_1 \neq E_2$ and let $L = E_1E_2$. At least one of the E_i/F is totally ramified, so $[L{:}F] = 4$ and the maximal unramified sub-extension E/F of L/F has degree 2. Set $\varkappa_i = \varkappa_{E_i/F}$. The representation $\rho = \rho_{\xi_i}$ is fixed under tensoring with $\varkappa_1$ and $\varkappa_2$, hence also with $\varkappa_{E/F}$. That is, $\rho \in \mathcal{G}_2^{\mathrm{nr}}(F)$. The lemma now implies that $E_1 = E_2 = E$, contrary to hypothesis. We conclude that the map (34.1.1) is injective.

We have to prove it is surjective. If $\rho \in \mathcal{G}_2^{\mathrm{nr}}(F)$, elementary considerations show that ρ is induced from a character ξ of $\mathcal{W}_E$, where E/F is unramified quadratic. The pair $(E/F, \xi)$ is then admissible and $\rho \cong \rho_\xi$.

We therefore assume ρ totally ramified; in particular, $p \neq 2$. Since $\dim \rho = 2$, the restriction of ρ to the pro p-group $\mathcal{P}_F$ decomposes as a sum of characters. It follows that there is a finite, tamely ramified Galois extension K/F such that $\rho \mid \mathcal{W}_K$ decomposes as a sum of characters,

$$\rho \mid \mathcal{W}_K = \theta \oplus \theta',$$

say. Suppose first that $\theta \neq \theta'$. The $\mathcal{W}_F$-stabilizer of θ is therefore $\mathcal{W}_L$, for some quadratic extension L/F contained in K. The natural representation of

$\mathcal{W}_L$ in the θ-isotypic subspace of ρ provides a character ξ of $\mathcal{W}_L$ such that $\rho = \mathrm{Ind}_{L/F}\,\xi$. We view ξ as a character of $L^\times$; we have to show that the pair $(L/F, \xi)$ is admissible. Let $\sigma \in \mathrm{Gal}(L/F)$, $\sigma \neq 1$. The character $\xi^\sigma \mid \mathcal{W}_K$ is θ', so $\xi \neq \xi^\sigma$ and ξ does not factor through $\mathrm{N}_{L/F}$. Consider $\xi \mid U_L^1$. The extension L/F is totally ramified, since ρ is totally ramified. Thus, if $\xi \mid U_L^1$ factors through $\mathrm{N}_{L/F}$, the character ξ^σ/ξ is trivial on $U_L = U_F U_L^1$, and is unramified: $\xi^\sigma = \chi_L \xi$, for some unramified character χ of $F^\times$, $\chi \neq 1$. Thus $\rho \cong \chi \otimes \rho$, and $\rho \in \mathcal{G}_2^{\mathrm{nr}}(F)$, contrary to hypothesis. We conclude that, in this case, the pair $(L/F, \xi)$ is admissible and $\rho \cong \rho_\xi$.

It remains only to exclude the possibility $\theta = \theta'$. Let L/F be the maximal unramified sub-extension of K/F. Since K/L is cyclic, θ admits extension to a character ξ of $\mathcal{W}_L$, such that ξ occurs in $\rho \mid \mathcal{W}_L$. In other words, we could have taken K/F unramified. But this would imply $\rho \in \mathcal{G}_2^{\mathrm{nr}}(F)$, contrary to hypothesis. □

34.2. For $p \neq 2$, we have canonical bijections

$$\begin{aligned} \mathbb{P}_2(F) &\longrightarrow \mathcal{A}_2^0, & \mathbb{P}_2(F) &\longrightarrow \mathcal{G}_2^0(F), \\ (E/F,\xi) &\longmapsto \pi_\xi, & (E/F,\xi) &\longmapsto \rho_\xi, \end{aligned}$$

given by 20.2 and 34.1 respectively. The implied bijection

$$\begin{aligned} \mathcal{G}_2^0(F) &\xrightarrow{\approx} \mathcal{A}_2^0(F), \\ \rho_\xi &\longmapsto \pi_\xi, \end{aligned} \tag{34.2.1}$$

is not the map π *demanded by* 33.4. For example, if $(E/F, \xi) \in \mathbb{P}_2(F)$, the representation ρ_ξ has determinant $\varkappa_{E/F} \otimes \xi|F^\times$ while (see 20.2) π_ξ has central character $\xi|F^\times$, contrary to the requirement of 33.4 Proposition. When $p = 2$, we have a parallel situation for unramified admissible pairs. We obtain the map π of 33.4 by systematically modifying the bijection (34.2.1).

34.3. If K/F is a finite separable extension and $\psi \in \widehat{F}$, $\psi \neq 1$, let

$$\lambda_{K/F}(\psi) = \frac{\varepsilon(R_{K/F}, s, \psi)}{\varepsilon(1_K, s, \psi_K)}$$

denote the Langlands constant, as in (30.4.1), 30.4 Corollary.

Proposition. *Let* $\psi \in \widehat{F}$ *have level one, and let* K/F *be a tamely ramified quadratic extension.*

(1) *If* K/F *is unramified, then* $\varkappa_{K/F}$ *is unramified of order* 2 *and*

$$\lambda_{K/F}(\psi) = -1.$$

(2) *If K/F is totally ramified, then $\varkappa_{K/F}$ is the non-trivial character of $F^\times/\mathrm{N}_{K/F}(K^\times)$ and*

$$\lambda_{K/F}(\psi) = \tau(\varkappa_{K/F}, \psi)/q^{1/2}.$$

In particular, $\lambda_{K/F}(\psi)^2 = \varkappa_{K/F}(-1)$.

Proof. We have

$$R_{K/F} = \mathrm{Ind}_{K/F}\, 1_K = 1_F \oplus \varkappa_{K/F}.$$

The declared formulæ are then given by the proposition and theorem of 23.5, the final assertion being (30.4.3). □

We observe that, in (2), $\varkappa_{K/F}$ is of order 2, and ramified of level zero. Thus $\varkappa_{K/F}(-1)$ is given by the Jacobi symbol:

$$\varkappa_{K/F}(-1) = \left(\frac{-1}{q}\right). \tag{34.3.1}$$

34.4. We take an admissible pair $(E/F, \xi) \in \mathbb{P}_2(F)$. We associate to this pair a character $\Delta = \Delta_\xi$ of $E^\times$ of level zero. First:

Definition. *Let $(E/F, \xi)$ be an admissible pair in which E/F is unramified. Define Δ_ξ to be unramified, of order* 2.

The ramified case is more involved. We recall that $\boldsymbol{\mu}_F$ denotes the group of roots of unity in F of order prime to p. Let E/F be a totally tamely ramified quadratic extension, let ϖ be a prime element of E, and let $\beta \in E^\times$. Since $U_E = \boldsymbol{\mu}_E U_E^1 = \boldsymbol{\mu}_F U_E^1$, there is a unique root of unity $\zeta(\beta, \varpi) \in \boldsymbol{\mu}_F$ such that

$$\beta\varpi^{-v_E(\beta)} \equiv \zeta(\beta, \varpi) \pmod{U_E^1}. \tag{34.4.1}$$

Proposition-Definition. Let $\psi \in \widehat{F}$ have level one.

(1) *Let $(E/F, \xi) \in \mathbb{P}_2(F)$ be a minimal pair such that E/F is totally ramified. Let n be the level of ξ and let $\alpha \in \mathfrak{p}_E^{-n}$ satisfy $\xi(1{+}x) = \psi_E(\alpha x)$, $x \in \mathfrak{p}_E^n$. There is a unique character $\Delta = \Delta_\xi$ of $E^\times$ such that:*

$$\begin{aligned} \Delta \mid U_E^1 = 1, \quad \Delta \mid F^\times &= \varkappa_{E/F}, \\ \Delta(\varpi) = \varkappa_{E/F}(\zeta(\alpha, \varpi))\, &\lambda_{E/F}(\psi)^n, \end{aligned} \tag{34.4.2}$$

for any prime element ϖ of E. The definition of Δ_ξ is independent of the choices of ψ and α.

(2) *Let $(E/F, \xi) \in \mathbb{P}_2(F)$ and suppose that E/F is totally ramified. Write $\xi = \xi' \cdot \chi_E$, for a minimal pair $(E/F, \xi')$ and a character χ of $F^\times$. Define*

$$\Delta_\xi = \Delta_{\xi'}. \tag{34.4.3}$$

The definition of Δ_ξ is independent of the choice of decomposition $\xi = \xi' \cdot \chi_E$.

Proof. There is certainly at most one character Δ of $E^\times$ satisfying the conditions (34.4.2). If we fix the character ψ, the definition of Δ depends only on the coset αU_E^1. If we replace ψ by another character $\psi' \in \widehat{F}$ of level one, then $\psi' = u\psi$, for some $u \in U_F$. The element α is replaced by $u^{-1}\alpha$, and

$$\varkappa_{E/F}(\zeta(u^{-1}\alpha,\varpi))\,\lambda_{E/F}(u\psi)^n = \varkappa_{E/F}(\zeta(\alpha,\varpi))\,\lambda_{E/F}(\psi)^n\,\varkappa_{E/F}(u)^{n-1},$$

while $\varkappa_{E/F}(u)^{n-1} = 1$, since n is odd. In other words, the character (assuming it exists) is defined independently of the choices of ψ and α within the permitted ranges.

To establish its existence, we have to check the consistency conditions

$$\begin{aligned} \Delta(u\varpi) &= \Delta(u)\Delta(\varpi), \quad u \in U_E, \\ \Delta(\varpi^2) &= \Delta(\varpi)^2. \end{aligned}$$

For the first, we write $u = \eta v$, with $\eta \in \boldsymbol{\mu}_F$ and $v \in U_E^1$. This gives $\zeta(\alpha, u\varpi) \equiv \eta\zeta(\alpha,\varpi) \pmod{\boldsymbol{\mu}_F^2}$ and

$$\Delta(u\varpi) = \varkappa_{E/F}(\eta)\,\varkappa_{E/F}(\zeta(\alpha,\varpi))\,\lambda_{E/F}(\psi)^n = \Delta(u)\Delta(\varpi).$$

For the second, we note that $\varpi^2 \equiv -\mathrm{N}_{E/F}(\varpi) \pmod{U_E^1}$, whence by (34.3.1),

$$\Delta(\varpi^2) = \Delta(-\mathrm{N}_{E/F}(\varpi)) = \varkappa_{E/F}(-1) = \Delta(\varpi)^2,$$

as required.

The final assertion is immediate. □

Lemma.

(1) *If $(E/F,\xi)$ is an admissible pair, the pair $(E/F,\Delta_\xi\xi)$ is admissible and its isomorphism class depends only on that of $(E/F,\xi)$. The character Δ_ξ satisfies $\Delta_\xi^2 = 1$,* except *when E/F is totally ramified and $q \equiv 3 \pmod 4$. In the exceptional case, Δ_ξ has order 4.*

(2) *The map*

$$\begin{aligned} \mathbb{P}_2(F) &\longrightarrow \mathbb{P}_2(F), \\ (E/F,\xi) &\longmapsto (E/F,\Delta_\xi\xi), \end{aligned}$$

is bijective.

Proof. (1) follows directly from the definitions, 34.3 Proposition and (34.3.1). Part (2) follows from the observation that Δ_ξ is tamely ramified, depending only on E/F and $\xi \mid U_E^1$. □

Tame Langlands Correspondence.

(1) *Suppose $p \neq 2$. For $\rho \in \mathcal{G}_2^0(F)$, define $\boldsymbol{\pi}(\rho) = \pi_{\Delta_\xi \xi}$ for any $(E/F, \xi) \in \mathbb{P}_2(F)$ such that $\rho \cong \rho_\xi$. The map*

$$\boldsymbol{\pi} : \mathcal{G}_2^0(F) \longrightarrow \mathcal{A}_2^0(F)$$

is a bijection satisfying

$$\varepsilon(\chi \otimes \rho, s, \psi) = \varepsilon(\chi \boldsymbol{\pi}(\rho), s, \psi), \tag{34.4.4}$$

for all characters χ of $F^\times$ and all $\psi \in \widehat{F}$, $\psi \neq 1$.

(2) *In the case $p = 2$, the map $\rho_\xi \mapsto \pi_{\Delta_\xi \xi}$ is a bijection*

$$\boldsymbol{\pi} : \mathcal{G}_2^{\mathrm{nr}}(F) \longrightarrow \mathcal{A}_2^{\mathrm{nr}}(F)$$

satisfying the analogue of (34.4.4).

(3) *In both cases, the map $\boldsymbol{\pi}$ satisfies*

$$\boldsymbol{\pi}(\chi \otimes \rho) = \chi\, \boldsymbol{\pi}(\rho) \quad \text{and} \quad \boldsymbol{\pi}(\check{\rho}) = \boldsymbol{\pi}(\rho)^\vee, \tag{34.4.5}$$

for all ρ and all characters χ of $F^\times$.

Remark. Because of the uniqueness properties (33.2), $\boldsymbol{\pi}$ is the Langlands correspondence when $p \neq 2$, and is the restriction to $\mathcal{G}_2^{\mathrm{nr}}(F)$ of the Langlands correspondence when $p = 2$.

Proof. First we note that $\boldsymbol{\pi}$ is bijective because of the lemma, 34.1 Theorem and 20.2 Theorem, while (34.4.5) follows by construction and 20.2 Theorem (3).

The construction also gives $\omega_{\boldsymbol{\pi}(\rho)} = \det \rho$, so it is enough to check (34.4.4) for one choice of $\psi \in \widehat{F}$. We take ψ to have level one.

Further, $n(\rho, \psi) = n(\boldsymbol{\pi}(\rho), \psi)$ (as follows readily from 20.2 Theorem (1)), so we have only to check the ε-relation at $s = \frac{1}{2}$. We divide into cases.

Case 1. *Let $\rho = \mathrm{Ind}_{E/F}\, \xi$, where $(E/F, \xi)$ is minimal and E/F totally ramified.*

Thus ξ has odd level $n = 2m+1$. We choose α such that $\xi(1+x) = \psi_E(\alpha x)$, $x \in \mathfrak{p}_E^{m+1}$. Using the standard reduction technique (23.6.4), we get

$$\begin{aligned} \varepsilon(\xi, \tfrac{1}{2}, \psi_E) &= \check{\xi}(\alpha)\psi_E(\alpha), \\ \varepsilon(\rho, \tfrac{1}{2}, \psi) &= \varepsilon(\xi, \tfrac{1}{2}, \psi_E)\, \lambda_{E/F}(\psi) \\ &= \check{\xi}(\alpha)\psi_E(\alpha)\, \lambda_{E/F}(\psi). \end{aligned}$$

Setting $\pi = \boldsymbol{\pi}(\rho)$, 25.5 Corollary gives

$$\varepsilon(\pi, \tfrac{1}{2}, \psi) = \check{\Delta}_\xi(\alpha)\check{\xi}(\alpha)\psi_A(\alpha).$$

We have to check that $\check{\Delta}_\xi(\alpha) = \lambda_{E/F}(\psi)$.

We choose a prime element ϖ of E and $\zeta \in \mu_F$ such that $\alpha\varpi^n \equiv \zeta \pmod{U_E^1}$. The definition of $\Delta = \Delta_\xi$ gives

$$\Delta(\alpha) = \varkappa_{E/F}(\zeta)\,\Delta(\varpi^{-n}).$$

Since n is odd and $\lambda_{E/F}(\psi)$ is a 4-th root of unity, we get

$$\varkappa_{E/F}(\zeta)\,\Delta(\varpi^{-n})\,\lambda_{E/F}(\psi) = \varkappa_{E/F}(\zeta)^{1-n}\,\lambda_{E/F}(\psi)^{1-n^2} = 1,$$

as required.

Case 2. *Let* $\rho = \mathrm{Ind}_{E/F}\,\xi$ *be as in* Case 1, *and let* χ *be a character of* $F^\times$ *of level* $l \geqslant 1$.

We consider the representation $\chi \otimes \rho$. This is obtained, by induction, from the admissible pair $(E/F, \xi\chi_E)$. If $l \leqslant m$, the pair $(E/F, \xi\chi_E)$ is minimal and there is nothing more to do. Otherwise, χ_E has level $2l > n$ and there exists $c \in F$ such that $\chi(1+x) = \psi(cx)$, $x \in \mathfrak{p}^{[l/2]+1}$. Thus $\chi_E(1+y) = \psi_E(cy)$, $y \in \mathfrak{p}_E^{l+1}$. Further, $\xi(1+x) = \psi_E(\alpha x)$, $x \in \mathfrak{p}_E^l$. From (23.6.4), we therefore get

$$\begin{aligned}
\varepsilon(\xi\chi_E, \tfrac12, \psi_E)
&= q^{-1/2} \sum_{y\in\mathfrak{p}_E^l/\mathfrak{p}_E^{l+1}} \check{\xi}\check{\chi}_E\big((\alpha+c)(1+y)\big)\psi_E\big((\alpha+c)(1+y)\big)\\
&= q^{-1/2}\,\check{\xi}(c+\alpha)\,\check{\chi}_E(1+c^{-1}\alpha)\,\psi_E(\alpha) \sum_y \check{\chi}_E(c(1+y))\,\psi_E(c(1+y))\\
&= \check{\xi}(c+\alpha)\,\check{\chi}_E(1+c^{-1}\alpha)\,\psi_E(\alpha)\,\varepsilon(\chi_E, \tfrac12, \psi_E).
\end{aligned}$$

Therefore, via (30.4.2) and 23.6,

$$\begin{aligned}
\varepsilon(\chi\otimes\rho, \tfrac12, \psi) &= \check{\xi}(c+\alpha)\,\check{\chi}_E(1+c^{-1}\alpha)\,\psi_E(\alpha)\,\varepsilon(\chi_E, \tfrac12, \psi_E)\,\lambda_{E/F}(\psi)\\
&= \check{\xi}(c+\alpha)\,\check{\chi}_E(1+c^{-1}\alpha)\,\psi_E(\alpha)\,\varepsilon(\chi, \tfrac12, \psi)\,\varepsilon(\chi\varkappa_{E/F}, \tfrac12, \psi)\\
&= \check{\xi}(c+\alpha)\,\check{\chi}_E(1+c^{-1}\alpha)\,\psi_E(\alpha)\,\varepsilon(\chi, \tfrac12, \psi)^2\,\varkappa_{E/F}(c).
\end{aligned}$$

On the other hand, with $\pi = \pi(\rho)$ as before, the same techniques ((23.6.2), (25.6.1) and 25.5 Corollary) yield

$$\varepsilon(\chi\pi, \tfrac12, \psi) = \check{\Delta}\check{\xi}(c+\alpha)\,\check{\chi}(\det(1+c^{-1}\alpha))\,\psi_A(\alpha)\,\varepsilon(\chi, \tfrac12, \psi)^2.$$

However, $\psi_A(\alpha) = \psi_E(\alpha)$ and, since Δ is tamely ramified, we have$\check{\Delta}(c+\alpha) = \check{\Delta}(c) = \varkappa_{E/F}(c)$. We deduce

$$\varepsilon(\chi\otimes\rho, \tfrac12, \psi) = \varepsilon(\chi\pi, \tfrac12, \psi),$$

as required.

Case 3. *Let* $\rho = \mathrm{Ind}_{E/F}\,\xi$, *where* $(E/F, \xi)$ *is minimal and* E/F *is unramified.*

If the level n of ξ is odd, the argument is a simpler version of Case 1, so we omit the details. We therefore assume n is even. The case $n = 0$ follows from the identity (25.4.1) and straightforward manipulations. We therefore assume $n = 2m > 0$. We have (23.6.2), (23.6.4),

$$\varepsilon(\xi, \tfrac{1}{2}, \psi_E) = q^{-1} \sum_{x \in \mathfrak{p}_E^m / \mathfrak{p}_E^{m+1}} \check{\xi}(\alpha(1+x))\, \psi_E(\alpha(1+x)),$$

for α chosen suitably, in the usual way. On the other hand (25.5 Corollary),

$$\varepsilon(\pi, \tfrac{1}{2}, \psi) = q^{-3} \sum_{y \in \mathfrak{p}^m \mathfrak{M} / \mathfrak{p}^{m+1} \mathfrak{M}} \operatorname{tr} \Lambda^{\vee}(\alpha(1+y))\, \psi_A(\alpha(1+y)).$$

(Here, we use the notation of the construction of π_ξ in 19.4.)

Lemma. *Let $x \in U_{\mathfrak{M}}^m$; then $\operatorname{tr} \Lambda(\alpha x) = 0$ unless αx is $U_{\mathfrak{M}}^m$-conjugate to an element of $\alpha U_E^m U_{\mathfrak{M}}^{m+1}$.*

This is a restatement of 22.3 Lemma. Taking account of the number of elements of the conjugacy classes, we get

$$\begin{aligned} \varepsilon(\pi, \tfrac{1}{2}, \psi) &= -q^{-1} \sum_{y \in \mathfrak{p}_E^m / \mathfrak{p}_E^{m+1}} \check{\xi}(\alpha(1+y))\, \psi_A(\alpha(1+y)). \\ &= \varepsilon(\rho, \tfrac{1}{2}, \psi), \end{aligned}$$

as desired.

The remaining cases, of representations $\chi \otimes \rho$, where ρ is unramified and given by a minimal pair, follow much the same course as Case 2, so we omit the details.

We have completed the proof of 34.4 Theorem. We have also proved 33.1 Theorem in the case $p \neq 2$. □

35. The ℓ-adic Correspondence

We return to the Langlands correspondence $\boldsymbol{\pi} : \mathcal{G}_2(F) \to \mathcal{A}_2(F)$ in the general setting of 33.1. For the purposes of this section, we assume that 33.1 Theorem has been proved in all cases.

35.1. We consider representations over the field $\overline{\mathbb{Q}}_\ell$, $\ell \neq p$, as in 32.4 *et seq.*

Let $\mathcal{G}_2(F, \overline{\mathbb{Q}}_\ell)$ denote the set of equivalence classes of semisimple, Deligne representations of $\mathcal{W}_F$, over $\overline{\mathbb{Q}}_\ell$ and of dimension 2. We recall (32.7) that $\mathcal{G}_2(F, \overline{\mathbb{Q}}_\ell)$ is in canonical bijection with the set of equivalence classes of 2-dimensional, Φ-semisimple, continuous representations of $\mathcal{W}_F$ over $\overline{\mathbb{Q}}_\ell$.

Let $\mathcal{A}_2(F, \overline{\mathbb{Q}}_\ell)$ denote the set of equivalence classes of irreducible smooth representations of $G = \mathrm{GL}_2(F)$ over $\overline{\mathbb{Q}}_\ell$.

The choice of an isomorphism $\iota : \mathbb{C} \cong \overline{\mathbb{Q}}_\ell$ gives bijections

$$\begin{aligned} \mathcal{G}_2(F) &\cong \mathcal{G}_2(F, \overline{\mathbb{Q}}_\ell), & \mathcal{A}_2(F) &\cong \mathcal{A}_2(F, \overline{\mathbb{Q}}_\ell), \\ \sigma &\mapsto {}^\iota\sigma & \pi &\mapsto {}^\iota\pi. \end{aligned}$$

The Langlands correspondence thus gives a bijection $\mathcal{G}_2(F, \overline{\mathbb{Q}}_\ell) \cong \mathcal{A}_2(F, \overline{\mathbb{Q}}_\ell)$.

However, such a bijection *is not canonical,* in that it depends on the choice of ι. Put another way, let $\mathrm{Aut}\,\mathbb{C}$ denote the group of (not necessarily continuous) automorphisms of the field $\mathbb{C}$. The constructions of 32.1, 32.2 define actions of $\mathrm{Aut}\,\mathbb{C}$ on the sets $\mathcal{G}_2(F)$, $\mathcal{A}_2(F)$. We denote these

$$\varphi : \eta \longmapsto {}^\varphi\eta,$$

for $\varphi \in \mathrm{Aut}\,\mathbb{C}$ and $\eta \in \mathcal{G}_2(F)$ or $\mathcal{A}_2(F)$.

The problem arises from the fact that the Langlands correspondence does not respect this action. To get a canonical correspondence for representations over $\overline{\mathbb{Q}}_\ell$, one has to modify the ideas slightly.

Definition. *For $\rho \in \mathcal{G}_2^{\mathrm{ss}}(F)$, define $\tilde{\rho} \in \mathcal{G}_2^{\mathrm{ss}}(F)$ by*

$$\tilde{\rho} : x \longmapsto \|x\|^{-\frac{1}{2}}\, \rho(x), \quad x \in \mathcal{W}_F.$$

For $\sigma = (\rho, V, \mathfrak{n}) \in \mathcal{G}_2(F)$, define $\tilde{\sigma} = (\tilde{\rho}, V, \mathfrak{n})$.

Theorem. *The map*

$$\begin{aligned} \Pi_{\mathbb{C}} : \mathcal{G}_2(F) &\longrightarrow \mathcal{A}_2(F), \\ \sigma &\longmapsto \boldsymbol{\pi}(\tilde{\sigma}), \end{aligned} \tag{35.1.1}$$

is a bijection which commutes with the natural actions of $\mathrm{Aut}\,\mathbb{C}$*:*

$$\Pi_{\mathbb{C}}({}^\varphi\rho) = {}^\varphi\Pi_{\mathbb{C}}(\rho), \quad \rho \in \mathcal{G}_2(F),\ \varphi \in \mathrm{Aut}\,\mathbb{C}.$$

Moreover, $\Pi_{\mathbb{C}}$ is the unique map $\mathcal{G}_2(F) \to \mathcal{A}_2(F)$ which satisfies

$$\begin{aligned} L(\chi\Pi_{\mathbb{C}}(\sigma), s) &= L(\chi \otimes \sigma, s - \tfrac{1}{2}), \\ \varepsilon(\chi\Pi_{\mathbb{C}}(\sigma), s, \psi) &= \varepsilon(\chi \otimes \sigma, s - \tfrac{1}{2}, \psi), \end{aligned} \tag{35.1.2}$$

for all characters χ of $F^\times$ and all $\psi \in \widehat{F}$, $\psi \neq 1$.

Note that the condition (35.1.2) specifies $\Pi_{\mathbb{C}}$ uniquely via the Converse Theorem, as in the standard case.

We sketch a proof of the theorem in the following paragraphs. Before passing on to that, we record the desired consequence:

ℓ-adic Langlands Correspondence. *There is a unique bijection*

$$\Pi_\ell : \mathcal{G}_2(F, \overline{\mathbb{Q}}_\ell) \longrightarrow \mathcal{A}_2(F, \overline{\mathbb{Q}}_\ell),$$

such that

$$\Pi_\ell({}^\iota\sigma) = {}^\iota\Pi_{\mathbb{C}}(\sigma),$$

for all $\sigma \in \mathcal{G}_2(F)$ and all field isomorphisms $\iota : \overline{\mathbb{Q}}_\ell \cong \mathbb{C}$.

35.2. We outline a proof of 35.1 Theorem.

We first have to investigate the effect of an automorphism $\varphi \in \operatorname{Aut}\mathbb{C}$ on the constructions of §23. We let φ act on the rings $\mathbb{C}((X))$, $\mathbb{C}(X)$, $\mathbb{C}(q^{-s})$ etc. via the coefficients, *fixing X and q^{-s}*. In the definition of $Z(\Phi, \chi, X)$, we choose a Haar measure μ^* on $F^\times$ for which $\mu^*(U_F) \in \mathbb{Q}$. The measures of compact open subgroups of $F^\times$ then are all rational. The map $\Phi \mapsto \varphi\circ\Phi$ is an additive automorphism of $C_c^\infty(F)$. For a character χ of $F^\times$ and $\Phi \in C_c^\infty(F)$, we then have

$$Z(\varphi\circ\Phi, {}^\varphi\chi, X) = {}^\varphi Z(\Phi, \chi, X).$$

Consequently

$$L({}^\varphi\chi, s) = {}^\varphi L(\chi, s). \tag{35.2.1}$$

Now take $\psi \in \widehat{F}$ of level k. The composite $\varphi\circ\psi$ is again a smooth homomorphism $F \to \mathbb{C}^\times$, and so lies in $\widehat{F}$. The character $\varphi\circ\psi$ has level k, and hence ψ, $\varphi\circ\psi$ give rise to the same self-dual Haar measure on F. However, to compare the Fourier transforms $\mathcal{F}_\psi$, $\mathcal{F}_{\varphi\circ\psi}$ on $C_c^\infty(F)$, we have to remember that this self-dual measure $\mu_\psi = \mu_{\varphi\circ\psi}$ does not necessarily take rational values on compact open subgroups. Indeed, μ_ψ is rational on compact open subgroups if and only if q is a square or if k is even. Otherwise, these values lie in $\sqrt{q}\,\mathbb{Q}$. We deduce

$$\mathcal{F}_{\varphi\circ\psi}(\varphi\circ\Phi) = \alpha(\varphi,\psi)\cdot\varphi\circ(\mathcal{F}_\psi(\Phi)), \quad \Phi \in C_c^\infty(F),$$

where

$$\alpha(\varphi,\psi) = ({}^\varphi\sqrt{q}/\sqrt{q})^k.$$

It follows that

$$\varepsilon({}^\varphi\chi, s, \varphi\circ\psi) = \alpha(\varphi,\psi)\,{}^\varphi\varepsilon(\chi, s, \psi).$$

35.3. The definitions of $L(\sigma, s)$, $\sigma \in \mathcal{G}_n(F)$, give

$$L({}^\varphi\sigma, s) = {}^\varphi L(\sigma, s).$$

The machinery of induction constants, especially the uniqueness properties, implies

$$\varepsilon({}^\varphi\sigma, s, \varphi\circ\psi) = \alpha(\varphi,\psi)^{\dim\sigma}\cdot{}^\varphi\varepsilon(\sigma, s, \psi).$$

Therefore

$$\varepsilon({}^\varphi\sigma, s, \varphi\circ\psi) = {}^\varphi\varepsilon(\sigma, s, \psi), \quad \sigma \in \mathcal{G}_2(F).$$

35.4. For $\pi \in \mathcal{A}_2(F)$, $f \in \mathcal{C}(\pi)$ and $\Phi \in C_c^\infty(A)$, we similarly get

$$Z(\varphi\circ\Phi, {}^{\varphi}f, X) = {}^{\varphi}Z(\Phi, f, X),$$

and

$$L({}^{\varphi}\pi, s-\tfrac{1}{2}) = {}^{\varphi}L(\pi, s-\tfrac{1}{2}).$$

In this case, the self-dual Haar measure μ_ψ^A on A, relative to any non-trivial $\psi \in \widehat{F}$, takes rational values on compact open subgroups of A. In notation analogous to that of 35.2, we therefore have

$$\mathcal{F}_{\varphi\circ\psi}(\varphi\circ\Phi) = \varphi\circ\mathcal{F}_\psi(\Phi), \quad \Phi \in C_c^\infty(A),$$

which implies

$$\varepsilon({}^{\varphi}\pi, s-\tfrac{1}{2}, \varphi\circ\psi) = {}^{\varphi}\varepsilon(\pi, s-\tfrac{1}{2}, \psi).$$

The theorem now follows from the characterization (35.1.2). □

Further reading. We briefly discuss the Langlands Correspondence in higher dimensions. To state it precisely, one needs the L-function and the local constant of a *pair* of irreducible smooth representations π_i of $\mathrm{GL}_{n_i}(F)$, $i = 1, 2$. These are denoted $L(\pi_1\times\pi_2, s)$, $\varepsilon(\pi_1\times\pi_2, s, \psi)$ respectively. As before, the L-function is of the form $P_{\pi_1\times\pi_2}(q^{-s})^{-1}$, where $P_{\pi_1\times\pi_2}(t)$ is a polynomial with constant term 1, and $\varepsilon(\pi_1\times\pi_2, s, \psi)$ is a monomial in q^{-s}. The original account is in [48], but see also [79]. The crucial property is that, if π is an irreducible cuspidal representation of $\mathrm{GL}_n(F)$ (notation: $\pi \in \mathcal{A}_n^0(F)$), then π is determined, up to equivalence, by the function $\sigma \mapsto \varepsilon(\sigma\times\pi, s, \psi)$, as σ ranges over the sets $\mathcal{A}_m^0(F)$, $m \leqslant n-1$ [42]. If $\dim\sigma = 1$, the definition gives $\varepsilon(\sigma\times\pi, s, \psi) = \varepsilon(\sigma\pi, s, \psi)$ in the sense we use.

The Langlands Conjecture states that, for each $n \geqslant 1$, there is a unique bijection $\boldsymbol{\pi}_n : \mathcal{G}_n^0(F) \to \mathcal{A}_n^0(F)$ such that

$$\varepsilon(\rho_1\otimes\rho_2, s, \psi) = \varepsilon(\pi_1\times\pi_2, s, \psi),$$

for all $\rho_i \in \mathcal{G}_{n_i}^0(F)$, $\pi_i = \boldsymbol{\pi}_{n_i}(\rho_i)$. As we have noted in the Introduction, the conjecture is proved in all dimensions: the positive characteristic case is in [58], the characteristic zero case in [38] (or [43], building on [37]).

The technique of §34 can likewise be used to give an explicit account of the Langlands correspondence in dimension $n \not\equiv 0 \pmod{p}$: see [17], [18] (but the results are not complete at the time of writing). The considerations of §35 likewise apply in higher dimension: see [15].

9

The Weil Representation

Having proved the existence of the Langlands correspondence in the "tame" case $p \neq 2$ (34.4), we now need only concern ourselves with the case where the residual characteristic p of F is 2. To get started on this, however, we use a general construction valid for all p, and there is no advantage in imposing any restrictions. In fact, there is some overlap between the implications of this chapter and what we have done already. We get another definition of the Langlands correspondence when p is odd, and another proof of the Converse Theorem (27.1) for cuspidal representations. However, much of what one approach yields easily is quite difficult in the other, and it seems preferable to have both.

We first return to the mirabolic subgroup $M = SN$ of G, encountered in §8, Chapter III. The reason is that, if π is an irreducible cuspidal representation of G, the restriction of π to M is the unique irreducible smooth representation of M of infinite dimension. This can be realized, in an obvious way, on the space $C_c^\infty(F^\times)$. Thus $C_c^\infty(F^\times)$ carries a representation of G, equivalent to π and restricting to M in the specified fashion. Via the Bruhat decomposition in G, the cuspidal representation π is determined by its central character ω_π and the operator $\pi(w)$ on $C_c^\infty(F^\times)$, where w is the standard Weyl element $\left(\begin{smallmatrix} 0 & 1 \\ -1 & 0 \end{smallmatrix}\right)$. The miracle is that the operator $\pi(w)$ can be described in terms of the local constants $\varepsilon(\chi\pi, s, \psi)$, as χ ranges over the characters of $F^\times$.

There is a parallel strand. One starts with a separable quadratic extension E/F and constructs, from $E^\times$ and G, a group $\mathcal{G} = \mathcal{G}_{E/F}$. This is a locally profinite group, which has a subgroup structure and a Bruhat decomposition analogous to those of G. One uses this to construct a standard representation of $\mathcal{G}$ on the space $C_c^\infty(E)$: the process imitates the action of G on $C_c^\infty(F^\times)$ given by a cuspidal representation of G, but the Weyl element w acts as a Galois-twisted Fourier transform.

The compact group $\mathcal{K}_{E/F} = \operatorname{Ker} \mathrm{N}_{E/F}$ is a central subgroup of $\mathcal{G}$, so one can decompose the standard representation according to characters of $\mathcal{K}_{E/F}$. If Θ is a character of $E^\times$, non-trivial on $\mathcal{K}_{E/F}$, the component of the standard representation corresponding to $\Theta \mid \mathcal{K}_{E/F}$ gives rise to an irreducible cuspidal representation $\pi_{E/F}(\Theta)$ of G. This is the *Weil representation* defined by the datum $(E/F, \Theta)$. Its detailed structure is accessible to the methods of the first part of the chapter.

The hypothesis $\Theta \mid \mathcal{K}_{E/F} \neq 1$ is equivalent to the representation $\rho_\Theta = \operatorname{Ind}_{E/F} \Theta$ of $\mathcal{W}_F$ being irreducible. The process $\rho_\Theta \mapsto \pi_{E/F}(\Theta)$ then gives a well-defined map from a subset of $\mathfrak{G}_2^0(F)$ to $\mathcal{A}_2^0(F)$, with the properties demanded of a Langlands correspondence. When the residual characteristic p of F is odd, it is defined on all of $\mathfrak{G}_2^0(F)$ and is the Langlands correspondence. (In light of the results of 34.4, we do not give the complete argument.)

In the more difficult case $p = 2$, the existence of this "partial correspondence" is the starting point for the work of the later chapters XI and XII. It is indeed only the existence which matters for that. There will be no reason to re-visit the methods of this chapter: we give a self-contained statement of the outcome in §40.

Our discussion of $\mathcal{G}_{E/F}$ and its representations is an adaptation of a small part of an extensive general theory. We say nothing of this, beyond allowing the broader context to influence the choice of vocabulary.

36. Whittaker and Kirillov Models

Let ϑ be a non-trivial character of the group N of upper triangular unipotent matrices in $G = \mathrm{GL}_2(F)$.

36.1. Let (π, V) be an irreducible smooth representation of G; we consider the space V_ϑ, as in 8.1.

If $\dim V = 1$, then $\pi = \phi \circ \det$, for some character ϕ of $F^\times$. Clearly, $V_\vartheta = 0$ in this case. Otherwise:

Theorem. *Let ϑ be a non-trivial character of N. If (π, V) is an irreducible smooth representation of G of infinite dimension, then* $\dim V_\vartheta = 1$.

Before proving this result, we derive some consequences. Frobenius Reciprocity (2.4) gives canonical isomorphisms

$$\mathrm{Hom}_N(V_\vartheta, \vartheta) \cong \mathrm{Hom}_M(V, \mathrm{Ind}_N^M \vartheta) \cong \mathrm{Hom}_G(V, \mathrm{Ind}_N^G \vartheta). \tag{36.1.1}$$

The theorem asserts that the first of these spaces has dimension one, so we deduce:

Corollary 1. *Let (π, V) be an irreducible smooth representation of G of infinite dimension. There is a non-zero G-homomorphism $\phi : V \to \mathrm{Ind}_N^G \vartheta$ which is, moreover, uniquely determined up to a constant scalar factor.*

The map ϕ is injective, since π is irreducible. It follows that there is a unique G-subspace $\boldsymbol{W}(\pi, \vartheta)$ of $\mathrm{Ind}_N^G \vartheta$ which is isomorphic to π. It is called the *Whittaker model* of π.

The other isomorphism in (36.1.1) yields the first assertion of:

Corollary 2. *Let (π, V) be an irreducible smooth representation of G of infinite dimension.*

(1) *There is a non-zero M-homomorphism $f : V \to \mathrm{Ind}_N^M \vartheta$ which is, moreover, uniquely determined up to a constant scalar factor.*
(2) *The map f is injective.*
(3) *The image of f contains c-$\mathrm{Ind}_N^M \vartheta$, and f induces an isomorphism*

$$V_N \cong \mathrm{Im}\, f / c\text{-}\mathrm{Ind}_N^M \vartheta.$$

In particular, π is cuspidal if and only if $f(V) = c$-$\mathrm{Ind}_N^M \vartheta$.

Proof. Part (1) has already been noted. Since $\dim V_\vartheta = 1$, the composite map $V \to \mathrm{Ind}_N^M \vartheta \to \vartheta$ induces an isomorphism $V_\vartheta \cong \vartheta$. It follows that $(\mathrm{Ker}\, f)_\vartheta = 0$. Therefore $(\mathrm{Ker}\, f)(N) = 0$ (8.1 Corollary 1) and so N acts trivially on $\mathrm{Ker}\, f$. Thus $\mathrm{Ker}\, f = 0$ (9.2 Exercise 1).

Since f is injective, the space $f(V(N))$ is non-zero, and

$$f(V(N)) \subset (\mathrm{Ind}_N^M \vartheta)(N) = c\text{-}\mathrm{Ind}_N^M \vartheta$$

(8.2 Proposition (1)). Since c-$\mathrm{Ind}_N^M \vartheta$ is irreducible (8.2 Corollary), we deduce that

$$f(V) \supset f(V(N)) = c\text{-}\mathrm{Ind}_N^M \vartheta,$$

Therefore f induces an M-isomorphism $V_N = V/V(N) \cong f(V)/c$-$\mathrm{Ind}_N^M \vartheta$, and the final assertion follows. □

Therefore:

Corollary-Definition. *Let (π, V) be an irreducible smooth representation of G of infinite dimension, and let ϑ be a non-trivial character of M.*

(1) *There is a unique M-subspace $\boldsymbol{K} = \boldsymbol{K}(\pi, \vartheta)$ of $\mathrm{Ind}_N^M \vartheta$ which is M-isomorphic to V.*
(2) *There is a unique homomorphism $\pi_{\boldsymbol{K}} : G \to \mathrm{Aut}_{\mathbb{C}}(\boldsymbol{K})$ such that*
   (a) *$\pi_{\boldsymbol{K}} \mid M$ induces the natural action of M on $\boldsymbol{K}$, and*
   (b) *$\pi_{\boldsymbol{K}} \cong \pi$.*

The representation $(\pi_{\boldsymbol{K}}, \boldsymbol{K}(\pi, \vartheta))$ is called the Kirillov model of π.

Restating Corollary 2, we have $\boldsymbol{K}(\pi, \psi) \supset c\text{-}\mathrm{Ind}_N^M \vartheta$, with equality if and only if π is cuspidal.

The canonical isomorphism of Frobenius Reciprocity

$$\mathrm{Hom}_G(V, \mathrm{Ind}_N^G \vartheta) \cong \mathrm{Hom}_M(V, \mathrm{Ind}_N^M \vartheta)$$

is given by restricting functions to M. We deduce:

Corollary 3. *Restriction of functions from G to M induces an M-isomorphism $\boldsymbol{W}(\pi, \vartheta) \cong \boldsymbol{K}(\pi, \vartheta)$, and hence a G-isomorphism*

$$\boldsymbol{W}(\pi, \vartheta) \cong (\pi_{\boldsymbol{K}}, \boldsymbol{K}(\pi, \vartheta)).$$

36.2. We start the proof of 36.1 Theorem with the case of a *principal series* representation. Let χ be a character of T and consider the induced representation $(\Sigma, X) = \mathrm{Ind}_B^G \chi$. Let $W = X(N)$ be the kernel of the canonical B-map $X \to X_N$: this is the same as the space $W = V(N)$ in 9.7. The space X_N has dimension 2 and N acts trivially, so $W_\vartheta = X_\vartheta$ (8.1.3). However, W is irreducible as representation of M (9.7) so $W \cong c\text{-}\mathrm{Ind}_N^M \vartheta$ (8.3 Corollary), whence $\dim W_\vartheta = \dim X_\vartheta = 1$ (8.2 Proposition).

Let (π, V) be a G-composition factor of (Σ, X), of infinite dimension. If $\pi = \Sigma$ is irreducible, we have $V_\vartheta = X_\vartheta$ and this has dimension 1. Otherwise, Σ has G-composition length 2: one factor is V and the other, call it Y, has dimension 1. Thus $Y_\vartheta = 0$ and $\dim V_\vartheta = \dim X_\vartheta = 1$, as required.

36.3. We now assume that (π, V) is *cuspidal*. Thus $V_N = 0$, whence (8.1 Corollary 1) $V_\vartheta \neq 0$, for any $\vartheta \neq 1$. We have to show that $\dim V_\vartheta = 1$.

This assertion is insensitive to twisting by characters of $F^\times$. We therefore assume $\ell(\pi) \leqslant \ell(\chi\pi)$, for all characters χ of $F^\times$. We take a cuspidal type $(\mathfrak{A}, J, \Lambda)$ occurring in π; thus $\pi = c\text{-}\mathrm{Ind}_J^G \Lambda$.

Lemma 1.

(1) *The natural map $(T \cap J)\backslash T \to J\backslash G/N$ is a surjection.*
(2) *Let $t \in T$. The dimension of the space*

$$\mathrm{Hom}_N\big(c\text{-}\mathrm{Ind}_{N\cap t^{-1}Jt}^N(\Lambda^t), \vartheta\big)$$

is the multiplicity in Λ of the character $\vartheta^{t^{-1}} \mid N \cap J$.

Proof. In all cases, there is a field extension E/F, of degree 2, such that $E^\times \subset J$. Let B be the group of upper triangular matrices in G, as always. In particular, $B = TN$ and B is the G-stabilizer of the line Fe_1, $e_1 = \left(\begin{smallmatrix}1\\0\end{smallmatrix}\right)$, in F^2. Let $g \in G$ and consider the element ge_1. The group $E^\times$ acts transitively on the set of non-zero elements of F^2, so there exists $x \in E^\times$ such that $xge_1 = e_1$. Therefore $xg \in B$, and we deduce that $G = E^\times B = E^\times TN$. Since $J \supset E^\times$, it follows that the natural map

$$\begin{aligned}(T \cap J)\backslash T &\longrightarrow J\backslash G/N,\\ t &\longmapsto JtN,\end{aligned}$$

is surjective, as required for (1).

In part (2), we note that $N \cap t^{-1}JT$ is an open subgroup of N; the result now follows directly from Frobenius Reciprocity (2.5.2). □

Working explicitly with the functions in c-Ind Λ, we find:

$$\pi \mid N = \bigoplus_{g \in J\backslash G/N} c\text{-Ind}_{N\cap g^{-1}Jg}^N \Lambda^g,$$

$$\operatorname{Hom}_G(\pi, \operatorname{Ind}_N^G \vartheta) = \operatorname{Hom}_N(\pi, \vartheta) = \prod_{g \in J\backslash G/N} \operatorname{Hom}_N(c\text{-Ind}_{N\cap g^{-1}Jg}^N \Lambda^g, \vartheta).$$

According to Lemma 1, we can take the coset representatives g from T. For $t \in T$, we have

$$\operatorname{Hom}_N(c\text{-Ind}_{N\cap t^{-1}Jt}^N \Lambda^t, \vartheta) \cong \operatorname{Hom}_N(c\text{-Ind}_{N\cap J}^N \Lambda, \vartheta^{t^{-1}}).$$

We therefore have to show:

Lemma 2. *Let ϑ_1 be a non-trivial character of N such that $\vartheta_1 \mid N \cap J$ occurs in Λ.*

(1) *The character $\vartheta_1 \mid N \cap J$ occurs in Λ with multiplicity one.*
(2) *If ϑ_2 is a non-trivial character of N such that $\vartheta_2 \mid N \cap J$ occurs in Λ, then ϑ_2 is conjugate to ϑ_1 under $T \cap J$.*

Proof. We shall use our standard notation

$$N_j = \begin{pmatrix}1 & \mathfrak{p}^j\\ 0 & 1\end{pmatrix}, \quad j \in \mathbb{Z}.$$

Suppose first that $\ell(\pi) = 0$. We may take $J = F^\times \mathrm{GL}_2(\mathfrak{o})$. We have $N \cap J = N_0$. The restriction of Λ to $\mathrm{GL}_2(\mathfrak{o})$ is the inflation of an irreducible cuspidal representation of $\mathrm{GL}_2(\boldsymbol{k})$. It follows (*cf.* (6.4.1)) that $\Lambda \mid N_0$ is the direct sum of the non-trivial characters of N_0 which are trivial on N_1, each occurring with multiplicity one. Any two characters of N which are trivial on N_1, but not on N_0, are conjugate under $T \cap \mathrm{GL}_2(\mathfrak{o})$ (*cf.* 1.7), and the result follows in this case.

We therefore assume $\ell(\pi) > 0$, and fix a character $\psi \in \widehat{F}$ of level one. There is a simple stratum $(\mathfrak{A}, n, \alpha)$, $n \geqslant 1$, such that $\Lambda \in C(\psi_\alpha, \mathfrak{A})$. Setting $E = F[\alpha]$, we have $J = E^\times U_{\mathfrak{A}}^{[(n+1)/2]}$. We will work explicitly, so we assume that $\mathfrak{A} = \mathfrak{M}$ or $\mathfrak{I}$.

Lemma 3.

(1) *We have* $N \cap J = N \cap U_{\mathfrak{A}}^{[(n+1)/2]}$.
(2) *Let* ϑ *be a character of* N *such that* $\vartheta \mid N \cap J$ *occurs in* $\Lambda \mid N \cap J$. *Then* $\vartheta \mid N \cap U_{\mathfrak{A}}^{[n/2]+1} = \psi_\alpha \mid U_{\mathfrak{A}}^{[n/2]+1}$.
(3) *The restriction of* ψ_α *to* $U_{\mathfrak{A}}^n \cap N$ *is non-trivial.*
(4) *Any two characters of* N*, agreeing with* ψ_α *on* $N \cap U_{\mathfrak{A}}^{[n/2]+1}$*, are conjugate under* $T \cap U_{\mathfrak{A}}^{[(n+1)/2]} \subset T \cap J$.

Proof. Write $\mathfrak{P} = \mathrm{rad}\,\mathfrak{A}$. Any unipotent element of J is contained in $U_{\mathfrak{A}} \cap J = U_E U_{\mathfrak{A}}^{[(n+1)/2]}$. Writing $N = 1+\mathcal{N}$, let $x \in \mathcal{N} \cap (\mathfrak{o}_E + \mathfrak{P}^m)$, where $m = [(n+1)/2]$. We accordingly write $x = y+z$, with $y \in \mathfrak{o}_E$, $z \in \mathfrak{P}^m$. We choose integers j, k maximal for the properties $y \in \mathfrak{p}_E^j$, $z \in \mathfrak{P}^k$. If $j \geqslant k$, there is nothing to do. Otherwise, $j < k$ and $x^2 = 0 \in \mathfrak{P}^{2j} \smallsetminus \mathfrak{P}^{1+2j}$, which is nonsense. This proves (1).

Assertion (2) simply re-iterates the defining fact that $\Lambda \mid U_{\mathfrak{A}}^{[n/2]+1}$ is a multiple of ψ_α, and part (3) is given by the explicit form of simple elements. If $j \geqslant 0$ is an integer, we have $N \cap U_{\mathfrak{M}}^j = N_j$ and $N \cap U_{\mathfrak{I}}^j = N_{[j/2]}$. Part (4) now follows from a simple calculation. □

Lemma 3 reduces us to checking that, if ϑ is a character of N, then $\vartheta \mid J \cap N$ occurs in $\Lambda \mid J \cap N$ with multiplicity at most one. If $\dim \Lambda = 1$, this is trivial. We therefore assume that $n = 2m$ is even, so $\mathfrak{A} = \mathfrak{M}$. We have $\dim \Lambda = q$, $J = E^\times U_{\mathfrak{M}}^m$, and $N \cap J = N_m$. If θ is a character of N_m occurring in $\Lambda \mid N_m$, there are q distinct conjugates θ^x, $x \in T \cap U_{\mathfrak{M}}^m$, all of which occur in $\Lambda \mid J \cap N$. The restriction $\Lambda \mid J \cap N$ is therefore multiplicity-free. □

This completes the proof of 36.1 Theorem. □

37. Manifestation of the Local Constant

Let (π, V) be an irreducible *cuspidal* representation of G. In this section, we give a description of the local constant $\varepsilon(\pi, s, \psi)$ of π (24.2) in terms of the Kirillov model of π. The arguments can be extended to apply to any irreducible smooth representation (π, V) of infinite dimension, but we shall have no need for the non-cuspidal case.

37.1. Let $\psi \in \widehat{F}$, $\psi \neq 1$, and define a character ϑ of N by

$$\psi(x) = \vartheta \begin{pmatrix} 1 & x \\ 0 & 1 \end{pmatrix}, \quad x \in F.$$

We henceforward write

$$\boldsymbol{W}(\pi,\psi) = \boldsymbol{W}(\pi,\vartheta), \quad \boldsymbol{K}(\pi,\psi) = \boldsymbol{K}(\pi,\vartheta).$$

Since (π, V) is cuspidal, we have $\boldsymbol{K}(\pi,\psi) = c\text{-Ind}_N^M \vartheta$. As in 8.2 Gloss, restriction of functions to the group

$$S = \left\{ \begin{pmatrix} x & 0 \\ 0 & 1 \end{pmatrix} : x \in F^\times \right\} \cong F^\times$$

induces a linear isomorphism $c\text{-Ind}\,\vartheta \cong C_c^\infty(F^\times)$. Combining with 36.1 Corollary 3, we have linear isomorphisms

$$\boldsymbol{W}(\pi,\psi) \cong \boldsymbol{K}(\pi,\psi) = c\text{-Ind}\,\vartheta \cong C_c^\infty(F^\times).$$

The composite map transfers the actions of G on $\boldsymbol{W}(\pi,\psi)$, $\boldsymbol{K}(\pi,\psi)$ to one on $C_c^\infty(F^\times)$. This is isomorphic to π (or $\pi_{\boldsymbol{K}}$), so we continue to denote it

$$\pi_{\boldsymbol{K}} : G \longrightarrow \mathrm{Aut}_{\mathbb{C}}(C_c^\infty(F^\times)).$$

Remark. One often also refers to the representation $(\pi_{\boldsymbol{K}}, C_c^\infty(F^\times))$ as the Kirillov model of π.

This version of the representation $\pi_{\boldsymbol{K}}$ satisfies the following identities. If $\phi \in C_c^\infty(F^\times)$, $a, b \in F^\times$ and $x \in F$, then

$$\begin{aligned} \pi_{\boldsymbol{K}} \begin{pmatrix} a & 0 \\ 0 & 1 \end{pmatrix} \phi &: b \longmapsto \phi(ab), \\ \pi_{\boldsymbol{K}} \begin{pmatrix} 1 & x \\ 0 & 1 \end{pmatrix} \phi &: b \longmapsto \psi(bx)\phi(b), \\ \pi_{\boldsymbol{K}} \begin{pmatrix} a & 0 \\ 0 & a \end{pmatrix} \phi &= \omega_\pi(a)\phi. \end{aligned} \tag{37.1.1}$$

The Bruhat decomposition (5.2)

$$G = B \cup BwN, \quad w = \begin{pmatrix} 0 & 1 \\ -1 & 0 \end{pmatrix},$$

taken together with (37.1.1), shows that the operators $\pi_{\boldsymbol{K}}(g)$, $g \in G$, can be expressed purely in terms of $\pi_{\boldsymbol{K}}(w)$ and $\pi_{\boldsymbol{K}} \mid B$. The representation $\pi_{\boldsymbol{K}}$ is therefore completely determined by ω_π and the operator $\pi_{\boldsymbol{K}}(w)$ on $C_c^\infty(F^\times)$.

We will need to know how the Kirillov model behaves with respect to twisting:

Lemma. *Let π be an irreducible cuspidal representation of G. Let χ be a character of $F^\times$ and define*

$$\begin{aligned} t_\chi : C_c^\infty(F^\times) &\longrightarrow C_c^\infty(F^\times), \\ f &\longmapsto \chi^{-1} f. \end{aligned}$$

The map t_χ is a G-isomorphism $(\chi\pi)_{\boldsymbol{K}} \to \chi\pi_{\boldsymbol{K}}$.

Proof. We define a map t_χ on $\mathrm{Ind}_N^G \vartheta$ in the analogous way. Let $\pi_{\boldsymbol{W}}$ denote the natural representation of G on the function space $\boldsymbol{W}(\pi, \psi)$. For $f \in \boldsymbol{W}(\chi\pi, \psi)$,

$$t_\chi \circ (\chi\pi)_{\boldsymbol{W}}(g) f : x \longmapsto \chi^{-1}(\det x) f(xg) = \chi(\det g)\, \pi_{\boldsymbol{W}}(g) \circ t_\chi f(x).$$

That is, $t_\chi \circ (\chi\pi)_{\boldsymbol{W}} = \chi\pi_{\boldsymbol{W}} \circ t_\chi$. The lemma follows on restricting functions, as in 36.1 Corollary 3. □

37.2. We show how to express the operator $\pi_{\boldsymbol{K}}(w)$ in terms of local constants.

We have to re-interpret the discussion of 24.4 *et seq.* in the present context. Set $A = \mathrm{M}_2(F)$. For $m \in \mathbb{Z}$, let G_m be the set of $g \in G$ such that $\upsilon_F(\det g) = m$. As in (24.4.3), we put

$$z_m(\Phi, f) = \int_{G_m} \Phi(x) f(x)\, d^*x,$$

where $\Phi \in C_c^\infty(A)$, $f \in \mathcal{C}(\pi)$, and d^*x is a fixed Haar measure on G. The series

$$Z(\Phi, f, X) = \sum_{m\in\mathbb{Z}} z_m(\Phi, f)\, X^m$$

is then a Laurent polynomial in X (24.5 Lemma (3)). By 24.5 Corollary, 24.2 Corollary, there is a monomial $c(\pi, X, \psi) = c(\pi, \psi) X^{n(\pi,\psi)}$ such that

$$Z(\hat{\hat{\Phi}}, \check{f}, 1/q^2 X) = c(\pi, X, \psi)\, Z(\Phi, f, X). \tag{37.2.1}$$

We have $c(\pi, q^{-\frac{1}{2}-s}, \psi) = \varepsilon(\pi, s, \psi)$ (24.6.3) .

37.3. For a character χ of $F^\times$ and $k \in \mathbb{Z}$, we define a function $\xi_{\chi,k} \in C_c^\infty(F^\times)$ by

$$\xi_{\chi,k}(x) = \begin{cases} \chi(x) & \text{if } \upsilon_F(x) = k, \\ 0 & \text{otherwise.} \end{cases}$$

The set of functions $\xi_{\chi,k}$ provides a basis of $C_c^\infty(F^\times)$.

Theorem. *Let χ be a character of $F^\times$ and $n \in \mathbb{Z}$; then*

$$\pi_{\boldsymbol{K}}(w)\, \xi_{\chi,k} = c(\chi^{-1}\pi, q^{-1}, \psi)\, \xi_{\chi^{-1}\omega_\pi, -n(\chi^{-1}\pi,\psi)-k}. \tag{37.3.1}$$

The proof of the theorem will occupy the rest of the section.

Remark. The equation (37.3.1) can be written in the form

$$\pi_{\boldsymbol{K}}(w)\,\xi_{\chi,k} = \varepsilon(\chi^{-1}\pi,\tfrac{1}{2},\psi)\,\xi_{\chi^{-1}\omega_\pi,-n(\chi^{-1}\pi,\psi)-k}. \tag{37.3.2}$$

Once one knows that the local constants $\varepsilon(\chi\pi,s,\psi)$ determine ω_π (from, for instance, 27.4 Lemma), the theorem yields another proof of the Converse Theorem 27.1 for cuspidal representations.

37.4. To prove the theorem, we apply the functional equation (37.2.1) to carefully chosen functions Φ and f.

Let H be a compact open subgroup of G. The map

$$W \longmapsto \int_H W(h)\,d^*h, \quad W \in \boldsymbol{W}(\pi,\psi),$$

is a smooth linear form. So, for $W \in \boldsymbol{W}(\pi,\psi)$, the function

$$f_{W,H} : g \longmapsto \mu^*(H)^{-1}\int_H W(hg)\,d^*h, \quad g \in G,$$

is a coefficient of π.

For $W \in \boldsymbol{W}(\pi,\psi)$ and $\Phi \in C_c^\infty(A)$, we choose a compact open subgroup H of G which fixes Φ under left and right translation, and set

$$Z(\Phi,W,X) = Z(\Phi,f_{W,H},X) = \sum_{m\in\mathbb{Z}} z_m(\Phi,f_{W,H})\,X^m.$$

In particular,

$$z_m(\Phi,f_{W,H}) = \int_{G_m} \Phi(x)W(x)\,d^*x.$$

This definition is independent of the choice of H. We can perform analogous constructions using the function $W^\vee : g \mapsto W(g^{-1})$ in place of W. In these terms, the functional equation (37.2.1) reads

$$Z(\hat{\Phi},W^\vee,1/q^2X) = c(\pi,X,\psi)\,Z(\Phi,W,X). \tag{37.4.1}$$

37.5. Let $\psi \in \widehat{F}$ have level ν, so that ψ is trivial on $\mathfrak{p}^\nu$ but not on $\mathfrak{p}^{\nu-1}$. Let n be a positive integer (to be chosen "sufficiently large"), and let Φ_n be the function defined by

$$\operatorname{supp}\Phi_n = S_n = \begin{pmatrix} \mathfrak{p}^{\nu-n} & \mathfrak{p}^{\nu-n} \\ \mathfrak{p}^n & 1+\mathfrak{p}^n \end{pmatrix},$$

and

$$\Phi_n\begin{pmatrix} a & b \\ c & d \end{pmatrix} = \bar{\psi}(b), \quad \begin{pmatrix} a & b \\ c & d \end{pmatrix} \in S_n.$$

The function Φ_n lies in $C_c^\infty(A)$.

Let N' denote the group of lower triangular unipotent matrices in G, and consider the "big cell"

$$BN' = \left\{ \left(\begin{smallmatrix} a & b \\ c & d \end{smallmatrix}\right) \in G : d \neq 0 \right\}.$$

The cell BN' is the image of $F \times F^\times \times F^\times \times F$ under the bijective map

$$(t, z, a, y) \longmapsto z \left(\begin{smallmatrix} 1 & t \\ 0 & 1 \end{smallmatrix}\right) \left(\begin{smallmatrix} a & 0 \\ 0 & 1 \end{smallmatrix}\right) \left(\begin{smallmatrix} 1 & 0 \\ y & 1 \end{smallmatrix}\right) = z \left(\begin{smallmatrix} a+ty & t \\ y & 1 \end{smallmatrix}\right), \tag{37.5.1}$$

Further, $S_n \cap G$ is contained in BN' and is the image (under (37.5.1)) of the set

$$\mathfrak{p}^{\nu-n} \times U_F^n \times (\mathfrak{p}^{\nu-n} \cap F^\times) \times \mathfrak{p}^n.$$

For $x \in S_n \cap G$, written according to the decomposition (37.5.1), we have

$$\Phi_n(x) = \bar{\psi}(zt) = \bar{\psi}(t).$$

Let $W \in \boldsymbol{W}(\pi, \psi)$ be fixed by $K_n = U_{\mathfrak{M}}^n$. For $x \in S_n \cap G$, written in the form (37.5.1), we have

$$W(x) = \psi(t)\, W\left(\begin{smallmatrix} a & 0 \\ 0 & 1 \end{smallmatrix}\right),$$
$$\Phi_n(x) W(x) = W\left(\begin{smallmatrix} a & 0 \\ 0 & 1 \end{smallmatrix}\right).$$

Corresponding to (37.5.1), the Haar measure d^*x on BN' decomposes thus:

$$d^*x = \|a\|^{-1}\, d\mu(t)\, d\mu^\times(z)\, d\mu^\times(a)\, d\mu(y),$$

for Haar measures μ on F and $\mu^\times$ on $F^\times$ (*cf.* 7.6 Exercises). We compute

$$Z(\Phi_n, W, X) = k \sum_{m \geqslant \nu-n} q^m X^m \int_{\varpi^m U_F} W\left(\begin{smallmatrix} a & 0 \\ 0 & 1 \end{smallmatrix}\right) d\mu^\times(a),$$

where ϖ is a prime element of F and

$$k = \mu(\mathfrak{p}^{\nu-n})\, \mu(\mathfrak{p}^n)\, \mu^\times(U_F^n). \tag{37.5.2}$$

37.6. We consider the Fourier transform $\hat{\Phi}_n$. This has support

$$\Sigma_n = \begin{pmatrix} \mathfrak{p}^n & \mathfrak{p}^{\nu-n} \\ 1+\mathfrak{p}^n & \mathfrak{p}^{\nu-n} \end{pmatrix},$$

and

$$\hat{\Phi}_n \left(\begin{smallmatrix} a & b \\ c & d \end{smallmatrix}\right) = \psi(d), \quad \left(\begin{smallmatrix} a & b \\ c & d \end{smallmatrix}\right) \in \Sigma_n.$$

We have to compute

$$Z(\hat{\Phi}_n, W^\vee, X) = \sum_{m\in\mathbb{Z}} X^m \int_{G_m} \hat{\Phi}_n(x) W(x^{-1})\, d^*x.$$

The set $G \cap \Sigma_n$ is the image of the map

$$\mathfrak{p}^{\nu-n} \times U_F^n \times (\mathfrak{p}^{\nu-n} \cap F^\times) \times \mathfrak{p}^n \longrightarrow G,$$

$$(t, z, a, y) \longmapsto w^{-1} z \begin{pmatrix} 1 & 0 \\ y & 1 \end{pmatrix} \begin{pmatrix} 1 & 0 \\ 0 & a \end{pmatrix} \begin{pmatrix} 1 & t \\ 0 & 1 \end{pmatrix} = z \begin{pmatrix} -y & -a-ty \\ 1 & t \end{pmatrix}.$$

For such an element x of $\Sigma_n \cap G$, we have $\hat{\Phi}_n(x) = \psi(t)$, while

$$W(x^{-1}) = \bar{\psi}(t)\, W\big(\big(\begin{smallmatrix} 1 & 0 \\ 0 & a^{-1} \end{smallmatrix}\big)\, w\big) = \bar{\psi}(t)\, \omega_\pi(a)^{-1} W\big(\big(\begin{smallmatrix} a & 0 \\ 0 & 1 \end{smallmatrix}\big)\, w\big),$$

giving

$$\hat{\Phi}_n(x) W(x^{-1}) = \omega_\pi(a)^{-1}\, W\big(\big(\begin{smallmatrix} a & 0 \\ 0 & 1 \end{smallmatrix}\big)\, w\big).$$

Therefore

$$Z(\hat{\Phi}_n, W^\vee, X) = k \sum_{m \geqslant \nu - n} q^m X^m \int_{\varpi^m U_F} \omega_\pi(a)^{-1}\, W\big(\big(\begin{smallmatrix} a & 0 \\ 0 & 1 \end{smallmatrix}\big)\, w\big)\, d\mu^\times(a),$$

with the same k as before (37.5.2).

37.7. For $W \in \boldsymbol{W}(\pi, \psi)$, we put

$$\Psi(W, X) = \sum_{m\in\mathbb{Z}} X^m \int_{\varpi^m U_F} W\big(\begin{smallmatrix} a & 0 \\ 0 & 1 \end{smallmatrix}\big)\, d\mu^\times(a),$$

$$\widetilde{\Psi}(W, X) = \sum_{m\in\mathbb{Z}} X^m \int_{\varpi^m U_F} \omega_\pi(a)^{-1} W\big(\begin{smallmatrix} a & 0 \\ 0 & 1 \end{smallmatrix}\big)\, d\mu^\times(a).$$

Since the function $a \mapsto W\big(\begin{smallmatrix} a & 0 \\ 0 & 1 \end{smallmatrix}\big)$ has compact support in $F^\times$, the expressions $\Psi(W, X)$, $\widetilde{\Psi}(W, X)$ are Laurent polynomials in X. For n sufficiently large, we have

$$\begin{aligned} Z(\Phi_n, W, X) &= k\, \Psi(W, qX), \\ Z(\hat{\Phi}_n, W^\vee, X) &= k\, \widetilde{\Psi}(\pi(w)W, qX). \end{aligned}$$

The functional equation (37.4.1) therefore reduces to

$$\widetilde{\Psi}(\pi(w)W, X) = c(\pi, 1/qX, \psi)\, \Psi(W, 1/X). \tag{37.7.1}$$

We now take W so that

$$W\big(\begin{smallmatrix} a & 0 \\ 0 & 1 \end{smallmatrix}\big) = \xi_{1,m}(a),$$

for some integer m, where 1 denotes the trivial character of $F^\times$. Then

$$\Psi(W, X) = \mu^\times(U_F)\, X^m.$$

We deduce from (37.7.1) that

$$\widetilde{\Psi}(\pi(w)W, X) = \mu^*(U_F)\, c(\pi, 1/q, \psi)\, X^{-m-n(\pi,\psi)}. \tag{37.7.2}$$

Under the action of the group $\left(\begin{smallmatrix} U_F & 0 \\ 0 & 1 \end{smallmatrix}\right)$, it is clear from the choice of W that

$$a \longmapsto \pi(w)W\left(\begin{smallmatrix} a & 0 \\ 0 & 1 \end{smallmatrix}\right)$$

transforms according to the character ω_π. It follows that $\pi_{\boldsymbol{K}}(w)\xi_{1,m}$ is a linear combination of the functions $\xi_{\omega_\pi,k}$, $k \in \mathbb{Z}$. The equation (37.7.2) implies

$$\pi_{\boldsymbol{K}}(w)\, \xi_{1,m} = c(\pi, 1/q, \psi)\, \xi_{\omega_\pi, -n(\pi,\psi)-m},$$

and the theorem follows in the case $\chi = 1$.

Applying this special case to the representation $\chi^{-1}\pi$, we get

$$(\chi^{-1}\pi)_{\boldsymbol{K}}(w)\, \xi_{1,m} = c(\chi^{-1}\pi, 1/q, \psi)\, \xi_{\chi^{-2}\omega_\pi, -n(\chi^{-1}\pi,\psi)-m}.$$

The theorem now follows from 37.1 Lemma. □

38. A Metaplectic Representation

Let E/F be a separable quadratic extension, with $\mathrm{Gal}(E/F) = \{1, \sigma\}$. In this section, we first attach to E/F a locally profinite group $\mathcal{G} = \mathcal{G}_{E/F}$. We then fix a character $\psi \in \widehat{F}$, $\psi \neq 1$. We construct a smooth representation $\eta_\psi^{\mathcal{G}}$ of the group $\mathcal{G}_{E/F}$ on the vector space $C_c^\infty(E)$.

38.1. Let $\varkappa = \varkappa_{E/F}$ be the character of $F^\times$ with kernel $\mathrm{N}_{E/F}(E^\times)$ and let $G_\varkappa$ denote the kernel of the map $g \mapsto \varkappa(\det g)$, $g \in G$. Let

$$\mathcal{K} = \mathcal{K}_{E/F} = \mathrm{Ker}\, \mathrm{N}_{E/F} \subset U_E.$$

We define

$$\mathcal{G} = \mathcal{G}_{E/F} = \left\{(g, h) : g \in G,\ h \in E^\times, \det g = \mathrm{N}_{E/F}(h)^{-1}\right\}.$$

This is a closed subgroup of $G \times E^\times$, and so is locally profinite.

Projecting on the two factors, we get exact sequences

$$1 \to \mathcal{K} \longrightarrow \mathcal{G} \longrightarrow G_\varkappa \to 1,$$
$$1 \to \mathrm{SL}_2(F) \longrightarrow \mathcal{G} \longrightarrow E^\times \to 1.$$

38.2. The group $\mathcal{G} = \mathcal{G}_{E/F}$ has distinguished subgroups similar to those of G. It has particular families of elements:

$$\begin{aligned} n(t) &= \left(\left(\begin{smallmatrix} 1 & t \\ 0 & 1 \end{smallmatrix}\right), 1\right), & t &\in F, \\ a(x) &= \left(\left(\begin{smallmatrix} \mathrm{N}_{E/F}(x) & 0 \\ 0 & 1 \end{smallmatrix}\right), x^{-\sigma}\right), & x &\in E^\times, \\ z(\zeta) &= \left(\left(\begin{smallmatrix} \zeta & 0 \\ 0 & \zeta \end{smallmatrix}\right), \zeta^{-1}\right), & \zeta &\in F^\times. \end{aligned} \tag{38.2.1}$$

(Here, $x^{-\sigma}$ means $(x^{-1})^\sigma = (x^\sigma)^{-1}$.) The sets

$$\mathcal{N} = \{n(t) : t \in F\}, \quad \mathcal{A} = \{a(x) : x \in E^\times\}, \quad \mathcal{Z} = \{z(\zeta) : \zeta \in F^\times\},$$

are then subgroups of $\mathcal{G}$. The set $\mathcal{ZA}$ is also a group; if we let $\mathcal{A}^1$ be the group of elements $a(x)$, $x \in \mathcal{K}$, then $\mathcal{A}^1 \cong \mathcal{K}$ and $\mathcal{ZA}^1$ is the centre of $\mathcal{G}$.

The product map

$$\mathcal{Z} \times \mathcal{A} \times \mathcal{N} \longrightarrow \mathcal{G}$$

is injective. Its image, denoted $\mathcal{B}$, is given by

$$\mathcal{B} = \mathcal{ZAN} = \{(b, x) \in B \times E^\times : \det(b)\,\mathrm{N}_{E/F}(x) = 1\};$$

$\mathcal{B}$ is a closed subgroup of $\mathcal{G}$.

We write

$$w = \left(\begin{smallmatrix} 0 & 1 \\ -1 & 0 \end{smallmatrix}\right).$$

Thus $(w, 1) \in \mathcal{G}$. We also denote this element w. The Bruhat decomposition for G implies:

Lemma. *The group $\mathcal{G}$ is the disjoint union $\mathcal{G} = \mathcal{B} \cup \mathcal{B}w\mathcal{N}$. The map*

$$\begin{aligned} \mathcal{B} \times \mathcal{N} &\longrightarrow \mathcal{B}w\mathcal{N}, \\ (b, n) &\longmapsto bwn, \end{aligned}$$

is a bijection.

38.3. The representation theory of the group $\mathcal{G}$ provides a link between representations of $G_\varkappa$ and characters of $E^\times$.

Let (π, V) be an irreducible smooth representation of the group $\mathcal{G} = \mathcal{G}_{E/F}$. Schur's Lemma (2.6) applies, so the centre $\mathcal{ZA}^1$ of $\mathcal{G}$ acts on V via a character θ. The restriction of θ to $\mathcal{A}^1 = \mathcal{K}$ can be extended to a character Θ of $E^\times$. The composition of Θ with the projection $\mathcal{G} \to E^\times$ defines a character $\Theta_\mathcal{G}$ of $\mathcal{G}$.

Consider the representation $\Theta_\mathcal{G}^{-1} \otimes \pi$. This is an irreducible representation of $\mathcal{G}$, in which $\mathcal{A}^1 = \mathcal{K}$ acts trivially. Thus $\Theta_\mathcal{G}^{-1} \otimes \pi$ is the inflation of an irreducible smooth representation π_0 of $\mathcal{G}/\mathcal{K} = G_\varkappa$. That is,

Proposition.

(1) *Let π be an irreducible smooth representation of $\mathcal{G}$. There is an irreducible smooth representation π_0 of $G_\varkappa$ and a character Θ of $E^\times$ such that π is equivalent to the representation*

$$\pi_0\Theta : (g,x) \longmapsto \Theta(x)\pi_0(g), \quad (g,x) \in \mathcal{G} \subset G \times E^\times.$$

(2) *For $i = 1,2$, let π_i be an irreducible smooth representation of $G_\varkappa$ and Θ_i a character of $E^\times$; then $\pi_1\Theta_1 \cong \pi_2\Theta_2$ if and only if there exists a character χ of $F^\times$ such that $\Theta_2 = \Theta_1 \otimes \chi_E$ and π_2 is equivalent to the representation $g \mapsto \chi(\det g)^{-1}\pi_1(g)$, $g \in G_\varkappa$.*

In (2) we have used the notation $\chi_E = \chi \circ \mathrm{N}_{E/F}$.

38.4. We fix a character $\psi \in \widehat{F}$, $\psi \neq 1$, and use it to define a representation $\eta_\psi^{\mathcal{B}}$ of $\mathcal{B}$ on $C_c^\infty(E)$. For $f \in C_c^\infty(E)$ and $y \in E$, we set:

$$\begin{aligned} n(t)f &: y \longmapsto \psi(t\,\mathrm{N}_{E/F}(y))f(y), && t \in F, \\ a(x)f &: y \longmapsto f(xy), && x \in E^\times, \\ z(\zeta)f &: y \longmapsto f(y), && \zeta \in F^\times. \end{aligned} \tag{38.4.1}$$

Proposition. *There is a unique representation $\eta_\psi^{\mathcal{B}}$ of $\mathcal{B}$ on $C_c^\infty(E)$ satisfying the relations* (38.4.1). *The representation $\eta_\psi^{\mathcal{B}}$ is smooth.*

Proof. In each case $b = n(t), a(x), z(\zeta)$, the function bf lies in $C_c^\infty(E)$. The only non-trivial relation among the generators (38.2.1) is

$$a(x)\,n(t)\,a(x)^{-1} = n(t\,\mathrm{N}_{E/F}(x)), \quad t \in F,\ x \in E^\times.$$

To show that (38.4.1) defines a representation of $\mathcal{B}$, we therefore need only check that

$$a(x)\,n(t)\,a(x)^{-1}f : y \longmapsto \psi(t\,\mathrm{N}_{E/F}(xy))f(y), \quad f \in C_c^\infty(E).$$

This is straightforward.

Let $f \in C_c^\infty(E)$; then f is surely fixed, under the representation $\eta_\psi^{\mathcal{B}}$, by $\mathcal{Z}$ and an open subgroup $\mathcal{A}_f$ of $\mathcal{A}$. Further, since f has compact support, it is fixed by an open subgroup $\mathcal{N}_f$ of $\mathcal{N}$. The set $\mathcal{Z}\mathcal{A}_f\mathcal{N}_f$ is open in $\mathcal{B}$, and it generates an open subgroup of $\mathcal{B}$ which fixes f. It follows that $\eta_\psi^{\mathcal{B}}$ is smooth. □

38.5. We consider the general problem of extending representations of $\mathcal{B}$ to representations of $\mathcal{G}$. The Bruhat decomposition (38.2 Lemma) gives:

Lemma. *For $i = 1, 2$, let (ρ_i, V) be a representation of $\mathcal{G}$, and suppose that $\rho_1(b) = \rho_2(b)$ for all $b \in \mathcal{B}$. Suppose also that $\rho_1(w) = \rho_2(w)$. We then have $\rho_1 = \rho_2$.*

Proposition. *Let (ρ, V) be a representation of $\mathcal{B}$, and let $A \in \mathrm{Aut}_{\mathbb{C}}(V)$. The following are equivalent:*

(1) *there is a representation $(\tilde\rho, V)$ of $\mathcal{G}$, satisfying $\tilde\rho(b) = \rho(b)$, $b \in \mathcal{B}$, and $\tilde\rho(w) = A$;*
(2) *the operator A satisfies*

$$\begin{aligned} A^2 &= \rho\big(z(-1)a(-1)\big), \\ A\,\rho\big(z(\zeta)\big) &= \rho\big(z(\zeta)\big)\,A, && \zeta \in F^\times, \\ A\,\rho\big(a(x)\big) &= \rho\big(z(\mathrm{N}_{E/F}\,x)a(x^{-\sigma})\big)\,A, && x \in E^\times, \\ A\,\rho\big(n(t)\big)\,A &= \rho\big(z(-t)a(-t^{-1})n(-t)\big)\,A\,\rho\big(n(-t^{-1})\big), && t \in F^\times. \end{aligned} \tag{38.5.1}$$

If ρ is smooth and these conditions hold, then $\tilde\rho$ is smooth.

Proof. In $\mathcal{G}$, we have the relations

$$\begin{aligned} w^2 &= z(-1)a(-1), \\ w\,z(\zeta) &= z(\zeta)\,w, && \zeta \in F^\times, \\ w\,a(x) &= z(\mathrm{N}_{E/F}\,x)\,a(x^{-\sigma})\,w, && x \in E^\times, \\ w\,n(t)\,w &= z(-t)\,a(-t^{-1})\,n(-t)\,w\,n(-t^{-1}), && t \in F^\times. \end{aligned}$$

The Bruhat decomposition shows that these relations, taken together with the obvious relations inside $\mathcal{B}$, give a presentation of $\mathcal{G}$. The first assertion follows immediately. For the second, take $v \in V$. There is an open subgroup $\mathcal{H}$ of $\mathcal{B}$ which fixes both v and Av. Thus v is fixed by the subgroup of $\mathcal{G}$ generated by $\mathcal{H}$ and $w\mathcal{H}w^{-1}$. This subgroup is open, so $\tilde\rho$ is smooth, as required. □

38.6. We now construct a representation $\eta_\psi^{\mathcal{G}}$ of the group $\mathcal{G} = \mathcal{G}_{E/F}$ on $C_c^\infty(E)$. We set $\psi_E = \psi \circ \mathrm{Tr}_{E/F}$, and write $f \mapsto \hat f$ for the Fourier transform operation on $C_c^\infty(E)$ defined using the Haar measure on E which is self-dual with respect to ψ_E. We recall that $\varkappa = \varkappa_{E/F}$ is the character of $F^\times$ with kernel $\mathrm{N}_{E/F}(E^\times)$. Let $\lambda_{E/F}(\psi)$ be the Langlands constant, as in 30.4.

Theorem. *There is a unique representation $\eta = \eta_\psi^{\mathcal{G}}$ of $\mathcal{G}$ on $C_c^\infty(E)$ satisfying:*

$$\begin{aligned} \eta\big(n(t)\big)f &: y \longmapsto \psi\big(t\,\mathrm{N}_{E/F}(y)\big)f(y), && t \in F, \\ \eta\big(a(x)\big)f &: y \longmapsto \|x\|_E^{1/2} f(xy), && x \in E^\times, \\ \eta\big(z(\zeta)\big)f &: y \longmapsto \varkappa(\zeta) f(y), && \zeta \in F^\times, \\ \eta(w)f &: y \longmapsto \lambda_{E/F}(\psi)\hat f(y^\sigma), \end{aligned} \tag{38.6.1}$$

for $f \in C_c^\infty(E)$ and $y \in E$. The representation $\eta_\psi^{\mathcal{G}}$ is smooth.

Proof. The first three relations in (38.6.1) define a representation η of $\mathcal{B}$ which is the twist, by a character, of the representation $\eta_\psi^{\mathcal{B}}$ of 38.4 Proposition. Explicitly,

$$\eta(g) = \varkappa(d)\|ad^{-1}\|_F^{1/2}\eta_\psi^{\mathcal{B}}(g), \quad g = \left(\left(\begin{smallmatrix} a & b \\ 0 & d \end{smallmatrix}\right), x\right) \in \mathcal{B}.$$

It is therefore enough to check the relations (38.5.1) for this representation η and the operator A defined by

$$Af : y \longmapsto \lambda_{E/F}(\psi)\, \hat{f}(y^\sigma), \quad f \in C_c^\infty(E),\ y \in E.$$

For $f \in C_c^\infty(E)$, the function $A^2 f$ is that given by

$$y \longmapsto \lambda_{E/F}(\psi)^2 f(-y) = \varkappa(-1)f(-y) = \eta\big(z(-1)a(-1)\big)f(y),$$

as required for the first relation in (38.5.1).

The second relation is immediate, so we turn to the third. For $f \in C_c^\infty(E)$ and $t \in E^\times$, let f^t temporarily denote the function $y \mapsto f(ty)$. We then have

$$\widehat{f^t}(y) = \|t\|_E^{-1}\hat{f}(t^{-1}y).$$

This gives us

$$A\eta\big(a(x)\big)f : y \longmapsto \lambda(E/F,\psi)\, \|x\|_E^{-1/2}\hat{f}(x^{-1}y^\sigma)$$

while, on the other hand,

$$\eta\big(z(\mathrm{N}_{E/F}\, x)a(x^{-\sigma})\big)Af : y \longmapsto \lambda(E/F,\psi)\, \|x\|_E^{-1/2}\hat{f}(x^{-1}y^\sigma),$$

since $\varkappa(\mathrm{N}_{E/F}\, x) = 1$. This proves the third relation in (38.5.1).

38.7. The fourth relation in (38.5.1) is a consequence of a property of the Fourier transform $f \mapsto \hat{f}$ on $C_c^\infty(E)$. This is a sort of twisted Fourier inversion formula, which we now develop. Set

$$\mathfrak{q} = \psi \circ \mathrm{N}_{E/F} : E \longrightarrow \mathbb{C}^\times$$

and

$$\langle x, y\rangle = \psi_E(xy^\sigma), \quad x, y \in E.$$

These functions satisfy the relation

$$\mathfrak{q}(x+y) = \mathfrak{q}(x)\,\mathfrak{q}(y)\,\langle x, y\rangle, \quad x, y \in E.$$

We also abbreviate

$$[f](y) = \hat{f}(y^\sigma), \quad f \in C_c^\infty(E),\ y \in E.$$

The function $\mathfrak{q}$ is locally constant on E, so $\mathfrak{q}f \in C_c^\infty(E)$ for any $f \in C_c^\infty(E)$.

Proposition. *For $f \in C_c^\infty(E)$ and a Haar measure dx on E, we have*

$$\int_E \mathfrak{q}(x)\,[f](x)\,dx = \lambda_{E/F}(\psi) \int_E \mathfrak{q}(x)^{-1} f(x)\,dx.$$

Proof. The assertion is independent of the choice of Haar measure dx, so we take dx self-dual with respect to ψ_E.

We first prove an approximate version of the proposition.

Lemma. *There is a constant $\gamma = \gamma_{E/F}(\psi) \in \mathbb{C}^\times$ such that*

$$\int_E \mathfrak{q}(x)\,[f](x)\,dx = \gamma \int_E \mathfrak{q}(x)^{-1} f(x)\,dx,$$

for $f \in C_c^\infty(E)$.

Proof. We define a linear form I on $C_c^\infty(E)$ by

$$I(f) = \int_E \mathfrak{q}(x)\,[\mathfrak{q}f](x)\,dx.$$

We fix $y \in E$ and consider the function $h = f_y : x \mapsto f(x+y)$. We have

$$\begin{aligned}
[\mathfrak{q}h](x) &= \int_E \mathfrak{q}h(t)\langle t, x\rangle\,dt \\
&= \int_E \mathfrak{q}(t) f(t+y)\langle t, x\rangle\,dt \\
&= \int_E \mathfrak{q}(t-y) f(t)\langle t-y, x\rangle\,dt,
\end{aligned}$$

so

$$\begin{aligned}
\mathfrak{q}(x)\,[\mathfrak{q}h](x) &= \int_E \mathfrak{q}(x)\mathfrak{q}(t-y) f(t)\langle t-y, x\rangle\,dt = \int_E \mathfrak{q}(x+t-y) f(t)\,dt \\
&= \mathfrak{q}(x-y) \int_E \mathfrak{q}(t) f(t)\langle x-y, t\rangle\,dt = \mathfrak{q}(x-y)\,[\mathfrak{q}f](x-y).
\end{aligned}$$

Integrating over $x \in E$, we get $I(f_y) = I(f)$, so I is a scalar multiple of Haar measure. The map $f \mapsto \mathfrak{q}\,[\mathfrak{q}f]$ is bijective on $C_c^\infty(E)$, so the functional I is nonzero and the lemma is proved. □

The proof of the proposition continues in the next paragraph.

38.8. It remains to show that

$$\gamma_{E/F}(\psi) = \lambda_{E/F}(\psi). \tag{38.8.1}$$

We first observe:

Lemma. *The constant* $\gamma = \gamma_{E/F}(\psi)$ *satisfies* $\gamma\bar{\gamma} = 1$.

Proof. A simple calculation shows that, for $f \in C_c^\infty(E)$, we have

$$\overline{[f](y)} = [\bar{f}]\,(-y).$$

Also, $\bar{\mathfrak{q}} = \mathfrak{q}^{-1}$ and $\mathfrak{q}(-x) = \mathfrak{q}(x)$. So, applying complex conjugation to the defining relation for γ, we get (in abbreviated notation)

$$\int_E \mathfrak{q}^{-1}\,[\bar{f}] = \bar{\gamma}\int_E \mathfrak{q}\bar{f} = \gamma^{-1}\int_E \mathfrak{q}\bar{f},$$

so $\bar{\gamma} = \gamma^{-1}$, as required. □

The Langlands constant $\lambda_{E/F}(\psi)$ is a fourth root of unity (30.4.3), so also $|\lambda_{E/F}(\psi)| = 1$. It is therefore enough to prove that

$$\gamma_{E/F}(\psi) = c\,\lambda_{E/F}(\psi),$$

for some $c > 0$. In the following argument, c denotes a positive real number, varying from line to line.

To compute γ, we let f be the characteristic function of $\mathfrak{p}_E^d$, for a large positive integer d. We let μ be the level of ψ_E. We have

$$\int_E \mathfrak{q}^{-1} f(x)\,dx = c,$$

so

$$\begin{aligned}\gamma &= c\int_{\mathfrak{p}_E^{\mu-d}} \psi(\mathrm{N}_{E/F}\,x)\,dx \\ &= c\sum_{m\geqslant\mu-d}\int_{\varpi_E^m U_E} \psi(\mathrm{N}_{E/F}\,x)\,\|x\|_E\,d^\times x,\end{aligned}$$

for a prime element ϖ_E of E and a Haar measure $d^\times x$ on $E^\times$. The integrand is constant on cosets of $\mathcal{K}_{E/F}$. Since $\|x\|_E = \|\mathrm{N}_{E/F}\,x\|_F$, we get

$$\begin{aligned}\gamma &= c\sum_{m\geqslant\mu-d}\int_{\mathrm{N}_{E/F}(\varpi_E^m U_E)} \psi(y)\,\|y\|_F\,d^\times y \\ &= c\sum_{m\geqslant k}\int_{\varpi_F^m U_F} (1{+}\varkappa)(y)\,\psi(y)\,\|y\|_F\,d^\times y,\end{aligned}$$

where $k = (\mu{-}d)f(E|F)$, ϖ_F is a prime element of F, and $d^\times y$ is a Haar measure on $F^\times$. The integral

$$\sum_{m\geqslant k}\int_{\varpi_F^m U_F} \psi(y)\,\|y\|_F\,d^\times y = \int_{\mathfrak{p}^k} \psi(y)\,dy$$

is zero for $k < \nu$, where ν is the level of ψ. It is therefore zero for d sufficiently large. This leaves

$$\gamma = c \int_{\mathfrak{p}^k} \varkappa(y)\psi(y)\, dy. \tag{38.8.2}$$

When E/F is ramified, the character $\varkappa$ is ramified and the integral gives the Gauss sum formula for $\varepsilon(\varkappa, 1, \psi)$ (23.5 Exercise). Since $\varepsilon(\varkappa, 1, \psi) = c\lambda_{E/F}(\psi)$, we have the result in this case.

Suppose, therefore, that E/F is unramified. The integral (38.8.2) is then $c\zeta(\hat{\Psi}, \varkappa, 1)$, where Ψ is the characteristic function of $1+\mathfrak{p}^{\mu-k}$. The function $\zeta(\Psi, \varkappa, s)$ is a positive constant (for d sufficiently large). The L-function $L(\varkappa, s)$ is $(1+q^{-s})^{-1}$, and takes only positive values on the real axis. The functional equation (23.4.4) therefore gives

$$\gamma = c\varepsilon(\varkappa, 1, \psi) = c\lambda_{E/F}(\psi).$$

This proves (38.8.1) and completes the proof of 38.7 Proposition. □

Corollary. *Let $f \in C_c^\infty(E)$ and put $g = \mathfrak{q}\,[f]$, $h = \mathfrak{q}^{-1}f$. We then have*

$$[g](y) = \lambda_{E/F}(\psi)\,\mathfrak{q}^{-1}(y)\,[h](-y), \quad y \in E.$$

Proof. When $y = 0$, this is given by the proposition. In general, we apply the proposition to the function $f_{-y} : x \mapsto f(x-y)$, and note that

$$[g](y) = \int_E \mathfrak{q}\,[f_{-y}],$$

$$[h](-y) = \mathfrak{q}(y) \int_E \mathfrak{q}^{-1} f_{-y}. \quad \square$$

38.9. We apply 38.8 Corollary to the task of verifying the fourth identity in (38.5.1) for the operator $A : f \mapsto \lambda_{E/F}(\psi)\,[f]$. In this paragraph, we treat the case $t = 1$. Taking $f \in C_c^\infty(E)$, $g = \mathfrak{q}\,[f]$, $h = \mathfrak{q}^{-1}f$, and abbreviating $\lambda = \lambda_{E/F}(\psi)$, we have

$$\eta(n(1))Af = \lambda\mathfrak{q}\,[f] = \lambda g,$$

and so

$$A\eta(n(1))Af = \lambda^2\,[g] = \varkappa(-1)\,[g].$$

On the other hand, $\eta(n(-1))f = \mathfrak{q}^{-1}f = h$ and $A\eta(n(-1))f = \lambda\,[h]$, so that

$$\eta(z(-1)a(-1)n(-1))A\eta(n(-1))f(y) = \varkappa(-1)\,\lambda\,\mathfrak{q}^{-1}\,[h](-y)$$

which, by 38.8 Corollary, is $\varkappa(-1)\,[g](y)$, as required.

38.10. The general case of the relation will follow from a variant of the same argument. For $f \in C_c^\infty(E)$, we have $Af = \lambda\,[f]$, so

$$\eta(n(t))Af = \lambda \mathfrak{q}_t\,[f],$$

where $\mathfrak{q}_t$ denotes the function $y \mapsto \psi(t\,\mathrm{N}_{E/F}(y))$. Thus

$$A\,\eta(n(t))Af = \lambda^2\,[\mathfrak{q}_t\,[f]] = \varkappa(-1)\,[\mathfrak{q}_t\,[f]].$$

On the other hand, $\eta(n(-t^{-1}))f = \mathfrak{q}_{-t^{-1}}f$, so

$$A\,\eta(n(-t^{-1}))f = \lambda\,[\mathfrak{q}_{-t^{-1}}f].$$

and

$$\eta(z(-t)a(-t^{-1})n(-t))\,A\,\eta(n(-t^{-1}))f:$$
$$y \longmapsto \varkappa(-t)\,\lambda\,\|t\|_E^{-1/2}\,\mathfrak{q}_{-t}(-t^{-1}y)\,[\mathfrak{q}_{-t^{-1}}f](-t^{-1}y).$$

We therefore want to prove

$$[\mathfrak{q}_t\,[f]](y) = \varkappa(t)\,\lambda\,\|t\|_E^{-1/2}\,\mathfrak{q}_{-t}(-t^{-1}y)\,[\mathfrak{q}_{-t^{-1}}f](-t^{-1}y).$$

We fix $t \in F^\times$ and consider the character $t\psi : x \mapsto \psi(tx)$ in place of ψ. We denote by $\mathcal{F}$ the normalized Fourier transform on $C_c^\infty(E)$ relative to the character $t\psi \circ \mathrm{Tr}_{E/F}$. Thus

$$\mathcal{F}(f)(y) = \|t\|_E^{1/2}\hat{f}(ty).$$

Defining $\widetilde{\mathcal{F}}(f)$ as the function $y \mapsto \mathcal{F}(f)(y^\sigma)$, we get

$$\widetilde{\mathcal{F}}(f)(y) = \|t\|_E^{1/2}\,[f](ty).$$

We apply 38.8 Corollary to $t\psi$ and $\widetilde{\mathcal{F}}$; for $\varphi \in C_c^\infty(E)$, we obtain

$$\widetilde{\mathcal{F}}(\mathfrak{q}_t\widetilde{\mathcal{F}}(\varphi))(y) = \lambda(E/F, t\psi)\,\mathfrak{q}_t^{-1}(y)\,\widetilde{\mathcal{F}}(\mathfrak{q}_t^{-1}\varphi)(-y).$$

We set $[\varphi]^t(y) = [\varphi](ty)$. Since $\lambda(E/F, t\psi) = \varkappa(t)\lambda(E/F, \psi)$, this last relation reads

$$\widetilde{\mathcal{F}}(\mathfrak{q}_t\,[\varphi]^t)(y) = \varkappa(t)\,\lambda\,\mathfrak{q}_{-t}(-y)\,[\mathfrak{q}_{-t}\varphi]^t(-y).$$

We define φ by $\varphi(y) = \|t\|_E f(ty)$, so that $[\varphi]^t = [f]$. We get

$$\|t\|_E^{1/2}[\mathfrak{q}_t\,[f]](ty) = \varkappa(t)\,\lambda\,\mathfrak{q}_{-t}(-y)\,[\mathfrak{q}_{-t}\varphi]^t(-y),$$

so it is enough to prove

$$[\mathfrak{q}_{-t}\varphi](-y) = [\mathfrak{q}_{-t^{-1}}f](-t^{-1}y).$$

This, however, follows from the definition of φ. $\square$

39. The Weil Representation

Again, let E/F be a separable quadratic extension with $\mathrm{Gal}(E/F) = \{1, \sigma\}$. We let Θ be a character of $E^\times$ which is E/F-*regular*, in the sense that $\Theta^\sigma \neq \Theta$. Using the representation $\eta_\psi^{\mathcal{G}}$ of $\mathcal{G} = \mathcal{G}_{E/F}$ defined in §38, we construct an irreducible cuspidal representation $\pi_{E/F}(\Theta)$ canonically associated with $(E/F, \Theta)$.

39.1. To start with, let θ be a *non-trivial* character of $\mathcal{K} = \mathcal{K}_{E/F}$. Let $C_c^\infty(E, \theta)$ be the space of functions $f \in C_c^\infty(E)$ such that

$$f(xy) = \theta(x) f(y), \quad x \in \mathcal{K},\ y \in E. \tag{39.1.1}$$

In the notation of (38.6.1), condition (39.1.1) can be re-written as:

$$\eta_\psi^{\mathcal{G}}(a(x)) f = \theta(x) f, \quad x \in \mathcal{K}. \tag{39.1.2}$$

The function f is constant on a neighbourhood of 0, so (39.1.1) implies $f(0) = 0$. That is:

$$C_c^\infty(E, \theta) \subset C_c^\infty(E^\times). \tag{39.1.3}$$

As $\mathcal{K}$ is contained in the centre of $\mathcal{G}$, $C_c^\infty(E, \theta)$ is a $\mathcal{G}$-subspace of $(\eta_\psi^{\mathcal{G}}, C_c^\infty(E))$. That is, $C_c^\infty(E, \theta)$ carries the representation of $\mathcal{G}$ restricted from $C_c^\infty(E)$.

We choose a character Θ of $E^\times$ such that $\Theta \mid \mathcal{K} = \theta$ and form the representation

$$\big(\xi(\Theta, \psi), C_c^\infty(E, \theta)\big) = \big(\Theta^{-1}\eta_\psi^{\mathcal{G}}, C_c^\infty(E, \theta)\big),$$

where $\Theta^{-1}\eta_\psi^{\mathcal{G}}$ is the representation $(g, x) \mapsto \Theta(x)^{-1}\eta_\psi^{\mathcal{G}}(g, x)$, $(g, x) \in \mathcal{G}$.

The representation $\xi(\Theta, \psi)$ is trivial on $\mathcal{K}$ and is therefore the inflation of a representation of $\mathcal{G}/\mathcal{K} = G_\varkappa$, for which we use the same notation.

39.2. Let $f \in C_c^\infty(E, \theta)$; the function $\Theta^{-1} f$ is constant on cosets of $\mathcal{K}$ and so is of the form $\Theta^{-1} f(x) = f^0(\mathrm{N}_{E/F}(x))$, for a uniquely determined function f^0 on the group $\mathrm{N}_{E/F}(E^\times) = \mathrm{Ker}\,\varkappa$, which we now denote $F_\varkappa^\times$. Indeed, $f \mapsto f^0$ gives a linear isomorphism $C_c^\infty(E, \theta) \cong C_c^\infty(F_\varkappa^\times)$ which we use to transport $\xi(\Theta, \psi)$ to a representation $\xi_\varkappa(\Theta, \psi)$ of $G_\varkappa$ on $C_c^\infty(F_\varkappa^\times)$.

Under the representation $\xi_\varkappa(\Theta, \psi)$, the group $B \cap G_\varkappa$ acts on $f \in C_c^\infty(F_\varkappa^\times)$ as follows:

$$\begin{pmatrix} 1 & t \\ 0 & 1 \end{pmatrix} f : y \longmapsto \psi(ty) f(y), \qquad t \in F,$$

$$\begin{pmatrix} \zeta & 0 \\ 0 & \zeta \end{pmatrix} f : y \longmapsto \varkappa(\zeta)\Theta(\zeta) f(y), \qquad \zeta \in F^\times,$$

$$\begin{pmatrix} x & 0 \\ 0 & 1 \end{pmatrix} f : y \longmapsto \|x\|_F^{1/2}\Theta(x) f(xy), \quad x \in F_\varkappa^\times.$$

The space $C_c^\infty(F_\varkappa^\times)$ admits a linear automorphism $\varphi \mapsto \varphi'$, where

$$\varphi'(x) = \|x\|_F^{1/2}\Theta(x)\varphi(x). \tag{39.2.1}$$

Composing with this, we obtain an equivalent representation $\pi_\varkappa = \pi_\varkappa(\Theta, \psi)$ of $G_\varkappa$ on $C_c^\infty(F_\varkappa^\times)$. It acts on $\varphi \in C_c^\infty(F_\varkappa^\times)$ as follows:

$$\begin{aligned} \pi_\varkappa \left(\begin{smallmatrix} 1 & t \\ 0 & 1 \end{smallmatrix}\right) \varphi &: y \longmapsto \psi(ty)\varphi(y), & t &\in F, \\ \pi_\varkappa \begin{pmatrix} \zeta & 0 \\ 0 & \zeta \end{pmatrix} \varphi &: y \longmapsto \varkappa(\zeta)\Theta(\zeta)\varphi(y), & \zeta &\in F^\times, \\ \pi_\varkappa \left(\begin{smallmatrix} x & 0 \\ 0 & 1 \end{smallmatrix}\right) \varphi &: y \longmapsto \varphi(xy), & x &\in F_\varkappa^\times. \end{aligned} \tag{39.2.2}$$

We define

$$\pi(\Theta, \psi) = \mathrm{Ind}_{G_\varkappa}^G \, \pi_\varkappa(\Theta, \psi),$$

where, as usual, Ind is the functor of smooth induction.

Note. The character θ of $\mathcal{K}$ determines Θ up to tensoring with $\chi_E = \chi \circ \mathrm{N}_{E/F}$, for a character χ of $F_\varkappa^\times$. In the opposite direction, a character Θ of $E^\times$ satisfies $\Theta \mid \mathcal{K} \neq 1$ if and only if Θ is E/F-regular. In particular, the representation $\pi(\Theta, \psi)$ is defined for any regular character Θ of $E^\times$.

Proposition. *Let Θ be a regular character of $E^\times$, and let $\psi \in \widehat{F}$, $\psi \neq 1$.*

(1) *The representation $\pi(\Theta, \psi)$ of G is irreducible and cuspidal.*
(2) *If χ is a character of $F^\times$, then*

$$\pi(\chi_E\Theta, \psi) \cong \chi\pi(\Theta, \psi).$$

In particular,

$$\varkappa_{E/F}\, \pi(\Theta, \psi) \cong \pi(\Theta, \psi).$$

(3) *The central character ω_π of $\pi(\Theta, \psi)$ is given by*

$$\omega_\pi = \varkappa_{E/F} \otimes \Theta \mid F^\times.$$

Proof. Parts (2) and (3) are direct consequences of the definitions.

Define a character ϑ of N by

$$\vartheta \begin{pmatrix} 1 & x \\ 0 & 1 \end{pmatrix} = \psi(x).$$

Write $M_\varkappa = M \cap G_\varkappa$. Writing $\pi = \pi(\Theta, \psi)$, the relation $\pi = \mathrm{Ind}_{G_\varkappa}^G \pi_\varkappa$ implies $\pi \mid M = \mathrm{Ind}_{M_\varkappa}^M (\pi_\varkappa | M_\varkappa)$. Comparing (39.2.2) with (37.1.1), we get $\pi \mid M \cong$ c-$\mathrm{Ind}_N^M \vartheta$. Thus indeed π is irreducible and cuspidal (36.1 Corollary 2). $\square$

39.3. We reach the crucial property of the representation $\pi(\Theta, \psi)$.

Theorem. *Let $\psi \in \widehat{F}$, $\psi \neq 1$. Let Θ be a regular character of $E^\times$ and set $\rho = \operatorname{Ind}_{E/F} \Theta$, $\pi = \pi(\Theta, \psi)$. Then π is an irreducible cuspidal representation of G satisfying*

$$\varepsilon(\chi\pi, s, \psi) = \varepsilon(\chi \otimes \rho, s, \psi),$$

for all characters χ of $F^\times$.

Proof. By 39.2 Proposition (2), it is enough to treat the case $\chi = 1$. We calculate $\varepsilon(\pi, s, \psi)$ using 37.3. We have therefore to track the action of the Weyl element $w \in G_\varkappa$ under the isomorphisms of 39.2. We abbreviate $\pi_\varkappa = \pi_\varkappa(\Theta, \psi)$.

We recall that $\psi_E = \psi \circ \operatorname{Tr}_{E/F}$. We have

$$\varepsilon(\Theta, s, \psi_E) = q_E^{(\frac12 - s)n(\Theta, \psi_E)}\, \varepsilon(\Theta, \tfrac12, \psi_E),$$

for an integer $n(\Theta, \psi_E)$ (23.4 Corollary 2).

Lemma. *Let $\varphi \in C_c^\infty(F_\varkappa^\times)$ be the characteristic function of $U_F \cap F_\varkappa^\times$. The support of $\pi_\varkappa(w)\varphi$ is the set of $x \in F_\varkappa^\times$ with $\upsilon_F(x) = -n(\Theta, \psi_E) f(E|F)$. If $\upsilon_F(x) = -n(\Theta, \psi_E) f(E|F)$, then*

$$\pi_\varkappa(w)\varphi(x) = \lambda_{E/F}(\psi)\, \varepsilon(\Theta, \tfrac12, \psi_E)\, \Theta(x). \tag{39.3.1}$$

Proof. Let $f \in C_c^\infty(E, \theta)$ be the function with support U_E such that $f(y) = \Theta^{-\sigma}(y)$, $y \in U_E$. The composite of the two isomorphisms of 39.2 takes f to the characteristic function φ of $U_F \cap F_\varkappa^\times$.

From (38.6.1), we get

$$\eta_\psi^{\mathcal{G}}(w) f(x) = \lambda_{E/F}(\psi)\, \hat{f}(x^\sigma).$$

Expanding, with Φ_0 denoting the characteristic function of U_E,

$$\begin{aligned}
\hat{f}(x^\sigma) &= \int_E f(y)\, \psi_E(yx^\sigma)\, dy \\
&= \int_E \Phi_0(y)\, \Theta^{-1}(y^\sigma)\, \psi_E(x^\sigma y)\, dy \\
&= \int_E \Phi_0(y)\, \Theta^{-1}(y)\, \psi_E(xy)\, dy \\
&= \int_E \Phi_0(y)\, \Theta^{-1}(y)\, \psi_E(xy)\, d^\times y \\
&= \Theta(x) \int_{U_E} \Theta^{-1}(xy)\, \psi_E(xy)\, d^\times y,
\end{aligned}$$

where $d^\times y = \|y\|_E^{-1}\,dy$. By 23.5 Exercise, the last integral vanishes unless $\upsilon_E(x) = -n(\Theta, \psi_E)$. In that case

$$\int_{U_E} \Theta^{-1}(xy)\,\psi_E(xy)\,d^\times y = \varepsilon(\Theta, 1, \psi_E),$$

and so

$$\eta_\psi^{\mathfrak{g}}(w)f(x) = \begin{cases} \lambda_{E/F}(\psi)\,\Theta(x)\,\varepsilon(\Theta,1,\psi_E) & \text{if } \upsilon_E(x) = -n(\Theta,\psi_E), \\ 0 & \text{otherwise.} \end{cases}$$

Tracking back to $C_c^\infty(F_\varkappa^\times)$, the lemma follows. □

Within the Kirillov model of $\pi = \pi(\Theta,\psi)$, the natural representation of $G_\varkappa$ on the subspace $C_c^\infty(F_\varkappa^\times)$ of $C_c^\infty(F^\times)$ is equivalent to $\pi_\varkappa$. Comparing the lemma with 37.3 Theorem, we get $n(\Theta,\psi_E)f(E|F) = n(\pi,\psi) = n$, say, and

$$\varepsilon(\pi, \tfrac{1}{2}, \psi)\,\omega_\pi(x) = \lambda_{E/F}(\psi)\,\varepsilon(\Theta, \tfrac{1}{2}, \psi_E)\,\Theta(x),$$

for any $x \in F_\varkappa^\times$ of valuation $-n$. For such x, we have $\omega_\pi(x) = \Theta(x)$ (39.2 Proposition (3)), and the result follows. □

39.4. We extract some complementary detail.

Corollary. *Let E/F be a separable quadratic extension, and let Θ be an E/F-regular character of $E^\times$. Let $\psi \in \widehat{F}$, $\psi \neq 1$, and put $\rho = \mathrm{Ind}_{E/F}\,\Theta$, $\pi = \pi(\Theta,\psi)$. We then have*

$$\varepsilon(\chi\pi, s, \psi') = \varepsilon(\chi\otimes\rho, s, \psi'),$$

for all characters χ of $F^\times$ and all $\psi' \in \widehat{F}$, $\psi' \neq 1$. Consequently, $\pi(\Theta,\psi) \cong \pi(\Theta,\psi')$.

Proof. Because of 39.2 Proposition (2), it is enough to treat the case $\chi = 1$. Part (3) of the same proposition gives $\omega_\pi = \det\rho$, and the first assertion follows from 24.3 Proposition and 29.4 Proposition (2). Setting $\pi' = \pi(\Theta,\psi')$, we have

$$\varepsilon(\chi\pi', s, \psi) = \varepsilon(\chi\otimes\rho, s, \psi) = \varepsilon(\chi\pi, s, \psi)$$

for all χ, whence (27.1) $\pi' \cong \pi$, as desired. □

Thus, if E/F is a separable quadratic extension and Θ is an E/F-regular character of $E^\times$, the isomorphism class of $\pi(\Theta,\psi)$ does not depend on ψ. We therefore denote it $\pi(\Theta)$, or $\pi_{E/F}(\Theta)$, and call it the *Weil representation* defined by $(E/F,\Theta)$.

40. A Partial Correspondence

We can now use the Weil representation, as defined in §39, to give a partial construction of the map $\mathcal{G}_2^0(F) \to \mathcal{A}_2^0(F)$ demanded by Theorem 33.4.

40.1. Let $\rho \in \mathcal{G}_2^0(F)$; one says that ρ is *imprimitive* if there exists a separable quadratic extension E/F and a character ξ of $E^\times$ such that $\rho \cong \operatorname{Ind}_{E/F} \xi$.

Let $\mathcal{G}_2^{\mathrm{im}}(F)$ denote the set of imprimitive equivalence classes $\rho \in \mathcal{G}_2^0(F)$.

Theorem. *Let $\psi \in \widehat{F}$, $\psi \neq 1$. There exists a unique map*

$$\boldsymbol{\pi} : \mathcal{G}_2^{\mathrm{im}}(F) \longrightarrow \mathcal{A}_2^0(F)$$

with the property

$$\varepsilon(\chi \otimes \rho, s, \psi) = \varepsilon(\chi\boldsymbol{\pi}(\rho), s, \psi) \tag{40.1.1}$$

for all $\rho \in \mathcal{G}_2^{\mathrm{im}}(F)$ and all characters χ of $F^\times$. Indeed,

$$\boldsymbol{\pi}(\chi \otimes \rho) \cong \chi\boldsymbol{\pi}(\rho),$$

for all ρ, χ, and (40.1.1) holds for all $\psi \in \widehat{F}$, $\psi \neq 1$.

Proof. Let $\rho \in \mathcal{G}_2^{\mathrm{im}}(F)$; choose a separable quadratic extension E/F and a character ξ of $E^\times$ such that $\rho \cong \operatorname{Ind}_{E/F} \xi$. Let $\pi = \pi_{E/F}(\xi) \in \mathcal{A}_2^0(F)$ be the Weil representation defined by $(E/F, \xi)$. This has the property $\varepsilon(\chi\pi, s, \psi) = \varepsilon(\chi \otimes \rho, s, \psi)$ for all χ and ψ (39.3). By the Converse Theorem 27.1, it is the only element of $\mathcal{A}_2^0(F)$ with this property. The assignment $\operatorname{Ind}_{E/F} \xi \mapsto \pi_{E/F}(\xi)$ thus gives a well-defined map $\mathcal{G}_2^{\mathrm{im}}(F) \to \mathcal{A}_2^0(F)$, and it is the only map with the desired properties. The next assertion is 39.2 Proposition (2), and the final one is 39.4 Corollary. □

Remark. When the residual characteristic p of F is odd, we have $\mathcal{G}_2^{\mathrm{im}}(F) = \mathcal{G}_2^0(F)$ (34.1 Theorem). The uniqueness property shows that $\boldsymbol{\pi}$ is the same as the tame Langlands correspondence $\boldsymbol{\pi}$ of 34.4. In this case, therefore, $\boldsymbol{\pi}$ gives a bijection $\mathcal{G}_2^0(F) \to \mathcal{A}_2^0(F)$. If $p = 2$, the same reasoning shows that $\boldsymbol{\pi}$ extends the bijection $\mathcal{G}_2^{\mathrm{nr}}(F) \cong \mathcal{A}_2^{\mathrm{nr}}(F)$ of 34.4. In the case $p = 2$, we have $\mathcal{G}_2^{\mathrm{nr}}(F) \subsetneq \mathcal{G}_2^{\mathrm{im}}(F) \subsetneq \mathcal{G}_2^0(F)$, and there is considerably more to do.

Further reading The proof of 36.1 Theorem (36.3) has been engineered to illustrate the relation with the structure of cuspidal representations. The standard source for such results in general is [70], but the method of [5] is more straightforward in our context. The rest of the chapter is a simplified version of [46], except for the treatment of the Kirillov model and its relation with the Godement-Jacquet functional equation: this owes something to [47].

The Weil representation occurs in a much more general context: see, for example, [62]. This generalization however does not lead to the Langlands correspondence in higher dimension. There is an alternative method of constructing the Langlands correspondence, in arbitrary dimension, on irreducible representations of $\mathcal{W}_F$ which are "imprimitive" in the right sense. It uses the technique of *automorphic induction* [44] and is reasonably transparent. It does, however, rely heavily on global constructions and so is not admissible here. The process is worked out fully in [40], using a slightly older version of the technology.

10

Arithmetic of Dyadic Fields

We assume, throughout this chapter, that the residual characteristic p of the base field F is 2. This chapter is devoted to a careful examination of the irreducible two-dimensional representations of $\mathcal{W}_F$. Our account is quite detailed, but we give only what we need for the construction of the Langlands correspondence in Chapter XII. In particular, we do not establish the existence of primitive representations. While not particularly hard, it would require a further digression, and the situation becomes completely transparent once we have the Langlands correspondence.

We also prove an analogue of the Converse Theorem of 27.1: an *imprimitive* representation $\rho \in \mathcal{G}_2^0(F)$ is determined, up to isomorphism, by the function $\chi \mapsto \varepsilon(\chi \otimes \rho, s, \psi)$ of characters χ of $F^\times$.

41. Imprimitive Representations

We fix a separable algebraic closure $\overline{F}/F$ (as required for the definition of $\mathcal{W}_F$) and, for the time being, consider only field extensions E/F with $E \subset \overline{F}$. In particular, E/F is always assumed to be separable.

41.1. We review some basic facts concerning the arithmetic of those quadratic field extensions E/F which are *totally wildly ramified.*

Lemma. *Let E/F be quadratic and totally ramified. Let ϖ_E be a prime element of E and let $f(X) = X^2+aX+b$ be the minimal polynomial of ϖ_E over F. Then:*

(1) $\upsilon_F(b) = 1$, $\upsilon_F(a) \geqslant 1$ *and* $\mathfrak{o}_E = \mathfrak{o}[\varpi_E]$;

(2) *the relative discriminant* $\mathfrak{d}_{E/F}$ *of the field extension* E/F *is of the form* $\mathfrak{d}_{E/F} = \mathfrak{p}^{d+1}$, *where*

$$d = \min\{2\upsilon_F(a)-1, \upsilon_F(4)\}.$$

In particular, $1 \leqslant d \leqslant \upsilon_F(4)$, *and* d *is odd except in the case where* F *has characteristic zero and* $d = \upsilon_F(4)$.

(3) *Let* ϖ *be a prime element of* E; *then*

$$\upsilon_F(\mathrm{Tr}_{E/F}\,\varpi) \begin{cases} = (d+1)/2 & \text{if } d \text{ is odd,} \\ \geqslant 1+\upsilon_F(2) & \text{if } d = \upsilon_F(4). \end{cases}$$

(4) *Let* ϖ *be any prime element of* E *and let* σ *generate* $\mathrm{Gal}(E/F)$; *then* $\upsilon_E(\varpi-\varpi^\sigma) = d+1$.

The proof is an elementary exercise. Observe that, if F has characteristic 2, then $\upsilon_F(4) = \infty$ and $d = 2\upsilon_F(a)-1$ is odd. (The case $a = 0$, $d = \infty$, arises exactly when E/F is purely inseparable, and is irrelevant to our purposes.)

We will normally use the notation

$$\mathfrak{d}_{E/F} = \mathfrak{p}^{d+1}, \quad d = d(E|F), \tag{41.1.1}$$

when E/F is a separable quadratic extension.

41.2. Let E/F be a totally ramified quadratic extension with $d(E|F) = d$. Let $\varphi = \varphi_{E/F}$ be the *inverse Herbrand function* for the extension E/F. Thus φ is the function of non-negative integers given by

$$\varphi(j) = \begin{cases} j & \text{if } 0 \leqslant j \leqslant d, \\ 2j-d & \text{if } j \geqslant d. \end{cases}$$

We then have the following properties:

Proposition.

(1) *Let* $k \in \mathbb{Z}$; *then*

$$\mathrm{Tr}_{E/F}(\mathfrak{p}_E^k) = \mathfrak{p}_F^l, \quad l = \left[\frac{k+d+1}{2}\right].$$

The trace map $\mathrm{Tr}_{E/F}$ *induces an isomorphism*

$$\mathfrak{p}_E^{2m-d}/\mathfrak{p}_E^{2m-d+1} \xrightarrow{\approx} \mathfrak{p}^m/\mathfrak{p}^{m+1},$$

for all $m \in \mathbb{Z}$.

(2) *Suppose* $j \geqslant 0$, $j \neq d$; *the field norm* $\mathrm{N}_{E/F}$ *induces an isomorphism* $U_E^{\varphi(j)}/U_E^{1+\varphi(j)} \cong U_F^j/U_F^{1+j}$.

(3) *The norm map* $\mathrm{N}_{E/F}$ *induces a homomorphism* $U_E^d/U_E^{1+d} \to U_F^d/U_F^{1+d}$ *with kernel and cokernel of order* 2.

Proof. The inverse different of E/F is $\mathfrak{D}_{E/F}^{-1} = \mathfrak{p}_E^{-(1+d)}$. By definition, this is the largest fractional ideal of $\mathfrak{o}_E$ with trace contained in $\mathfrak{o}$, and (1) follows easily.

Under the canonical inclusion $U_F \to U_E$, we get $U_F/U_F^1 = U_E/U_E^1 \cong \boldsymbol{k}^\times$. The map $\boldsymbol{k}^\times \to \boldsymbol{k}^\times$, induced by the norm, is $x \mapsto x^2$ and so it is bijective. This proves (2) in the case $j = 0$.

Let $1 \leqslant j \leqslant d{-}1$, so that $j = \varphi(j)$. For $x \in \mathfrak{p}_E^j$, we have

$$\begin{aligned}\mathrm{N}_{E/F}(1{+}x) &= 1 + \mathrm{Tr}_{E/F}(x) + \mathrm{N}_{E/F}(x)\\ &\equiv 1 + \mathrm{N}_{E/F}(x) \pmod{\mathfrak{p}^{j+1}}.\end{aligned}$$

The elements $u\varpi^j$, $u \in \mathfrak{o}/\mathfrak{p}$, provide a set of representatives for $\mathfrak{p}_E^j/\mathfrak{p}_E^{j+1}$, and their norms $u^2\mathrm{N}_{E/F}(\varpi)^j$ give a set of representatives for $\mathfrak{p}^j/\mathfrak{p}^{j+1}$. This proves part (2) of the proposition when $1 \leqslant j \leqslant d{-}1$.

For $j \geqslant d{+}1$ and $x \in \mathfrak{p}_E^{\varphi(j)}$, we get

$$\begin{aligned}\mathrm{N}_{E/F}(1{+}x) &= 1 + \mathrm{Tr}_{E/F}(x) + \mathrm{N}_{E/F}(x)\\ &\equiv 1 + \mathrm{Tr}_{E/F}(x) \pmod{\mathfrak{p}^{j+1}},\end{aligned}$$

and this case of part (2) follows from part (1).

Let $\sigma \in \mathrm{Gal}(E/F)$, $\sigma \neq 1$. The well-known "Hilbert's Theorem 90" says that the kernel of the norm map $\mathrm{N}_{E/F}$ consists of all elements x/x^σ, $x \in E^\times$. If ϖ is a prime element of E, then $\varpi^\sigma/\varpi \equiv 1 \pmod{\mathfrak{p}_E^d}$, by 41.1 Lemma (4). On the other hand, if $x \in U_E = U_F U_E^1$, then $x^\sigma/x \in U_E^{d+1}$.

In (3), we certainly have $\mathrm{N}_{E/F}(U_E^d) \subset U_F^d$. If $\mathcal{K}$ denotes the kernel of $\mathrm{N}_{E/F}$, then $\mathcal{K} \subset U_E^d$ and the group $\mathcal{K}/\mathcal{K} \cap U_E^{d+1} \cong \mathcal{K}U_E^{d+1}/U_E^{d+1}$ has order 2: the non-trivial element of this group is the coset of ϖ/ϖ^σ, for a prime element ϖ of E. On the other hand, it follows from (2) that $\mathrm{N}_{E/F}(U_E^{d+1}) = U_F^{d+1}$. So, if $x \in U_E^d$ satisfies $\mathrm{N}_{E/F}(x) \in U_F^{d+1}$, we may write $x = yz$, with $z \in U_E^{d+1}$ and $y \in \mathcal{K}$. The kernel of the norm map $U_E^d/U_E^{d+1} \to U_F^d/U_F^{d+1}$ thus has order 2, whence so does its cokernel. □

For later use, we record a more detailed comment on the structure of the group $\mathrm{Ker}\,\mathrm{N}_{E/F}$.

Lemma. *Let* $m \geqslant 0$, $m \in \mathbb{Z}$. *The map* $\widetilde{N} : x \mapsto x/x^\sigma$ *induces an isomorphism*

$$U_E^{2m+1}/U_E^{2m+2} \longrightarrow U_E^{2m+d+1}/U_E^{2m+d+2}.$$

Proof. Let ϖ be a prime element of E. The elements $x = 1+y\varpi$, $y \in \mathfrak{p}^m/\mathfrak{p}^{m+1}$, give a set of coset representatives for U_E^{2m+1}/U_E^{2m+2}. As in 41.1 Lemma (4), $\upsilon_E(\varpi-\varpi^\sigma) = d+1$, so

$$x/x^\sigma \equiv 1 + y(\varpi-\varpi^\sigma) \pmod{U_E^{2m+d+2}}.$$

Since $U_E^{2m+2} = U_F^{m+1}U_E^{2m+3}$, we have $\widetilde{N}(U_E^{2m+2}) \subset U_F^{2m+d+2}$ and the lemma follows. □

41.3. Let E/F be a separable quadratic extension and ξ a character of $\mathcal{W}_E$. The representation $\mathrm{Ind}_{E/F}\,\xi$ of $\mathcal{W}_F$ is then irreducible if and only if $\xi \neq \xi^\sigma$, where σ generates $\mathrm{Gal}(E/F)$. If we view ξ as a character of $E^\times$ via class field theory, this irreducibility property is equivalent (by Hilbert 90) to ξ not factoring through the norm map $\mathrm{N}_{E/F}$.

Definition. *Let $\rho \in \mathfrak{G}_2^0(F)$; define $\mathfrak{T}(\rho)$ to be the group of characters χ of $\mathcal{W}_F$ such that $\chi \otimes \rho \cong \rho$.*

Comparing determinants, we see that $\chi \in \mathfrak{T}(\rho)$ implies $\chi^2 = 1$. In particular, $\mathfrak{T}(\rho)$ is an elementary abelian 2-group. In practice, it will be more convenient to view the elements of $\mathfrak{T}(\rho)$ as characters of $F^\times$, via class field theory.

If E/F is a separable quadratic extension, then $\varkappa_{E/F}$ (29.2.2) is the unique non-trivial character of $F^\times$ trivial on $\mathrm{N}_{E/F}(E^\times)$. Thus $\varkappa_{E/F}$ is unramified if and only if E/F is unramified. Otherwise, $\varkappa_{E/F}$ has level $d(E|F)$ (as follows from 41.2 Proposition).

Class field theory (29.1) shows that, if χ is a character of $F^\times$ of order 2, there is a unique quadratic extension E/F (inside $\overline{F}$) such that $\chi = \varkappa_{E/F}$.

Proposition. *For any $\rho \in \mathfrak{G}_2^0(F)$, the group $\mathfrak{T}(\rho)$ has order dividing* 4.

Proof. We start with:

Lemma. *Let $\rho \in \mathfrak{G}_2^0(F)$ and let E/F be a separable quadratic extension. The following are equivalent:*

(1) *$\varkappa_{E/F} \in \mathfrak{T}(\rho)$;*
(2) *there is a character ξ of $\mathcal{W}_E$ such that $\rho \cong \mathrm{Ind}_{E/F}\,\xi$.*

If condition (2) holds, the character ξ is uniquely determined by ρ, up to conjugation by $\mathrm{Gal}(E/F)$.

Proof. All assertions follow from standard Clifford theory. □

Suppose that $|\mathfrak{T}(\rho)| \neq 1$, and let $\varkappa = \varkappa_{E/F} \in \mathfrak{T}(\rho)$. Choose a character ξ of $E^\times$ such that $\rho \cong \mathrm{Ind}_{E/F}\,\xi$.

If $|\mathfrak{T}(\rho)| > 2$, choose $\chi \in \mathfrak{T}(\rho)$, $\chi \neq 1, \varkappa$. The equation $\rho \cong \chi \otimes \rho$ implies $\rho \cong \mathrm{Ind}_{E/F}(\chi_E \otimes \xi)$, where $\chi_E = \chi \circ \mathrm{N}_{E/F}$ (or $\chi \mid \mathcal{W}_E$). It follows (Lemma) that $\chi_E \otimes \xi = \xi$ or ξ^σ, where $\sigma \in \mathrm{Gal}(E/F)$, $\sigma \neq 1$. By hypothesis, $\chi_E \neq 1$, so $\chi_E \otimes \xi = \xi^\sigma$, that is, $\chi_E = \xi^\sigma/\xi$. There are, at most, two characters χ of $F^\times$ satisfying $\chi_E = \xi^\sigma/\xi$, namely χ itself and $\varkappa\chi$, so $\mathfrak{T}(\rho) = \{1, \varkappa, \chi, \varkappa\chi\}$. □

We will make repeated use of one intermediate step in the preceding argument. We therefore exhibit it explicitly:

Corollary. *Let $\rho = \mathrm{Ind}_{E/F}\,\xi \in \mathcal{G}_2^0(F)$. A character ϕ of $F^\times$ lies in $\mathfrak{T}(\rho)$ if and only if $\phi_E = 1$ or $\phi_E = \xi/\xi^\sigma$, where σ generates* $\mathrm{Gal}(E/F)$.

One says that ρ is

$$\left.\begin{array}{r}\text{primitive}\\ \text{simply imprimitive}\\ \text{triply imprimitive}\end{array}\right\} \quad \textit{if } |\mathfrak{T}(\rho)| = \quad \left\{\begin{array}{l}1,\\ 2,\\ 4.\end{array}\right.$$

In particular, an irreducible representation $\mathrm{Ind}_{E/F}\,\xi$ is triply imprimitive if and only if ξ/ξ^σ factors through $\mathrm{N}_{E/F}$.

41.4. We recall that $\mathcal{G}_2^{\mathrm{nr}}(F)$ denotes the set of classes of representations $\rho \in \mathcal{G}_2^0(F)$ such that $\mathfrak{T}(\rho)$ contains an unramified character $\chi \neq 1$ (just as in 34.1). If $\rho \in \mathcal{G}_2^0(F) \smallsetminus \mathcal{G}_2^{\mathrm{nr}}(F)$, then ρ is called *totally ramified.*

Let E/F be a separable, totally ramified, quadratic extension. Let ξ be a character of $E^\times$, of level $l(\xi) \geqslant 1$. We say that ξ is *minimal over F* if $l(\xi) \leqslant l(\chi_E\,\xi)$, for all characters χ of $F^\times$.

Equivalently, ξ is minimal over F if and only if $\xi \mid U_E^{l(\xi)}$ does not factor through $\mathrm{N}_{E/F}$. In particular, if ξ is minimal over F, the representation $\mathrm{Ind}_{E/F}\,\xi$ is irreducible.

If ξ is a character of $E^\times$ which does not factor through $\mathrm{N}_{E/F}$, then clearly $\xi = \chi_E\,\xi_0$, for a character χ of $F^\times$ and some ξ_0 which is minimal over F. When analyzing induced representations $\rho = \mathrm{Ind}_{E/F}\,\xi$ or the groups $\mathfrak{T}(\rho)$, therefore, it is enough to treat the case of ξ minimal.

Lemma. *Let $d(E|F) = d \geqslant 1$, and let ξ be a character of $E^\times$.*

(1) *If ξ is minimal over F, then $l(\xi) \geqslant d$.*
(2) *Suppose that $l(\xi) \geqslant 1{+}d$; then ξ is minimal over F if and only if $l(\xi) \not\equiv d$* (mod 2).

Proof. If ξ has level l, $1 \leqslant l < d$, then $\xi \mid U_E^l$ certainly factors through the norm (41.1 Proposition (2)), whence follows (1). Part (2) is a direct consequence of 41.1 Proposition (2). □

Using the same notation as in the lemma:

Proposition. *Suppose that ξ is minimal of level l, and put $\rho = \mathrm{Ind}_{E/F}\,\xi$.*

(1) *The representation ρ lies in $\mathcal{G}_2^{\mathrm{nr}}(F)$ if and only if $l = d$.*
(2) *If $d < l < 2d$, the representation ρ is triply imprimitive, and $\mathfrak{T}(\rho)$ contains a character of level $l-d$.*

Proof. Let σ be the generator of $\mathrm{Gal}(E/F)$. If ξ has level d, the character ξ/ξ^σ is trivial on U_E^1 (41.2 Lemma) and hence on $U_E = U_F U_E^1$. It is therefore unramified (and non-trivial). Thus $\xi/\xi^\sigma = \chi_E$, where χ is a non-trivial unramified character of $F^\times$. By 41.3 Corollary, $\chi \in \mathfrak{T}(\rho)$ and so $\rho \in \mathcal{G}_2^{\mathrm{nr}}(F)$.

Suppose now that ξ has level $l > d$. Since $l \not\equiv d \pmod 2$, any $x \in U_E^l$ satisfies $x \equiv y^\sigma/y \pmod{U_E^{l+1}}$, for some $y \in U_E^{l-d}$ (41.2 Lemma). We deduce that ξ/ξ^σ has level $l-d \geqslant 1$. In particular, ξ/ξ^σ cannot be of the form χ_E, for an unramified character χ of $F^\times$. Therefore ρ is totally ramified, and we have proved (1).

If, however, $d < l < 2d$, the character ξ/ξ^σ has level $l-d < d$ and so it factors through $\mathrm{N}_{E/F}$, say $\xi/\xi^\sigma = \phi \circ \mathrm{N}_{E/F}$, for a character ϕ of $F^\times$. The level of ϕ is $l-d$, so $\phi \neq \varkappa_{E/F}$ while $\phi \in \mathfrak{T}(\rho)$. Thus ρ is triply imprimitive, as required. □

41.5. We can phrase this differently, using the Langlands-Deligne local constant. Let $\rho \in \mathcal{G}_2^0(F)$ and $\psi \in \widehat{F}$, $\psi \neq 1$. Recall that

$$\varepsilon(\rho, s, \psi) = q^{n(\rho,\psi)(\frac12 - s)}\,\varepsilon(\rho, \tfrac12, \psi),$$

for some $n(\rho,\psi) \in \mathbb{Z}$ (29.4 Proposition). For ρ fixed, $n(\rho,\psi)$ depends only on the level of ψ *loc. cit.* We write $\ell(\rho) = n(\rho,\psi)$, for some ψ of level one: we call $\ell(\rho)$ the *level* of ρ. We say ρ is *minimal* if $\ell(\rho) \leqslant \ell(\chi \otimes \rho)$, for every character χ of $F^\times$.

Lemma. *Let $\rho \in \mathcal{G}_2^0(F)$ be totally ramified and imprimitive. Then ρ is minimal if and only if $\ell(\rho)$ is odd.*

Proof. Write $\rho = \mathrm{Ind}_{E/F}\,\xi$, $d = d(E|F)$, and let $\psi \in \widehat{F}$ have level 1. It follows, from the inductive properties of the local constant and 30.4 Corollary, that $n(\xi, \psi_E) = n(\rho, \psi)$. The character $\psi_E = \psi \circ \mathrm{Tr}_{E/F}$ has level $1-d$ by 41.1 Proposition (1). Thus ξ has level $\ell(\rho) - d$ (23.5), and the lemma follows from 41.4 Lemma. □

Theorem. *Let $\rho \in \mathcal{G}_2^0(F)$ be totally ramified, minimal and imprimitive. The group $\mathfrak{T}(\rho)$ then contains a character χ of level $d \geqslant 1$ satisfying $3d \leqslant \ell(\rho)$.*

Proof. Set $\ell = \ell(\rho)$ and take $\chi \in \mathfrak{T}(\rho)$, $\chi \neq 1$. Let d be the level of χ, so that $\chi = \varkappa_{E/F}$ for a quadratic extension E/F with $d(E|F) = d$. We can write $\rho = \operatorname{Ind}_{E/F} \xi$, for a character ξ of $E^\times$. The character ξ has level $l(\xi) = \ell - d$. The minimality of ρ implies that ξ is minimal over F.

If $l(\xi) \geqslant 2d$, there is nothing more to do. We therefore assume the contrary, that is, $d < l(\xi) < 2d$. By 41.4 Proposition, $\mathfrak{T}(\rho)$ contains a character ϕ of level $d' = l(\xi) - d < d$. We have $\ell(\rho) = l(\xi) + d = d' + 2d > 3d'$, as required. □

It is sometimes useful to have a little more detail. Suppose ρ is triply imprimitive, so that $\mathfrak{T}(\rho)$ is non-cyclic of order 4. There are two possibilities: either all non-trivial $\chi \in \mathfrak{T}(\rho)$ have the same level $l(\chi) = d \geqslant 1$, and the characters $\chi \mid U_F^d$ are distinct, or else there is a unique non-trivial element with minimal level $l(\chi)$. The proof of the theorem gives:

Corollary. *Let ρ be totally ramified, triply imprimitive and minimal, of level n. Let χ_i, $1 \leqslant i \leqslant 3$, be the non-trivial elements of $\mathfrak{T}(\rho)$ and let $l(\chi_i)$ be the level of χ_i. The characters may be numbered so that either*

(1) $l(\chi_1) < l(\chi_2) = l(\chi_3)$ *and* $3l(\chi_2) > n > 3l(\chi_1)$, *or else*
(2) $l(\chi_1) = l(\chi_2) = l(\chi_3)$ *and* $n = 3l(\chi_i)$.

In the second case, setting $l = l(\chi_i)$, the characters $\chi_i \mid U_F^l$ are distinct.

Exercise. *Suppose, for the moment, that $p \neq 2$. Let $\rho \in \mathfrak{G}_2^0(F)$ be minimal and triply imprimitive. Show that ρ has level zero. Classify the minimal triply imprimitive elements of $\mathfrak{G}_2^0(F)$.*

42. Primitive Representations

We give a preliminary analysis of the *primitive* 2-dimensional representations of $\mathcal{W}_F$. By definition, a primitive representation $\rho \in \mathfrak{G}_2^0(F)$ is totally ramified.

42.1. Let $\rho \in \mathfrak{G}_2^0(F)$. If K/F is a finite field extension, with $K \subset \overline{F}$ as always, we put $\rho_K = \rho \mid \mathcal{W}_K$. Observe that if χ is a character of $\mathcal{W}_F$, then $\chi \mid \mathcal{W}_K$ corresponds to $\chi_K = \chi \circ \mathrm{N}_{K/F}$, via class field theory.

Proposition. *Let $\rho \in \mathfrak{G}_2^0(F)$ be totally ramified, and let K/F be a finite, tamely ramified field extension.*

(1) *The representation ρ_K is irreducible and totally ramified.*
(2) *The canonical map*

$$\begin{aligned} \mathfrak{T}(\rho) &\longrightarrow \mathfrak{T}(\rho_K), \\ \chi &\longmapsto \chi_K, \end{aligned}$$

is injective.

(3) *If* $\rho' \in \mathcal{G}_2^0(F)$ *satisfies* $\rho'_K \cong \rho_K$*, then* $\rho' \cong \chi \otimes \rho$*, for a uniquely determined character* χ *of* $F^\times$ *which is trivial on* $\mathrm{N}_{K/F}(K^\times)$*.*

Proof. In (1), we suppose for a contradiction that there exists a finite, tamely ramified extension such that ρ_K is reducible. Since ρ_K is semisimple (28.7 Lemma), it is a sum of characters, $\rho_K = \theta_1 \oplus \theta_2$. We may as well assume that K/F is Galois. We are then in the situation of the proof of 34.1 Theorem. As there, if $\theta_1 \neq \theta_2$, then ρ is induced from a character of $\mathcal{W}_L$, where L/F is quadratic and tamely ramified, hence unramified. If $\theta_1 = \theta_2$, the same argument as in 34.1 shows that $\rho \in \mathcal{G}_2^{\mathrm{nr}}(F)$. Either of these outcomes is contrary to hypothesis, so ρ_K is irreducible.

Let L/K be unramified quadratic. If ρ_K were not totally ramified, then ρ_L would be a sum of characters (41.3 Lemma), contrary to the first argument.

In part (2), let $\chi \in \mathfrak{T}(\rho)$, $\chi \neq 1$. By hypothesis, χ is not unramified; since $\chi^2 = 1$, it is not tamely ramified and so has level $\geqslant 1$. The norm map $\mathrm{N}_{K/F} : U_K^1 \to U_F^1$ is surjective (18.1 Lemma (2)), so $\chi_K \neq 1$.

We first prove (3) under the assumption that K/F is Galois. We can assume that ρ, ρ' act on the same vector space V and, replacing ρ' by a conjugate if necessary, that $\rho(h) = \rho'(h)$, $h \in \mathcal{W}_K$. For $h \in \mathcal{W}_K$ and $g \in \mathcal{W}_F$, we have

$$\begin{aligned}\rho(g)\rho(h)\rho(g)^{-1} &= \rho(ghg^{-1}) = \rho'(ghg^{-1}) \\ &= \rho'(g)\rho'(h)\rho'(g)^{-1} = \rho'(g)\rho(h)\rho'(g)^{-1}.\end{aligned}$$

Thus $\rho(g)^{-1}\rho'(g)$ commutes with all the operators $\rho(h)$. Schur's Lemma implies that $\rho(g)^{-1}\rho'(g) = \chi(g)$, for a scalar $\chi(g) \in \mathbb{C}^\times$. The map $g \mapsto \chi(g)$ is a character of $\mathcal{W}_F$ trivial on $\mathcal{W}_K$, and $\rho' \cong \chi \otimes \rho$. The character χ is tamely ramified, and uniquely determined modulo $\mathfrak{T}(\rho)$. By hypothesis, $\mathfrak{T}(\rho)$ contains no tamely ramified character other than 1, so χ is uniquely determined. This proves (3) for a Galois extension K/F.

In the general case, let L/F be the normal closure of K/F. Surely $\rho_L \cong \rho'_L$, so $\rho' \cong \rho \otimes \chi$, for a character χ of $\mathcal{W}_F$ trivial on $\mathcal{W}_L$. Therefore $\rho'_K \cong \rho_K \otimes \chi_K$, and hence $\chi_K \in \mathfrak{T}(\rho_K)$. The quadratic character χ_K is tamely ramified, therefore unramified. Since ρ_K is totally ramified, we have $\chi_K = 1$. Viewing χ as a character of $F^\times$, this is equivalent to χ vanishing on norms from K. □

42.2. We can now unpick the detailed structure of the 2-dimensional primitive representations of $\mathcal{W}_F$.

Theorem. *Let* $\rho \in \mathcal{G}_2^0(F)$ *be primitive. There exists a cubic extension* K/F *such that* ρ_K *is imprimitive. Moreover:*

(1) *If* K/F *is Galois (hence cyclic), the representation* ρ_K *is triply imprimitive.*

(2) *Suppose K/F is not cyclic, let L/F be the normal closure of K/F and let E/F be the maximal unramified sub-extension of L/F. The representation ρ_K is simply imprimitive, ρ_L is triply imprimitive, and ρ_E is primitive.*

Proof. We start with a more general analysis.

Lemma 1. *Let $\rho \in \mathcal{G}_2^0(F)$ be primitive, and let L/F be a finite, tamely ramified, Galois extension such that ρ_L is imprimitive. Then:*

(1) *The Galois group* $\mathrm{Gal}(L/F)$ *acts on $\mathfrak{T}(\rho_L)$ with only the trivial fixed point.*
(2) *The representation ρ_L is triply imprimitive.*
(3) *If $F \subset E \subset L$ and* $\mathrm{Gal}(L/E)$ *acts trivially on $\mathfrak{T}(\rho_L)$, the canonical map $\mathfrak{T}(\rho_E) \to \mathfrak{T}(\rho_L)$ is an isomorphism.*

Proof. Let E/F be the maximal unramified sub-extension of L/F. Suppose $\xi \in \mathfrak{T}(\rho_L)$ is fixed by $\mathrm{Gal}(L/F)$, $\xi \neq 1$. In particular, ξ is fixed by $\mathrm{Gal}(L/E)$. Thus $\xi = \chi_L$, for a character χ of $E^\times$. Since $(\chi \otimes \rho_E)_L \cong \rho_L$, we have $\chi \otimes \rho_E \cong \theta \otimes \rho_E$, for a character θ of $E^\times$ trivial on norms from $L^\times$ (42.1 Proposition (3)). That is, we can replace χ by $\theta^{-1}\chi$ and assume $\chi \in \mathfrak{T}(\rho_E)$.

If σ generates $\mathrm{Gal}(E/F)$, we have $(\chi^\sigma)_L = \chi_L$, so χ^σ/χ is trivial on norms from $L^\times$. In particular, χ^σ/χ is trivial on U_E^1. On the other hand, $\chi^\sigma \in \mathfrak{T}(\rho_E)$, so $\mathfrak{T}(\rho_E)$ contains the tamely ramified character χ^σ/χ. Since ρ_E is totally ramified (42.1 Proposition (1)), this implies $\chi^\sigma = \chi$. We deduce that χ is of the form τ_E, for some character τ of $F^\times$. Going through the same argument, we get $\tau \in \mathfrak{T}(\rho)$, which is impossible. This proves (1).

If ρ_L were simply imprimitive, the action of $\mathrm{Gal}(L/F)$ on $\mathfrak{T}(\rho_L)$ would necessarily be trivial, contradicting (1) and proving (2).

In (3), it is enough to treat the case where L/E is cyclic. A character $\chi \in \mathfrak{T}(\rho_L)$ is of the form $\chi = \xi_L$, for some character ξ of $E^\times$. Arguing as in the proof of (1), we may choose ξ to lie in $\mathfrak{T}(\rho_E)$. The map $\mathfrak{T}(\rho_E) \to \mathfrak{T}(\rho_L)$ is therefore surjective. It is injective by 42.1 Proposition (2). □

Next, we prove:

Lemma 2. *There is a finite, tamely ramified extension K/F such that the representation ρ_K is imprimitive.*

Proof. By 28.6 Proposition, we can assume that ρ is a representation of Ω_F. We view ρ as a homomorphism $\Omega_F \to \mathrm{GL}_2(\mathbb{C})$, and we write $G = \rho(\Omega_F)$. We can identify G with $\mathrm{Gal}(E/F)$, for some finite Galois extension E/F. Let $\overline{G}$ denote the image of G in the projective linear group $\mathrm{PGL}_2(\mathbb{C}) = \mathrm{GL}_2(\mathbb{C})/\mathbb{C}^\times$.

The group $Z = G \cap \mathbb{C}^\times$ is finite cyclic, and central in G. We have an exact sequence:

$$1 \to Z \longrightarrow G \longrightarrow \overline{G} \to 1. \tag{42.2.1}$$

Let $\overline{H}$ be a 2-Sylow subgroup of $\overline{G}$, with inverse image H in G. The index $(\overline{G} : \overline{H}) = (G : H)$ is odd, so the field extension E^H/F is tamely ramified. By 42.1 Proposition, the restriction $\rho \mid H$ is irreducible. The group H is a cyclic central extension of a finite 2-group. Thus H is *nilpotent* and so any irreducible representation of H is induced from a one-dimensional representation. The desired field K is then E^H. □

So, by Lemma 2, there exists a finite, tamely ramified, Galois extension L/F such that ρ_L is imprimitive. Lemma 1 (2) implies that ρ_L is triply imprimitive. The action of $\mathrm{Gal}(L/F)$ on $\mathfrak{T}(\rho_L)$ is given by a homomorphism

$$\mathrm{Gal}(L/F) \longrightarrow \mathrm{Aut}(\mathfrak{T}(\rho_L)) \cong \mathrm{GL}_2(\mathbb{F}_2) \cong S_3, \tag{42.2.2}$$

where S_3 is the symmetric group on 3 letters. The image of the map (42.2.2) acts without fixed point on the non-trivial elements of $\mathfrak{T}(\rho_L)$, so it is either S_3 or the cyclic alternating group A_3. By Lemma 1 (3), we may choose L/F so that (42.2.2) is injective. That is, $\mathrm{Gal}(L/F)$ is either S_3 or A_3.

If $\mathrm{Gal}(L/F) \cong A_3$, we are in the first case of the theorem with $K = L$. Otherwise, a non-trivial involution $\sigma \in \mathrm{Gal}(L/F)$ has a non-trivial fixed point $\chi \in \mathfrak{T}(\rho_L)$; if K is the fixed field of σ we have $\chi = \xi_L$, for some $\xi \in \mathfrak{T}(\rho_K)$. The extension K/F is totally ramified and cubic, while the representation ρ_K is simply imprimitive (Lemma 1 (3)). We are in case (2) of the theorem, and the final assertion follows from Lemma 1 (1). □

Remark. Suppose $\rho \in \mathfrak{G}_2^0(F)$ is primitive, and let K/F be a tamely ramified, Galois extension such that ρ_K is triply imprimitive. The group $\mathrm{Gal}(K/F)$ permutes the non-trivial elements of $\mathfrak{T}(\rho_K)$ transitively, and so they all have the same level. Corollary 41.5 implies that this level is $n_K/3$, where n_K is the level of ρ_K. In particular, n_K is divisible by 3.

42.3. We can work out the structure of the group $\overline{G}$ in (42.2.1). Take K as in the theorem, with L/F the normal closure of K/F. There are totally ramified quadratic extensions E_i/L, $i = 1, 2, 3$, such that the characters $\varkappa_{E_i/L}$ are the non-trivial elements of $\mathfrak{T}(\rho_L)$. For each i, there is a character ξ_i of $E_i^\times$ such that $\rho_L = \mathrm{Ind}_{E_i/L}\, \xi_i$.

Let E/L be the composite of the extensions E_i/L; then E/L is totally ramified (42.2 Remark and 41.5 Corollary) and $\mathrm{Gal}(E/L) = V_4$, the non-cyclic group of order 4. The character $\xi = \xi_{i,E}$ of $E^\times$ is fixed by $\mathrm{Gal}(E/F)$ and ρ is effectively a representation of a cyclic central extension of $\mathrm{Gal}(E/F)$. We thus have $\overline{G} = \mathrm{Gal}(E/F)$.

In the first case of the theorem, we have an exact sequence

$$1 \to V_4 \longrightarrow \overline{G} \longrightarrow A_3 \to 1,$$

and A_3 acts non-trivially on the subgroup V_4. Elementary group-theoretic arguments give $\overline{G} \cong A_4$. In case (2), we have an exact sequence

$$1 \to V_4 \longrightarrow \overline{G} \longrightarrow S_3 \to 1,$$

giving $\overline{G} \cong S_4$.

Comment. It is a "well-known fact, equivalent to the classification of the Platonic solids", that the finite subgroups of $\mathrm{PGL}_2(\mathbb{C})$ are the cyclic groups C_n of order n, $n \geqslant 1$, the dihedral groups D_n of order $2n$, $n \geqslant 2$, and the permutation groups A_4, S_4 and A_5. The case $\overline{G} = C_n$ cannot arise (because ρ would be reducible), neither can $\overline{G} = D_n$ (since ρ would then be imprimitive). Since $\overline{G}$ is soluble, the case $\overline{G} = A_5$ cannot happen. This leaves the alternatives $\overline{G} = A_4$ or S_4, just as in the theorem. This background explains why the first case is often called "tetrahedral" and the second "octahedral".

There is one aspect of this analysis to which we shall return.

Proposition. *Let $\rho \in \mathfrak{G}_2^0(F)$ be primitive.*

(1) *There is a cubic extension K/F such that ρ_K is imprimitive, and this condition determines K uniquely, up to F-isomorphism.*
(2) *Let E/F be a finite, tamely ramified field extension. The representation ρ_E is imprimitive if and only if E contains some F-conjugate of K.*

Proof. The existence of K/F is given by 42.2 Theorem. Let L/F be the normal closure of K/F; if $K = L$ we set $L_0 = F$ while, if $K \neq L$, we let L_0/F be the maximal unramified sub-extension of L/F. It is enough to show that if E/F is a finite, tamely ramified Galois extension such that ρ_E is imprimitive, then $E \supset L$. We assume the contrary. It follows that $E \cap L = L_0$ or F.

By Lemma 1 (2), the representations ρ_L, ρ_E are both triply imprimitive. The canonical maps $\mathfrak{T}(\rho_L) \to \mathfrak{T}(\rho_{EL})$, $\mathfrak{T}(\rho_E) \to \mathfrak{T}(\rho_{EL})$ are both isomorphisms. The group $\mathrm{Gal}(EL/E\cap L) = \mathrm{Gal}(L/E\cap L) \times \mathrm{Gal}(E/E\cap L)$ therefore acts trivially on $\mathfrak{T}(\rho_{EL})$. Lemma 1 (3) now implies that $\rho_{E\cap L}$ is triply imprimitive. If $E \cap L = F$, this contradicts the hypothesis. Otherwise, $E\cap L/F$ is unramified quadratic, and 42.2 Theorem (2) implies $\rho_{E\cap L}$ primitive.

These contradictions prove the proposition. □

42.4. Let K/F be a cubic extension, and let $\tau \in \mathfrak{G}_2^0(K)$. The question naturally arises as to whether τ is of the form ρ_K, for some $\rho \in \mathfrak{G}_2^0(F)$.

Proposition. *Let K/F be a cubic extension, and let $\tau \in \mathfrak{G}_2^0(K)$ be totally ramified.*

(1) *Suppose K/F is cyclic. There exists $\rho \in \mathcal{G}_2^0(F)$, such that $\tau \cong \rho_K$, if and only if $\tau \cong \tau^\sigma$, where σ generates* $\mathrm{Gal}(K/F)$.
(2) *Suppose K/F is not cyclic, and let L/F be the normal closure of K/F. There exists $\rho \in \mathcal{G}_2^0(F)$ such that $\tau \cong \rho_K$ if and only if there exists $\rho' \in \mathcal{G}_2^0(F)$ such that $\tau_L \cong \rho'_L$.*
(3) *In the situation of (2), let $\vartheta \in \mathcal{G}_2^0(L)$ be totally ramified. There exists $\rho \in \mathcal{G}_2^0(F)$ such that $\vartheta \cong \rho_L$ if and only if $\vartheta^\gamma \cong \vartheta$ for all $\gamma \in$* $\mathrm{Gal}(L/F)$.

Proof. Part (1) is elementary. In part (2), the necessity of the condition is clear. As for sufficiency, we observe that K/F is totally ramified while L/K is unramified and quadratic. We have $\tau_L \cong (\rho'_K)_L$, so $\tau = \phi \otimes \rho'_K$, for an unramified character ϕ of $K^\times$ such that $\phi^2 = 1$ (42.1 Proposition (3)). Necessarily $\phi = \chi_K$, where χ is the unramified character of $F^\times$ of the same order as ϕ. Thus $\tau = (\chi \otimes \rho')_K$, as required.

In part (3), the necessity of the condition is clear. As for sufficiency, let E/F be the maximal unramified sub-extension of L/F. Thus L/E is totally ramified and cubic. In particular, ϑ is of the form μ_L, for some $\mu \in \mathcal{G}_2^0(E)$. Further, if $\mu' \in \mathcal{G}_2^0(E)$, then $\mu'_L \cong \vartheta$ if and only if $\mu' \cong \mu \otimes \chi$, for some character χ of $E^\times$ trivial on norms from L.

These three representations μ of $\mathcal{W}_E$, which satisfy $\mu_L \cong \vartheta$, are mutually inequivalent as they have different determinants. The group $\mathrm{Gal}(E/F)$ acts on this set and it has a fixed point. In other words, we may choose μ to satisfy $\mu^\sigma \cong \mu$, where σ generates $\mathrm{Gal}(E/F)$. It follows that $\mu \cong \rho_E$, for some $\rho \in \mathcal{G}_2^0(F)$. □

43. A Converse Theorem

In this section, we prove an analogue of the Converse Theorem (27.1) for *imprimitive* representations $\rho \in \mathcal{G}_2^0(F)$.

43.1. We establish notation for a preliminary result. We let E/F be a separable, totally ramified quadratic extension. We put $d(E|F) = d$ and $\mathrm{Gal}(E/F) = \{1, \sigma\}$.

We let ξ be a character of $E^\times$, which is minimal over F, of level $n-d > d$. The integer n is odd (41.4 Lemma (2)): we put $n = 2m+1$. The representation $\rho = \mathrm{Ind}_{E/F}\,\xi$ is irreducible, totally ramified, and of level n.

For an integer $j \geqslant 0$, we let Γ_j denote the group of characters of $F^\times$ which are trivial on U_F^j. We put

$$F(n) = \{x \in F : \upsilon_F(x) = -n\},$$
$$E(n) = \{y \in E : \upsilon_E(y) = -n\}.$$

For $x \in E(n)$, we have $\mathrm{N}_{E/F}(x) \in F(n)$ and $\mathrm{Tr}_{E/F}(x) \in \mathfrak{p}^{-m+[d/2]} \subset \mathfrak{p}^{-m}$. If $u \in U_E^{n-d+1}$, then

$$\mathrm{N}_{E/F}(xu) \equiv \mathrm{N}_{E/F}(x) \pmod{U_F^{m+1}},$$

$$\mathrm{Tr}_{E/F}(xu) \equiv \mathrm{Tr}_{E/F}(x) \pmod{\mathfrak{p}},$$

so we have a canonical map

$$\begin{aligned} \Psi_E : \mathrm{Gal}(E/F)\backslash E(n)/U_E^{n-d+1} &\longrightarrow F(n)/U_F^{m+1} \times \mathfrak{p}^{-m}/\mathfrak{p}, \\ x &\longmapsto (\mathrm{N}_{E/F}(x), \mathrm{Tr}_{E/F}(x)). \end{aligned} \tag{43.1.1}$$

Lemma. *The map Ψ_E is injective.*

Proof. Let $\gamma_1, \gamma_2 \in E(n)$, and suppose that $\Psi_E(\gamma_1) = \Psi_E(\gamma_2)$. The relation $\mathrm{N}_{E/F}(\gamma_1^{-1}\gamma_2) \in U_F^1$ implies $\gamma_1 = \gamma_2(1{+}x)$, for some $x \in \mathfrak{p}_E$. Since $\mathrm{N}_{E/F}(1{+}x) \in U_F^{m+1} \subset U_F^{d+1}$, we have $1{+}x \in \mathrm{Ker}\,\mathrm{N}_{E/F} \cdot U_E^{d+1}$. Replacing γ_2 by γ_2^g if necessary, we can assume $1{+}x \in U_E^{d+1}$.

Since $\mathrm{Tr}_{E/F}(\gamma_1) \equiv \mathrm{Tr}_{E/F}(\gamma_2) \pmod{\mathfrak{p}}$, we have $v_F(\mathrm{Tr}_{E/F}(\gamma_2 x)) \geqslant 1$. Temporarily set $k = v_E(\gamma_2 x)$ and suppose first that $k \equiv d \pmod 2$. Parts (2) and (3) of 41.1 Lemma show that $v_F(\mathrm{Tr}_{E/F}(\gamma_2 x)) = (k{+}d)/2$. This implies $v_E(x) \geqslant n{-}d{+}1$, as required. Suppose therefore that $k \not\equiv d \pmod 2$ or, equivalently, $v_E(x) \equiv d \pmod 2$. In the same way, $v_F(\mathrm{Tr}_{E/F}(x)) = (v_E(x){+}d)/2$. The conditions $\mathrm{N}_{E/F}(\gamma_1\gamma_2^{-1}) \in U_F^{m+1}$, $x \in \mathfrak{p}_E^{d+1}$, then translate into $(v_E(x){+}d)/2 \geqslant m{+}1$. The desired relation $v_E(x) \geqslant n{+}1{-}d$ again follows. □

Let χ range over Γ_{m+1} and μ over the characters of F which are trivial on $\mathfrak{p}$. We define a function $\Upsilon(\chi, \mu)$ of such characters by

$$\Upsilon(\chi, \mu) = \begin{cases} 0 & \text{if } \mu \mid \mathfrak{o} = 1, \\ \varepsilon(\chi_E \xi, \tfrac{1}{2}, \mu_E) & \text{otherwise.} \end{cases}$$

If $\mu \in \widehat{F}$ has level one, the local constant $\varepsilon(\chi_E\xi, s, \mu_E)$, for $\chi \in \Gamma_{m+1}$, depends only on $\mu \mid \mathfrak{p}^{-m}$. We therefore regard Υ as a function of $\chi \in \Gamma_{m+1}$ and $\mu \in (\mathfrak{p}^{-m}/\mathfrak{p})\widehat{\ }$.

We fix $\psi \in \widehat{F}$ of level one. The map $c \mapsto c\psi$ then gives an isomorphism $\mathfrak{o}/\mathfrak{p}^{m+1} \to (\mathfrak{p}^{-m}/\mathfrak{p})\widehat{\ }$, under which the characters non-trivial on $\mathfrak{o}/\mathfrak{p}$ correspond to units $c \in U_F$.

Let $a \in F(n)$; then $\varepsilon(\chi_E\xi, \frac{1}{2}, c\psi_E)\chi(a)$ depends only on a modulo U_F^{m+1} and the coset $\chi\Gamma_0$.

For $a \in F(n)/U_F^{m+1}$, $b \in \mathfrak{p}^{-m}/\mathfrak{p}$, consider the function

$$\begin{aligned}\Phi(a,b) &= \sum_{\substack{\chi\in\Gamma_{m+1}/\Gamma_0,\\ c\in U_F/U_F^{m+1}}} \varepsilon(\chi_E\xi, \tfrac{1}{2}, c\psi_E)\,\chi(a)\,\psi(-cb) \\ &= \sum_{\substack{\chi\in\Gamma_{m+1}/\Gamma_0,\\ c\in\mathfrak{o}/\mathfrak{p}^{m+1}}} \Upsilon(\chi, c\psi)\,\chi(a)\,\psi(-cb).\end{aligned} \tag{43.1.2}$$

Thus Φ is essentially the Fourier transform of Υ on the finite group $\Gamma_{m+1}/\Gamma_0 \times \mathfrak{o}/\mathfrak{p}^{m+1}$. In particular, we recover Υ as

$$\Upsilon(\chi, c\psi) = (q-1)^{-1}q^{-(2m+1)}\sum_{a,b}\Phi(a,b)\,\chi(a)^{-1}\,\psi(cb),$$

where a runs through $F(n)/U_F^{m+1}$ and b through $\mathfrak{p}^{-m}/\mathfrak{p}$.

On the other hand, for $x \in E(n)$, the quantity $\check{\xi}(x)+\check{\xi}^\sigma(x)$ depends only on $\Psi_E(x)$. Thus we can define a function Φ' on $F(n)/U_F^{m+1} \times \mathfrak{p}^{-m}/\mathfrak{p}$, with support in $\Psi_E(E(n))$, such that

$$\Phi'(\mathrm{N}_{E/F}(x), \mathrm{Tr}_{E/F}(x)) = \check{\xi}(x)+\check{\xi}^\sigma(x), \quad x \in E(n).$$

Proposition. *There is a constant $k > 0$, depending only on n, such that $\Phi = k\Phi'$.*

Proof. By Fourier inversion, it is enough to prove

$$\Upsilon(\chi, \mu) = k'\sum_{a,b}\Phi'(a,b)\,\chi(a)^{-1}\,\mu(b),$$

for all characters μ of $\mathfrak{p}^{-m}$ trivial on $\mathfrak{p}$ and all $\chi \in \Gamma_{m+1}$, where k' is a positive constant depending only on n. The right hand side is

$$\begin{aligned}&k' \sum_{x\in E(n)/U_E^{n-d+1}} (\check{\xi}(x)+\check{\xi}^\sigma(x))\,\chi_E(x)^{-1}\mu_E(x) \\ &= k'\int_{E(n)} (\check{\xi}(x)+\check{\xi}^\sigma(x))\,\chi_E(x)^{-1}\mu_E(x)\,d\nu^*(x),\end{aligned}$$

where ν^* is the Haar measure on $E^\times$ for which $\nu^*(U_E^{n-d+1}) = 1$. If μ has level $\leqslant 0$, then μ_E has level $\leqslant -d$ and the integral vanishes (as follows easily from 23.5 Lemma). If μ has level 1, then μ_E has level $1-d$ and, as in the proof of 23.5 Theorem, the integral reduces to $k''\varepsilon(\chi_E\xi, \frac{1}{2}, \mu_E)$, for some $k'' > 0$ depending only on n. □

Let us re-state the proposition in more transparent form:

Corollary. *There exists a constant $k > 0$, depending only on n, such that*

$$\check{\xi}(x)+\check{\xi}^\sigma(x) = k \sum_{\substack{\chi\in\Gamma_{m+1}/\Gamma_0,\\ c\in U_F/U_F^{m+1}}} \varepsilon(\chi_E\xi, \tfrac{1}{2}, c\psi)\,\chi(\mathrm{N}_{E/F}(x))\,\psi(-c\mathrm{Tr}_{E/F}(x)),$$

for all $x \in E(n)$.

43.2. The result we seek is:

Galois Converse Theorem. *Let $\rho_1, \rho_2 \in \mathfrak{G}_2^0(F)$ be imprimitive and suppose that ρ_1 is totally ramified. Suppose also that*

$$\varepsilon(\chi\otimes\rho_1, s, \psi) = \varepsilon(\chi\otimes\rho_2, s, \psi),$$

for all characters χ of $F^\times$ and some $\psi \in \widehat{F}$, $\psi \neq 1$. We then have $\rho_1 \cong \rho_2$.

We first note that, using the Stability Theorem (29.4 Proposition (4)), the hypothesis implies $\det\rho_1 = \det\rho_2$. The hypothesis is therefore independent of ψ. We may therefore choose ψ of level one. Further, we can assume that the ρ_i are both minimal over F.

43.3. The proof of 43.2 Theorem proceeds in two steps. We first treat the case where $\mathfrak{T}(\rho_1)\cap\mathfrak{T}(\rho_2)$ is non-trivial. Let $\varkappa_{E/F} \in \mathfrak{T}(\rho_1)\cap\mathfrak{T}(\rho_2)$. The extension E/F is then totally ramified. We set $d(E|F) = d$, and use the other notation of 43.1. The integer $n = n(\rho_1,\psi) = n(\rho_2,\psi)$ is then $> 2d$ and so ρ_2 is also totally ramified (41.4 Proposition (1)).

By 41.3 Lemma, we can choose characters ξ_1, ξ_2 of $E^\times$ such that $\rho_i = \mathrm{Ind}_{E/F}\,\xi_i$, $i = 1, 2$. Our hypothesis implies

$$\varepsilon(\chi_E\xi_1, \tfrac{1}{2}, \psi_E) = \varepsilon(\chi_E\xi_2, \tfrac{1}{2}, \psi_E),$$

for all characters χ of $F^\times$. Corollary 43.1 yields

$$\check{\xi}_1(x) + \check{\xi}_1^\sigma(x) = \check{\xi}_2(x) + \check{\xi}_2^\sigma(x), \tag{43.3.1}$$

for all $x \in E(n)$. The relation $\det\rho_1 = \det\rho_2$ implies $\xi_1 \mid F^\times = \xi_2 \mid F^\times$, so the two sides of (43.3.1) agree on the set $F^\times E(n) = \{y \in E^\times : v_E(y) \equiv 1 \pmod 2\}$. Thus, if ϕ denotes the unramified character of $E^\times$ of order 2, we have

$$(\xi_1{-}\phi\xi_1) + (\xi_1^\sigma{-}\phi\xi_1^\sigma) = (\xi_2{-}\phi\xi_2) + (\xi_2^\sigma{-}\phi\xi_2^\sigma),$$

as functions on $E^\times$. Since ρ_i is totally ramified, we have $\xi_i \neq \phi\xi_i^\sigma$. Linear independence of characters of $E^\times$ now implies that $\xi_1 = \xi_2$ or ξ_2^σ. Therefore $\rho_1 \cong \rho_2$, as desired.

43.4. We prove 43.2 Theorem in the general case. Let $\varkappa = \varkappa_{K/F} \in \mathfrak{T}(\rho_2)$. For any character χ of $F^\times$, we have

$$\varepsilon(\chi\varkappa \otimes \rho_1, s, \psi) = \varepsilon(\chi\varkappa \otimes \rho_2, s, \psi) = \varepsilon(\chi \otimes \rho_2, s, \psi) = \varepsilon(\chi \otimes \rho_1, s, \psi).$$

Since ρ_1 is totally ramified and $\mathfrak{T}(\rho_1) = \mathfrak{T}(\varkappa \otimes \rho_1)$, we can apply the special case of 43.3 to deduce $\varkappa \otimes \rho_1 \cong \rho_1$. Thus $\mathfrak{T}(\rho_1) \cap \mathfrak{T}(\rho_2) \neq \{1\}$, and so $\rho_1 \cong \rho_2$ by the first case. □

Remark. The theorem also holds if both $\rho_i \in \mathcal{G}_2^{\mathrm{nr}}(F)$, because of the uniqueness of the tame Langlands correspondence (34.4).

Further reading. Primitive 2-dimensional representations of $\mathcal{W}_F$ are classified in Weil [88], using a somewhat different approach from that here. A full classification of primitive representations of $\mathcal{W}_F$, in arbitrary dimension and residual characteristic, is given by Koch [49]. The idea of writing the inducing character of an imprimitive representation in terms of local constants seems to have first been employed in a series of papers of Gérardin and Li, for example [34].

11

Ordinary Representations

We return to the partial correspondence

$$\boldsymbol{\pi} : \mathcal{G}_2^{\mathrm{im}}(F) \longrightarrow \mathcal{A}_2^0(F),$$

defined on the set $\mathcal{G}_2^{\mathrm{im}}(F)$ of equivalence classes of irreducible *imprimitive* two-dimensional representations of $\mathcal{W}_F$, as in 40.1. The image $\boldsymbol{\pi}(\mathcal{G}_2^{\mathrm{im}}(F))$ comprises the cuspidal Weil representations of §39. The mode of construction of the map $\boldsymbol{\pi}$ is no longer relevant: we rely solely on its formal properties. The object of this chapter is to give an independent characterization of the representations $\pi \in \boldsymbol{\pi}(\mathcal{G}_2^{\mathrm{im}}(F))$, in terms of the classification theory of Chapter IV.

We know from the Tame Parametrization Theorem (34.4) that the set $\mathcal{A}_2^{\mathrm{nr}}(F)$ of equivalence classes of unramified cuspidal representations is contained in $\boldsymbol{\pi}(\mathcal{G}_2^{\mathrm{im}}(F))$ and also, when $p \neq 2$, that $\mathcal{G}_2^{\mathrm{im}}(F) = \mathcal{G}_2^0(F)$ and $\boldsymbol{\pi}$ is bijective. So, we assume henceforth that the residual characteristic p of F is 2, and concentrate on totally ramified representations. (That said, the argument works unchanged when $p \neq 2$, provided we stick to *totally ramified* representations. Some variation is needed to deal with representations $\rho \in \mathcal{G}_2^{\mathrm{nr}}(F)$ using this approach.)

44. Ordinary Representations and Strata

Let $\pi \in \mathcal{A}_2^0(F)$ satisfy $\ell(\pi) \leqslant \ell(\chi\pi)$, for all characters χ of $F^\times$: we shall now say that such representations are *minimal.* Suppose also that π is *totally ramified.*

There is a ramified simple stratum $(\mathfrak{A}, n, \alpha)$ in $A = \mathrm{M}_2(F)$ such that π contains the character ψ_α of $U_{\mathfrak{A}}^{[n/2]+1}$. As we shall discover in this section, it is the stratum $(\mathfrak{A}, n, \alpha)$ which determines whether or not π lies in the image of the partial correspondence $\boldsymbol{\pi} \mid \mathcal{G}_2^{\mathrm{im}}(F)$.

44.1. We gather together those properties of the partial correspondence on which we shall rely.

Imprimitive Langlands correspondence. *Let $\psi \in \widehat{F}$, $\psi \neq 1$. There is a unique map*

$$\boldsymbol{\pi} = \boldsymbol{\pi}_F : \mathcal{G}_2^{\mathrm{im}}(F) \longrightarrow \mathcal{A}_2^0(F)$$

such that

$$\varepsilon(\chi\boldsymbol{\pi}(\rho), s, \psi) = \varepsilon(\chi \otimes \rho, s, \psi) \tag{44.1.1}$$

for all characters χ of $F^\times$.

The map $\boldsymbol{\pi}$ has the further properties:

(1) *The relation* (44.1.1) holds for all $\psi \in \widehat{F}$, $\psi \neq 1$.
(2) *The map $\boldsymbol{\pi}$ is injective.*
(3) *For any $\rho \in \mathcal{G}_2^{\mathrm{im}}(F)$ and any character χ of $F^\times$, we have*

$$\boldsymbol{\pi}(\chi \otimes \rho) \cong \chi\boldsymbol{\pi}(\rho).$$

(4) *If $\rho \in \mathcal{G}_2^0(F)$ and $\pi = \boldsymbol{\pi}(\rho)$, then $\omega_\pi =$* $\det \rho$.

Proof. The existence and uniqueness of $\boldsymbol{\pi}$ are taken from 40.1 Theorem, as are (1) and (3). Property (4) is 39.2 Proposition (3), 29.2 Proposition. Injectivity (2) is implied by the Galois Converse Theorem of 43.2. □

Definition. *Let $\pi \in \mathcal{A}_2^0(F)$; say π is* ordinary *if $\pi = \boldsymbol{\pi}(\rho)$, for some $\rho \in \mathcal{G}_2^{\mathrm{im}}(F)$. Otherwise, call π* exceptional.

Observe that unramified cuspidal representations π are invariably ordinary.

44.2 We make a brief return to matrix calculations. Let $(\mathfrak{A}, n, \alpha)$ be a ramified simple stratum with $n = 2m+1 \geqslant 1$. We can replace $\mathfrak{A}$ by a G-conjugate, when convenient, and identify it with the standard chain order $\mathfrak{J}$:

$$\mathfrak{A} = \mathfrak{J} = \begin{pmatrix} \mathfrak{o} & \mathfrak{o} \\ \mathfrak{p} & \mathfrak{o} \end{pmatrix}.$$

We choose a prime element ϖ of F. The element α is then conjugate (in G or $\mathcal{K}_{\mathfrak{A}}$) to the matrix

$$\begin{pmatrix} 0 & -\zeta\varpi^{-(m+1)} \\ \varpi^{-m} & t \end{pmatrix},$$

where $\zeta \in U_F$ (so $\det\alpha = \zeta\varpi^{-n}$) and $t = \operatorname{tr}\alpha \in \mathfrak{p}^{-m}$ (*cf.* 27.5).

Lemma. *Let $(\mathfrak{A}, n, \alpha)$, $(\mathfrak{A}, n, \alpha')$ be ramified simple strata with $n = 2m+1 \geqslant 1$. Let $\psi \in \widehat{F}$ have level one. The following conditions are equivalent:*

(1) *the characters* ψ_α, $\psi_{\alpha'}$ *of* $U_{\mathfrak{A}}^{m+1}$ *intertwine in* G;
(2) *the characters* ψ_α, $\psi_{\alpha'}$ *of* $U_{\mathfrak{A}}^{m+1}$ *are conjugate in* $\mathcal{K}_{\mathfrak{A}}$;
(3) *the cosets* $\alpha U_{\mathfrak{A}}^{m+1}$, $\alpha' U_{\mathfrak{A}}^{m+1}$ *are* $\mathcal{K}_{\mathfrak{A}}$*-conjugate;*
(4) *the elements* α, α' *satisfy*

$$\det\alpha \equiv \det\alpha' \pmod{U_F^{[m/2]+1}},$$
$$\operatorname{tr}\alpha \equiv \operatorname{tr}\alpha' \pmod{\mathfrak{p}^{-[m/2]}}.$$

Proof. The equivalence of (1) and (2) is the Conjugacy Theorem 15.2, and (3) is the dual version of (2). The equivalence of (3) and (4) follows from the "normal form" above. □

44.3. We fix a character $\psi \in \widehat{F}$ of level one, to use the classification theory of Chapter IV, and standardize our notation for the rest of the section.

Let $(\mathfrak{A}, n, \alpha)$, $n \geqslant 1$, be a ramified simple stratum in $A = \mathrm{M}_2(F)$. By definition, the integer n is odd, so we write $n = 2m{+}1$. We let $\mathcal{A}_2^0(F;\alpha)$ denote the set of equivalence classes of representations $\pi \in \mathcal{A}_2^0(F)$ which contain the character $\psi_\alpha : 1{+}x \mapsto \psi_A(\alpha x) = \psi(\operatorname{tr}(\alpha x))$ of $U_{\mathfrak{A}}^{m+1}$.

Note 1. The Classification Theorem (15.5) and 15.6 Proposition 1 imply that we have a bijection

$$C(\psi_\alpha, \mathfrak{A}) \longrightarrow \mathcal{A}_2^0(F;\alpha),$$
$$\Lambda \longmapsto c\text{-}\mathrm{Ind}_J^G\,\Lambda,$$

where $J = F[\alpha]^\times U_{\mathfrak{A}}^{m+1}$.

Note 2. Every $\pi \in \mathcal{A}_2^0(F;\alpha)$ has $\ell(\pi) = n/2$. If $(\mathfrak{A}', n', \alpha')$ is some other ramified simple stratum, with $n' \geqslant 1$, the Conjugacy Theorem (15.2) implies that either $\mathcal{A}_2^0(F;\alpha') = \mathcal{A}_2^0(F;\alpha)$ or $\mathcal{A}_2^0(F;\alpha') \cap \mathcal{A}_2^0(F;\alpha) = \emptyset$.

If L/F is a finite field extension, $L(n)$ will denote the set of $x \in L$ for which $v_L(x) = -n$.

If E/F is a separable, totally ramified quadratic extension, we set $\mathfrak{d}_{E/F} = \mathfrak{p}^{d+1}$ and $d = d(E|F)$. The unique non-trivial character of $F^\times$ trivial on $\mathrm{N}_{E/F}(E^\times)$ will be denoted $\varkappa_{E/F}$. We choose $\delta_{E/F} \in \mathfrak{p}^{-d}$ so that $\varkappa_{E/F}(1{+}x) = \psi(x\delta_{E/F})$, $x \in \mathfrak{p}^{1+[d/2]}$.

The section is devoted to proving the following result.

Theorem. *Let* $(\mathfrak{A}, n, \alpha)$, $n \geqslant 1$, *be a ramified simple stratum, and write* $n = 2m{+}1$. *The following are equivalent:*

(1) *The set* $\mathcal{A}_2^0(F;\alpha)$ *contains an ordinary representation.*
(2) *All representations* $\pi \in \mathcal{A}_2^0(F;\alpha)$ *are ordinary.*
(3) *There exist a totally ramified, separable quadratic extension* E/F *and an element* $\beta \in E(n)$ *such that*

(a) $n \geqslant 3d(E|F)$, *and*
(b) *the following congruences are satisfied:*

$$\begin{aligned} \det\alpha &\equiv \mathrm{N}_{E/F}(\beta) \pmod{U_F^{[m/2]+1}}, \\ \operatorname{tr}\alpha &\equiv \delta_{E/F} + \mathrm{Tr}_{E/F}(\beta) \pmod{\mathfrak{p}^{-[m/2]}}. \end{aligned} \tag{44.3.1}$$

The proof occupies the rest of the section.

Remark. If we regard E/F and β as given, the congruences (44.3.1) determine the $\mathcal{K}_{\mathfrak{A}}$-conjugacy class of the character $\psi_\alpha \mid U_{\mathfrak{A}}^{m+1}$ completely (44.2). If, however, we start with ψ_α, they do not determine E/F or β with any precision: we have to analyze this phenomenon below.

Note also that the conditions in the theorem are independent of the choice of $\psi \in \widehat{F}$ of level one.

44.4. We start the proof of 44.3 Theorem with a calculation.

Let $\rho \in \mathcal{G}_2^{\mathrm{im}}(F)$. We assume that ρ is totally ramified and *minimal*, in the sense that $n(\rho,\psi) \leqslant n(\chi\otimes\rho,\psi)$, for all characters χ of $F^\times$ (*cf.* 41.5). Thus there is a separable quadratic extension E/F, and a character ξ of $E^\times$, such that $\rho \cong \mathrm{Ind}_{E/F}\,\xi$. By 41.5 Theorem, we may choose E so that

$$n = n(\rho,\psi) \geqslant 3d(E|F). \tag{44.4.1}$$

The character ξ is minimal, of level $n-d$, $d = d(E|F)$. The integer n is invariably odd (41.5 Lemma), so we write $n = 2m+1$.

The character $\psi_E = \psi\circ\mathrm{Tr}_{E/F}$ has level $1-d$ (as follows from 41.2 Proposition (1)), so there exists $\beta \in E(n)$ such that

$$\xi(1+x) = \psi_E(\beta x), \quad v_E(x) \geqslant 1+[(n-d)/2]. \tag{44.4.2}$$

This condition determines the coset $\beta U_E^{[(n-d+1)/2]}$.

The representation $\pi = \boldsymbol{\pi}(\rho)$ is minimal of normalized level $\ell(\pi) = n/2$, and $n = n(\pi,\psi) = n(\rho,\psi)$. Therefore there is a ramified simple stratum $(\mathfrak{A}, n, \alpha)$ such that $\pi \in \boldsymbol{\mathcal{A}}_2^0(F;\alpha)$. Our calculation will show:

Lemma. *The elements* α, β *satisfy the congruences* (44.3.1).

Proof. We first exploit the relation

$$\omega_\pi = \det\rho = \varkappa_{E/F}\otimes\xi \mid F^\times.$$

Let $1+x \in U_{\mathfrak{A}}^{m+1}\cap F^\times = U_F^{[m/2]+1}$. Thus $\omega_\pi(1+x) = \psi(x\,\mathrm{tr}(\alpha))$. On the other hand, the hypothesis $n \geqslant 3d$ implies $[m/2] \geqslant [d/2]$, so

$$\det\rho(1+x) = \psi\big((\delta_{E/F}+\mathrm{Tr}_{E/F}(\beta))x\big),$$

whence

$$\operatorname{tr}\alpha \equiv \delta_{E/F}+\mathrm{Tr}_{E/F}(\beta) \pmod{\mathfrak{p}^{-[m/2]}}. \tag{44.4.3}$$

44.5. We describe π in terms of the classification theory of §15. The representation π contains a character Λ of the group $J = F[\alpha]^\times U_{\mathfrak{A}}^{m+1}$ such that $\Lambda \mid U_{\mathfrak{A}}^{m+1} = \psi_\alpha$. Thus $\Lambda \in C(\psi_\alpha, \mathfrak{A})$ and $(\mathfrak{A}, J, \Lambda)$ is a cuspidal type in π. The character Λ induces an irreducible representation Ξ of $\mathcal{K}_{\mathfrak{A}}$: the pair $(\mathfrak{A}, \Xi)$ is a cuspidal inducing datum in π.

As before, we let Γ_r denote the group of characters of $F^\times$ trivial on U_F^r, $r \geqslant 0$. Let $\mathcal{K}(n)$ denote the set of $g \in \mathcal{K}_{\mathfrak{A}}$ with $v_F(\det g) = -n$. From (27.7.1), we have the formula

$$\operatorname{tr} \check{\Xi}(g) = k \sum_{\substack{\chi \in \Gamma_{m+1}/\Gamma_0, \\ c \in U_F/U_F^{m+1}}} \varepsilon(\chi\pi, \tfrac{1}{2}, c\psi)\, \chi(\det g)\, \psi(-c\operatorname{tr}(g))$$

$$= k \sum_{\substack{\chi \in \Gamma_{m+1}/\Gamma_0, \\ c \in U_F/U_F^{m+1}}} \varepsilon(\chi\pi, \tfrac{1}{2}, \psi)\, \chi^2\omega_\pi(c)\, \chi(\det g)\, \psi(-c\operatorname{tr}(g)),$$

for $g \in \mathcal{K}(n)$, where k is a positive constant depending only on n.

Lemma. *Let du be the Haar measure on $U_{\mathfrak{A}}^{m+1}$ such that $\int_{U_{\mathfrak{A}}^{m+1}} du = 1$.*

(1) *Let $\gamma \in \mathcal{K}(n)$. The function of $g \in \mathcal{K}(n)$, defined by*

$$g \longmapsto \operatorname{tr} \check{\Xi} * \psi_\gamma^{-1}(g) = \int_{U_{\mathfrak{A}}^{m+1}} \operatorname{tr} \check{\Xi}(gu)\, \psi_\gamma(u)\, du,$$

is identically zero unless the cosets $\gamma U_{\mathfrak{A}}^{m+1}$, $\alpha U_{\mathfrak{A}}^{m+1}$ are $\mathcal{K}_{\mathfrak{A}}$-conjugate.

(2) *For $g \in J \cap \mathcal{K}(n)$, we have*

$$\operatorname{tr} \check{\Xi} * \psi_\alpha^{-1}(g) = \check{\Lambda}(g).$$

Proof. Set $\mathfrak{P} = \operatorname{rad} \mathfrak{A}$. By definition, the representation $\Xi \mid U_{\mathfrak{A}}^{m+1}$ is the direct sum of the characters ψ_γ, $\gamma \in \mathcal{K}(n)$, such that the coset $\gamma + \mathfrak{P}^{-m} = \gamma U_{\mathfrak{A}}^{m+1}$ is $\mathcal{K}_{\mathfrak{A}}$-conjugate to $\alpha U_{\mathfrak{A}}^{m+1}$, whence follows (1). The natural representation of J on the ψ_α-isotypic space in Ξ is the character Λ, and this implies (2). □

In particular,

$$\operatorname{tr} \check{\Xi} * \psi_\alpha^{-1}(\alpha) \neq 0. \tag{44.5.1}$$

The relation $\pi = \boldsymbol{\pi}(\rho)$ gives

$$\operatorname{tr} \check{\Xi}(g) = k \sum_{\substack{\chi \in \Gamma_{m+1}/\Gamma_0, \\ c \in U_F/U_F^{m+1}}} \varepsilon(\chi \otimes \rho, \tfrac{1}{2}, \psi)\, \chi^2 \det \rho(c)\, \chi(\det g)\, \psi(-c\operatorname{tr}(g)),$$

for $g \in \mathcal{K}(n)$. Taking $u \in U_{\mathfrak{A}}^{m+1}$, we have $\chi(\det \alpha u) = \chi(\det \alpha)\chi(\det u)$ and $\psi(-c\,\mathrm{tr}(\alpha u)) = \psi(-c\,\mathrm{tr}(\alpha))\,\psi_{-c\alpha}(u)$. Therefore

$$\begin{aligned} &\mathrm{tr}\,\check{\Xi} * \psi_\alpha^{-1}(\alpha) \\ &\qquad = k \sum_{\chi, c} \varepsilon(\chi \otimes \rho, \tfrac{1}{2}, \psi)\, \chi^2 \det \rho(c)\, \chi(\det \alpha)\, \psi(-c\,\mathrm{tr}(\alpha)) \\ &\qquad\qquad \int_{U_{\mathfrak{A}}^{m+1}} \chi(\det u)\, \psi_{-c\alpha}(u)\, \psi_\alpha(u)\, du. \end{aligned}$$

We choose $\delta_\chi \in \mathfrak{p}^{-m}$ so that $\chi(1+x) = \psi(\delta_\chi x)$, $x \in \mathfrak{p}^{[m/2]+1}$. Thus $\chi(\det u) = \psi_{\delta_\chi}(u)$, $u \in U_{\mathfrak{A}}^{m+1}$. So, writing $\mathfrak{P} = \mathrm{rad}\,\mathfrak{A}$, the inner integral is non-zero if and only if

$$\delta_\chi + (1-c)\alpha \equiv 0 \pmod{\mathfrak{P}^{-m}}. \tag{44.5.2}$$

(When the integral is non-zero, its value is 1.) We could replace $\mathfrak{P}$ by $\mathfrak{p}_{F[\alpha]}$ here. In the field $F[\alpha]$, the element δ_χ has even valuation while $(c-1)\alpha$ has odd valuation (or is zero). The congruence (44.5.2) is therefore equivalent to

$$\delta_\chi \equiv (1-c)\alpha \equiv 0 \pmod{\mathfrak{P}^{-m}}.$$

A character χ thus contributes to the sum only if $\delta_\chi \in \mathfrak{P}^{-m} \cap F = \mathfrak{p}^{-[m/2]}$: that is, χ has level $\leqslant [m/2]$. A unit $c \in U_F$ contributes to the sum only if $c \in U_{\mathfrak{A}}^{m+1} \cap F = U_F^{[m/2]+1}$. For such χ and c we have $\chi(c) = 1$. Therefore, invoking (44.4.3),

$$\begin{aligned} &\mathrm{tr}\,\check{\Xi} * \psi_\alpha^{-1}(\alpha) \\ &\qquad = k' \sum_{\substack{\chi \in \Gamma_{[m/2]+1}/\Gamma_0, \\ c \in U_F^{[m/2]+1}/U_F^{m+1}}} \varepsilon(\chi \otimes \rho, \tfrac{1}{2}, \psi)\, \det \rho(c)\, \chi(\det \alpha)\, \psi(-c\,\mathrm{tr}(\alpha)) \\ &\qquad = k'' \sum_{\chi \in \Gamma_{[m/2]+1}/\Gamma_0} \varepsilon(\chi \otimes \rho, \tfrac{1}{2}, \psi)\, \chi(\det \alpha)\, \psi(-\mathrm{tr}(\alpha)), \end{aligned}$$

for positive constants k', k''. For $\chi \in \Gamma_{[m/2]+1}$, we have $\varepsilon(\chi \otimes \rho, \frac{1}{2}, \psi) = \lambda_{E/F}(\psi)\, \varepsilon(\chi_E \xi, \frac{1}{2}, \psi_E)$. Writing $\varphi = \varphi_{E/F}$ (*cf.* 44.2), the character χ_E has level $\leqslant \varphi([m/2]) < (n-d)/2$. We can therefore apply 23.8 to get

$$\varepsilon(\chi \otimes \rho, \tfrac{1}{2}, \psi) = \chi_E(\beta)^{-1}\, \varepsilon(\rho, \tfrac{1}{2}, \psi),$$

We achieve

$$\mathrm{tr}\,\check{\Xi} * \psi_\alpha^{-1}(\alpha) = k''\varepsilon(\rho, \tfrac{1}{2}, \psi)\, \psi(-\mathrm{tr}(\alpha)) \sum_\chi \chi(\det \alpha / \mathrm{N}_{E/F}(\beta)),$$

with χ ranging over $\Gamma_{[m/2]+1}/\Gamma_0$. This sum is non-zero if and only if

$$\det \alpha \equiv \mathrm{N}_{E/F}(\beta) \pmod{U^{[m/2]+1}}, \tag{44.5.3}$$

so completing the proof of 44.4 Lemma. □

44.6. We now prove (1) $\Leftrightarrow$ (3) in 44.3 Theorem. Let $(\mathfrak{A}, n, \alpha)$, $n = 2m{+}1 \geqslant 1$, be a ramified simple stratum. Let $\pi \in \mathcal{A}_2^0(F; \alpha)$ be ordinary. Let $\rho \in \mathcal{G}_2^{\mathrm{im}}(F)$ satisfy $\pi = \boldsymbol{\pi}(\rho)$. We write $\rho = \mathrm{Ind}_{E/F}\, \xi$ as in (44.4.2). We have $n = n(\rho, \psi)$ and ρ is minimal. The congruences (44.3.1) are given by 44.4 Lemma.

Conversely, let $(\mathfrak{A}, n, \alpha)$ satisfy (3), relative to an extension E/F and an element $\beta \in E(n)$. We choose a character ξ of $E^\times$ satisfying (44.4.2). Since n is odd, the character ξ is minimal over F, the representation $\rho = \mathrm{Ind}_{E/F}\, \xi$ is irreducible, and 44.4 Lemma shows that $\boldsymbol{\pi}(\rho) \in \mathcal{A}_2^0(F; \alpha)$. Thus $(\mathfrak{A}, n, \alpha)$ satisfies (1).

44.7. At this point, it becomes convenient to introduce some new vocabulary and summarize the position. Let $\overline{F}/F$ be the separable algebraic closure of F used to define $\mathcal{W}_F$.

Definition 1. *An* admissible wild triple $(E/F, n, \beta)$ *consists of*

(a) *a totally ramified quadratic extension* E/F, $E \subset \overline{F}$,
(b) *an odd integer* $n = 2m + 1 \geqslant 1$, *and*
(c) *an element* $\beta \in E(n)$,

satisfying the condition $n \geqslant 3d(E|F)$.

Definition 2. *Let* $(E/F, n, \beta)$ *be an admissible wild triple. Define* $\mathcal{G}_2^0(F; \beta)$ *to be the set of classes of representations* $\rho \in \mathcal{G}_2^0(F)$ *of the form* $\rho = \mathrm{Ind}_{E/F}\, \xi$, *where*

$$\xi(1{+}x) = \psi_E(\beta x), \quad x \in \mathfrak{p}_E^{1+[(n-d)/2]}, \tag{44.7.1}$$

and $d = d(E|F)$.

Observe here that ξ has level $n{-}d$ and $n(\rho, \psi) = n$. Moreover, ρ and ξ are minimal, so $\boldsymbol{\pi}(\rho) \in \mathcal{A}_2^0(F; \alpha)$, for some ramified simple stratum $(\mathfrak{A}, n, \alpha)$. Lemma 44.4 shows that the stratum must satisfy the conditions (3) of 44.3 Theorem. To summarize:

Proposition. *Let* $(E/F, n, \beta)$ *be an admissible wild triple,* $n = 2m{+}1$. *There exists a ramified simple stratum* $(\mathfrak{A}, n, \alpha)$ *satisfying*

$$\begin{aligned} \det \alpha &\equiv \mathrm{N}_{E/F}(\beta) && \pmod{U_F^{[m/2]+1}}, \\ \operatorname{tr} \alpha &\equiv \delta_{E/F} + \mathrm{Tr}_{E/F}(\beta) && \pmod{\mathfrak{p}^{-[m/2]}}. \end{aligned}$$

These conditions determine the G*-conjugacy class of the character* $\psi_\alpha \mid U_{\mathfrak{A}}^{m+1}$ *uniquely, and*

$$\boldsymbol{\pi}(\mathcal{G}_2^0(F; \beta)) \subset \mathcal{A}_2^0(F; \alpha).$$

44.8. To prove (1) ⇒ (2) in 44.3 Theorem, we rely on a counting argument. This will occupy the rest of the section.

The group Γ_1 acts on the set $\mathcal{A}_2^0(F;\alpha)$ by $(\chi, \pi) \mapsto \chi\pi$.

Lemma 1. *Let* $(\mathfrak{A}, n, \alpha)$ *be a ramified simple stratum,* $n = 2m+1 \geqslant 1$*; then*

$$|\Gamma_1 \backslash \mathcal{A}_2^0(F;\alpha)| = q^m.$$

Proof. Set $J = F[\alpha]^\times U_{\mathfrak{A}}^{m+1}$. The elements π of $\mathcal{A}_2^0(F;\alpha)$ are the induced representations $\pi_\Lambda = c\text{-Ind}_J^G \Lambda$, where Λ ranges over the set $C(\psi_\alpha, \mathfrak{A})$ of characters of J satisfying $\Lambda \mid U_{\mathfrak{A}}^{m+1} = \psi_\alpha$. Indeed, $\Lambda \mapsto \pi_\Lambda$ is a bijection between $C(\psi_\alpha, \mathfrak{A})$ and $\mathcal{A}_2^0(F;\alpha)$ (15.6 Proposition 1). Thus two representations π_{Λ_i} lie in the same Γ_1-orbit if and only the characters Λ_i agree on $J \cap U_{\mathfrak{A}}^1 = U_{F[\alpha]}^1 U_{\mathfrak{A}}^{m+1}$. Therefore

$$|\Gamma_1 \backslash \mathcal{A}_2^0(F;\alpha)| = (J \cap U_{\mathfrak{A}}^1 : U_{\mathfrak{A}}^{m+1}) = (U_{F[\alpha]}^1 : U_{F[\alpha]}^{m+1}) = q^m,$$

as required. □

We need a corresponding result on the Galois side. If $(E/F, n, \beta)$ is an admissible wild triple, then Γ_1 acts on $\mathcal{G}_2^0(F;\beta)$ by $(\chi, \rho) \mapsto \chi \otimes \rho$.

Lemma 2. *Let* $(E/F, n, \beta)$ *be an admissible wild triple. With* $d = d(E|F)$*, we have*

$$|\Gamma_1 \backslash \mathcal{G}_2^0(F;\beta)| = \begin{cases} q^{[(n-d)/2]} & \text{if } n > 3d, \\ \frac{1}{2} q^{[(n-d)/2]} & \text{if } n = 3d. \end{cases}$$

Proof. Let $\sigma \in \mathrm{Gal}(E/F)$, $\sigma \neq 1$. There are precisely $q^{[(n-d)/2]}$ characters ξ of U_E^1 satisfying (44.7.1). For any such ξ, the representation $\mathrm{Ind}_{E/F}\,\xi$ is irreducible. These characters ξ are Galois-conjugate in pairs if the characters $\eta : 1+x \mapsto \psi_E(\beta x)$, $\eta^\sigma : 1+x \mapsto \psi_E(\beta^\sigma x)$ of $U_E^{1+[(n-d)/2]}$ are the same. Otherwise, they give rise to distinct elements of $\Gamma_1 \backslash \mathcal{G}_2^0(F;\beta)$.

The characters η, η^σ of $U_E^{1+[(n-d)/2]}$ are the same if and only if

$$\beta/\beta^\sigma \in U_E^{[(n-d+1)/2]}. \tag{44.8.1}$$

We have $\upsilon_E(\beta/\beta^\sigma - 1) = d$ (41.1 Lemma), so (44.8.1) holds if and only if $[(n-d+1)/2] \leqslant d$. This translates as $n \leqslant 3d$, that is, $n = 3d$ and the lemma follows. □

44.9. In this paragraph, we analyze the set of admissible wild triples giving rise to the same simple stratum in 44.7. To do this, we fix an admissible wild triple $(E/F, n, \beta)$, $n = 2m+1$.

Proposition. *Let E'/F be a totally ramified, separable quadratic extension, and let $\beta' \in E'(n)$. Suppose that*

$$\begin{aligned} \mathrm{N}_{E'/F}(\beta') &\equiv \mathrm{N}_{E/F}(\beta) \pmod{U_F^{[m/2]+1}}, \\ \mathrm{Tr}_{E'/F}(\beta') &\equiv \mathrm{Tr}_{E/F}(\beta) \pmod{\mathfrak{p}^{-[m/2]}}. \end{aligned} \tag{44.9.1}$$

We then have $d(E'|F) = d(E|F)$ and $\delta_{E'/F} \equiv \delta_{E/F} \pmod{\mathfrak{p}^{-[m/2]}}$. In particular, $(E'/F, n, \beta')$ is an admissible wild triple.

Proof. Let us abbreviate $d = d(E|F)$, $d' = d(E'|F)$. By 41.1 Lemma,

$$v_F(\mathrm{Tr}_{E/F}(\beta)) \geqslant -m + [d/2],$$

with equality if and only if d is odd, i.e., $d < v_F(4)$. This translates into

$$d = \min\{2v_F(\mathrm{Tr}_{E/F}(\beta)){+}2m{+}1, v_F(4)\}.$$

Likewise,

$$d' = \min\{2v_F(\mathrm{Tr}_{E'/F}(\beta')){+}2m{+}1, v_F(4)\}.$$

The assumption $n \geqslant 3d$ implies $-m{+}[d/2] < -[m/2]$, so the congruence of traces implies

$$v_F(\mathrm{Tr}_{E/F}(\beta)) = v_F(\mathrm{Tr}_{E'/F}(\beta')).$$

We deduce that $d = d' \leqslant n/3$, whence $(E'/F, n, \beta')$ is admissible.

The second assertion of the proposition is equivalent to demanding that the characters $\varkappa = \varkappa_{E/F}$, $\varkappa' = \varkappa_{E'/F}$ agree on the group $U_F^{[m/2]+1}$. These characters both have level d and they satisfy $\varkappa^2 = \varkappa'^2 = 1$. It is therefore enough to show that $\varkappa'$ vanishes on any element of $U_F^{[m/2]+1}$ which is a norm from E. If $[m/2] \geqslant d$, there is nothing to do, since both $\varkappa$ and $\varkappa'$ are trivial on $U_F^{[m/2]+1} \subset U_F^{d+1} \subset \mathrm{Ker}\,\varkappa \cap \mathrm{Ker}\,\varkappa'$. We therefore assume $[m/2] < d$.

Any element of $E = F[\beta]$ can be written uniquely in the form $x{+}y\beta$, $x, y \in F$. The element $x{+}y\beta$ lies in U_E^1 if and only if $x \in U_F^1$ and $y\beta \in \mathfrak{p}_E$. That is, $U_E^1 = U_F^1(1{+}\mathfrak{p}^{m+1}\beta)$. Since $\varkappa'^2 = 1$, this reduces us to showing that $\varkappa'(\mathrm{N}_{E/F}(1{+}x\beta)) = 1$ whenever $\mathrm{N}_{E/F}(1{+}x\beta) \in U_F^{[m/2]+1}$.

Let $x \in \mathfrak{p}^{m+1}$; then $\mathrm{N}_{E/F}(1{+}x\beta) \in U_F^{[m/2]+1}$ if and only if $1{+}x\beta \in U_E^{[m/2]+1}$. Equivalently, $2v_F(x) \geqslant n{+}[m/2]{+}1$. This condition is unchanged on replacing β by β'. For such x, we have

$$\begin{aligned} \mathrm{N}_{E/F}(1{+}x\beta) &= 1 + x\mathrm{Tr}_{E/F}(\beta) + x^2\mathrm{N}_{E/F}(\beta) \\ &\equiv 1 + x\mathrm{Tr}_{E'/F}(\beta') + x^2\mathrm{N}_{E'/F}(\beta') \pmod{\mathfrak{p}^{d+1}}, \end{aligned}$$

whence $\varkappa'(\mathrm{N}_{E/F}(1{+}x\beta)) = \varkappa'(\mathrm{N}_{E'/F}(1{+}x\beta')) = 1$, as required. □

Consequently:

Corollary. *Let $(E/F, n, \beta)$ be an admissible wild triple. Let $(\mathfrak{A}, n, \alpha)$ be a ramified simple stratum such that $\boldsymbol{\pi}(\mathfrak{G}_2^0(F;\beta)) \subset \mathcal{A}_2^0(F;\alpha)$. If (E', n, β') is a triple satisfying (44.9.1), then $\boldsymbol{\pi}(\mathfrak{G}_2^0(F;\beta')) \subset \mathcal{A}_2^0(F;\alpha)$.*

44.10. We let $(E/F, n, \beta)$, $(E'/F, n, \beta')$ be admissible wild triples satisfying the conditions (44.9.1), namely

$$\begin{aligned} \mathrm{N}_{E'/F}(\beta') &\equiv \mathrm{N}_{E/F}(\beta) \pmod{U_F^{[m/2]+1}}, \\ \mathrm{Tr}_{E'/F}(\beta') &\equiv \mathrm{Tr}_{E/F}(\beta) \pmod{\mathfrak{p}^{-[m/2]}}, \end{aligned}$$

where $n = 2m+1$. We set $d = d(E|F) = d(E'|F)$.

Proposition. *Suppose, with the hypotheses above, that $\mathcal{G}_2^0(F;\beta) \cap \mathcal{G}_2^0(F;\beta') \neq \emptyset$. We then have*

$$\begin{aligned} &\mathrm{N}_{E'/F}(\beta') \equiv \mathrm{N}_{E/F}(\beta) \pmod{\mathrm{N}_{E/F}(U_E^{[(1+n-d)/2]})}, \\ &\mathrm{Tr}_{E'/F}(\beta') \equiv \mathrm{Tr}_{E/F}(\beta) \pmod{\mathrm{Tr}_{E/F}(\mathfrak{p}_E^{-n+[(1+n-d)/2]})}. \end{aligned} \tag{44.10.1}$$

In this case, $\mathcal{G}_2^0(F;\beta) = \mathcal{G}_2^0(F;\beta')$ and $E = E'$.

Proof. Denote by $\psi_{E,\beta}$ the character $1+x \mapsto \psi_E(\beta x)$ of $U_E^{1+[(n-d)/2]}$, and similarly define $\psi_{E',\beta'}$.

We take $\rho \in \mathcal{G}_2^0(F;\beta) \cap \mathcal{G}_2^0(F;\beta')$ and accordingly write $\rho = \mathrm{Ind}_{E/F}\,\xi = \mathrm{Ind}_{E'/F}\,\xi'$, for characters ξ of $E^\times$, ξ' of $E'^\times$ such that

$$\xi \mid U_E^{1+[(n-d)/2]} = \psi_{E,\beta}, \quad \xi' \mid U_{E'}^{1+[(n-d)/2]} = \psi_{E,\beta'}.$$

Consider first the possibility $E = E'$. We then have $\xi' = \xi^\tau$, for some $\tau \in \mathrm{Gal}(E/F)$ (41.3 Lemma). Consequently,

$$\beta' \equiv \beta^\tau \pmod{U_E^{[(1+n-d)/2]}}. \tag{44.10.2}$$

Directly from the definition, therefore, $\mathcal{G}_2^0(F;\beta') = \mathcal{G}_2^0(F;\beta)$ and the congruence of norms holds.

In additive notation, the congruence (44.10.2) says

$$\beta' \equiv \beta^\tau \pmod{\mathfrak{p}_E^{-n+[(1+n-d)/2]}}.$$

The congruence of traces is satisfied and the result therefore holds in the case $E = E'$.

If $n > 3d$, 41.5 Corollary implies that $E = E'$. If $n = 3d$ and ρ is triply imprimitive, there are three fields E_i/F from which ρ is induced. However, no two of the elements $\delta_{E_i/F}$ are congruent modulo $\mathfrak{p}^{1-d}$ *loc. cit.* Since $n = 3d$, we have $1-d \leqslant -[m/2]$, so 44.9 Proposition again implies $E = E'$. □

44.11. Let $(E/F, n, \beta)$ be an admissible wild triple and $(\mathfrak{A}, n, \alpha)$ a simple stratum such that $\pi(\mathcal{G}_2^0(F;\beta)) \subset \mathcal{A}_2^0(F;\alpha)$. The triple determines the character ψ_α of $U_{\mathfrak{A}}^{[n/2]+1}$ rather than the field $F[\alpha]$. However, one can extract some useful information.

Proposition. *Let $(\mathfrak{A}, n, \alpha)$ be a ramified simple stratum and let $(E/F, n, \beta)$ be a wild admissible triple. Suppose that $\pi(\mathcal{G}_2^0(F;\beta)) \subset \mathcal{A}_2^0(F;\alpha)$. Then*

$$n > 3d(E|F) \iff n > 3d(F[\alpha]|F)$$
$$n = 3d(E|F) \iff n \leqslant 3d(F[\alpha]|F).$$

When the first alternative holds, we have $d(F[\alpha]|F) = d(E|F)$.

Proof. We abbreviate $d = d(E|F)$, $d_\alpha = d(F[\alpha]|F)$ and work case by case. Assume to start with that $n > 3d$ and d is odd. We have

$$\upsilon_F(\operatorname{Tr}_{E/F}\beta) = -m + (d-1)/2 < -d,$$

since $3d < n$. For the same reason, this valuation is $< -[m/2]$, so

$$\upsilon_F(\operatorname{tr}\alpha) = \upsilon_F(\operatorname{Tr}_{E/F}(\beta) + \delta_{E/F}) = -m + (d-1)/2.$$

If d_α is odd, we get $d_\alpha = d < n/3$, as required. This valuation is $< -m+\upsilon_F(2)$, so the case of $d_\alpha = \upsilon_F(4)$ cannot arise.

We take the case $n > 3d$, $d = \upsilon_F(4)$. We then have

$$\upsilon_F(\operatorname{Tr}_{E/F}(\beta) + \delta_{E/F}) \geqslant \min\{-m+\upsilon_F(2), -\upsilon_F(4)\}.$$

If d_α were odd, so $d_\alpha < d = \upsilon_F(4)$, we would have

$$-m + (d_\alpha-1)/2 \geqslant \min\{-m+\upsilon_F(2), -\upsilon_F(4)\},$$

which is impossible. Thus $d_\alpha = d = \upsilon_F(4)$ and $n > 3d_\alpha$.

We therefore assume $n = 3d$. In particular, $d < \upsilon_F(4)$, d is odd and $n-d$ is even. The element $\operatorname{Tr}_{E/F}(\beta)$ has valuation $-m + (d-1)/2 = -d < -[m/2]$. Thus $\upsilon_F(\operatorname{tr}\alpha) \geqslant -d$. If $d_\alpha = \upsilon_F(4)$, then surely $n < 3d_\alpha$, so we assume d_α is odd. Therefore

$$\upsilon_F(\operatorname{tr}\alpha) = -m + (d_\alpha-1)/2 \geqslant -d = -n/3.$$

This yields $d_\alpha \geqslant n/3$, as desired. □

Remark. Suppose we are in the case $n \leqslant 3d(F[\alpha]|F)$. If $\mathcal{A}_2^0(F;\alpha)$ has an ordinary element, there is a wild admissible triple $(E/F, n, \beta)$ satisfying the congruences (44.3.1). The proposition gives $n = 3d(E|F)$ whence $n \equiv 0 \pmod 3$. Put another way, if $(\mathfrak{A}, n, \alpha)$ is a ramified simple stratum with $n \leqslant 3d(F[\alpha]|F)$ and $n \equiv \pm 1 \pmod 3$, every representation $\pi \in \mathcal{A}_2^0(F;\alpha)$ is exceptional.

44.12. We now prove the implication (1) ⇒ (2) in 44.3 Theorem.

By hypothesis, there is an admissible wild triple $(E/F, n, \beta)$ with $\boldsymbol{\pi}(\mathcal{G}_2^0(F;\beta))$ contained in $\mathcal{A}_2^0(F;\alpha)$. In particular, the elements α, β satisfy the congruences (44.3.1). Set

$$r = \big(U_F^{[m/2]+1} : \mathrm{N}_{E/F}(U_E^{[(n-d+1)/2]})\big)\big(\mathfrak{p}^{-[m/2]} : \mathrm{Tr}_{E/F}(\mathfrak{p}_E^{-n+[(n-d+1)/2]})\big).$$

Proposition 44.10 gives admissible wild triples $(E/F, n, \beta_i)$, $1 \leqslant i \leqslant r$, satisfying

$$\begin{aligned} \boldsymbol{\pi}(\mathcal{G}_2^0(F;\beta_i)) \subset \mathcal{A}_2^0(F;\alpha), \quad & 1 \leqslant i \leqslant r, \\ \mathcal{G}_2^0(F;\beta_i) \cap \mathcal{G}_2^0(F;\beta_j) = \emptyset, \quad & i \neq j. \end{aligned}$$

Each set $\Gamma_1\backslash\mathcal{G}_2^0(F;\beta_i)$ has $q^{[(n-d)/2]}$ elements if $n > 3d$, or $q^{[(n-d)/2]}/2$ elements if $n = 3d$ (44.8 Lemma 2). The partial correspondence $\boldsymbol{\pi}$ is injective (44.1 Theorem (2)), so Lemma 1 of 44.8 reduces us to proving

$$q^m/r = \begin{cases} q^{[(n-d)/2]} & \text{if } n > 3d, \\ q^{[(n-d)/2]}/2 & \text{if } n = 3d. \end{cases}$$

This follows from a routine calculation, and we have completed the proof of 44.3 Theorem. □

Remark. The list $(E_i/F, n, \beta_i)$, $1 \leqslant i \leqslant r$, in the last proof does not exhaust the set of admissible wild triples $(E'/F, n, \beta')$ for which $\boldsymbol{\pi}(\mathcal{G}_2^0(F;\beta')) \subset \mathcal{A}_2^0(F;\alpha)$: it counts only those satisfying (44.9.1) relative to a fixed triple $(E/F, n, \beta)$. These are enough.

We have no further use for this fact, but it is worthy of a formal record:

Corollary. *Let $(\mathfrak{A}, n, \alpha)$, $n = 2m+1$, be a ramified simple stratum and let $(E/F, n, \beta)$ be an admissible wild triple satisfying*

$$\begin{aligned} \det\alpha &\equiv \mathrm{N}_{E/F}(\beta) \pmod{U_F^{[m/2]+1}}, \\ \mathrm{tr}\,\alpha &\equiv \delta_{E/F} + \mathrm{Tr}_{E/F}(\beta) \pmod{\mathfrak{p}^{-[m/2]}}. \end{aligned}$$

Let $\pi \in \mathcal{A}_2^0(F;\alpha)$. There exists an admissible wild triple $(E'/F, n, \beta')$, satisfying

$$\begin{aligned} \mathrm{N}_{E'/F}(\beta') &\equiv \mathrm{N}_{E/F}(\beta) \pmod{U_F^{[m/2]+1}}, \\ \mathrm{Tr}_{E'/F}(\beta') &\equiv \mathrm{Tr}_{E/F}(\beta) \pmod{\mathfrak{p}^{-[m/2]}}, \end{aligned}$$

such that $\pi = \boldsymbol{\pi}(\rho)$, for some $\rho \in \mathcal{G}_2^0(F;\beta')$.

45. Exceptional Representations and Strata

Let $(\mathfrak{A}, n, \alpha)$, $n \geqslant 1$, be a ramified simple stratum. We shall say that it is *ordinary* if $\mathcal{A}_2^0(F;\alpha)$ contains an ordinary representation. By 44.3 Theorem, this is equivalent to $\mathcal{A}_2^0(F;\alpha) \subset \pi(\mathcal{G}_2^{\mathrm{im}}(F))$. The purpose of this section is to give a simple criterion for a stratum to be ordinary.

45.1. We start with an elementary observation:

Lemma. *There exists $\psi \in \widehat{F}$ such that*

(a) *ψ has level one, and*
(b) *$\psi(x+x^2) = 1$, $x \in \mathfrak{o}$.*

A character $\psi' \in \widehat{F}$ has properties (a) *and* (b) *if and only if $\psi' = u\psi$, for some $u \in U_F^1$.*

Proof. Let η be the non-trivial character of $\mathbb{F}_2$, and set $\psi_0 = \eta \circ \mathrm{Tr}_{\boldsymbol{k}/\mathbb{F}_2}$. The Frobenius substitution $x \mapsto x^2$ generates $\mathrm{Gal}(\boldsymbol{k}/\mathbb{F}_2)$, so $\psi_0(y^2) = \psi_0(y)$, $y \in \mathbb{K}$, whence $\psi_0(y+y^2) = 1$. Moreover, ψ_0 is the unique non-trivial character of $\boldsymbol{k}$ with this property.

We take $\psi \in \widehat{F}$ so that $\psi \mid \mathfrak{o}$ is the inflation of ψ_0: this character has the desired properties (a), (b), and its uniqueness follows from that of ψ_0. □

We call *special* a character $\psi \in \widehat{F}$ satisfying the conditions of the lemma.

45.2. We prove:

Theorem. *Let $(\mathfrak{A}, n, \alpha)$, $n = 2m+1 \geqslant 1$, be a ramified simple stratum. The following are equivalent:*

(1) *The stratum $(\mathfrak{A}, n, \alpha)$ is ordinary.*
(2) *Either*
   (a) *$n > 3d(F[\alpha]|F)$, or*
   (b) *the polynomial*

$$C_\alpha(X) = X^3 - \mathrm{tr}(\alpha)X^2 + \det(\alpha)$$

has a root in F.

Remark. The reducibility properties of the polynomial $C_\alpha(X)$ depend only on the character ψ_α of $U_{\mathfrak{A}}^{[n/2]+1}$, not on the stratum and the choice of character ψ. For, if we replace ψ by $u^{-1}\psi$, with $u \in U_F$, we have $\psi_\alpha = \psi'_{u\alpha}$, and $C_{u\alpha}(X)$ splits over F if and only if $C_\alpha(X)$ does. This enables us to make convenient choices of ψ.

45.3. Suppose that $(\mathfrak{A}, n, \alpha)$ is ordinary, and choose an admissible wild triple $(E/F, n, \beta)$ which satisfies the congruences (44.3.1) relative to α. Set $d = d(E|F)$, $d_\alpha = d(F[\alpha]|F)$. Suppose that $n \leqslant 3d_\alpha$, so that $n = 3d$, $d = d(E|F)$ (44.11 Proposition). It follows that d is odd and $\mathrm{Tr}_{E/F}(\beta)$ has valuation $-d$. Thus $\mathrm{Tr}_{E/F}(\beta)^2\,\mathrm{N}_{E/F}(\beta)^{-1}$ has valuation d.

It will now be convenient to assume that ψ is *special*, as in 45.1. We take $y \in \mathfrak{o}$ and set $x = y\mathrm{Tr}_{E/F}(\beta)/\mathrm{N}_{E/F}(\beta)$. Thus $1{+}x\beta \in U_E^d$ and

$$\begin{aligned} 1 = \varkappa_{E/F}(\mathrm{N}_{E/F}(1{+}x\beta)) &= \varkappa_{E/F}(1 + x\,\mathrm{Tr}_{E/F}(\beta) + x^2\,\mathrm{N}_{E/F}(\beta)) \\ &= \psi(\delta_{E/F}(x\,\mathrm{Tr}_{E/F}(\beta) + x^2\,\mathrm{N}_{E/F}(\beta)) \\ &= \psi(\delta_{E/F}\mathrm{Tr}_{E/F}(\beta)^2\,\mathrm{N}_{E/F}(\beta)^{-1}(y{+}y^2)). \end{aligned}$$

This holds for all $y \in \mathfrak{o}$. By 45.1 Lemma, we get

$$\delta_{E/F}\,\mathrm{Tr}_{E/F}(\beta)^2\,\mathrm{N}_{E/F}(\beta)^{-1} \equiv 1 \pmod{\mathfrak{p}}.$$

Setting $t = \mathrm{Tr}_{E/F}(\beta)$, the congruences (44.3.1) imply

$$C_\alpha(t) = t^3 - \mathrm{tr}(\alpha)t^2 + \det(\alpha) \equiv 0 \pmod{\mathfrak{p}^{1-n}}.$$

We have $v_F(\det\alpha) = -n = -3d$, so we write $\det\alpha = ua^{-3}$, for some $a \in F^\times$, $u \in U_F$. Setting $s = at$, we have

$$c_\alpha(s) = s^3 - a\mathrm{tr}(\alpha)s^2 + u \equiv 0 \pmod{\mathfrak{p}}.$$

The element $\mathrm{tr}(\alpha)$ has valuation $\geqslant -d$, so $a\mathrm{tr}(\alpha) \in \mathfrak{o}$. The reduction modulo $\mathfrak{p}$ of the polynomial $c_\alpha(X)$ is separable, so Hensel's Lemma applies to show that $c_\alpha(X)$ has a root in F. Thus C_α has a root in F, and (1) $\Rightarrow$ (2) in 45.2 Theorem.

45.4. We turn to the converse. Let $(\mathfrak{A}, n, \alpha)$ be a ramified simple stratum with $n = 2m{+}1 \geqslant 1$.

Suppose first that $n \leqslant 3d_\alpha$, $d_\alpha = d(F[\alpha]|F)$, and that $C_\alpha(X)$ has a root in F. If $n \not\equiv 0 \pmod 3$, elementary changes of variable transform $C_\alpha(X)$ into an Eisenstein polynomial, so C_α is irreducible over F in this case. We deduce that $n \equiv 0 \pmod 3$.

Elementary Lemma. *Let n be an integer satisfying $n \geqslant 1$, $n \equiv 0 \pmod 3$. Let $a, b \in F$ satisfy $v_F(b) = -n$, $v_F(a) \geqslant -n/3$. Any root $\gamma \in F$ of the polynomial $c(X) = X^3 + aX^2 + b$ then satisfies $v_F(\gamma) = -n/3$.*

Proof. Extending the base field, we can assume that $c(X)$ splits completely in F, without changing the relative conditions on the coefficients. If $c(X)$ has a root of valuation $< -n/3$, the vanishing of the linear term in $c(X)$ implies it is the only root with this property. This, however, is inconsistent with the condition on the coefficient a. □

Since $\upsilon_F(\operatorname{tr}\alpha) \geqslant -n/3$, we deduce that any root $\gamma \in F$ of $C_\alpha(X)$ has $\upsilon_F(\gamma) = -n/3$.

Lemma 1. *Let $\epsilon \in F$, $\upsilon_F(\epsilon) \geqslant -m$. There exists $\beta = \beta_\epsilon \in \mathcal{K}_{\mathfrak{A}}$ such that*

$$\det\beta = \det\alpha, \qquad \operatorname{tr}\beta = \epsilon.$$

The algebra $E = F[\beta]$ is a field satisfying $E^\times \subset \mathcal{K}_{\mathfrak{A}}$.

Proof. This is immediate, as a consequence of the normal form used in 44.2. □

We also need a refinement of 45.1 Lemma.

Lemma 2. *Let j be an integer, $0 \leqslant j \leqslant [(\upsilon_F(2)-1)/2]$. There exists a character $\psi \in \widehat{F}$, of level one, such that*

$$\psi(y+y^2) = 1, \quad y \in \mathfrak{p}^{-j}.$$

Proof. Consider the map $\wp : \mathfrak{p}^{-j}/\mathfrak{p} \to F/\mathfrak{p}$ given by $y \mapsto y+y^2$. The condition on j implies that $\wp$ is additive, but its image does not contain $\mathfrak{o}/\mathfrak{p}$. □

We now assume ψ has been chosen to satisfy the conditions of Lemma 2 for the largest possible value of j.

Set $d = n/3$ (which is odd) and let ϵ be a root of $C_\alpha(X)$ in F. We define β as in Lemma 1 for this ϵ and set $E = F[\beta]$. This gives $d(E|F) = d = n/3$ and $(E/F, n, \beta)$ is an admissible wild triple.

We put $\delta = \operatorname{tr}(\alpha)-\epsilon$. Thus $\upsilon_F(\delta) \geqslant -d$. We show that

$$\delta \equiv \delta_{E/F} \pmod{\mathfrak{p}^{-[m/2]}}.$$

This will imply that $(E/F, n, \beta)$ is an admissible wild triple satisfying (44.3.1) and that $(\mathfrak{A}, n, \alpha)$ is ordinary, as required.

To do this, it is enough to show that the character $\varkappa : 1+x \to \psi(\delta x)$ of $U_F^{[m/2]+1}$ is trivial on norms from $U_E^{[m/2]+1}$. This reduces to showing that $\varkappa(\mathrm{N}_{E/F}(1+x\beta)) = 1$ when $x \in F$ and $x\beta \in \mathfrak{p}_E^{[m/2]+1}$. For such x, we can write $x = y\mathrm{Tr}_{E/F}(\beta)/\mathrm{N}_{E/F}(\beta) = y\epsilon/\det\alpha$, where

$$\begin{aligned} 2\upsilon_F(y) &= 2\upsilon_F(x) - 2n + 2d \geqslant \left[\tfrac{m}{2}\right] - n + 2d + 1 \\ &= \left[\tfrac{m}{2}\right] - d + 1 \geqslant 1 - \upsilon_F(2). \end{aligned}$$

We now repeat the calculation of 45.3:

$$\begin{aligned} \varkappa(\mathrm{N}_{E/F}(1+x\beta)) &= \psi(\delta(x\,\mathrm{Tr}_{E/F}(\beta) + x^2\,\mathrm{N}_{E/F}(\beta)) \\ &= \psi(\delta\,\mathrm{Tr}_{E/F}(\beta)^2\,\mathrm{N}_{E/F}(\beta)^{-1}(y+y^2)). \end{aligned}$$

Since

$$\delta\,\mathrm{Tr}_{E/F}(\beta)^2\,\mathrm{N}_{E/F}(\beta)^{-1} = (\mathrm{tr}\,\alpha - \epsilon)\,\epsilon^2\,\det\alpha^{-1} = 1,$$

this reduces to

$$\varkappa(\mathrm{N}_{E/F}(1{+}x\beta)) = \psi(y{+}y^2) = 1,$$

as required.

45.5. It remains to show that a ramified simple stratum $(\mathfrak{A}, n, \alpha)$, which satisfies $n > 3d$, $d = d(F[\alpha]|F)$, is necessarily ordinary.

The case $n > 4d$ is trivial: $(F[\alpha]/F, n, \alpha)$ is an admissible wild triple which satisfies the requirements of (44.3.1). We therefore assume $3d < n < 4d$. This implies

$$[m/2] < d.$$

We consider an admissible wild triple $(E/F, n, \beta)$, with $E^\times \subset \mathcal{K}_{\mathfrak{A}}$, in which $d(E|F) = d$.

Lemma. *Let v be an integer satisfying*

$$d \geqslant v > [m/2].$$

Let $\beta' \in \mathcal{K}_{\mathfrak{A}}$ satisfy

$$\det\beta' = \det\beta, \qquad \mathrm{tr}(\beta') \equiv \mathrm{tr}(\beta) \pmod{\mathfrak{p}^{-v}}.$$

Writing $E' = F[\beta']$, we then have

$$\delta_{E'/F} \equiv \delta_{E/F} \pmod{\mathfrak{p}^{n-2d-2v}}.$$

Proof. The condition on the trace of β' implies that $d(E'|F) = d$.

Write $\varkappa = \varkappa_{E/F}$, $\varkappa' = \varkappa_{E'/F}$. The assertion of the lemma is equivalent to $\varkappa^{-1}\varkappa'$ being trivial on $(1{+}2d{+}2v{-}n)$-units in F. Since $1{+}2d{+}2v{-}n \leqslant d$, and $\varkappa^2 = \varkappa'^2 = 1$, it is enough to show that $\varkappa$, $\varkappa'$ agree on norms of $(1{+}2d{+}2v{-}n)$-units of E of the form $1{+}x\beta$, with $x \in F$. The condition $1{+}x\beta \in U_E^{1+2d+2v-n}$ is equivalent to $2v_F(x) \geqslant 1{+}2d{+}2v$, or $v_F(x) \geqslant 1{+}d{+}v$. Writing out,

$$\begin{aligned}\mathrm{N}_{E/F}(1{+}x\beta) &= 1 + x\,\mathrm{tr}(\beta) + x^2\det(\beta)\\ &\equiv 1 + x\,\mathrm{tr}(\beta') + x^2\det(\beta') \pmod{\mathfrak{p}^{d+1}}.\end{aligned}$$

The character $\varkappa'$ has level d, so $\varkappa'(\mathrm{N}_{E/F}(1{+}x\beta)) = \varkappa'(\mathrm{N}_{E'/F}(1{+}x\beta')) = 1$, and symmetrically. The lemma is proved. □

We return to the stratum $(\mathfrak{A}, n, \alpha)$, with $d = d(F[\alpha]|F)$, $3d < n < 4d$. We construct a sequence of quadratic field extensions $E_i = F[\beta_i]/F$, with $\delta_i = \delta_{E_i/F}$, as follows.

We put $\beta_0 = \alpha$. For $i \geqslant 0$, we set

$$\det \beta_{i+1} = \det \alpha, \qquad \mathrm{tr}(\beta_{i+1}) = \mathrm{tr}(\alpha) - \delta_i.$$

In particular, $\mathrm{tr}(\beta_1) \equiv \mathrm{tr}(\beta_0) \pmod{\mathfrak{p}^{-d}}$, so we can apply the lemma to get

$$\delta_1 \equiv \delta_0 \pmod{\mathfrak{p}^{n-4d}}.$$

Since $n-4d > -d$, this amounts to $\delta_1 \equiv \delta_0 \pmod{\mathfrak{p}^{1-d}}$. Therefore

$$\mathrm{tr}(\beta_2) = \mathrm{tr}(\alpha) - \delta_1 = \mathrm{tr}(\beta_1) - \delta_1 + \delta_0 \equiv \mathrm{tr}(\beta_1) \pmod{\mathfrak{p}^{1-d}}.$$

The condition $4d > n$ implies $d-1 > [m/2]$, we can apply the lemma to get

$$\delta_2 \equiv \delta_1 \pmod{\mathfrak{p}^{n-4d+2}}.$$

In particular, $\delta_2 \equiv \delta_1 \pmod{\mathfrak{p}^{2-d}}$. Iterating, we get

$$\delta_{i+1} \equiv \delta_i \pmod{\mathfrak{p}^{i+1-d}}$$

provided $i < d-[m/2]$. Setting $k = d-[m/2]$, $(E_k/F, n, \beta_k)$ is an admissible wild triple satisfying the conditions (44.3.1):

$$\det \beta_k = \det \alpha, \qquad \mathrm{tr}\, \beta_k + \delta_k \equiv \mathrm{tr}\, \alpha \pmod{\mathfrak{p}^{-[m/2]}}.$$

By 44.3 Theorem, the stratum $(\mathfrak{A}, n, \alpha)$ is ordinary, as required. □

45.6. For ease of reference, we display the formal consequence of the theorem. We call the ramified simple stratum $(\mathfrak{A}, n, \alpha)$ *exceptional* if it is not ordinary. Equivalently, any $\pi \in \mathcal{A}_2^0(F;\alpha)$ is exceptional.

Corollary. *Let* $(\mathfrak{A}, n, \alpha)$, $n \geqslant 1$, *be a ramified simple stratum. The following are equivalent:*

(1) *the stratum* $(\mathfrak{A}, n, \alpha)$ *is exceptional;*
(2) $n \leqslant 3d(F[\alpha]|F)$ *and the polynomial* $C_\alpha(X) = X^3 - \mathrm{tr}(\alpha)X^2 + \det(\alpha)$ *is irreducible over* F.

Remark. Let $(\mathfrak{A}, n, \alpha)$ be a ramified simple stratum, $1 \leqslant n \leqslant 3d_\alpha$. We have already observed (45.3) that $C_\alpha(X)$ is irreducible over F in the case $n \not\equiv 0 \pmod 3$. So, according to the theorem, $(\mathfrak{A}, n, \alpha)$ cannot be ordinary: *cf.* 44.11 Remark.

12

The Dyadic Langlands Correspondence

In the sole remaining case $p = 2$, the Langlands correspondence has only been defined (44.1) on the set $\mathcal{G}_2^{\mathrm{im}}(F)$ of classes of imprimitive representations $\rho \in \mathcal{G}_2^0(F)$. We have to define $\pi_F(\rho)$ when ρ is *primitive,* and verify that the map $\pi_F : \mathcal{G}_2^0(F) \to \mathcal{A}_2^0(F)$ so obtained satisfies the requirements of the statement in 33.1.

Any primitive $\rho \in \mathcal{G}_2^0(F)$ is totally ramified, that is, $\rho \in \mathcal{G}_2^{\mathrm{wr}}(F) = \mathcal{G}_2^0(F) \smallsetminus \mathcal{G}_2^{\mathrm{nr}}(F)$. We correspondingly put $\mathcal{A}_2^{\mathrm{wr}}(F) = \mathcal{A}_2^0(F) \smallsetminus \mathcal{A}_2^{\mathrm{nr}}(F)$. Let $\rho \in \mathcal{G}_2^{\mathrm{wr}}(F)$; the definition of $\pi = \pi_F(\rho)$ is given in 50.3: after a little preparation, it is explicit and direct. At first pass, however, it only gives the relation $\varepsilon(\rho, s, \psi)^3 = \varepsilon(\pi, s, \psi)^3$. The process of extracting the cube root, to get the required equality of local constants, depends on a further analysis of these objects and their properties relative to certain sorts of base field extension (§48, §49).

That achieved, we have a map $\pi_F : \mathcal{G}_2^{\text{wr}}(F) \to \mathcal{A}_2^{\text{wr}}(F)$ satisfying the defining (and determining) property of the Langlands correspondence. It remains only to prove that it is bijective. This requires a different family of methods.

The strong uniqueness properties of the correspondence, expressed via the Converse Theorems, imply that it is compatible with automorphisms of the base field F (§47). The argument revolves around the ability to recognize fixed points for certain automorphisms of F.

On the Galois side, this has already been clarified in Chapter X. If K/F is a finite, cyclic, *tamely ramified* field extension, the set of Galois-fixed points $\mathcal{G}_2^{\text{wr}}(K)^{\text{Gal}(K/F)}$ is the image of $\mathcal{G}_2^{\text{wr}}(F)$ under restriction of representations (from $\mathcal{W}_F$ to $\mathcal{W}_K$):

$$\text{Res}_{K/F} : \mathcal{G}_2^{\text{wr}}(F) \longrightarrow \mathcal{G}_2^{\text{wr}}(K).$$

On the automorphic side, there is a similar map

$$\text{Lft}_{K/F} : \mathcal{A}_2^{\text{wr}}(F) \longrightarrow \mathcal{A}_2^{\text{wr}}(K),$$

defined for any finite, tamely ramified extension K/F, and called *tame lifting.* Its properties are exactly parallel to those of $\text{Res}_{K/F}$ and, in particular, it gives the same description of Galois-fixed points. If F contains a primitive cube root of unity, then any primitive $\rho \in \mathcal{G}_2^0(F)$ is *tetrahedral,* in the sense of 42.3 Comment. In this case, the fixed-point properties of the tame lifting map imply the bijectivity of π_F with little difficulty (51.3). The general case (§52) requires a more delicate analysis of Galois-fixed points, and of the manner in which two representations are distinguished by the local constants of their twists.

It is known that the operations of restriction and tame lifting are related via the Langlands correspondence: $\pi_K \circ \text{Res}_{K/F} = \text{Lft}_{K/F} \circ \pi_F$, for any finite tamely ramified extension K/F. However, we have not used this fact: the only known proof relies on global methods and is beyond the scope of this book: we discuss the matter further at the end of the chapter.

Except where the contrary is explicitly allowed, we assume that $p = 2$ throughout this chapter.

46. Tame Lifting

Recall that $\mathcal{A}_2^{\text{nr}}(F)$ denotes the subset of $\mathcal{A}_2^0(F)$ consisting of the classes of unramified representations (20.1). We set

$$\mathcal{A}_2^{\text{wr}}(F) = \mathcal{A}_2^0(F) \smallsetminus \mathcal{A}_2^{\text{nr}}(F).$$

We call representations $\pi \in \mathcal{A}_2^{\text{wr}}(F)$ *totally ramified.*

This corresponds to another dissection of the set $\mathcal{A}_2^0(F)$: a cuspidal representation π is unramified if and only if contains a cuspidal inducing datum of the form $(\mathrm{M}_2(\mathfrak{o}), \Xi)$ (20.3 Lemma). Thus π is totally ramified if and only if it contains a cuspidal inducing datum $(\mathfrak{A}, \Xi)$ in which $e_{\mathfrak{A}} = 2$.

The object of this section is to define a canonical map

$$\mathrm{Lft}_{K/F} : \mathcal{A}_2^{\mathrm{wr}}(F) \longrightarrow \mathcal{A}_2^{\mathrm{wr}}(K),$$

for any finite, tamely ramified field extension K/F. The definition is straightforward in principle, but it proceeds via the classification theory of Chapter IV. Some effort, therefore, is required to show it is independent of choices: this is a tortuous process extending from the statement of 46.3 Proposition to 46.5 Definition, and could be omitted at a first reading.

46.1. We start by defining a notion of lifting for ramified simple strata.

Let $(\mathfrak{A}, n, \alpha)$, $n = 2m{+}1 \geqslant 1$, be a ramified simple stratum in $A = \mathrm{M}_2(F)$. In particular, $E = F[\alpha]$ is a field and $\mathfrak{A}$ is the unique $\mathfrak{o}$-chain order in A with $\alpha \in \mathcal{K}_{\mathfrak{A}}$ (*cf.* 13.5 Proposition).

Let K/F be a finite, tamely ramified field extension and abbreviate $e_K = e(K|F)$. (In particular, e_K is *odd*.) We set $A_K = \mathrm{M}_2(K)$ and $G_K = \mathrm{GL}_2(K)$. We regard A as embedded in A_K in the obvious way. If K/F happens to be Galois, we can make $\mathrm{Gal}(K/F)$ act on A_K coefficient by coefficient, and then A is the set of $\mathrm{Gal}(K/F)$-fixed points in A_K. Similarly for G and G_K.

Since E/F is totally wildly ramified and K/F is tamely ramified, the algebra $EK = E \otimes_F K$ is a field, naturally identified with the subfield $K[\alpha]$ of A_K. The extension EK/K is quadratic and totally ramified. The valuation $\upsilon_{EK}(\alpha) = -e_K n$ is odd, whence α is minimal over K.

Proposition. *Let $(\mathfrak{A}, n, \alpha)$ be a ramified simple stratum in $A = \mathrm{M}_2(F)$. Let K/F be a finite, tamely ramified field extension with $e_K = e(K|F)$.*

(1) *There is a unique $\mathfrak{o}_K$-chain order $\mathfrak{A}_K$ in $A_K = \mathrm{M}_2(K)$ such that $\alpha \in \mathcal{K}_{\mathfrak{A}_K}$. The triple $(\mathfrak{A}_K, n_K, \alpha)$ is a ramified simple stratum in A_K, where $n_K = e_K n$.*

(2) *Write* $\mathrm{rad}\,\mathfrak{A} = \mathfrak{P}$ *and* $\mathrm{rad}\,\mathfrak{A}_K = \mathfrak{P}_K$*; then*

$$\mathfrak{P}_K^{1+j} \cap A = \mathfrak{P}^{1+[j/e_K]}, \quad j \in \mathbb{Z},$$

and, in particular,

$$\mathfrak{A} = \mathfrak{A}_K \cap A, \quad \mathfrak{P} = \mathfrak{P}_K \cap A.$$

Proof. Since α is minimal over K, part (1) follows from 13.5 Proposition.

We may choose bases so that $\mathfrak{A}$ is the standard chain order $\mathfrak{I}$ and α is a matrix of the form $\left(\begin{smallmatrix} 0 & * \\ * & * \end{smallmatrix}\right)$. The order $\mathfrak{A}_K$ is then given by

$$\mathfrak{A}_K = \begin{pmatrix} \mathfrak{o}_K & \mathfrak{p}_K^{(1-e_K)/2} \\ \mathfrak{p}_K^{(1+e_K)/2} & \mathfrak{o}_K \end{pmatrix}. \tag{46.1.1}$$

Since K/F is tamely ramified, we have

$$\mathrm{Tr}_{K/F}(\mathfrak{p}_K^{j+1}) = \mathfrak{p}_K^{j+1} \cap F = \mathfrak{p}^{1+[j/e_K]}, \tag{46.1.2}$$

for all $j \in \mathbb{Z}$ (18.1 Lemma). All assertions now follow from simple calculations. □

The group $\mathcal{K}_{\mathfrak{A}}$ is generated by $U_{\mathfrak{A}}$, $F^\times$, and the element α. The G_K-normalizer $\mathcal{K}_{\mathfrak{A}_K}$ of $\mathfrak{A}_K$ thus contains $\mathcal{K}_{\mathfrak{A}}$. The proposition shows:

Gloss. *Let $\mathfrak{A}$ be a chain order in A with $e_{\mathfrak{A}} = 2$. There is a unique $\mathfrak{o}_K$-chain order $\mathfrak{A}_K$ in A_K such that $\mathcal{K}_{\mathfrak{A}} \subset \mathcal{K}_{\mathfrak{A}_K}$. If $\alpha \in \mathcal{K}_{\mathfrak{A}}$ is minimal over F, then $\mathfrak{A}_K$ is the unique chain order in A_K such that $\alpha \in \mathcal{K}_{\mathfrak{A}_K}$.*

46.2. With K/F as in 46.1, define

$$\begin{aligned} T : A_K = K \otimes_F A &\longrightarrow A, \\ x \otimes a &\longmapsto \mathrm{Tr}_{K/F}(x)a. \end{aligned}$$

In matrix terms,

$$T : \begin{pmatrix} a & b \\ c & d \end{pmatrix} \longmapsto \begin{pmatrix} \mathrm{Tr}_{K/F}\, a & \mathrm{Tr}_{K/F}\, b \\ \mathrm{Tr}_{K/F}\, c & \mathrm{Tr}_{K/F}\, d \end{pmatrix}.$$

Let $(\mathfrak{A}, n, \alpha)$, $n = 2m+1 \geqslant 1$, be a ramified simple stratum in A, and define $(\mathfrak{A}_K, n_K, \alpha)$ as in 46.1. Writing $n_K = e(K|F)\, n = 2m_K+1$, we have

$$T(\mathfrak{P}_K^{n_K+1}) = \mathfrak{P}^{n+1}, \quad T(\mathfrak{P}_K^{m_K+1}) = \mathfrak{P}^{m+1}.$$

Thus there is a surjective group homomorphism

$$\begin{aligned} T^* : U_{\mathfrak{A}_K}^{m_K+1}/U_{\mathfrak{A}_K}^{n_K+1} &\longrightarrow U_{\mathfrak{A}}^{m+1}/U_{\mathfrak{A}}^{n+1}, \\ 1+x &\longmapsto 1+T(x). \end{aligned}$$

Let $\psi \in \widehat{F}$ have level one. The character $\psi_K = \psi \circ \mathrm{Tr}_{K/F} \in \widehat{K}$ then has level one. Consider the characters

$$\begin{aligned} \psi_\alpha &: 1+x \longmapsto \psi(\mathrm{tr}_A(\alpha x)), & 1+x \in U_{\mathfrak{A}}^{m+1}, \\ \psi_\alpha^K &: 1+y \longmapsto \psi_K(\mathrm{tr}_{A_K}(\alpha y)), & 1+y \in U_{\mathfrak{A}_K}^{m_K+1}. \end{aligned} \tag{46.2.1}$$

Immediately:

$$\psi_\alpha^K = \psi_\alpha \circ T^*. \tag{46.2.2}$$

This process preserves the intertwining (or conjugacy, *cf.* 15.2) structures in which we are interested:

Lemma. *Let* $(\mathfrak{B}, n, \beta)$ *be a ramified simple stratum in* A*; then* $\psi_\beta \mid U_{\mathfrak{B}}^{m+1}$ *intertwines in* G *with* $\psi_\alpha \mid U_{\mathfrak{A}}^{m+1}$ *if and only if* $\psi_\beta^K \mid U_{\mathfrak{B}_K}^{m_K+1}$ *intertwines with* $\psi_\alpha^K \mid U_{\mathfrak{A}_K}^{m_K+1}$ *in* G_K.

Proof. The G-conjugacy class of the character ψ_α is determined by the pair of cosets

$$\det(\alpha)\, U_F^{[m/2]+1}, \quad \mathrm{tr}_A(\alpha)+\mathfrak{p}^{-[m/2]},$$

(44.2 Lemma). Similarly for ψ_α^K. We have

$$U_K^{[m_K/2]+1} \cap F = U_F^{[m/2]+1}, \quad \mathfrak{p}_K^{-[m_K/2]} \cap F = \mathfrak{p}^{-[m/2]},$$

whence the lemma follows. □

The lemma says that the G_K-conjugacy class of the character $\psi_\alpha^K \mid U_{\mathfrak{A}_K}^{m_K+1}$ and the G-conjugacy class of $\psi_\alpha \mid U_{\mathfrak{A}}^{m+1}$ determine each other uniquely (*cf.* 15.2).

46.3. With $(\mathfrak{A}, n, \alpha)$, K/F and ψ as before, let $E = F[\alpha]$ and $J = E^\times U_{\mathfrak{A}}^{m+1}$. Thus J is the set of elements of G which intertwine the character ψ_α of $U_{\mathfrak{A}}^{m+1}$ (15.1). Likewise, $J_K = EK^\times U_{\mathfrak{A}_K}^{m_K+1}$ is the set of $g \in G_K$ which intertwine ψ_α^K.

Recall that $C(\psi_\alpha, \mathfrak{A})$ is the set of characters Λ of J such that $\Lambda \mid U_{\mathfrak{A}}^{m+1} = \psi_\alpha$ (15.3 Definition). The set $C(\psi_\alpha^K, \mathfrak{A}_K)$ is defined similarly.

Consider the character $\varkappa_{K/F} = \det \mathrm{Ind}_{K/F}\, 1_K$, where 1_K denotes the trivial character of $\mathcal{W}_K$. We view $\varkappa_{K/F}$ as a character of $F^\times$. It satisfies $\varkappa_{K/F}^2 = 1$ and it is tamely ramified. It is therefore unramified. We define

$$\epsilon_{K/F} = \varkappa_{K/F}(\varpi) = \pm 1, \tag{46.3.1}$$

for any prime element ϖ of F. Using this notation, we prove:

Proposition. *Let* $\Lambda \in C(\psi_\alpha, \mathfrak{A})$. *There exists a unique character*

$$\Lambda_K = \mathcal{L}_{K/F}^\alpha(\Lambda) \in C(\psi_\alpha^K, \mathfrak{A}_K)$$

such that

$$\Lambda_K(y) = \epsilon_{K/F}^{v_{EK}(y)}\, \Lambda(\mathrm{N}_{EK/E}(y)), \quad y \in EK^\times. \tag{46.3.2}$$

Further:

(1) *Let* $\Lambda, \Lambda' \in C(\psi_\alpha, \mathfrak{A})$*; then* $\Lambda'_K = \Lambda_K$ *if and only if* $\Lambda' = \chi \circ \det \otimes \Lambda$, *for a character* χ *of* $F^\times$ *which is trivial on* $\mathrm{N}_{K/F}(K^\times)$.

(2) *If* L/K *is a finite, tamely ramified field extension, then*

$$\mathcal{L}_{L/F}^\alpha = \mathcal{L}_{L/K}^\alpha \circ \mathcal{L}_{K/F}^\alpha.$$

(3) *If* $(\mathfrak{A}, n, \beta)$ *is a simple stratum such that* $\psi_\beta = \psi_\alpha$ *on* $U_{\mathfrak{A}}^{m+1}$, *then* $C(\psi_\beta, \mathfrak{A}) = C(\psi_\alpha, \mathfrak{A})$, $C(\psi_\beta^K, \mathfrak{A}_K) = C(\psi_\alpha^K, \mathfrak{A}_K)$ *and*

$$\mathcal{L}_{K/F}^{\beta} = \mathcal{L}_{K/F}^{\alpha}.$$

Proof. We take T as in 46.2. For $x \in EK$, we have $T(x) = \mathrm{Tr}_{EK/E}(x)$. So, if $u \in EK^\times \cap U_{\mathfrak{A}_K}^{m_K+1} = U_{EK}^{m_K+1}$, then $\psi_\alpha^K(u) = \psi_\alpha(\mathrm{N}_{EK/E}(u))$. The formula (46.3.2) therefore specifies a unique character Λ_K of J_K with $\Lambda_K \in C(\psi_\alpha^K, \mathfrak{A}_K)$.

In (1), the norm map $\mathrm{N}_{EK/E} : U_{EK}^1 \to U_E^1$ is surjective, since EK/E is tamely ramified. Thus, if $\Lambda'_K = \Lambda_K$, the characters Λ, Λ' agree on $U_E^1 U_{\mathfrak{A}}^{m+1}$. Therefore $\Lambda' = \chi \circ \det \otimes \Lambda$, for a tamely ramified character χ of $F^\times$. The character χ_{EK} of $EK^\times$ is trivial; since EK/K is totally wildly ramified, this implies $\chi_K = 1$. This proves assertion (1).

Assertion (2) is immediate from the definition.

In (3), the equality $C(\psi_\beta^K, \mathfrak{A}_K) = C(\psi_\alpha^K, \mathfrak{A}_K)$ follows from 46.2 Lemma and the Conjugacy Theorem 15.2. To prove the main assertion, we first assume that the extension K/F is *cyclic*, of degree d say. We set $\Gamma = \mathrm{Gal}(K/F)$ and choose a generator γ of Γ. We let Γ act, coefficient by coefficient, on A_K. The group J_K is then stable under Γ, and we can form the semi-direct product $\Gamma \ltimes J_K$. For $x \in J_K$, the element

$$\mathbf{N}_\gamma(x) = (\gamma x)^d$$

surely lies in J_K.

Lemma. *Let* $x \in U_{\mathfrak{A}_K}^{m_K+1}$; *then* $\mathbf{N}_\gamma(x) \in U_{\mathfrak{A}}^{m+1} U_{\mathfrak{A}_K}^{n_K+1}$, *and* $\mathbf{N}_\gamma$ *induces a surjective homomorphism*

$$\overline{\mathbf{N}}_\gamma : U_{\mathfrak{A}_K}^{m_K+1}/U_{\mathfrak{A}_K}^{n_K+1} \longrightarrow U_{\mathfrak{A}}^{m+1}/U_{\mathfrak{A}}^{n+1}$$

satisfying

$$\psi_\alpha^K(x) = \psi_\alpha(\overline{\mathbf{N}}_\gamma(x)), \quad x \in U_{\mathfrak{A}_K}^{m_K+1}.$$

Proof. Write $x = 1+a$, $a \in \mathfrak{P}_K^{m_K+1}$; then

$$\mathbf{N}_\gamma(x) \equiv 1 + \sum_{j=0}^{d-1} \gamma^{j}(a) \pmod{\mathfrak{P}_K^{n_K+1}}.$$

The right hand side is $1 + T(a) = T^*(x)$, in the notation of 46.2. As in 46.2, the map $T^* : U_{\mathfrak{A}_K}^{m_K+1}/U_{\mathfrak{A}_K}^{n_K+1} \to U_{\mathfrak{A}}^{m+1}/U_{\mathfrak{A}}^{n+1}$ is surjective, and the lemma follows. □

Write $X = \operatorname{Ker}\psi_\alpha$, $X_K = \operatorname{Ker}\psi_\alpha^K$. It follows from the lemma that $\mathfrak{N}_\gamma$ induces an isomorphism

$$\mathfrak{N}_\gamma : U_{\mathfrak{A}_K}^{m_K+1}/X_K \cong U_{\mathfrak{A}}^{m+1}/X.$$

For β as in (3), we have $J = F[\beta]^\times U_{\mathfrak{A}}^{m+1}$ and $J_K = K[\beta]^\times U_{\mathfrak{A}_K}^{m_K+1}$. So, if $x \in J_K$, we can write $x = zu$, $z \in K[\beta]^\times$, $u \in U_{\mathfrak{A}_K}^{m_K+1}$. Since both γ and conjugation by elements of $K[\beta]^\times$ fix ψ_α^K, we get

$$\mathfrak{N}_\gamma(x) \equiv \mathrm{N}_{K[\beta]/F[\beta]}(z)\,\mathfrak{N}_\gamma(u) \pmod{X_K}.$$

Thus $\mathfrak{N}_\gamma$ induces a group homomorphism

$$\mathfrak{N}_\gamma : J_K/X_K \longrightarrow J/X,$$

which is visibly independent of the choice of β. For $y \in J_K$, we have

$$\Lambda_K(y) = \epsilon_{K/F}^{v_K(\det y)}\,\Lambda(\mathfrak{N}_\gamma(y)),$$

and we have proved part (3) of the proposition when K/F is cyclic.

The transitivity property (2) implies (3) in the case where K/F is Galois.

46.4. To deal with the general case, we use a rather different sort of method:

Lemma. *Let K/F be a finite, tamely ramified field extension. Let $\Lambda \in C(\psi_\alpha, \mathfrak{A})$ and put $\Lambda_K = \mathcal{L}_{K/F}^\alpha(\Lambda)$. Set $\pi =$ c-$\mathrm{Ind}_J^G\,\Lambda$ and $\varrho =$ c-$\mathrm{Ind}_{J_K}^{G_K}\,\Lambda_K$. We then have*

$$\varepsilon(\varrho, s, \psi_K) = \epsilon_{K/F}\,\varepsilon(\pi, s, \psi)^{[K:F]}. \tag{46.4.1}$$

Proof. Combining formula (25.2.1) with 25.5 Corollary, we get

$$\varepsilon(\pi, s, \psi) = q^{n(\frac{1}{2}-s)}\,\Lambda(\alpha)^{-1}\,\psi(\mathrm{tr}_A\,\alpha),$$

where, we recall, $q = |\boldsymbol{k}|$ is the size of the residue field of F. Similarly,

$$\varepsilon(\varrho, s, \psi_K) = q_K^{n_K(\frac{1}{2}-s)}\,\Lambda_K(\alpha)^{-1}\,\psi_K(\mathrm{tr}_{A_K}\,\alpha),$$

where $q_K = |\boldsymbol{k}_K| = q^{f(K|F)}$ and $n_K = e(K|F)n$. Also

$$\psi_K(\mathrm{tr}_{A_K}\,\alpha) = \psi_K(\mathrm{tr}_A\,\alpha) = \psi(\mathrm{tr}_A\,\alpha)^{[K:F]},$$

so the result follows from the definition of Λ_K. □

We finish the proof of 46.3 Proposition. The result for cyclic extensions and the transitivity properties reduce us to the case in which K/F is *totally tamely ramified,* and hence of odd degree.

Let $C_1(\psi_\alpha, \mathfrak{A})$ denote the set of characters of $J^1 = J \cap U^1_{\mathfrak{A}} = U^1_E U^{m+1}_{\mathfrak{A}}$ which extend ψ_α on $U^{m+1}_{\mathfrak{A}}$. Use similar notation for other base fields.

For any tame extension K'/F, the norm map $\mathrm{N}_{EK'/E} : U^1_{EK'} \to U^1_E$ is surjective, since EK'/E is tamely ramified. It follows that $\mathcal{L}^\alpha_{K'/F}$ induces an injective map $\mathcal{L}^{1,\alpha}_{K'/F} : C_1(\psi_\alpha, \mathfrak{A}) \to C_1(\psi^{K'}_\alpha, \mathfrak{A}_{K'})$.

If L/F is the normal closure of K/F, then $\Lambda_K \mid J^1_K$ is the unique element of $C_1(\psi^K_\alpha, \mathfrak{A}_K)$ with image $\mathcal{L}^{1,\alpha}_{L/F}(\Lambda \mid J^1_L)$ under $\mathcal{L}^{1,\alpha}_{L/K}$. Thus $\Lambda_K \mid J^1_K$ is determined independently of the choice of α. We have $\Lambda_K \mid K^\times = (\Lambda \mid F^\times) \circ \mathrm{N}_{K/F}$. The character $\mathcal{L}^\beta_{K/F}(\Lambda)$ is therefore either Λ_K or $\chi \circ \det \otimes \Lambda_K$, where χ is unramified of order 2. The second possibility is excluded by the lemma. □

We are therefore justified in henceforward writing $\mathcal{L}_{K/F} = \mathcal{L}^\alpha_{K/F}$.

46.5. To be useful, the lifting operation $\Lambda \mapsto \Lambda_K$ must respect the relations of twisting and intertwining (or conjugacy). We check this.

Lemma 1. *Let $(\mathfrak{A}, n, \alpha)$, $(\mathfrak{B}, n, \beta)$ be ramified simple strata in A, and set $n = 2m+1 \geqslant 1$. Suppose that the characters $\psi_\alpha \mid U^{m+1}_{\mathfrak{A}}$, $\psi_\beta \mid U^{m+1}_{\mathfrak{B}}$ intertwine in $G = \mathrm{GL}_2(F)$. Let $\Lambda \in C(\psi_\alpha, \mathfrak{A})$, and let Λ' be the unique element of $C(\psi_\beta, \mathfrak{B})$ which intertwines with Λ in G.*

Let K/F be a finite, tamely ramified field extension. The characters $\mathcal{L}_{K/F}(\Lambda)$, $\mathcal{L}_{K/F}(\Lambda')$ then intertwine in $\mathrm{GL}_2(K)$.

Proof. Set $J = F[\alpha]^\times U^{m+1}_{\mathfrak{A}}$, $J' = F[\beta]^\times U^{m+1}_{\mathfrak{B}}$. The intertwining hypothesis says there exists $g \in G$ such that $\mathfrak{B} = g^{-1}\mathfrak{A}g$ and, on $U^{m+1}_{\mathfrak{B}}$, $\psi_\beta = \psi^g_\alpha = \psi_{g^{-1}\alpha g}$. We then have $J' = g^{-1}Jg$, and $\Lambda' = \Lambda^g$.

Reverting briefly to our earlier notation, the definition gives

$$\mathcal{L}^{g^{-1}\alpha g}_{K/F}(\Lambda^g) = \big(\mathcal{L}^\alpha_{K/F}(\Lambda)\big)^g.$$

The maps $\mathcal{L}^{g^{-1}\alpha g}_{K/F}$, $\mathcal{L}^\beta_{K/F}$ on $C(\psi_\beta, \mathfrak{B})$ are the same (46.3 Proposition (3)), so the lemma is proven. □

Directly from the original definition (46.3.2), we get:

Lemma 2. *Let $(\mathfrak{A}, n, \alpha)$ be a ramified simple stratum in A, $n = 2m+1 \geqslant 1$. Let χ be a character of $F^\times$ of level $l < n/2$, and let $c \in F$ satisfy $\chi(1+x) = \psi(cx)$, $x \in \mathfrak{p}^{[l/2]+1}$, or, if $l = 0$, set $c = 0$.*

Let $\Lambda \in C(\psi_\alpha, \mathfrak{A})$. The character

$$\chi\Lambda = (\chi \circ \det) \otimes \Lambda$$

of $F[\alpha]^\times U^{m+1}_{\mathfrak{A}}$ lies in $C(\psi_{\alpha+c}, \mathfrak{A})$ and, if K/F is a finite, tamely ramified field extension, then

$$(\chi\Lambda)_K = \chi_K \Lambda_K,$$

where $\chi_K = \chi \circ \mathrm{N}_{K/F}$.

We can finally make the desired:

Definition. *Let $\pi \in \mathcal{A}_2^{\mathrm{wr}}(F)$, and let K/F be a finite, tamely ramified field extension. Define a representation*

$$\pi_K = \mathrm{Lft}_{K/F}(\pi) \in \mathcal{A}_2^{\mathrm{wr}}(K)$$

as follows.

(1) *If π is minimal, choose a ramified simple stratum $(\mathfrak{A}, n, \alpha)$ in $\mathrm{M}_2(F)$ such that π contains a character $\Lambda \in C(\psi_\alpha, \mathfrak{A})$. The representation π_K is then the unique element of $\mathcal{A}_2^{\mathrm{wr}}(K)$ containing the character $\Lambda_K \in C(\psi_\alpha^K, \mathfrak{A}_K)$ of* (46.3.2).

(2) *Otherwise, choose a decomposition $\pi = \chi\pi'$, where χ is a character of $F^\times$ and $\pi' \in \mathcal{A}_2^{\mathrm{wr}}(F)$ is minimal. Define π'_K by* (1) and put $\pi_K = \chi_K \pi'_K$.

The representation π_K is called the *K/F-lift* of π. The lemmas show that the definition of π_K is independent of the various choices. It is also independent of the choice of character ψ. We therefore have a well-defined map

$$\begin{aligned} \mathrm{Lft}_{K/F} : \mathcal{A}_2^{\mathrm{wr}}(F) &\longrightarrow \mathcal{A}_2^{\mathrm{wr}}(K), \\ \pi &\longmapsto \pi_K, \end{aligned} \tag{46.5.1}$$

which we call *tame lifting* (relative to K/F). Summarizing its basic properties:

Proposition. *Let K/F be a finite, tamely ramified field extension and let $\pi \in \mathcal{A}_2^{\mathrm{wr}}(F)$.*

(1) *The central characters ω_π, ω_{π_K} are related by*

$$\omega_{\pi_K} = \omega_\pi \circ \mathrm{N}_{K/F}.$$

(2) *If χ is a character of $F^\times$, then $(\chi\pi)_K = \chi_K\pi_K$.*

(3) *If L/K is a finite, tamely ramified field extension, then*

$$\mathrm{Lft}_{L/F} = \mathrm{Lft}_{L/K} \circ \mathrm{Lft}_{K/F}.$$

(4) *If $\pi, \pi' \in \mathcal{A}_2^{\mathrm{wr}}(F)$, then $\pi_K \cong \pi'_K$ if and only if $\pi' \cong \chi\pi$, for a character χ of $F^\times$ trivial on $\mathrm{N}_{K/F}(K^\times)$.*

Proof. Part (1) is immediate from the definition (46.3.2). Parts (2) and (3) have already been noted. Assertion (4) follows from the definition, 46.2 Lemma and 46.3 Proposition (1). □

As a direct consequence of 45.2 Theorem, we get:

Corollary. *Let $\pi \in \mathcal{A}_2^{\mathrm{wr}}(F)$; there exists a cubic extension K/F such that π_K is ordinary.*

46.6. Suppose, for the moment, that K/F is a finite Galois extension. The natural action of $\mathrm{Gal}(K/F)$ on $G_K = \mathrm{GL}_2(K)$ induces an action on the sets $\mathcal{A}_2^0(K)$, $\mathcal{A}_2^{\mathrm{wr}}(K)$, which we denote $\pi \mapsto \pi^\gamma$, $\gamma \in \mathrm{Gal}(K/F)$. Clearly, if K/F is also tamely ramified, then

$$\mathrm{Lft}_{K/F}\big(\mathcal{A}_2^{\mathrm{wr}}(F)\big) \subset \mathcal{A}_2^{\mathrm{wr}}(K)^{\mathrm{Gal}(K/F)}.$$

Proposition. *Let K/F be a finite, tamely ramified, cyclic extension and set $\mathrm{Gal}(K/F) = \Sigma$. If $\tau \in \mathcal{A}_2^{\mathrm{wr}}(K)^\Sigma$, there exists $\pi \in \mathcal{A}_2^{\mathrm{wr}}(F)$ such that $\tau = \pi_K$. The representation π is determined up to twisting with a character of $F^\times$ trivial on $\mathrm{N}_{K/F}(K^\times)$.*

Proof. Suppose first that τ is minimal. We choose a ramified simple stratum $(\mathfrak{B}, s, \beta)$ in A_K, $s = 2r{+}1$, such that τ contains the character ψ_β^K of $U_{\mathfrak{B}}^{r+1}$. For $\sigma \in \Sigma$, the characters ψ_β^K, $(\psi_\beta^K)^\sigma = \psi_{\beta^\sigma}^K$ intertwine in G_K, since they both occur in $\tau = \tau^\sigma$. We conclude:

$$\det\beta \equiv \det\beta^\sigma \pmod{U_K^{[r/2]+1}}, \quad \mathrm{tr}\,\beta \equiv \mathrm{tr}\,\beta^\sigma \pmod{\mathfrak{p}_K^{-[r/2]}}.$$

The first of these congruences implies that the coset $\det\beta\, U_K^1$ has a Σ-fixed point, whence $e(K|F)$ divides $v_K(\det\beta) = -s$. Set $n = s/e(K|F) = 2m{+}1$. We then have

$$U_K^{[r/2]+1} \cap F = U_F^{[m/2]+1}, \quad \mathfrak{p}_K^{-[r/2]} \cap F = \mathfrak{p}^{-[m/2]}.$$

Since K/F is tamely ramified, it is easy to show that the cohomology groups $H^1(\Sigma, U_K^{[r/2]+1})$ and $H^1(\Sigma, \mathfrak{p}_K^{-[r/2]})$ are both trivial. It follows that there exist $x \in F^\times$ with $v_F(x) = -n$ and $y \in \mathfrak{p}^{-m}$ satisfying

$$x \equiv \det\beta \pmod{U_K^{[r/2]+1}}, \qquad y \equiv \mathrm{tr}\,\beta \pmod{\mathfrak{p}_K^{-[r/2]}}.$$

There is a ramified simple stratum $(\mathfrak{A}, n, \alpha)$ in A with $\det\alpha = x$, $\mathrm{tr}\,\alpha = y$. The characters $\psi_\beta^K \mid U_{\mathfrak{B}}^{r+1}$, $\psi_\alpha^K \mid U_{\mathfrak{A}_K}^{r+1}$ are then conjugate in G_K (44.2 Lemma). Consequently, we can replace $(\mathfrak{B}, s, \beta)$ by $(\mathfrak{A}_K, n_K, \alpha)$ (in the notation of 46.1).

The character ψ_α^K of $U_{\mathfrak{A}_K}^{m_K+1}$ is fixed by Σ, whence the same applies to the character $\Theta \in C(\psi_\alpha^K, \mathfrak{A}_K)$ which occurs in τ. Thus $\Theta \mid K[\alpha]^\times$ is fixed under the natural action of Σ on $K[\alpha]$ as $\mathrm{Gal}(K[\alpha]/F[\alpha])$. Therefore $\Theta \mid K[\alpha]^\times$ factors through $\mathrm{N}_{K[\alpha]/F[\alpha]}$ and so $\Theta = \Lambda_K$, for a character $\Lambda \in C(\psi_\alpha, \mathfrak{A})$. Setting $\pi = c\text{-}\mathrm{Ind}_J^G\,\Lambda$, we get $\tau = \pi_K$.

In general, we write $\tau = \chi\tau'$, for a character χ of $K^\times$ and τ' minimal. Let $(\mathfrak{B}, s, \beta)$ be a ramified simple stratum such that $\tau' \in \mathcal{A}_2^0(K; \beta)$. This only determines the isomorphism class of the $\mathfrak{o}_K$-chain order $\mathfrak{B}$. We may therefore take $\mathfrak{B} = \mathfrak{A}_K$ (notation of 46.1), for some $\mathfrak{o}_F$-chain order $\mathfrak{A}$ in A. That is, we may choose $\mathfrak{B}$ to be Σ-stable. If χ has level t, then $2t > s$, and the characters

$\chi \circ \det \mid U_{\mathfrak{B}}^{2t}$, $\chi^\sigma \circ \det \mid U_{\mathfrak{B}}^{2t}$ intertwine in G_K. It follows that χ^σ, χ, agree on U_K^t, whence $\chi \mid U_K^t$ factors through $\mathrm{N}_{K/F}$. In other words, we can reduce the level of τ by twisting with a character lifted from F. Iterating, this gets us back to the first case.

The final assertion follows directly from 46.3 Proposition (1). □

We shall mainly use the proposition in the following context:

Corollary. *Let K/F be a cubic extension, let L/F be the normal closure of K/F and write $\Sigma = \mathrm{Gal}(L/F)$. Let $\tau \in \mathcal{A}_2^{\mathrm{wr}}(L)^\Sigma$. There exists $\pi \in \mathcal{A}_2^{\mathrm{wr}}(F)$ such that $\pi_L = \tau$. The representation π is determined up to twisting with a character of $F^\times$ trivial on $\mathrm{N}_{L/F}(L^\times)$.*

Proof. If K/F is cyclic, we are in a special case of the proposition. We assume, therefore, that K/F is not cyclic. Let E/F be the maximal unramified sub-extension of L/F. Thus $[E{:}F] = 2$ while L/E is totally ramified and cyclic of degree 3.

By the proposition, there exist precisely three representations $\zeta \in \mathcal{A}_2^{\mathrm{wr}}(E)$ such that $\zeta_L = \tau$. They form a single orbit under twisting by characters of $E^\times/\mathrm{N}_{L/E}(L^\times)$, and so are distinguished by their central characters. They are permuted by $\mathrm{Gal}(E/F)$, with a single fixed point ζ^0, say. There are precisely two representations $\pi \in \mathcal{A}_2^{\mathrm{wr}}(F)$ satisfying $\pi_E = \zeta^0$; they are twists of each other by the non-trivial character of $F^\times/\mathrm{N}_{E/F}(E^\times)$. Since E/F is the maximal abelian sub-extension of L/F, we have $\mathrm{N}_{E/F}(E^\times) = \mathrm{N}_{L/F}(L^\times)$ (*cf.* 29.1), and the result follows. □

47. Interior Actions

The strong uniqueness properties of the partial Langlands Correspondence of 44.1 translate into a selection of naturality properties with respect to isomorphisms of the base field. We give an account of the matter now, since it will be useful in the sections to follow.

The arguments centre on properties of local constants. Everything in this section applies in arbitrary residual characteristic.

47.1. Let $\phi : F \to F'$ be an isomorphism of local fields. It extends to an isomorphism $\phi : \overline{F} \to \overline{F}'$ of separable algebraic closures, yielding an isomorphism $\phi : \Omega_F \to \Omega_{F'}$ of absolute Galois groups, determined by the original map $\phi : F \to F'$ up to an inner automorphism of Ω_F. Explicitly, for $\sigma \in \Omega_F$, the automorphism $\phi\sigma$ of $\overline{F}'$ is $\phi \circ \sigma \circ \phi^{-1}$.

The map $\phi : \Omega_F \to \Omega_{F'}$ on Galois groups induces a topological isomorphism $\phi : \mathcal{W}_F \to \mathcal{W}_{F'}$ of Weil groups, determined up to inner automorphism of

Ω_F. It gives a bijection

$$\begin{aligned} \mathcal{G}^0(F') &\longrightarrow \mathcal{G}^0(F), \\ \rho &\longmapsto \rho^\phi, \end{aligned}$$

where $\mathcal{G}^0(F)$ denotes the set of equivalence classes of irreducible smooth representations of $\mathcal{W}_F$ of arbitrary dimension.

This map extends to a bijection $\mathcal{G}^{ss}(F') \to \mathcal{G}^{ss}(F)$ on the sets of equivalence classes of finite-dimensional semisimple smooth representations of $\mathcal{W}_F$. (See also 28.7 Remark.)

In particular, the isomorphism $\phi : F \cong F'$ induces an isomorphism $\mathcal{W}_F^{ab} \cong \mathcal{W}_{F'}^{ab}$ which relates the Artin Reciprocity maps $\boldsymbol{a}_F$, $\boldsymbol{a}_{F'}$ (29.1) in the obvious way, namely $\phi \circ \boldsymbol{a}_F = \boldsymbol{a}_{F'} \circ \phi$.

47.2. Likewise, ϕ induces a bijection $\mathcal{A}_2^0(F') \to \mathcal{A}_2^0(F)$ via the obvious isomorphism $\mathrm{GL}_2(F) \to \mathrm{GL}_2(F')$ given by ϕ. We denote this by $\pi \mapsto \pi^\phi = \pi \circ \phi$.

Proposition. *Let $\rho \in \mathcal{G}_2^{\mathrm{im}}(F')$; then $\boldsymbol{\pi}_F(\rho^\phi) = \boldsymbol{\pi}_{F'}(\rho)^\phi$.*

Proof. Let E/F be a finite extension, $E \subset \overline{F}$, and set $E' = \phi(E)$. If $\eta \in \widehat{E'}$, $\eta \neq 1$, then $\eta^\phi = \eta \circ \phi \in \widehat{E}$ and ϕ transports the self-dual Haar measure on E' (relative to η) to that on E relative to η^ϕ. If χ is a character of $E'^\times$, then $\chi^\phi = \chi \circ \phi$ is a character of $E^\times$ and, following through the definitions, we get

$$\varepsilon(\chi^\phi, s, \eta^\phi) = \varepsilon(\chi, s, \eta).$$

The maps on equivalence classes of representations of Weil groups, induced by ϕ, commute with induction. We deduce that $\varepsilon(\rho^\phi, s, \eta^\phi) = \varepsilon(\rho, s, \eta)$ for any $\rho \in \mathcal{G}^{ss}(E')$.

Similarly, if $\pi \in \mathcal{A}_2^0(F')$ and if $\psi \in \widehat{F'}$, $\psi \neq 1$, then $\varepsilon(\pi^\phi, s, \psi^\phi) = \varepsilon(\pi, s, \psi)$. Surely $(\chi\pi)^\phi = \chi^\phi\pi^\phi$, for any character χ of $F'^\times$.

Let $\rho \in \mathcal{G}_2^{\mathrm{im}}(F')$, and let χ be a character of $F'^\times$. We have

$$\begin{aligned} \varepsilon(\chi\boldsymbol{\pi}_{F'}(\rho), s, \psi) &= \varepsilon(\chi^\phi\boldsymbol{\pi}_{F'}(\rho)^\phi, s, \psi^\phi), \\ \varepsilon(\chi \otimes \rho, s, \psi) &= \varepsilon(\chi^\phi \otimes \rho^\phi, s, \psi^\phi), \\ \varepsilon(\chi \otimes \rho, s, \psi) &= \varepsilon(\chi\boldsymbol{\pi}_{F'}(\rho), s, \psi), \end{aligned}$$

whence

$$\varepsilon(\chi^\phi\boldsymbol{\pi}_{F'}(\rho)^\phi, s, \psi^\phi) = \varepsilon(\chi^\phi \otimes \rho^\phi, s, \psi^\phi),$$

for all characters χ of $F'^\times$ and all $\psi \in \widehat{F'}$, $\psi \neq 1$. The Converse Theorem 27.1 implies that $\boldsymbol{\pi}_{F'}(\rho)^\phi = \boldsymbol{\pi}_F(\rho^\phi)$, as required. □

In particular:

Corollary. *Let K/F be a finite Galois extension. If $\rho \in \mathcal{G}_2^{\text{im}}(K)$ and $\gamma \in \text{Gal}(K/F)$, then*

$$\pi_K(\rho^\gamma) = \pi_K(\rho)^\gamma.$$

47.3. The operation of tame lifting likewise behaves properly with regard to isomorphisms of fields. Suppose we have finite, tamely ramified extensions K/F, K'/F', together with an isomorphism $\phi : K \to K'$ such that $\phi(F) = F'$. The original definitions (46.3.2), 46.5 yield straightaway:

Proposition. *Let $\pi \in \mathcal{A}_2^{\text{wr}}(F')$; then $\text{Lft}_{K/F}(\pi^\phi) = \text{Lft}_{K'/F'}(\pi)^\phi$.*

48. The Langlands-Deligne Local Constant modulo Roots of Unity

We have to make connections between the operation of tame lifting and the process of restriction of representations of Weil groups. We do this via two parallel series of properties of the local constants of Langlands-Deligne and of Godement-Jacquet. In both cases, we get a stability theorem in a direction opposite to that of our earlier ones 25.7, 29.4 Proposition (4). This leads to very useful descriptions of the local constants via multiplicative congruences.

In this section, we deal with the Langlands-Deligne constant. As in §30, we have to work in arbitrary dimension and use the machinery of induction constants.

48.1. The first of these results concerns the effect on the local constant of twisting with a tamely ramified character of $F^\times$. It holds for arbitrary residual characteristic p.

Recall that $\mathcal{P}_F$ denotes the wild inertia subgroup of $\mathcal{W}_F$.

Theorem. *Let $\rho \in \mathcal{G}^{\text{ss}}(F)$, and suppose that $\rho \mid \mathcal{P}_F$ does not contain the trivial character. Let $\psi \in \widehat{F}$, $\psi \neq 1$.*

(1) *There exists $c_\rho = c(\rho, \psi) \in F^\times$ such that*

$$\varepsilon(\chi \otimes \rho, s, \psi) = \chi(c_\rho)^{-1} \varepsilon(\rho, s, \psi), \tag{48.1.1}$$

for all tamely ramified characters χ of $F^\times$. This property determines the coset $c_\rho U_F^1$ uniquely.

(2) *If ϕ is a tamely ramified character of $\mathcal{W}_F$, then $c(\phi \otimes \rho, \psi) = c(\rho, \psi)$.*

(3) *Let K/F be a finite, tamely ramified field extension, and set $\rho_K = \rho \mid \mathcal{W}_K$, $\psi_K = \psi \circ \text{Tr}_{K/F}$. We have*

$$c(\rho_K, \psi_K) \equiv c(\rho, \psi) \pmod{U_K^1}.$$

Proof. We first observe that if the result holds for $\rho_1, \rho_2 \in \mathfrak{G}^{ss}(F)$, then it holds for $\rho_1 \oplus \rho_2$ with

$$c(\rho_1 \oplus \rho_2, \psi) = c(\rho_1, \psi)\, c(\rho_2, \psi).$$

We may as well, therefore, start with the assumption that ρ is *irreducible.*

Twisting ρ with an unramified character of $\mathcal{W}_F$ amounts to a translation of the variable s and so changes nothing of relevance here. We therefore assume that ρ is an irreducible representation of $\Gamma = \mathrm{Gal}(E/F)$, for some finite Galois extension E/F. Let F_1/F be the maximal tamely ramified sub-extension of E/F. By hypothesis, $\rho \mid \mathrm{Gal}(E/F_1)$ does not contain the trivial character. The finite p-group $\mathrm{Gal}(E/F_1)$ has non-trivial centre, Z say. The group Z is a normal subgroup of Γ and the restriction of ρ to Z is a sum of characters, all conjugate under Γ. We can assume that none of these characters is trivial: otherwise, they all are and we can replace E by E^Z. Thus $\rho \mid Z$ is a sum of characters ϕ^γ, where ϕ is non-trivial and $\gamma \in \Gamma$.

We need a variant of the Brauer induction theorem:

Lemma. *There is an expression*

$$\rho = \sum_{i=1}^{r} n_i \,\mathrm{Ind}_{\Delta_i}^{\Gamma}\, \xi_i, \tag{48.1.2}$$

where Δ_i is a subgroup of Γ containing Z, ξ_i is a character of Δ_i such that $\xi_i \mid Z = \phi$, and $n_i \in \mathbb{Z}$.

Proof. Let Γ_0 be the Γ-centralizer of ϕ, and let ρ_0 be the natural representation of Γ_0 on the ϕ-isotypic subspace of ρ. The representation ρ_0 is then irreducible and $\rho = \mathrm{Ind}_{\Gamma_0}^{\Gamma}\, \rho_0$. It is therefore enough to prove the lemma under the assumption that $\Gamma = \Gamma_0$. We can further factor out the kernel of ϕ, and assume that ϕ is a faithful character of the central subgroup Z. We use the standard Brauer induction theorem to write

$$\rho = \sum_{i=1}^{r} n_i \,\mathrm{Ind}_{\Delta_i}^{\Gamma}\, \xi_i,$$

where ξ_i is a linear character of the subgroup Δ_i. The ϕ-isotypic component of the term $\mathrm{Ind}_{\Delta_i}^{\Gamma}\, \xi_i$ is zero unless $\xi_i \mid \Delta_i \cap Z = \phi \mid \Delta_i \cap Z$. If this condition is satisfied, we extend ξ_i to a character $\tilde{\xi}_i$ of $\Delta_i Z$ so that $\tilde{\xi}_i \mid Z = \phi$. We so get the required expression

$$\rho = \sum_{i} \mathrm{Ind}_{\Delta_i Z}^{\Gamma}\, \tilde{\xi}_i,$$

including only those terms for which $\xi_i \mid \Delta_i \cap Z = \phi \mid \Delta_i \cap Z$. □

Consider a typical term $\theta = \operatorname{Ind}_\Delta^\Gamma \xi$ in the expression (48.1.2). Let $L = E^\Delta$ and view ξ as a character of $L^\times$ via class field theory. Let k be the level of ξ; then $k \geqslant 1$ and there exists $\alpha \in L^\times$ such that $\xi(1+x) = \psi_L(\alpha x)$, $x \in \mathfrak{p}_L^k$. Assertion (1) then holds for θ with $c(\theta, \psi) = \mathrm{N}_{L/F}(\alpha)$ (*cf.* 23.6 Proposition). It therefore holds for ρ. Explicitly, if $L_i = E^{\Delta_i}$ and α_i is defined by analogy with α, (48.1.2) gives

$$c(\rho, \psi) = \prod_{i=1}^{r} \mathrm{N}_{L_i/F}(\alpha_i)^{n_i}.$$

The uniqueness assertion is immediate, and this proves part (1).

Abbreviating $c_\rho = c(\rho, \psi)$, in part (2) we have

$$\varepsilon(\chi\phi \otimes \rho, s, \psi) = \chi\phi(c_\rho)\,\varepsilon(\rho, s, \psi) = \chi(c_\rho)\,\varepsilon(\phi \otimes \rho, s, \psi),$$

whence $\chi(c_\rho) = \chi(c_{\phi\otimes\rho})$, for all tamely ramified characters χ of $F^\times$. Thus $c(\phi \otimes \rho, \psi) \equiv c(\rho, \psi) \pmod{U_F^1}$, as required.

To prove (3), let K/F be a finite, tamely ramified extension. Enlarging E if necessary, we can assume that $K \subset E$. Set $\mathrm{Gal}(E/K) = \Sigma$. The expression (48.1.2) yields

$$\rho_K = \sum_{i=1}^{r} n_i \operatorname{Ind}_{\Delta_i}^\Gamma \xi_i \mid \Sigma.$$

Taking a typical term as before,

$$\theta \mid \Sigma = \operatorname{Ind}_\Delta^\Gamma \xi \mid \Sigma = \sum_{\gamma \in \Delta\backslash\Gamma/\Sigma} \operatorname{Ind}_{\Delta^\gamma\cap\Sigma}^\Sigma \xi^\gamma.$$

Translating this in terms of fields

$$\theta \mid \Sigma = \sum_\gamma \operatorname{Ind}_{L^\gamma K/K}(\xi^\gamma \circ \mathrm{N}_{L^\gamma K/L^\gamma}).$$

The extension $L^\gamma K/L^\gamma$ is tamely ramified, so we get

$$c(\theta \mid \Sigma, \psi_K) = \prod_{\gamma \in \Delta\backslash\Gamma/\Sigma} \mathrm{N}_{L^\gamma K/K}(\alpha^\gamma) = \mathrm{N}_{L/F}(\alpha).$$

Assertion (3) therefore holds for θ, so it holds for ρ. □

48.2. We need a mild extension of the defining property (48.1.1) of $c_\rho = c(\rho, \psi)$.

Corollary. *Let $\rho \in \mathcal{G}_n^0(F)$; suppose that $\rho \mid \mathcal{P}_F$ is irreducible and non-trivial. Let $\theta \in \mathcal{G}^{\mathrm{ss}}(F)$ be trivial on $\mathcal{P}_F$. We then have*

$$\varepsilon(\theta \otimes \rho, s, \psi) = \det\theta(c_\rho)^{-1}\varepsilon(\rho, s, \psi)^{\dim\theta}.$$

Proof. We use the language and techniques of §30.

We can assume that θ is a representation of $G = \mathrm{Gal}(K/F)$, for a finite, tamely ramified Galois extension K/F. Consider the set $\widetilde{\Gamma}(G)$ of pairs (E,χ), where E is a field, $F \subset E \subset K$, and χ is a character of $\mathrm{Gal}(K/E)$. (We regard χ as a character of $E^\times$, of level zero, via local class field theory.) In the language of §30, the map

$$(E,\chi) \longmapsto \frac{\varepsilon(\chi\otimes\rho_E, s, \psi_E)}{\varepsilon(\rho_E, s, \psi_E)} = \chi(c_\rho)^{-1},$$

where $c_\rho = c(\rho,\psi)$, is a division on G. The maps

$$(E,\chi) \longmapsto \begin{cases} \varepsilon(\chi\otimes\rho_E, s, \psi_E), \\ \varepsilon(\rho_E, s, \psi_E), \end{cases} \qquad (E,\chi)\in\widetilde{\Gamma}(G),$$

are pre-inductive divisions on G, being the boundaries, respectively, of the induction constants

$$(E,\theta) \longmapsto \begin{cases} \varepsilon(\theta\otimes\rho_E, s, \psi_E) \\ \varepsilon(\rho_E, s, \psi_E)^{\dim\theta}. \end{cases} \qquad (E,\theta)\in\widetilde{K}_0G.$$

On the other hand, the division $(E,\chi) \mapsto \chi(c_\rho)^{-1}$ is the boundary of the induction constant $(E,\theta) \mapsto \det\theta(c_\rho)^{-1}$. An induction constant is determined by its boundary (30.1 Lemma 1). The corollary now follows. □

Remark. 48.1 Theorem and 48.2 Corollary apply, in particular, to representations $\rho\in\mathcal{G}_2^{\mathrm{wr}}(F)$.

48.3. For the remainder of this section, our blanket hypothesis $p = 2$ is essential.

Let K/F be a finite, tamely ramified extension. We set (as before) $R_{K/F} = \mathrm{Ind}_{K/F} 1_K$ and $\varkappa_{K/F} = \det R_{K/F}$: thus $\varkappa_{K/F}$ is unramified of order at most 2. Let $\lambda_{K/F}(\psi)$ be the Langlands constant, as in 30.4. We need the following particular consequence of 48.2 Corollary.

Proposition. *Let $\rho\in\mathcal{G}_2^{\mathrm{wr}}(F)$ and write $c_\rho = c(\rho,\psi)$. If K/F is a finite, tamely ramified extension, then*

$$\varepsilon(\rho_K, s, \psi_K) = \lambda_{K/F}(\psi)^{-2}\,\varkappa_{K/F}(c_\rho)^{-1}\varepsilon(\rho, s, \psi)^{[K:F]}. \tag{48.3.1}$$

Proof. We have $\mathrm{Ind}_{K/F}(\rho_K) = \rho\otimes R_{K/F}$; the inductive property of the local constant and 48.2 Corollary give

$$\lambda_{K/F}(\psi)^2\,\varepsilon(\rho_K, s, \psi_K) = \varepsilon(\rho\otimes R_{K/F}, s, \psi) = \varkappa_{K/F}(c_\rho)^{-1}\,\varepsilon(\rho, s, \psi)^{[K:F]},$$

as required. □

48.4. The invariant $c(\rho, \psi)$ has a more subtle rôle. Let $\mu_{\mathbb{C}}(2^\infty)$ denote the group of roots of unity in $\mathbb{C}$ of order a power of 2.

Theorem. *Let $\rho \in \mathcal{G}_2^{\mathrm{wr}}(F)$. Let $\psi \in \widehat{F}$, $\psi \neq 1$, and write $c_\rho = c(\rho, \psi)$. Then:*

$$\varepsilon(\rho, \tfrac{1}{2}, \psi) \equiv \det \rho(c_\rho)^{-1/2} \pmod{\mu_{\mathbb{C}}(2^\infty)}.$$

Proof. Suppose first that ρ is imprimitive. We can write $\rho = \mathrm{Ind}_{E/F}\,\xi$, for a ramified quadratic extension E/F and a character ξ of $E^\times$ of level $k \geqslant 1$, say. We then have $c_\rho = \mathrm{N}_{E/F}(\alpha)$, for any $\alpha \in E^\times$ such that $\xi(1+x) = \psi_E(\alpha x)$, $x \in \mathfrak{p}_E^k$.

Lemma. *We have*

$$\varepsilon(\xi, \tfrac{1}{2}, \psi_E) \equiv \xi(\alpha)^{-1} \pmod{\mu_{\mathbb{C}}(2^\infty)}.$$

Let us assume the lemma for the moment: we prove it in the next paragraph. We have $\varepsilon(\rho, \frac{1}{2}, \psi)/\varepsilon(\xi, \frac{1}{2}, \psi_E) = \lambda_{E/F}(\psi)$, which is a 4-th root of unity (30.4.3). Also, $\det \rho = \varkappa_{E/F} \otimes \xi \mid F^\times$, and $\varkappa_{E/F}$ takes values ± 1. Thus

$$\det \rho(c_\rho) \equiv \xi(\mathrm{N}_{E/F}(\alpha)) \pmod{\mu_{\mathbb{C}}(2^\infty)}.$$

If $\sigma \in \mathrm{Gal}(E/F)$, $\sigma \neq 1$, we have $\alpha^\sigma \equiv \alpha \pmod{U_E^1}$, with the result that

$$\xi(\mathrm{N}_{E/F}(\alpha)) \equiv \xi(\alpha)^2 \pmod{\mu_{\mathbb{C}}(2^\infty)},$$

implying the theorem in this case.

We therefore assume that ρ is *primitive*, and use the analysis of such representations in §42. We can assume that ρ is a representation of $\Gamma = \mathrm{Gal}(\widetilde{E}/F)$, for some finite extension $\widetilde{E}/F$. The centre Z of Γ is cyclic and Γ/Z is A_4 or S_4. Write $E = \widetilde{E}^Z$ and let F_1/F be the maximal tamely ramified subextension of E/F. Thus $\mathrm{Gal}(F_1/F)$ is cyclic of order 3 or dihedral of order 6, and F_1/F has a cubic sub-extension K/F such that ρ_K is imprimitive. The extension F_1/K is either trivial or unramified quadratic. The extension E/F_1 is totally wildly ramified with non-cyclic Galois group of order 4. It is the composite of three quadratic extensions $E^{(j)}/F_1$ and the exponents $d(E^{(j)}|F_1) = d$, say, are all the same (*cf.* 42.2 Remark).

There is a ramified quadratic extension K_1/K such that $\rho_K = \mathrm{Ind}_{K_1/K}\,\phi$, for some character ϕ of $K_1^\times$. The field K_1F_1 is one of the $E^{(j)}$, with the result that $d(K_1|K) = d$. Moreover, the character ϕ has level $l(K_1) \geqslant 2d$. If we write $\phi(1+x) = \psi_{K_1}(\alpha x)$, $x \in \mathfrak{p}_{K_1}^{l(K_1)}$, then

$$c(\rho_K, \psi_K) \equiv c(\rho, \psi) \equiv \mathrm{N}_{K_1/K}(\alpha) \pmod{U_{K_1}^1},$$

as in 48.1.

The restriction $\rho \mid Z$ is of the form $\chi \oplus \chi$, for some character χ of $E^\times$. Clearly $\chi = \phi \circ \mathrm{N}_{E/K_1}$, and χ has level $l(E) \geqslant 2d$. Indeed, $\chi(1+x) = \psi_E(\alpha x)$, $x \in \mathfrak{p}_E^{l(E)}$.

We now use 48.1 Lemma to write

$$\rho = \sum_i n_i \operatorname{Ind}_{L_i/F} \xi_i,$$

for various fields $L_i \subset E$. Each character ξ_i satisfies $\xi_i \circ \mathrm{N}_{E/L_i} = \chi$. If l_i is the level of ξ_i and we choose $\alpha_i \in L_i$ such that $\xi_i(1+x) = \psi_{L_i}(\alpha_i x)$, $x \in \mathfrak{p}_{L_i}^{l_i}$, we have $\alpha_i \equiv \alpha \pmod{U_E^1}$. Using $\sim$ to denote congruence modulo $\boldsymbol{\mu}_{\mathbb{C}}(2^\infty)$, it follows that

$$\varepsilon(\rho, \tfrac{1}{2}, \psi) \sim \prod_i \varepsilon(\operatorname{Ind}_{L_i/F} \xi_i, \tfrac{1}{2}, \psi_{L_i})^{n_i},$$

since the Langlands constants $\lambda_{L_i/F}(\psi)$ are 4-th roots of unity (30.4 Remark). The lemma now gives:

$$\varepsilon(\rho, \tfrac{1}{2}, \psi) \sim \prod_i \xi_i^{-n_i}(\alpha_i).$$

We extend each ξ_i, somehow, to a smooth character of the group generated by α and U_E^1. Writing $c_\rho = c(\rho, \psi)$, we get

$$\begin{aligned} \det \rho(c_\rho) &\sim \prod_i \xi_i^{n_i}(\mathrm{N}_{K_1/K}(\alpha)) \\ &\sim \prod_i \xi_i^{2n_i}(\alpha) \sim \prod_i \xi_i^{2n_i}(\alpha_i) \\ &\sim \varepsilon(\rho, \tfrac{1}{2}, \psi)^{-2}, \end{aligned}$$

whence the theorem follows. □

48.5. We have to prove 48.4 Lemma. Re-normalizing our notation, this amounts to[1]:

Lemma. *Let $\psi \in \widehat{F}$, $\psi \neq 1$. Let χ be a character of $F^\times$ of level $l \geqslant 1$. Let $\delta \in F^\times$ satisfy $\chi(1+x) = \psi(\delta x)$, $x \in \mathfrak{p}^l$. The quantity $\chi(\delta)\,\varepsilon(\chi, \frac{1}{2}, \psi)$ is then a root of unity of order a power of 2.*

Proof. The assertion is independent of the choice of ψ, so it will be convenient to assume ψ has level one. Thus

$$\varepsilon(\chi, \tfrac{1}{2}, \psi) = \tau(\chi, \psi)/q^{(l+1)/2}.$$

[1] The global argument used in the proof to follow is a variation of part of the standard proof of the unit theorem for global fields: see [54] or [31], for instance.

The condition imposed on δ determines only the coset δU_F^1, but $\chi(U_F^1) \subset \boldsymbol{\mu}_{\mathbb{C}}(2^\infty)$, so the assertion is independent of the choice of δ.

We have

$$\tau(\chi, \psi) = q \sum_{y \in U_F^1/U_F^l} \chi(\delta y)^{-1} \psi(\delta y),$$

so $\chi(\delta)\,\tau(\chi, \psi)$ lies in the field $\mathbb{F} = \mathbb{Q}[\zeta]$ generated by a 2^j-th root of unity ζ, for some $j \in \mathbb{Z}$. We take j sufficiently large to ensure that all values of $\chi \mid U_F^1$ and all values of $\psi \mid \mathfrak{p}^{-l}$ lie in $\mathbb{F}$. As we may take $j \geqslant 3$, we can further assume that $\mathbb{F}$ contains a square root of q.

Using $z \mapsto \bar{z}$ to denote complex conjugation, a simple manipulation gives

$$\overline{\tau(\chi, \psi)} = \chi(-1)\,|\chi(\delta)|^{-2} \tau(\check{\chi}, \psi).$$

The relation (23.6.2) and the functional equation (23.4.2) yield

$$\tau(\chi, \psi)\,\tau(\check{\chi}, \psi) = \chi(-1)\,q^{l+1}, \tag{48.5.1}$$

implying that

$$|\chi(\delta)\,\tau(\chi, \psi)| = q^{(l+1)/2}.$$

Let us now write

$$\xi(\chi, \psi) = \chi(\delta)\,\tau(\chi, \psi),$$

and take $\sigma \in \mathrm{Gal}(\mathbb{F}/\mathbb{Q})$. Thus $\psi^\sigma \mid \mathfrak{p}^{-l}$ is the restriction of a character of F of level one. Since $\chi(U_F^l) \subset \{\pm 1\}$, we also have $\chi^\sigma(1+x) = \psi^\sigma(\delta x)$ for $x \in \mathfrak{p}^l$. We conclude that

$$\xi(\chi, \psi)^\sigma \equiv \xi(\chi^\sigma, \psi^\sigma) \pmod{\boldsymbol{\mu}_{\mathbb{C}}(2^\infty)},$$

and hence that

$$|\xi(\chi, \psi)^\sigma| = q^{(l+1)/2}, \quad \sigma \in \mathrm{Gal}(\mathbb{F}/\mathbb{Q}).$$

Put another way,

$$\|q^{-(l+1)/2} \xi(\chi, \psi)\|_v = 1,$$

for all Archimedean places v of the field $\mathbb{F}$.

Let v_2 be the unique non-Archimedean place of $\mathbb{F}$ lying over 2. Let v be a non-Archimedean place of $\mathbb{F}$, $v \neq v_2$. The quantity $\xi(\chi, \psi)/q^{(l+1)/2}$ is then integral at v and the functional equation (48.5.1) implies that

$$\|\xi(\chi, \psi)/q^{(l+1)/2}\|_v = 1$$

However, the product formula for normalized valuations of $\mathbb{F}$,

$$\prod_v \|x\|_v = 1, \quad x \in \mathbb{F}^\times,$$

now implies that $\|\xi(\chi,\psi)/q^{(l+1)/2}\|_v = 1$ for all v. In particular, $\xi(\chi,\psi)/q^{(l+1)/2}$ is a unit in $\mathbb{F}$.

Set

$$\mathbb{F}_\infty = \prod_{v|\infty} \mathbb{F}_v \cong \mathbb{C}^{[\mathbb{F}:\mathbb{Q}]/2},$$

where v ranges over all Archimedean places of $\mathbb{F}$. The canonical map $\mathbb{F} \to \mathbb{F}_\infty$ identifies $\mathfrak{o}_\mathbb{F}^\times$ with a discrete subgroup of $\mathbb{F}_\infty^\times$. The multiplicative group generated by $\xi(\chi,\psi)/q^{(l+1)/2}$ is therefore discrete, but it is contained in the compact subgroup of vectors $(x_v)_{v|\infty}$ such that $|x_v| = 1$ for all $v|\infty$. It is therefore finite.

We deduce that $\xi(\chi,\psi)/q^{(l+1)/2}$ is a root of unity in $\mathbb{F}$, and so it lies in $\mu_\mathbb{C}(2^\infty)$, as required. □

49. The Godement-Jacquet Local Constant and Lifting

We seek some properties of the Godement-Jacquet local constant, analogous to those of §48. In this section, it will be enough to treat the case where $\psi \in \widehat{F}$ has level one, and is the character used to relate simple strata and characters.

49.1. The first step is extremely simple:

Lemma. *Let* $\pi \in \mathcal{A}_2^{\mathrm{wr}}(F)$.

(1) *There exists* $c_\pi = c(\pi,\psi)$ *such that*

$$\varepsilon(\chi\pi, s, \psi) = \chi(c_\pi)^{-1}\,\varepsilon(\pi, s, \psi) \qquad (49.1.1)$$

for all tamely ramified characters χ *of* $F^\times$. *This condition determines the coset* $c_\pi U_F^1$ *uniquely.*

(2) *Let* K/F *be a finite, tamely ramified field extension; then*

$$c(\pi_K, \psi_K) \equiv c(\pi,\psi) \pmod{U_K^1}. \qquad (49.1.2)$$

Proof. Suppose first that π is minimal: there exists a ramified simple stratum $(\mathfrak{A}, n, \alpha)$ in A such that $\pi \in \mathcal{A}_2^0(F;\alpha)$ (notation of 44.3). The Gauss sum formula (25.5 Corollary) for the local constant then gives (49.1.1) with $c_\pi = \det\alpha$. The second assertion is immediate and part (2) follows from the definition of tame lifting.

Otherwise, we write $\pi = \xi\pi_0$, where ξ is a character of $F^\times$ and $\ell(\pi_0) < \ell(\pi)$. Let ξ have level $k \geqslant 1$, and choose $\delta_\xi \in F$ so that $\xi(1{+}x) = \psi(\delta_\xi x)$, $x \in \mathfrak{p}^k$. The same Gauss sum formula then yields the result with $c_\pi = \delta_\xi^2$. □

It will be useful to record the specific values given by the proof of the lemma:

Gloss.

(1) *Suppose there exists a ramified simple stratum* $(\mathfrak{A}, n, \alpha)$ *such that* $\pi \in \mathcal{A}_2^0(F;\alpha)$*; then* $c(\pi,\psi) = \det\alpha$.
(2) *Suppose there exists a character* ξ *of* $F^\times$*, of level* $l \geqslant 1$*, such that* $\ell(\xi^{-1}\pi) < \ell(\pi)$*; then* $c(\pi,\psi) = \delta^2$*, where* $\delta \in \mathfrak{p}^{-l}$ *satisfies* $\xi(1+x) = \psi(\delta x)$*,* $x \in \mathfrak{p}^l$.

49.2. We can now work out the behaviour of the local constant under tame lifting. We only need a very special case:

Proposition. *Let* $\pi \in \mathcal{A}_2^{\mathrm{wr}}(F)$*, and let* K/F *be a cubic extension; then*

$$\varepsilon(\pi_K, s, \psi_K) = \varkappa_{K/F}(c_\pi)^{-1}\,\varepsilon(\pi, s, \psi)^3, \tag{49.2.1}$$

where $c_\pi = c(\pi,\psi)$ *and* $\lambda_{K/F} = \lambda_{K/F}(\psi)$.

Proof. Suppose first that $\pi \in \mathcal{A}_2^0(F;\alpha)$, for some ramified simple stratum $(\mathfrak{A}, n, \alpha)$. Immediately,

$$\left(q^{2\ell(\pi)}\right)^3 = q_K^{2\ell(\pi_K)},$$

so in this case the Gauss sum formula and the definition of π_K give

$$\varepsilon(\pi_K, s, \psi_K) = \epsilon_{K/F}\,\varepsilon(\pi, s, \psi)^3.$$

We have $\varkappa_{K/F}(\det\alpha) = \epsilon_{K/F}$, and the result follows in this case.

The next step is to prove:

$$\lambda_{K/F}(\psi)^2 = 1. \tag{49.2.2}$$

If K/F is unramified, 23.5 Proposition gives $\lambda_{K/F}(\psi) = 1$. Suppose next that K/F is totally ramified and cyclic. Let ϕ be a non-trivial character of $F^\times$ vanishing on norms from K. By (23.4.2),

$$\lambda_{K/F}(\psi) = \varepsilon(\phi, s, \psi)\,\varepsilon(\phi^{-1}s, \psi) = \phi(-1).$$

Since ϕ has order 3, we have $\phi(-1) = 1 = \lambda_{K/F}(\psi)$. This leaves only the case where K/F is non-cyclic, hence totally ramified. Let E/F be unramified quadratic. Thus EK/E is cyclic and cubic. We have

$$\begin{aligned}\lambda_{EK/F}(\psi) &= \lambda_{EK/K}(\psi_K)\,\lambda_{K/F}(\psi)^2\\ &= \lambda_{EK/E}(\psi_E)\,\lambda_{E/F}(\psi)^3.\end{aligned}$$

By the cyclic cubic case, $\lambda_{EK/E}(\psi_E) = 1$ while, for the quadratic unramified extensions E/F, EK/K, 23.5 Proposition gives

$$\lambda_{EK/K}(\psi_K) = \lambda_{E/F}(\psi) = -1.$$

Thus $\lambda_{K/F}(\psi)^2 = 1$, as desired.

We return to the proof of the Proposition. If π is not minimal, there is a ramified simple stratum $(\mathfrak{A}, n, \alpha)$, a character χ of $F^\times$ of level $l > n/2$, and a representation $\pi' \in \boldsymbol{A}_2^0(F;\alpha)$ such that $\pi = \chi\pi'$. The representation π' contains a character $\Lambda' \in C(\psi_\alpha, \mathfrak{A})$. Here, $\ell(\pi) = l$ and we have $q^{3\ell(\pi)} = q_K^{\ell(\pi_K)}$, so it is again enough to check (49.2.1) at $s = \frac{1}{2}$.

We choose $\delta \in \mathfrak{p}^{-l}$ such that $\chi(1+x) = \psi(\delta x)$, $x \in \mathfrak{p}^{[l/2]+1}$. This gives

$$\varepsilon(\pi, \tfrac{1}{2}, \psi) = q^{-1} \sum_x \chi\Lambda'((\alpha+\delta)x)^{-1}\psi_A((\alpha+\delta)x),$$

where $\psi_A = \psi \circ \mathrm{tr}_A$ and the sum is taken over $x \in U_{\mathfrak{A}}^l/U_{\mathfrak{A}}^{l+1}$ (25.5 Corollary). For such x, we have $\Lambda'(\alpha x)^{-1}\psi_A(\alpha x) = \Lambda'(\alpha)^{-1}\psi_A(\alpha)$, so this expression reduces to

$$\varepsilon(\pi, \tfrac{1}{2}, \psi) = q^{-1}\, \psi_A(\alpha)\, \Lambda'(\alpha+\delta)^{-1}\, \chi(\det(1+\alpha\delta^{-1}))^{-1} \sum_x \chi(\det \delta x)^{-1}\psi_A(\delta x).$$

Using the notation of 25.6, the identity (25.6.1) yields

$$\sum_x \chi(\det \delta x)^{-1}\psi_A(\delta x) = q^{-l}\, \tau(\chi, \psi)^2,$$

where

$$\tau(\chi, \psi) = \sum_{y \in U_F/U_F^{l+1}} \chi(\delta y)^{-1}\psi(\delta y).$$

That is,

$$\varepsilon(\pi, \tfrac{1}{2}, \psi) = \psi(\mathrm{tr}_A\, \alpha)\, \Lambda'(\alpha+\delta)^{-1}\, \chi(\det(1+\alpha\delta^{-1}))^{-1}\, \tau(\chi, \psi)^2/q^{(l+1)}.$$

The corresponding expression for π_K is

$$\varepsilon(\pi_K, \tfrac{1}{2}, \psi_K) = \psi_K(\mathrm{tr}_{A_K}\, \alpha)\, \Lambda_K'(\alpha+\delta)^{-1}\, \chi_K(\det(1+\alpha\delta^{-1}))^{-1}\, \tau(\chi_K, \psi_K)^2/q_K^{e_K l+1},$$

where $e_K = e(K|F)$. The definition of Λ_K' gives

$$\Lambda_K'(\alpha+\delta) = \Lambda'(\alpha+\delta)^3.$$

We have $c_\pi = \delta^2$, so we are reduced to proving

$$\tau(\chi_K, \psi_K)^2/q_K^{e_K l+1} = \tau(\chi, \psi)^6/q^{3(l+1)}. \tag{49.2.3}$$

To this end we use the identities

$$\begin{aligned}\tau(\chi,\psi) &= q^{(l+1)/2}\,\varepsilon(\chi,\tfrac{1}{2},\psi),\\ \tau(\chi_K,\psi_K) &= q_K^{(e(K|F)l+1)/2}\,\varepsilon(\chi_K,\tfrac{1}{2},\psi_K).\end{aligned}$$

The inductive property of the local constant gives

$$\varepsilon(\chi_K,\tfrac{1}{2},\psi_K) = \varepsilon(\chi\otimes R_{K/F},\tfrac{1}{2},\psi)\,\lambda_{K/F}(\psi)^{-1}.$$

A trivial instance of 48.1 gives $c_\chi = \delta$ and

$$\varepsilon(\chi\otimes R_{K/F},\tfrac{1}{2},\psi) = \varkappa_{K/F}(\delta)^{-1}\varepsilon(\chi,\tfrac{1}{2},\psi)^3.$$

Since $\varkappa_{K/F}^2 = 1$, the desired identity (49.2.3) now follows from (49.2.2). □

50. The Existence Theorem

In this section, we construct a map $\boldsymbol{\pi}_F : \mathcal{G}_2^0(F) \to \mathcal{A}_2^0(F)$ satisfying the local constant relation (33.4.1). That relation determines the map uniquely, so $\boldsymbol{\pi}_F$ is the Langlands correspondence.

Until further notice, $\psi \in \widehat{F}$ is of level one. We use it to make the connection between simple strata and cuspidal types, as in Chapter IV. Our blanket hypothesis $p = 2$ remains in force. We recall that $\mathcal{G}_2^{\mathrm{wr}}(F) = \mathcal{G}_2^0(F) \smallsetminus \mathcal{G}_2^{\mathrm{nr}}(F)$ is the set of totally ramified $\rho \in \mathcal{G}_2^0(F)$.

50.1. We extend some of the elementary observations of §41 to the class of primitive representations. Let $\rho \in \mathcal{G}_2^0(F)$; 29.4 Proposition (1) gives an integer $n(\rho,\psi)$ such that

$$\varepsilon(\rho,s,\psi) = q^{(\frac{1}{2}-s)n(\rho,\psi)}\,\varepsilon(\rho,\tfrac{1}{2},\psi).$$

We call $n(\rho,\psi)$ the *level* of ρ. We say that ρ is *minimal* if $n(\rho,\psi) \leqslant n(\chi\otimes\rho,\psi)$ for all characters χ of $F^\times$.

Proposition. *Let* $\rho \in \mathcal{G}_2^{\mathrm{wr}}(F)$.

(1) *The representation* ρ *is minimal if and only if* $n(\rho,\psi)$ *is odd.*
(2) *Suppose that* ρ *is minimal and let* χ *be a character of* $F^\times$ *of level* $l \geqslant 0$. *The representation* $\chi\otimes\rho$ *is minimal if and only if* $2l < n(\rho,\psi)$. *If* $2l > n(\rho,\psi)$, *then* $n(\chi\otimes\rho,\psi) = 2l$.

Proof. In the case where ρ is imprimitive, all assertions follow easily from the discussions in 41.4, 41.5.

Suppose therefore that ρ is primitive, and choose a cubic extension K/F such that ρ_K is imprimitive (42.2). We have $n(\rho_K,\psi_K) = e(K|F)\,n(\rho,\psi)$.

So, if $n(\rho, \psi)$ is odd, then $n(\rho_K, \psi_K)$ is odd and ρ_K is consequently minimal. Thus ρ is minimal.

Conversely, suppose that $n = n(\rho, \psi)$ is even, and let L/F be the normal closure of K/F: in particular, L/K is unramified of degree $\leqslant 2$. The representation ρ_L has level $n_L = e(K|F)n$ and so is not minimal over L. We write $\rho_L = \chi \otimes \tau$, for a character χ of $L^\times$ and a representation τ with $n(\tau, \psi_L) < n_L$. The character χ thus has level $n_L/2$. For $\gamma \in \mathrm{Gal}(L/F)$, we have $\rho_L^\gamma \cong \rho_L$, so $\tau \cong \chi^{-1}\chi^\gamma \otimes \tau^\gamma$. Thus $\chi^{-1}\chi^\gamma$ has level $< n_L/2$, and the restriction of χ to $U_L^{n_L/2}$ factors through $\mathrm{N}_{L/F}$. In other words, we could have taken $\chi = \phi_L$, for some character ϕ of $F^\times$. This character satisfies $n(\phi^{-1} \otimes \rho, \psi) < n(\rho, \psi)$, and ρ is not minimal. This completes the proof of (1) and (2) now follows readily. □

50.2. Let $\rho \in \mathcal{G}_2^{\mathrm{wr}}(F)$ be minimal. If ρ is imprimitive, then $\pi = \boldsymbol{\pi}_F(\rho)$ has been defined. The representation π is minimal, and 44.7 Proposition specifies a simple stratum $(\mathfrak{A}, n, \alpha)$ such that $\pi \in \boldsymbol{\mathcal{A}}_2^0(F; \alpha)$. We generalize this construction to the case of ρ primitive, by an indirect route.

Proposition. *Let $\rho \in \mathcal{G}_2^{\mathrm{wr}}(F)$ be minimal over F.*

(1) *There is a ramified simple stratum $(\mathfrak{A}, n, \alpha)$ in $\mathrm{M}_2(F)$, $n = 2m+1$, with the following property: if K/F is a finite, tamely ramified field extension such that ρ_K is imprimitive, then $\boldsymbol{\pi}_K(\rho_K) \in \boldsymbol{\mathcal{A}}_2^0(K; \alpha)$.*
(2) *Condition (1) determines the G-conjugacy class of the character $\psi_\alpha \mid U_{\mathfrak{A}}^{m+1}$, and hence the set $\boldsymbol{\mathcal{A}}_2^0(F; \alpha)$, uniquely.*

Proof. In the case where ρ is imprimitive, all assertions in (1) are immediate consequences of 44.7.

We therefore assume that ρ is primitive, and choose a cubic extension K/F such that ρ_K is imprimitive. We apply the first case to ρ_K to get a simple stratum $(\mathfrak{A}', n', \alpha')$ in $A_K = \mathrm{M}_2(K)$ with the property (1) relative to the base field K.

Let L/F be the normal closure of K/F and consider the representation $\zeta = \boldsymbol{\pi}_L(\rho_L)$. This lies in $\boldsymbol{\mathcal{A}}_2^0(L; \alpha')$ and is fixed by $\mathrm{Gal}(L/F)$. By 46.6 Corollary, there is $\pi \in \boldsymbol{\mathcal{A}}_2^{\mathrm{wr}}(F)$ such that $\zeta = \pi_L$. The representation π satisfies $\ell(\pi) = e(L|F)^{-1}\ell(\zeta) \notin \mathbb{Z}$, so π is minimal. There exists a simple stratum $(\mathfrak{A}, n, \alpha)$ in A such that $\pi \in \boldsymbol{\mathcal{A}}_2^0(F; \alpha)$, whence $\zeta \in \boldsymbol{\mathcal{A}}_2^0(L; \alpha)$. That is, $\boldsymbol{\mathcal{A}}_2^0(L; \alpha) = \boldsymbol{\mathcal{A}}_2^0(L; \alpha')$. Lemma 46.2 now implies $\boldsymbol{\mathcal{A}}_2^0(K; \alpha) = \boldsymbol{\mathcal{A}}_2^0(K; \alpha')$, so we could have taken $(\mathfrak{A}', n', \alpha') = (\mathfrak{A}_K, n_K, \alpha)$ (in the notation of 46.1)).

If E/F is a finite tame extension such that ρ_E is imprimitive, then E contains some F-conjugate of K (42.3 Proposition) and the first case implies $\boldsymbol{\pi}_E(\rho_E) \in \boldsymbol{\mathcal{A}}_2^0(E; \alpha)$.

Property (2) is implied by 46.2 Lemma. □

50.3. We come to the central result of the section, which establishes the existence of the Langlands correspondence. We continue with our fixed $\psi \in \widehat{F}$ of level one.

Existence Theorem. *Let $\rho \in \mathcal{G}_2^0(F)$. There exists a unique representation $\pi = \boldsymbol{\pi}_F(\rho) \in \mathcal{A}_2^0(F)$ such that*

$$\varepsilon(\chi\pi, s, \psi) = \varepsilon(\chi \otimes \rho, s, \psi), \tag{50.3.1}$$

for all characters χ of $F^\times$. Further:

1. *$\omega_{\boldsymbol{\pi}_F(\rho)} = \det \rho$, for all $\rho \in \mathcal{G}_2^0(F)$;*
2. *$\boldsymbol{\pi}_F(\chi \otimes \rho) \equiv \chi\boldsymbol{\pi}_F(\rho)$, for all $\rho \in \mathcal{G}_2^0(F)$ and all characters χ of $F^\times$;*
3. *the relation (50.3.1) holds for all $\psi \in \widehat{F}$, $\psi \neq 1$.*

Proof. The uniqueness assertion is a direct consequence of (50.3.1) and the Converse Theorem 27.1. Parts (1) and (2) will be seen as consequences of the construction (50.6.2), (50.6.3). Given these, part (3) follows from 24.3 Proposition and 29.4 Proposition (2).

If ρ is imprimitive, we set $\pi = \boldsymbol{\pi}_F(\rho)$, as in 44.1: this has all the required properties. We assume henceforth that ρ is *primitive*.

To start with, we also assume ρ is *minimal*. We choose a cubic extension K/F such that ρ_K is imprimitive. We take a ramified simple stratum $(\mathfrak{A}, n, \alpha)$ in $\mathrm{M}_2(F)$, as in 50.2 Proposition: thus $\zeta = \boldsymbol{\pi}_K(\rho_K) \in \mathcal{A}_2^0(K;\alpha)$. In particular, there is a unique character $\Lambda_\zeta \in C(\psi_\alpha^K, \mathfrak{A}_K)$ which occurs in ζ. Let ν be the unramified character of $K[\alpha]^\times$ of the same order as $\varkappa_{K/F}$, and write $n = 2m+1$.

Lemma 1. *There exists a unique character $\Lambda \in C(\psi_\alpha, \mathfrak{A})$ such that $\Lambda \mid F^\times = \det \rho$ and*

$$\Lambda^3 = \nu \otimes \Lambda_\zeta \mid F[\alpha]^\times U_{\mathfrak{A}}^{m+1}.$$

Proof. The restriction of Λ_ζ to $U_{\mathfrak{A}}^{m+1}$ is simply $\psi_{3\alpha}$. The restriction of Λ_ζ to $U_{F[\alpha]}^1$ has values in the group $\boldsymbol{\mu}_{\mathbb{C}}(2^\infty)$ of roots of unity in $\mathbb{C}$ of 2-power order. Thus $\Lambda_\zeta \mid U_{F[\alpha]}^1 U_{\mathfrak{A}}^{m+1}$ is of the form λ^3, for a unique character λ of $U_{F[\alpha]}^1 U_{\mathfrak{A}}^{m+1}$ such that $\lambda \mid U_{\mathfrak{A}}^{m+1} = \psi_\alpha$. We also have $\Lambda_\zeta \mid F^\times = (\det \rho)^3$, so we extend λ to a character Λ_0 of $F^\times U_{F[\alpha]}^1 U_{\mathfrak{A}}^{m+1}$ by deeming that $\Lambda_0 \mid F^\times = \det \rho$. The group $F^\times U_{F[\alpha]}^1 U_{\mathfrak{A}}^{m+1}$ has index 2 in $F[\alpha]^\times U_{\mathfrak{A}}^{m+1}$, so there is a unique character Λ, extending Λ_0 and with the required properties. □

Taking Λ as in Lemma 1, we set $\pi = c\text{-}\mathrm{Ind}_J^G \Lambda$, where $J = F[\alpha]^\times U_{\mathfrak{A}}^{m+1}$. We define

$$\boldsymbol{\pi}_F(\rho) = \pi. \tag{50.3.2}$$

By construction, we have

$$\omega_\pi = \det\rho, \tag{50.3.3}$$

as required for statement (1) of the theorem.

Lemma 2. *Let $\rho \in \mathcal{G}_2^0(F)$ be primitive and minimal over F, and define $\pi = \boldsymbol{\pi}_F(\rho)$, as in (50.3.2). Let χ be a character of $F^\times$ such that $\chi\otimes\rho$ is minimal over F. We then have $\boldsymbol{\pi}_F(\chi\otimes\rho) = \chi\pi$.*

Proof. This is immediate from 15.9 and the construction of π. □

We highlight a feature of the construction:

Proposition. *Let $\rho \in \mathcal{G}_2^{\mathrm{wr}}(F)$ be minimal.*

(1) *There is a ramified simple stratum $(\mathfrak{A}, n, \alpha)$ in A such that $\boldsymbol{\pi}_F(\rho) \in \mathcal{A}_2^0(F;\alpha)$.*
(2) *If E/F is a finite, tamely ramified extension, then $\boldsymbol{\pi}_E(\rho_E) \in \mathcal{A}_2^0(E;\alpha)$.*

Proof. If ρ is imprimitive, all assertions are given by 50.2 Proposition. Suppose, therefore, that ρ is primitive, and choose K/F cubic such that ρ_K is imprimitive. By construction, we have a ramified simple stratum $(\mathfrak{A}, n, \alpha)$ such that $\boldsymbol{\pi}_F(\rho) \in \mathcal{A}_2^0(F;\alpha)$ and $\boldsymbol{\pi}_K(\rho_K) \in \mathcal{A}_2^0(K;\alpha)$. Proposition 50.2 gives (2) in the case where ρ_E is imprimitive. Otherwise, we use 50.2 to get a ramified simple stratum $(\mathfrak{B}, r, \beta)$ in A_E such that $\boldsymbol{\pi}_L(\rho_L) \in \mathcal{A}_2^0(L;\beta)$ for any finite tame extension L/E with ρ_L imprimitive. However, we can assume $K \subset L$ and then 50.2 Proposition (2) gives $\mathcal{A}_2^0(L;\beta) = \mathcal{A}_2^0(L;\alpha)$. It follows from 46.2 Lemma that $\mathcal{A}_2^0(E;\beta) = \mathcal{A}_2^0(E;\alpha)$, as required for (2). □

Remark. We do *not* assert that $\boldsymbol{\pi}_K(\rho_K)$ is the K/F-lift π_K of π (although, in fact, it is: see 52.9 below).

50.4. We start the process of verifying the local constant relation (50.3.1) for the representation $\pi = \boldsymbol{\pi}_F(\rho)$.

Proposition. *Let $\rho \in \mathcal{G}_2^0(F)$ be primitive and minimal over F, and define $\pi = \boldsymbol{\pi}_F(\rho)$ as in (50.3.2). Then $\varepsilon(\pi, s, \psi) = \varepsilon(\rho, s, \psi)$.*

Proof. We use the notation of the construction of π, as in 50.3. By definition, $\varepsilon(\zeta, s, \psi_K) = \varepsilon(\rho_K, s, \psi)$. In particular,

$$e(K|F)\, n(\pi,\psi) = n(\zeta,\psi_K) = n(\rho_K,\psi_K) = e(K|F)\, n(\rho,\psi),$$

so $n(\pi,\psi) = n(\rho,\psi)$. We need only, therefore, check the desired relation at the point $s = \frac{1}{2}$.

If ϕ is a tamely ramified character of $K^\times$, then

$$\begin{aligned}\varepsilon(\phi\otimes\rho_K,s,\psi_K) &= \phi(c(\rho,\psi))^{-1}\,\varepsilon(\rho_K,s,\psi_K),\\ \varepsilon(\phi\zeta,s,\psi_K) &= \phi(\det\alpha)^{-1}\,\varepsilon(\zeta,s,\psi_K),\end{aligned}$$

by, respectively, 48.1 Theorem (3) and 49.1 Gloss. The defining relation (48.1.1), with χ unramified, gives $v_F(c(\rho,\psi)) = n(\rho,\psi)$, and this is odd (50.1 Proposition). From 48.3 Proposition and (49.2.2), we therefore get

$$\varepsilon(\rho_K,s,\psi_K) = \epsilon_{K/F}\,\varepsilon(\rho,s,\psi)^3,$$

while, by definition, $\varepsilon(\rho_K,s,\psi_K) = \varepsilon(\zeta,s,\psi_K)$. The Gauss sum formula 25.5 Corollary and the definition of π give $\varepsilon(\pi,\frac12,\psi)^3 = \epsilon_{K/F}\,\varepsilon(\zeta,\frac12,\psi_K)$, with the result that

$$\varepsilon(\pi,\tfrac12,\psi)^3 = \varepsilon(\rho,\tfrac12,\psi)^3. \tag{50.4.2}$$

We need to extract the cube root. Certainly

$$\varepsilon(\pi,\tfrac12,\psi) \equiv \Lambda(\alpha)^{-1} \equiv \omega_\pi(\det\alpha)^{-1/2} \pmod{\boldsymbol{\mu}_{\mathbb{C}}(2^\infty)}. \tag{50.4.3}$$

Invoking 48.4 Theorem, we get $\varepsilon(\pi,\frac12,\psi)/\varepsilon(\rho,\frac12,\psi) \in \boldsymbol{\mu}_{\mathbb{C}}(2^\infty)$, so (50.4.2) implies $\varepsilon(\pi,\frac12,\psi) = \varepsilon(\rho,\frac12,\psi)$, as required. □

Combining the proposition with 50.3 Lemma 2, we get:

Corollary. *The identity* (50.3.1) *holds for any χ such that $\chi\otimes\rho$ is minimal over F.*

50.5. We now have to check (50.3.1) for characters χ such that $\chi\otimes\rho$ is not minimal over F. Thus χ has level $l > n/2$, $n = n(\pi,\psi) = n(\rho,\psi)$. We choose $\delta \in \mathfrak{p}^{-l}$ such that $\chi(1+x) = \psi(\delta x)$, $x \in \mathfrak{p}^{[l/2]+1}$. Straightforward manipulations and 48.1 Theorem (3) give

$$c_\rho \equiv c_\pi \equiv \delta^2 \pmod{U_F^1}. \tag{50.5.1}$$

The standard Gauss sum formula and 50.1 Proposition (2) give

$$n(\chi\otimes\rho,\psi) = n(\chi\pi,\psi) = 2l.$$

From 48.3 Proposition and (49.2.2), we get

$$\varepsilon(\chi_K\otimes\rho_K,s,\psi_K) = \varepsilon(\chi\otimes\rho,s,\psi)^3,$$

while 49.2 Proposition and (49.2.2) give

$$\varepsilon(\chi_K\pi_K,s,\psi_K) = \varepsilon(\chi\pi,s,\psi)^3.$$

In all, this yields

$$\varepsilon(\chi\otimes\rho,\tfrac{1}{2},\psi)^3 = \varepsilon(\chi\pi,\tfrac{1}{2},\psi)^3,$$

whence

$$\det(\chi\otimes\rho)(\delta)^3\,\varepsilon(\chi\otimes\rho,\tfrac{1}{2},\psi)^3 = \omega_{\chi\pi}(\delta)^3\,\varepsilon(\chi\pi,\tfrac{1}{2},\psi)^3.$$

The left hand side lies in $\boldsymbol{\mu}_{\mathbb{C}}(2^\infty)$. The result is therefore implied by the following analogue of 48.4 Theorem:

Proposition. *The quantity $\omega_{\chi\pi}(\delta)\,\varepsilon(\chi\pi,\frac{1}{2},\psi)$ is a 2-power root of unity.*

Proof. The Gauss sum formula 25.5 Corollary gives

$$\varepsilon(\chi\pi,\tfrac{1}{2},\psi) = q^{-1}\sum_x \chi\Lambda((\alpha{+}\delta)x)^{-1}\psi_A((\alpha{+}\delta)x),$$

where $\psi_A = \psi\circ\mathrm{tr}_A$ and the sum is taken over $x\in U^l_{\mathfrak{A}}/U^{l+1}_{\mathfrak{A}}$. For such x, we have $\Lambda(\alpha x)^{-1}\psi_A(\alpha x) = \Lambda(\alpha)^{-1}\psi_A(\alpha)$, so this expression reduces to

$$\begin{aligned}\varepsilon(\chi\pi,\tfrac{1}{2},\psi)&\\ &= q^{-1}\,\psi_A(\alpha)\;\Lambda(\alpha{+}\delta)^{-1}\,\chi(\det(1{+}\alpha\delta^{-1}))^{-1}\sum_x \chi(\det\delta x)^{-1}\psi_A(\delta x).\end{aligned}$$

Using the notation of 25.6, the identity (25.6.1) yields

$$\sum_x \chi(\det\delta x)^{-1}\psi_A(\delta x) = q^{-l}\,\tau(\chi,\psi)^2,$$

where

$$\tau(\chi,\psi) = \sum_{y\in U_F/U_F^{l+1}} \chi(\delta y)^{-1}\psi(\delta y).$$

That is,

$$\varepsilon(\chi\pi,\tfrac{1}{2},\psi) = \psi_A(\alpha)\,\Lambda(\alpha{+}\delta)^{-1}\,\chi(\det(1{+}\alpha\delta^{-1}))^{-1}\,\tau(\chi,\psi)^2/q^{(l+1)}.$$

In this formula, surely $\psi_A(\alpha)\in\boldsymbol{\mu}_{\mathbb{C}}(2^\infty)$. We have $\delta^{-1}\alpha\in\mathfrak{p}_{F[\alpha]}$, therefore

$$\omega_\pi(\delta)\,\varepsilon(\chi\pi,\tfrac{1}{2},\psi) \equiv \tau(\chi,\psi)^2/q^{l+1} \pmod{\boldsymbol{\mu}_{\mathbb{C}}(2^\infty)},$$

or, equivalently,

$$\omega_{\chi\pi}(\delta)\,\varepsilon(\chi\pi,\tfrac{1}{2},\psi) \equiv \chi(\delta)^2\,\tau(\chi,\psi)^2/q^{l+1} \pmod{\boldsymbol{\mu}_{\mathbb{C}}(2^\infty)}.$$

However, the quantity $\chi(\delta)\,\tau(\chi,\psi)/q^{(l+1)/2}$ is a root of unity, of order a power of 2, by 48.5 Lemma. This completes the proof of the proposition. □

We have also completed the proof of the Existence Theorem 50.3 when the representation ρ is minimal.

50.6. It remains only to treat the case where ρ is primitive but not minimal. We choose a character ϕ of $F^\times$ such that $\rho' = \phi^{-1} \otimes \rho$ is minimal. We set

$$\boldsymbol{\pi}_F(\rho) = \phi\, \boldsymbol{\pi}_F(\rho'). \tag{50.6.1}$$

This is independent of the choice of ϕ, by 50.3 Lemma 2. This definition gives (*cf.* (50.3.3))

$$\omega_{\boldsymbol{\pi}_F(\rho)} = \det \rho, \tag{50.6.2}$$

and also

$$\boldsymbol{\pi}_F(\chi \otimes \rho) = \chi\, \boldsymbol{\pi}_F(\rho), \quad \rho \in \mathcal{G}_2^0(F), \tag{50.6.3}$$

for all characters χ of $F^\times$. The local constant relation (50.3.1) follows from (50.6.3) and the case already done. The proof of the Existence Theorem is complete. □

51. Some Special Cases

In preparation for the final result in the next section, we examine further the Langlands correspondence $\boldsymbol{\pi}_F$ of 50.3 in some special cases.

51.1. The map $\boldsymbol{\pi}_F$ of 50.3 is the only map $\mathcal{G}_2^0(F) \to \mathcal{A}_2^0(F)$ with the property (50.3.1). From an argument identical to the proof of 47.2 Proposition, we obtain:

Proposition. *Let $\phi : F \to F'$ be an isomorphism of fields; we then have*

$$\boldsymbol{\pi}_F(\rho^\phi) = \boldsymbol{\pi}_{F'}(\rho)^\phi,$$

for all $\rho \in \mathcal{G}_2^0(F')$.

51.2. In one case, it is easy to establish a connection between restriction and tame lifting.

Proposition. *Let $\rho \in \mathcal{G}_2^{\mathrm{wr}}(F)$, let K/F be a cyclic cubic extension, and suppose that ρ_K is imprimitive. We then have*

$$\boldsymbol{\pi}_K(\rho_K) = \boldsymbol{\pi}_F(\rho)_K.$$

Proof. Write $\Sigma = \mathrm{Gal}(K/F)$ and $\zeta = \boldsymbol{\pi}_K(\rho_K)$. Thus $\zeta \in \mathcal{A}_2^0(K)^\Sigma$, so $\zeta = \pi_K$, for some $\pi \in \mathcal{A}_2^{\mathrm{wr}}(F)$ (46.6 Proposition). We have $\omega_\zeta = \det \rho_K$, so there is a unique choice of π such that $\omega_\pi = \det \rho$. The character $\varkappa_{K/F}$ is trivial, so 49.2 Proposition gives us

$$\varepsilon(\chi_K \zeta, s, \psi_K) = \varepsilon(\chi\pi, s, \psi)^3,$$

for all characters χ of $F^\times$. On the other hand,

$$\varepsilon(\chi_K\zeta, s, \psi_K) = \varepsilon(\chi_K \otimes \rho_K, s, \psi_K) = \varepsilon(\chi \otimes \rho, \psi)^3$$

by (48.3.1), (49.2.2). So, if $\kappa = \boldsymbol{\pi}_F(\rho)$, we have

$$\varepsilon(\chi\kappa, s, \psi)^3 = \varepsilon(\chi\pi, s, \psi)^3,$$

for all characters χ of $F^\times$. Further, $\omega_\kappa = \omega_\pi$ while $c_{\chi\pi} \equiv c_{\chi\kappa} \equiv c_{\chi\otimes\rho}$ (mod $U_F^{\frac{1}{F}}$). We next observe:

Lemma. *The quantities* $\omega_{\chi\pi}(c_{\chi\pi})^{1/2}\varepsilon(\chi\pi, \frac{1}{2}, \psi)$, $\omega_{\chi\kappa}(c_{\chi\kappa})^{1/2}\varepsilon(\chi\kappa, \frac{1}{2}, \psi)$ *are 2-power roots of unity.*

Proof. The two cases are the same, so we deal only with π. If the representation $\chi\pi$ is not minimal, the assertion is given by 50.5 Proposition. Otherwise, it follows directly from the formula in 25.5 Corollary. □

So, $\varepsilon(\chi\kappa, \frac{1}{2}, \psi) = \varepsilon(\chi\pi, \frac{1}{2}, \psi)$, for all χ. Surely $n(\pi, \psi) = n(\kappa, \psi)$; it follows that $\varepsilon(\chi\kappa, s, \psi) = \varepsilon(\chi\pi, s, \psi)$, and hence $\pi \cong \kappa$, as required. □

51.3. In many cases, we can now complete the picture:

Proposition. *Suppose that the field F satisfies the condition $q \equiv 1 \pmod 3$. The map $\boldsymbol{\pi}_F : \mathcal{G}_2^0(F) \to \mathcal{A}_2^0(F)$ is then bijective.*

Proof. The effect of the hypothesis is that any cubic extension K/F is *cyclic*.

We first show that $\boldsymbol{\pi}_F$ is injective. Let $\rho, \rho' \in \mathcal{G}_2^0(F)$ satisfy $\boldsymbol{\pi}_F(\rho) = \boldsymbol{\pi}_F(\rho')$. If ρ and ρ' are both imprimitive, 44.1 gives $\rho = \rho'$.

Suppose next that ρ, ρ' are both *primitive*. Set $\pi = \boldsymbol{\pi}_F(\rho) = \boldsymbol{\pi}_F(\rho')$. In particular, we have $\det\rho = \det\rho' = \omega_\pi$. Take cubic extensions K/F, K'/F such that ρ_K and $\rho'_{K'}$ are both imprimitive.

If $K = K'$, 51.2 Proposition gives $\boldsymbol{\pi}_K(\rho_K) = \pi_K = \boldsymbol{\pi}_K(\rho'_K)$. Therefore $\rho_K \cong \rho'_K$, whence $\rho' \cong \chi \otimes \rho$, for some χ with $\chi^3 = 1$ (42.1 Proposition (3)). The relation $\det\rho = \det\rho'$ implies $\chi = 1$.

If $K \neq K'$, set $L = KK'$, so that L/F is abelian of degree 9. Applying 51.2 Proposition twice, we get

$$\boldsymbol{\pi}_L(\rho_L) = \pi_L = \boldsymbol{\pi}_L(\rho'_L),$$

and so $\rho_L \cong \rho'_L$. Surely $\det\rho_K = \det\rho'_K$, so we deduce $\rho_K \cong \rho'_K$ and then $\rho \cong \rho'$, using 42.1 again.

If ρ is primitive while ρ' is imprimitive, the same argument again leads to the conclusion $\rho = \rho'$. Thus $\boldsymbol{\pi}_F$ is injective.

We show that $\boldsymbol{\pi}_F$ is surjective. Let $\pi \in \mathcal{A}_2^0(F)$. If π is ordinary then, by definition, it lies in $\boldsymbol{\pi}_F(\mathcal{G}_2^{\mathrm{im}}(F))$. We therefore assume π is exceptional.

By 45.2 Theorem, there is a cubic extension K/F such that π_K is ordinary: we have $\pi_K = \boldsymbol{\pi}_K(\tau)$, for some $\tau \in \mathcal{G}_2^{\mathrm{wr}}(K)$. The representation τ is fixed by $\mathrm{Gal}(K/F)$, hence of the form $\tau = \rho_K$, for some $\rho \in \mathcal{G}_2^{\mathrm{wr}}(F)$. We may choose ρ such that $\det \rho = \omega_\pi$. Proposition 51.2 now implies $\pi = \boldsymbol{\pi}_F(\rho)$. □

51.4. We return to the general case $q \equiv \pm 1 \pmod 3$.

Let $(\mathfrak{A}, n, \alpha)$ be a ramified simple stratum in $A = \mathrm{M}_2(F)$; we define

$$\mathcal{G}_2^0(F)_\alpha = \{\rho \in \mathcal{G}_2^0(F) : \boldsymbol{\pi}_F(\rho) \in \mathcal{A}_2^0(F;\alpha)\}.$$

Proposition. *Let $(\mathfrak{A}, n, \alpha)$ be an ordinary, ramified simple stratum in $A = \mathrm{M}_2(F)$. The set $\mathcal{G}_2^0(F)_\alpha$ contains no primitive representation. In particular, the map $\boldsymbol{\pi}_F$ induces a bijection*

$$\mathcal{G}_2^0(F)_\alpha \xrightarrow{\approx} \mathcal{A}_2^0(F;\alpha). \tag{51.4.1}$$

Proof. The map $\boldsymbol{\pi}_F$ induces a bijection $\mathcal{G}_2^0(F)_\alpha \cap \mathcal{G}_2^{\mathrm{im}}(F) \to \mathcal{A}_2^0(F;\alpha)$, by 44.3 Theorem and 43.2 Theorem. So, $\boldsymbol{\pi}_F : \mathcal{G}_2^0(F)_\alpha \to \mathcal{A}_2^0(F;\alpha)$ is bijective if and only if $\mathcal{G}_2^0(F)_\alpha$ contains no primitive representation.

If $q \equiv 1 \pmod 3$, the bijectivity property is given by 51.3 Proposition, so we take $q \equiv -1 \pmod 3$. Let E/F be the unramified quadratic extension and $\Sigma = \mathrm{Gal}(E/F)$. Thus $q_E = q^2 \equiv 1 \pmod 3$. If $\rho \in \mathcal{G}_2^0(F)_\alpha$ is primitive, then ρ_E is primitive (42.2 Theorem (2)). It follows from 50.2 Proposition and 46.2 Lemma that $\rho_E \in \mathcal{G}_2^0(E)_\alpha$, contradicting what we have just said. □

51.5. We highlight a consequence of 51.4.

Let $\pi \in \mathcal{A}_2^0(F)$; by analogy with the group $\mathfrak{T}(\rho)$ defined in 41.3, let $\mathfrak{T}(\pi)$ denote the group of characters ϕ of $F^\times$ such that $\phi\pi \cong \pi$.

Proposition.

(1) *Let $\rho \in \mathcal{G}_2^0(F)$ be imprimitive and set $\pi = \boldsymbol{\pi}_F(\rho)$. We then have $\mathfrak{T}(\pi) = \mathfrak{T}(\rho)$.*

(2) *Let $(\mathfrak{A}, n, \alpha)$ be an exceptional, ramified simple stratum in $\mathrm{M}_2(F)$. Every $\pi \in \mathcal{A}_2^0(F;\alpha)$ then satisfies $\mathfrak{T}(\pi) = 1$.*

Proof. For any $\rho \in \mathcal{G}_2^0(F)$, part (2) of the Existence Theorem (50.3) implies that $\mathfrak{T}(\rho) \subset \mathfrak{T}(\boldsymbol{\pi}_F(\rho))$. The injectivity of $\boldsymbol{\pi}_F$ on imprimitive representations yields (1). Likewise, if $q \equiv 1 \pmod 3$, the bijectivity of $\boldsymbol{\pi}_F$ (51.3) gives $\mathfrak{T}(\boldsymbol{\pi}_F(\rho)) = \mathfrak{T}(\rho)$ in all cases. In particular, $\pi \in \mathcal{A}_2^0(F)$ is ordinary if and only if $\mathfrak{T}(\pi) \neq 1$.

We therefore pass to the case $q \equiv -1 \pmod 3$, and take $\pi \in \mathcal{A}_2^0(F;\alpha)$. Suppose, for a contradiction, that $\chi \in \mathfrak{T}(\pi)$, $\chi \neq 1$. The representation π is totally ramified, so χ is not unramified. Let E/F be the unramified quadratic extension. Thus $\chi_E \neq 1$ and $\chi_E \in \mathfrak{T}(\pi_E)$. However, $\pi_E \in \mathcal{A}_2^0(E;\alpha)$ and the polynomial $C_\alpha(X)$ (as in 45.2) remains irreducible over E. Thus π_E is exceptional, implying $\chi_E = 1$ by the first case. This contradiction completes the proof of (2). □

We can re-phrase the proposition:

Corollary. *An irreducible cuspidal representation of G is ordinary if and only if $\mathfrak{T}(\pi) \neq 1$.*

Or, in the language of §39, π is a Weil representation $\pi_{E/F}(\Theta)$ if and only if $\varkappa_{E/F} \in \mathfrak{T}(\pi)$.

51.6. Although we will not use it in the proofs, there is one further special case which is worthy of mention.

Let $(\mathfrak{A}, n, \alpha)$, $n = 2m+1 \geqslant 1$, be a ramified simple stratum in $A = \mathrm{M}_2(F)$. Write $d(F[\alpha]|F) = d_\alpha$.

Proposition. *Suppose that $(\mathfrak{A}, n, \alpha)$ is exceptional and $n = 3d_\alpha$. The splitting field of $C_\alpha(X)$ is then unramified over F.*

Proof. The hypothesis $n = 3d_\alpha$ implies that $\mathrm{tr}(\alpha) = \mathrm{Tr}_{F[\alpha]/F}(\alpha)$ has valuation $-d_\alpha$. Let ϖ be a prime element of F. The polynomial

$$c_\alpha(Y) = \varpi^{3d_\alpha}\, C_\alpha(\varpi^{-d_\alpha} Y) = Y^3 - aY^2 + b$$

has coefficients $a, b \in U_F$ and is irreducible over F. Its image in $\boldsymbol{k}[Y]$ is also irreducible, by Hensel's Lemma, whence the result follows. □

52. Octahedral Representations

In this section, we prove:

52.1 Bijectivity Theorem. *The Langlands correspondence*

$$\pi_F : \mathcal{G}_2^0(F) \longrightarrow \mathcal{A}_2^0(F)$$

is bijective.

After the discussion of imprimitive representations and the special cases of §51, all the remaining difficulties centre on primitive representations $\rho \in \mathcal{G}_2^0(F)$ which become imprimitive only over a *non-cyclic* cubic extension K/F. In the terminology of 42.3 Comment, these are the *octahedral* representations of $\mathcal{W}_F$.

52.2. Before embarking on the proof, we need to investigate some rather delicate properties of wildly ramified cuspidal types. We take a ramified simple stratum $(\mathfrak{A}, n, \alpha)$ in A, with $n = 2m+1 \geqslant 1$, and impose a "metric" on the set $\mathcal{A}_2^0(F;\alpha)$.

Set $J = F[\alpha]^\times U_{\mathfrak{A}}^{m+1}$. Let $\pi_1, \pi_2 \in \mathcal{A}_2^0(F;\alpha)$, so that $\pi_i = c\text{-Ind}_J^G \Lambda_i$, for a unique $\Lambda_i \in C(\psi_\alpha, \mathfrak{A})$. Let k denote the level of the character $\Lambda_1^{-1}\Lambda_2 \mid U_{F[\alpha]}^1$. We put

$$\|\pi_1 - \pi_2\| = k.$$

Thus $0 \leqslant \|\pi_1 - \pi_2\| \leqslant m$. The condition $\|\pi_1 - \pi_2\| = 0$ is equivalent to the existence of a tamely ramified character χ of $F^\times$ such that $\pi_2 = \chi\pi_1$.

One can determine this "distance" between representations in terms of local constants:

Lemma. *Let $(\mathfrak{A}, n, \alpha)$, $n \geqslant 1$, be a ramified simple stratum. Let $\pi, \pi' \in \mathcal{A}_2^0(F;\alpha)$ satisfy*

$$\omega_\pi = \omega_{\pi'} \quad \textit{and} \quad \|\pi - \pi'\| > 0.$$

The integer $\nu = \|\pi - \pi'\|$ is then odd, and $(n-\nu)/2$ is the minimum of the levels of characters χ of $F^\times$ such that

$$\varepsilon(\pi, \tfrac{1}{2}, \psi)/\varepsilon(\chi\pi, \tfrac{1}{2}, \psi) \neq \varepsilon(\pi', \tfrac{1}{2}, \psi)/\varepsilon(\chi\pi', \tfrac{1}{2}, \psi). \tag{52.2.1}$$

Proof. For any $j \geqslant 1$, we have $U_{F[\alpha]}^{2j} = U_F^j U_{F[\alpha]}^{2j+1}$; the condition $\omega_\pi = \omega_{\pi'}$ so implies that ν is odd.

Write $n = 2m+1$, $J = F[\alpha]^\times U_{\mathfrak{A}}^{m+1}$. Let $\Lambda, \Lambda' \in C(\psi_\alpha, \mathfrak{A})$ satisfy $\pi = c\text{-Ind}_J^G \Lambda$, $\pi' = c\text{-Ind}_J^G \Lambda'$. Take χ of level $k \leqslant m$, and let $\delta_\chi \in \mathfrak{p}^{-k}$ satisfy $\chi(\det 1+x) = \psi_{\delta_\chi}(1+x)$, $1+x \in U_{\mathfrak{A}}^{m+1}$. We then have (25.5 Corollary)

$$\varepsilon(\pi, \tfrac{1}{2}, \psi)/\varepsilon(\chi\pi, \tfrac{1}{2}, \psi) = \chi(\det(\alpha+\delta_\chi))\, \Lambda(1+\alpha^{-1}\delta_\chi)\, \psi_A(-\delta_\chi),$$

and similarly for π' in place of π. The inequality (52.2.1) is therefore equivalent to

$$\Lambda(1+\alpha^{-1}\delta_\chi) \neq \Lambda'(1+\alpha^{-1}\delta_\chi),$$

and the result follows. □

52.3. There is a complementary property, concerning the stability of a set $\mathcal{A}_2^0(F;\alpha)$ under twisting by characters of comparatively small level. This gives an alternative description of the elements of $\mathcal{A}_2^0(F;\alpha)$ sufficiently close to a fixed representation π.

Proposition. *Let $(\mathfrak{A}, n, \alpha)$, $n = 2m+1 \geqslant 1$, be a ramified simple stratum. Put $d_\alpha = d(F[\alpha]|F)$, suppose that $n \leqslant 3d_\alpha$, and write $c = [n/3]$. Let $\pi \in \mathcal{A}_2^0(F;\alpha)$; then $\chi\pi \in \mathcal{A}_2^0(F;\alpha)$ for every character χ of $F^\times$ of level $\leqslant c$.*

Proof. Set $\operatorname{rad}\mathfrak{A} = \mathfrak{P}$. Let χ have level $k \leqslant c$; if $k = 0$, there is nothing to prove, so we assume $k \geqslant 1$ (and hence $n \geqslant 3$). We choose $\delta \in \mathfrak{p}^{-k}$ such that $\chi(\det 1{+}x) = \psi_\delta(x)$, $x \in \mathfrak{P}^{m+1}$. Let $\Lambda \in C(\psi_\alpha, \mathfrak{A})$ occur in π. We consider the character

$$\chi\Lambda : x \longmapsto \chi(\det x)\, \Lambda(x), \quad x \in J.$$

The character $\chi\Lambda$ lies in $C(\psi_{\alpha+\delta}, \mathfrak{A})$ (15.9) and occurs in $\chi\pi$.

Lemma. *The cosets $\alpha U_{\mathfrak{A}}^{m+1}$, $(\alpha{+}\delta)U_{\mathfrak{A}}^{m+1}$ are conjugate in $\mathcal{K}_{\mathfrak{A}}$.*

Proof. By 44.2 Lemma, we have to show that $\det 1{+}\alpha^{-1}\delta \in U_F^{[m/2]+1}$ and $\upsilon_F(2\delta) \geqslant -[m/2]$.

Starting with the trace condition,

$$\upsilon_F(2\delta) \geqslant \frac{d_\alpha}{2} - k \geqslant \frac{d_\alpha}{2} - \left[\frac{n}{3}\right] \geqslant \frac{n}{6} - \left[\frac{n}{3}\right].$$

This is $\geqslant -[m/2]$ in all cases except $n = 3$. If $n = 3$, we have $k \leqslant 1$ and $\upsilon_F(2\delta) \geqslant 0 = -[m/2]$.

On the other hand, $1{+}\alpha^{-1}\delta \in U_{F[\alpha]}^{n-2k}$. We have $n{-}2k \geqslant n/3$ and $n/3 \leqslant d_\alpha$; therefore

$$\det 1{+}\alpha^{-1}\delta = \mathrm{N}_{F[\alpha]/F}(1{+}\alpha^{-1}\delta) \in U_F^{[n/3]}.$$

The lemma then follows, provided $[n/3] \leqslant [m/2]{+}1$. This holds for all odd integers $n \geqslant 3$ except $n = 5$. When $n = 5$, both $n{-}2k$ and d_α are $\geqslant 2$, and $\mathrm{N}_{F[\alpha]/F}(U_{F[\alpha]}^2) \subset U_F^2 = U_F^{[m/2]+1}$, as required. □

So, we choose $g \in \mathcal{K}_{\mathfrak{A}}$ such that $g\alpha g^{-1} \equiv \alpha{+}\delta \pmod{U_{\mathfrak{A}}^{m+1}}$: in terms of characters of $U_{\mathfrak{A}}^{m+1}$, this means $\psi_\alpha = \psi_{\alpha+\delta}^g$. The $\mathcal{K}_{\mathfrak{A}}$-normalizer of $\psi_{\alpha+\delta} \mid U_{\mathfrak{A}}^{m+1}$ is again J, so g normalizes J. It follows that the character $\Lambda' = (\chi\Lambda)^g$ of J lies in $C(\psi_\alpha, \mathfrak{A})$. It is contained in $\chi\pi$, so $\chi\pi \in \boldsymbol{\mathcal{A}}_2^0(F;\alpha)$, as required. □

Continuing in the same vein, we now sharpen our hypotheses and prove:

Corollary. *Let $(\mathfrak{A}, n, \alpha)$ be exceptional. Let $\pi_1, \pi_2 \in \boldsymbol{\mathcal{A}}_2^0(F;\alpha)$ satisfy*

$$\|\pi_1{-}\pi_2\| \leqslant c = [n/3].$$

There then exists a character χ of $F^\times$, of level $\leqslant c$, such that $\pi_2 = \chi\pi_1$.

Proof. Since $(\mathfrak{A}, n, \alpha)$ is exceptional, we have $n \leqslant 3d_\alpha$. We carry on from the end of the proof of the proposition, and push the calculation a little further. The element $g \in \mathcal{K}_{\mathfrak{A}}$ from the proof of the proposition, satisfying $g\alpha g^{-1} \equiv \alpha{+}\delta \pmod{\mathfrak{P}^{-m}}$, is only determined modulo $F[\alpha]^\times$. Since $\mathcal{K}_{\mathfrak{A}} = F[\alpha]^\times U_{\mathfrak{A}}$, we may as well take $g \in U_{\mathfrak{A}}$. It satisfies

$$\begin{aligned} g\alpha g^{-1} &\equiv \alpha \pmod{\mathfrak{P}^{-2k}}, \quad \text{or} \\ \alpha^{-1} g\alpha &\equiv g \pmod{\mathfrak{P}^{n-2k}}. \end{aligned}$$

By 16.2 Lemma, we can take $g = 1{+}x$, $x \in \mathfrak{P}^{n-2k}$.

We consider the restriction of the character $\Lambda' = (\chi\Lambda)^{1+x}$ to the group $U^{c+1}_{F[\alpha]}$. Taking $y \in \mathfrak{p}^{c+1}_{F[\alpha]}$, we get

$$(1{+}x)(1{+}y)(1{+}x)^{-1} \equiv 1{+}y{+}xy{-}yx \pmod{\mathfrak{P}^{n+1}},$$

giving

$$(1{+}x)(1{+}y)(1{+}x)^{-1}(1{+}y)^{-1} \equiv 1 + xy{-}yx \pmod{\mathfrak{P}^{n+1}}.$$

Certainly $xy{-}yx \in \mathfrak{P}^{m+1}$ so, in the obvious commutator notation,

$$\chi\Lambda[1{+}x, 1{+}y] = \psi_{\alpha+\delta}(1{+}xy{-}yx) = 1.$$

Since $k \leqslant c \leqslant d_\alpha$, the character $\chi \circ \det \mid U^1_{F[\alpha]}$ has level $\leqslant k$. Therefore

$$\Lambda' \mid U^{c+1}_{F[\alpha]} = (\chi\Lambda)^{1+x} \mid U^{c+1}_{F[\alpha]} = \chi\Lambda \mid U^{c+1}_{F[\alpha]} = \Lambda \mid U^{c+1}_{F[\alpha]}.$$

Setting $\pi' = \chi\pi$, we therefore have $\|\pi{-}\pi'\| \leqslant c$.

Let χ range over the characters of $F^\times$ of level $\leqslant c$; since $(\mathfrak{A}, n, \alpha)$ is exceptional, the representations $\chi\pi$ are distinct (51.5 Proposition). Counting, the proposition shows that they account for all $\pi' \in \boldsymbol{\mathcal{A}}^0_2(F;\alpha)$ with $\|\pi'{-}\pi\| \leqslant c$. □

52.4. We return to the Bijectivity Theorem to make a preliminary reduction. We know (Tame Langlands Correspondence, 34.4) that $\boldsymbol{\pi}_F$ induces a bijection $\mathcal{G}^{\mathrm{nr}}_2(F) \to \boldsymbol{\mathcal{A}}^{\mathrm{nr}}_2(F)$. We must therefore show that $\boldsymbol{\pi}_F$ restricts to a bijection $\mathcal{G}^{\mathrm{wr}}_2(F) \to \boldsymbol{\mathcal{A}}^{\mathrm{wr}}_2(F)$.

Let $k \geqslant 0$ be an integer; define $\mathcal{G}^{\mathrm{wr}}_2(F)_k$ to be the set of classes $\rho \in \mathcal{G}^{\mathrm{wr}}_2(F)$ for which $n(\rho,\psi) \leqslant k$, where $\psi \in \widehat{F}$ has level one. Define $\boldsymbol{\mathcal{A}}^{\mathrm{wr}}_2(F)_k$ in exactly the same way. Thus $\boldsymbol{\pi}_F\big(\mathcal{G}^{\mathrm{wr}}_2(F)_k\big) \subset \boldsymbol{\mathcal{A}}^{\mathrm{wr}}_2(F)_k$. We write $\boldsymbol{\pi}_{F,k}$ for the map $\mathcal{G}^{\mathrm{wr}}_2(F)_k \to \boldsymbol{\mathcal{A}}^{\mathrm{wr}}_2(F)_k$ induced by $\boldsymbol{\pi}_F$.

Proposition. *The following are equivalent:*

(1) *the map* $\boldsymbol{\pi}_F : \mathcal{G}^{\mathrm{wr}}_2(F) \to \boldsymbol{\mathcal{A}}^{\mathrm{wr}}_2(F)$ *is bijective;*
(2) *the map* $\boldsymbol{\pi}_{F,k}$ *is bijective for all* $k \geqslant 0$*;*
(3) *the map* $\boldsymbol{\pi}_{F,k}$ *is injective for all* $k \geqslant 0$*;*
(4) *the map* $\boldsymbol{\pi}_F : \mathcal{G}^{\mathrm{wr}}_2(F) \to \boldsymbol{\mathcal{A}}^{\mathrm{wr}}_2(F)$ *is injective.*

Proof. Clearly (1) $\Leftrightarrow$ (2) $\Rightarrow$ (3) $\Leftrightarrow$ (4). We show that (3) $\Rightarrow$ (2).

We also observe that all assertions hold when $q \equiv 1 \pmod 3$, by 51.3 Proposition.

In general, the set $\Gamma_1\backslash\mathcal{A}_2^{\mathrm{wr}}(F)_k$ is finite where, we recall, Γ_1 denotes the group of characters of $F^\times/U_F^1$. It follows that, when $q \equiv 1 \pmod 3$, we have

$$\left|\Gamma_1\backslash\mathcal{G}_2^{\mathrm{wr}}(F)_k\right| = \left|\Gamma_1\backslash\mathcal{A}_2^{\mathrm{wr}}(F)_k\right| < \infty.$$

Suppose therefore that $q \equiv -1 \pmod 3$. Let E/F be the unramified quadratic extension and put $\Sigma = \mathrm{Gal}(E/F)$. The canonical maps

$$\begin{aligned} \mathcal{G}_2^{\mathrm{wr}}(F) &\longrightarrow \mathcal{G}_2^{\mathrm{wr}}(E), & \mathcal{A}_2^{\mathrm{wr}}(F) &\longrightarrow \mathcal{A}_2^{\mathrm{wr}}(E), \\ \rho &\longmapsto \rho_E, & \pi &\longmapsto \pi_E, \end{aligned}$$

take $\mathcal{G}_2^{\mathrm{wr}}(F)_k$ to $\mathcal{G}_2^{\mathrm{wr}}(E)_k$ and $\mathcal{A}_2^{\mathrm{wr}}(F)_k$ to $\mathcal{A}_2^{\mathrm{wr}}(E)_k$. Thus, writing $\Gamma_1(E)$ for the group of characters of $E^\times/U_E^1$, we have canonical maps

$$\begin{aligned} \Gamma_1\backslash\mathcal{G}_2^{\mathrm{wr}}(F)_k &\longrightarrow \left(\Gamma_1(E)\backslash\mathcal{G}_2^{\mathrm{wr}}(E)_k\right)^\Sigma, \\ \Gamma_1\backslash\mathcal{A}_2^{\mathrm{wr}}(F)_k &\longrightarrow \left(\Gamma_1(E)\backslash\mathcal{A}_2^{\mathrm{wr}}(E)_k\right)^\Sigma. \end{aligned} \tag{52.4.1}$$

Lemma. *The maps* (52.4.1) *are both bijective.*

Proof. In the first case, representations $\rho, \rho' \in \mathcal{G}_2^{\mathrm{wr}}(F)$ lie in the same Γ_1-orbit if and only if $\rho \mid \mathcal{P}_F \cong \rho' \mid \mathcal{P}_F$. The same applies over E and, since E/F is unramified, we have $\mathcal{P}_E = \mathcal{P}_F$. The first map is therefore injective.

Let $\sigma \in \Sigma$, $\sigma \neq 1$. If $\tau \in \mathcal{G}_2^{\mathrm{wr}}(E)$ defines an orbit

$$\Gamma_1(E) \otimes \tau \in \left(\Gamma_1(E)\backslash\mathcal{G}_2^{\mathrm{wr}}(E)\right)^\Sigma,$$

then $\tau^\sigma \cong \chi \otimes \tau$, for some $\chi \in \Gamma_1(E)$. It follows that $\chi\chi^\sigma = 1$. This implies $\chi \mid \mu_E$ to be of the form $\lambda^{-1}\lambda^\sigma$, for some character $\lambda \in \Gamma_1(E)$. Replacing τ by $\lambda^{-1}\otimes\tau$, we can assume that χ is unramified. It follows that $\chi^2 = 1$. Suppose for a contradiction that $\chi \neq 1$. Thus $\chi = \phi_E$, where ϕ is an unramified character of $F^\times$ of order 4. The representation $\rho = \mathrm{Ind}_{E/F}\,\tau$ is irreducible and satisfies $\rho \cong \phi \otimes \rho$. Thus ρ is induced from the quartic unramified extension of F and τ cannot be totally ramified. Therefore $\chi = 1$ and $\tau \in \mathcal{G}_2^{\mathrm{wr}}(E)^\Sigma$. Therefore $\tau = \rho_E$, for some $\rho \in \mathcal{G}_2^{\mathrm{wr}}(F)$ and the first map in (52.4.1), we deduce, is surjective.

Let $\pi, \pi' \in \mathcal{A}_2^{\mathrm{wr}}(F)$ satisfy $\pi'_E = \chi\pi_E$, for some $\chi \in \Gamma_1(E)$. This implies $\chi^\sigma\pi_E = \chi\pi_E$ and, since π_E is totally ramified, it follows that $\chi = \chi^\sigma$. Thus $\chi = \phi_E$, for some $\phi \in \Gamma_1$. From 46.6 Proposition, we deduce that $\pi' = \phi'\pi$, for some $\phi' \in \Gamma_1$. The second map of (52.4.1) is therefore injective.

Now let $\tau \in \mathcal{A}_2^{\mathrm{wr}}(E)$ satisfy $\tau^\sigma = \chi\tau$, for some $\chi \in \Gamma_1(E)$. We deduce that $\chi\chi^\sigma = 1$. Exactly as in the preceding paragraph, we reduce to the case where χ is unramified and $\chi^2 = 1$. An argument identical to the proof of 46.6 Proposition gives a representation $\pi \in \mathcal{A}_2^{\mathrm{wr}}(F)$ such that $\pi_E = \phi\tau$, where ϕ is unramified and $\phi^2 = 1$. Therefore the orbit $\Gamma_1(E) \cdot \tau$ meets the image of $\mathcal{A}_2^{\mathrm{wr}}(F)$, as required to show that the second map (52.4.1) is surjective. □

We thus have canonical bijections

$$\begin{aligned} \Gamma_1\backslash\mathcal{G}_2^{\mathrm{wr}}(F)_k &\longleftrightarrow \Gamma_1(E)\backslash\mathcal{G}_2^{\mathrm{wr}}(E)_k^{\Sigma}, \\ \Gamma_1\backslash\mathcal{A}_2^{\mathrm{wr}}(F)_k &\longleftrightarrow \Gamma_1(E)\backslash\mathcal{A}_2^{\mathrm{wr}}(E)_k^{\Sigma}, \end{aligned}$$

while $\boldsymbol{\pi}_{E,k}$ induces a bijection

$$\Gamma_1(E)\backslash\mathcal{G}_2^{\mathrm{wr}}(E)_k^{\Sigma} \longleftrightarrow \Gamma_1(E)\backslash\mathcal{A}_2^{\mathrm{wr}}(E)_k^{\Sigma}.$$

If $\boldsymbol{\pi}_{F,k}$ is injective, it induces an injection $\Gamma_1\backslash\mathcal{G}_2^{\mathrm{wr}}(F)_k \to \Gamma_1\backslash\mathcal{A}_2^{\mathrm{wr}}(F)_k$. This is necessarily bijective, whence $\boldsymbol{\pi}_{F,k}$ is bijective. □

52.5. We now prove the Bijectivity Theorem 52.1. We take $\rho, \rho' \in \mathcal{G}_2^{\mathrm{wr}}(F)$, and assume that $\boldsymbol{\pi}_F(\rho) = \boldsymbol{\pi}_F(\rho')$. We show that $\rho = \rho'$: in the light of 52.4 Proposition, this will be enough.

Our assumption that $\boldsymbol{\pi}_F(\rho) = \boldsymbol{\pi}_F(\rho')$ carries the implication

$$\varepsilon(\chi \otimes \rho, s, \psi) = \varepsilon(\chi \otimes \rho', s, \psi), \tag{52.5.1}$$

for all χ. We deduce that, in particular,

$$\det \rho = \det \rho' \quad \text{and} \quad n(\rho, \psi) = n(\rho', \psi).$$

We can further assume that ρ, ρ' are both minimal.

The minimality hypothesis implies that there is a ramified simple stratum $(\mathfrak{A}, n, \alpha)$ such that $\boldsymbol{\pi}_F(\rho) = \boldsymbol{\pi}_F(\rho') \in \mathcal{A}_2^0(F;\alpha)$. If either of ρ, ρ' is imprimitive, the stratum $(\mathfrak{A}, n, \alpha)$ is ordinary and 51.4 Proposition implies both imprimitive. It follows that $\rho = \rho'$ in this case. We therefore assume that ρ, ρ' are both *primitive*. The stratum $(\mathfrak{A}, n, \alpha)$ is therefore exceptional.

Let K/F be a cubic extension such that ρ_K is imprimitive. If $\rho'_K = \rho_K$, then $\rho' \cong \chi\otimes\rho$, for a tame character χ of $F^\times$ (42.1); our assumption $\boldsymbol{\pi}_F(\rho') = \boldsymbol{\pi}_F(\rho)$ implies $\chi = 1$ (51.5 Proposition) and $\rho = \rho'$.

We therefore assume $\rho_K \neq \rho'_K$. However, by Propositions 50.2 and 50.3, we have $\boldsymbol{\pi}_K(\rho_K), \boldsymbol{\pi}_K(\rho'_K) \in \mathcal{A}_2^0(K;\alpha)$. As ρ_K is imprimitive, the stratum $(\mathfrak{A}_K, e(K|F)n, \alpha)$ in $\mathrm{M}_2(K)$ is ordinary, and so ρ'_K is imprimitive (51.4 Proposition).

If K/F is cyclic, 51.2 Proposition implies

$$\boldsymbol{\pi}_K(\rho_K) = \boldsymbol{\pi}_F(\rho)_K = \boldsymbol{\pi}_F(\rho')_K = \boldsymbol{\pi}_K(\rho'_K),$$

whence $\rho_K = \rho'_K$, contrary to hypothesis.

52.6. We are left, therefore, with the case where K/F is not cyclic. Since the stratum $(\mathfrak{A}, n, \alpha)$ is exceptional, we have $n \leqslant 3d(F[\alpha]|F)$, by 45.2 Theorem.

Set $n = 2m{+}1$. Let L/F be the normal closure of K/F, and put $\Sigma = \mathrm{Gal}(L/F)$. Set $\tau = \boldsymbol{\pi}_L(\rho_L)$, $\tau' = \boldsymbol{\pi}_L(\rho'_L)$. The stratum $(\mathfrak{A}_L, 3n, \alpha)$ in $\mathrm{M}_2(L)$ is ordinary and $\tau, \tau' \in \boldsymbol{\mathcal{A}}^0_2(L;\alpha)^\Sigma$. Moreover, $\omega_\tau = \omega_{\tau'}$. Define

$$\nu_L = \|\tau{-}\tau'\|.$$

This satisfies $0 \leqslant \nu_L \leqslant 3m{+}2$. If $\nu_L = 0$, there is a tamely ramified character ϕ of $L^\times$ such that $\tau' = \phi\tau$. We deduce that $\rho'_L \cong \phi \otimes \rho_L$. The relation $\det \rho'_L = \det \rho_L$ implies $\phi^2 = 1$ and hence ϕ is unramified. Therefore $\rho' \cong \chi{\otimes}\rho$, for a character χ of $F^\times$ of the form $\chi = \chi'\chi''$, where χ' is unramified (with $\chi'_L = \phi$) and χ'' is trivial on norms from L. In particular, χ'' is unramified; thus χ is also unramified and the relation $\boldsymbol{\pi}_F(\rho') = \boldsymbol{\pi}_F(\rho)$ implies $\chi = 1$.

We therefore assume $\nu_L > 0$. There exist representations $\pi, \pi' \in \boldsymbol{\mathcal{A}}^0_2(F;\alpha)$ such that $\tau = \pi_L$, $\tau' = \pi'_L$ (46.6 Corollary). Set $\nu_F = \|\pi{-}\pi'\|$; we have

$$\nu_L = 3\nu_F \tag{52.6.1}$$

from the definition of tame lifting. If $\nu_L \leqslant n$, then $\nu_F \leqslant [n/3]$, so there is a character χ of $F^\times$ such that $\pi' = \chi\pi$ (52.3 Corollary). This implies $\tau' = \chi_L\tau$ and, since $(\mathfrak{A}_L, 3n, \alpha)$ is ordinary, $\rho'_L = \chi_L \otimes \rho_L$ by 51.4 Proposition. Thus there is a character χ' of $F^\times$ such that $\rho' = \chi' \otimes \rho$ (42.1 Proposition (3)). The relation $\boldsymbol{\pi}_F(\rho') = \chi'\, \boldsymbol{\pi}_F(\rho) = \boldsymbol{\pi}_F(\rho)$ yields $\chi' = 1$ and $\rho' = \rho$.

This leaves only the possibility $\nu_L > n$. We consider the imprimitive representations ρ_K, ρ'_K. There are admissible wild triples $(E/K, 3n, \beta)$, $(E'/K, 3n, \beta')$ such that $\rho_K \in \mathfrak{G}^0_2(K;\beta)$ and $\rho'_K \in \mathfrak{G}^0_2(K;\beta')$. By definition, we have $\upsilon_E(\beta) = -3n$ while $n = d(E|K) = d(EL|L)$ (44.11 Proposition). Likewise for β'.

By definition, there is a character ξ of $E^\times$, of level $2n$, such that $\rho_K = \mathrm{Ind}_{E/K}\,\xi$. Thus $\rho_L = \mathrm{Ind}_{EL/L}\,\xi_L$, where $\xi_L = \xi \circ \mathrm{N}_{EL/E}$.

By 52.2 Lemma, there is a character χ of $L^\times$, of level $l = (3n{-}\nu_L)/2 < n$ such that

$$\varepsilon(\tau, \tfrac{1}{2}, \psi_L)/\varepsilon(\chi\tau, \tfrac{1}{2}, \psi_L) \neq \varepsilon(\tau', \tfrac{1}{2}, \psi_L)/\varepsilon(\chi\tau', \tfrac{1}{2}, \psi_L).$$

The level of χ is $l < n = d(EL|L)$, so χ_{EL} has level l. Therefore

$$\frac{\varepsilon(\tau, \frac{1}{2}, \psi_L)}{\varepsilon(\chi\tau, \frac{1}{2}, \psi_L)} = \frac{\varepsilon(\rho_L, \frac{1}{2}, \psi_L)}{\varepsilon(\chi \otimes \rho_L, \frac{1}{2}, \psi_L)} = \frac{\varepsilon(\xi_L, \frac{1}{2}, \psi_{EL})}{\varepsilon(\chi_{EL}\xi_L, \frac{1}{2}, \psi_{EL})}$$

$$= \chi_{EL}(\beta) = \chi(\mathrm{N}_{E/K}(\beta)),$$

using 23.6 Proposition. Likewise,

$$\frac{\varepsilon(\rho'_L, \frac{1}{2}, \psi_L)}{\varepsilon(\chi \otimes \rho'_L, \frac{1}{2}, \psi_L)} = \chi(\mathrm{N}_{E'/K}(\beta')).$$

Thus, by 52.2 Lemma,

$$\frac{3n - \nu_L}{2} = \upsilon_K\big(\mathrm{N}_{E/K}(\beta)/\mathrm{N}_{E'/K}(\beta') - 1\big). \tag{52.6.2}$$

52.7. We turn to the representations $\zeta = \boldsymbol{\pi}_K(\rho_K)$ and $\zeta' = \boldsymbol{\pi}_K(\rho'_K)$. We can do the same calculation over K, using the same elements β, β'. We conclude that

$$\|\zeta - \zeta'\| = \nu_L = 3\nu_F.$$

Let $\varUpsilon, \varUpsilon' \in C(\psi_\alpha^K, \mathfrak{A}_K)$ be such that $\zeta = c\text{-Ind}\,\varUpsilon$, $\zeta' = c\text{-Ind}\,\varUpsilon'$. The character $\varUpsilon^{-1}\varUpsilon'$ of $U^1_{K[\alpha]}$ has level ν_L. The norm $\mathrm{N}_{K[\alpha]/F[\alpha]}$ induces an isomorphism

$$U^{\nu_L}_{K[\alpha]}/U^{1+\nu_L}_{K[\alpha]} \longrightarrow U^{\nu_F}_{F[\alpha]}/U^{1+\nu_F}_{F[\alpha]},$$

which restricts to the isomorphism

$$\begin{aligned} U^{\nu_F}_{F[\alpha]}/U^{1+\nu_F}_{F[\alpha]} &\longrightarrow U^{\nu_F}_{F[\alpha]}/U^{1+\nu_F}_{F[\alpha]}, \\ x &\longmapsto x^3. \end{aligned}$$

It follows that $\varUpsilon^{-1}\varUpsilon' \mid U^1_{F[\alpha]} \neq 1$. Looking back at the construction in 50.3 (especially Lemma 1), the hypothesis $\boldsymbol{\pi}_F(\rho) = \boldsymbol{\pi}_F(\rho')$ would imply $\varUpsilon^{-1}\varUpsilon' \mid U^1_{F[\alpha]}$ trivial. This case also cannot arise therefore, and we have completed the proof of the Bijectivity Theorem 52.1. □

We have completed the proof of 33.1 Theorem and our discussion of the Langlands Correspondence for GL(2). ■

52.8. The Langlands correspondence gives a bijection between imprimitive representations of $\mathcal{W}_F$ and ordinary representations of $\mathrm{GL}_2(F)$. It therefore gives a bijection between primitive representations of $\mathcal{W}_F$ and exceptional representations of $\mathrm{GL}_2(F)$.

It is easy to produce exceptional representations of $\mathrm{GL}_2(F)$, using 45.2 Theorem: if $(\mathfrak{A}, n, \alpha)$ is a ramified simple stratum with $1 \leqslant n \leqslant 3d(F[\alpha]|F)$ and $n \not\equiv 0 \pmod 3$, any $\pi \in \mathcal{A}^0_2(F;\alpha)$ is exceptional (45.6 Remark). The cubic extension K/F which renders such representations π ordinary is totally ramified. On the other hand, it is easy to produce exceptional strata $(\mathfrak{A}, n, \alpha)$ with $n = 3d(F[\alpha]|F)$; in this case, K/F is unramified (51.6).

In particular, while all primitive representations $\rho \in \mathcal{G}^0_2(F)$ are tetrahedral in the case $q \equiv 1 \pmod 3$, both tetrahedral and octahedral cases arise when $q \equiv -1 \pmod 3$.

52.9 Comment. The Langlands correspondence is known to satisfy

$$\boldsymbol{\pi}_E(\rho_E) = \boldsymbol{\pi}_F(\rho)_E, \quad \rho \in \mathcal{G}_2^{\mathrm{wr}}(F), \tag{52.9.1}$$

for any finite, tamely ramified field extension E/F. However, no proof of this fact, using purely local methods, is known to the authors.

The central case is that where ρ is imprimitive and E/F is quadratic unramified. If we assume this fact, we get a quicker construction of the Langlands correspondence on primitive representations and an easier proof that it is bijective. To sketch the argument, let $\rho \in \mathcal{G}_2^0(F)$ be primitive. Choose K/F cubic so that ρ_K is imprimitive. Let L/F be the normal closure of K/F, and put $\Sigma = \mathrm{Gal}(L/F)$. Set $\tau = \boldsymbol{\pi}_K(\rho_K)$. By (52.9.1), we have $\tau_L = \boldsymbol{\pi}_L(\rho_L)$, so $\tau_L \in \mathcal{A}_2^{\mathrm{wr}}(L)^{\Sigma}$. Thus (as follows readily from 46.6 Proposition) there exists $\pi \in \mathcal{A}_2^0(F)$ such that $\tau_L = \pi_L$. There is a unique choice of π for which $\omega_\pi = \det\rho$. We set $\boldsymbol{\pi}_F(\rho) = \pi$. The local constant relations are established much as before. Bijectivity is easily proved, in the manner of 51.3 Proposition.

Exercise. *Assume that* (52.9.1) *holds for E/F quadratic unramified and ρ imprimitive. Show that it holds for all $\rho \in \mathcal{G}_2^{\mathrm{wr}}(F)$ and all tame extensions E/F.*

Further reading The notion of "tame lifting" has a quite general version [13], [14]. Likewise the twisting and congruence properties of the Langlands-Deligne local constant [28], [39].

Kutzko's original proof [52] (plus [53], covered here in Chapter XI) uses a different local constant technology, and derives (52.9.1) (for ρ imprimitive and E/F unramified quadratic) from Langlands' base change [57], via the preliminary summary [33]. From that point, Kutzko's argument is as summarized in 52.9 above.

The argument, both here and in Chapter XI, relies heavily on counting. This has been a persistent feature of attempts to prove the Langlands conjecture. Both approaches to general dimension in characteristic zero ([38], or [37] plus [43]), the proof in positive characteristic [58], and the attempt at an "explicit" version [14], all depend on the general counting arguments of [41].

13

The Jacquet-Langlands Correspondence

We return to a general non-Archimedean local field F, for which the residual characteristic p is arbitrary. The field F possesses a central division algebra D of dimension 4 and, up to isomorphism, only one. The group $D^\times$ is locally profinite and is compact modulo its centre $F^\times$. We consider the irreducible smooth representations of $D^\times$. One can classify these by a method parallel to that of Chapter IV, the principal series being effectively absent for $D^\times$. There is a straightforward theory of L-functions and local constants attached to irreducible representations of $D^\times$, rather similar to that of §23 for characters of $F^\times$.

Let $\mathcal{A}_1(D)$ denote the set of equivalence classes of irreducible smooth representations of $D^\times$. On the other hand, let $\mathcal{A}_2^\diamondsuit(F)$ denote the set of equivalence classes of irreducible smooth representations of $G = \mathrm{GL}_2(F)$ which are *essentially square-integrable modulo centre,* as in 17.4. This chapter is concerned with a canonical bijection, between $\mathcal{A}_2^\diamondsuit(F)$ and $\mathcal{A}_1(D)$, called the *Jacquet-Langlands correspondence.* We specify it in terms of L-functions and local constants. It relates the explicit classification of representations on either side.

We give virtually no proofs in this chapter. Once the basic structure of D has been described, all arguments are simpler versions of those we have already seen. The reader is invited to explore the topic as a sequence of exercises.

53. Division Algebras

Let D be an F-algebra which is a division ring, with centre F and dimension 4. We review[1] the basic structure of D as an F-algebra, and say something of the structure of the locally profinite group $D^\times$ of non-zero elements of D.

53.1. The relevant elements of the structure of D as F-algebra can be summed up as follows.

Proposition. *Let E/F be a separable quadratic field extension.*

(1) *There exists an embedding $E \to D$ of F-algebras. Any two such embeddings are conjugate by an element of $D^\times$.*
(2) *The E-algebra $B = E \otimes_F D$ is isomorphic to $\mathrm{M}_2(E)$.*
(3) *Let $a \in D$; then $\mathrm{tr}_{B/E}(1 \otimes a) \in F$ and, if $a \neq 0$, then $\det_B(1 \otimes a) \in F^\times$.*

In the context of the proposition, the quantities $\mathrm{tr}_B(1 \otimes a)$, $\det_B(1 \otimes a)$ actually depend only on a, not on E. One denotes them $\mathrm{Trd}_D\, a$, $\mathrm{Nrd}_D\, a$ respectively. The *reduced trace* Trd_D is a F-linear map $D \to F$ and the *reduced norm* Nrd_D is a homomorphism $D^\times \to F^\times$.

Let $a \in D$, $a \notin F$. The F-algebra $F[a]$ is then a field; since D is an $F[a]$-vector space, we must have $[F[a]{:}F] = 2$. Further,

$$\mathrm{Trd}_D \big| F[a] = \mathrm{Tr}_{F[a]/F}, \quad \mathrm{Nrd}_D \big| F[a]^\times = \mathrm{N}_{F[a]/F}.$$

The polynomial $X^2 - \mathrm{Trd}_D(a)X + \mathrm{Nrd}_D(a) \in F[X]$ is called the *reduced characteristic polynomial* of a.

Remark. Suppose, for the moment, that F is an arbitrary field. If D is a central F-division algebra of dimension 4, there exists a separable quadratic extension E/F such that E admits an F-embedding in D. The proposition then holds relative to that field E. From this general point of view, non-Archimedean local fields exhibit two idiosyncrasies. First, the algebra D is unique up to isomorphism, although that will not intervene in what we do. More importantly, *any* quadratic field extension E/F can be embedded in D: this has a profound effect.

53.2. The map

$$\upsilon_D : a \longmapsto \upsilon_F(\mathrm{Nrd}_D\, a), \quad a \in D,$$

is a valuation on D, in that it satisfies

$$\upsilon_D(ab) = \upsilon_D(a) + \upsilon_D(b)$$

[1]For a full discussion of division algebras over local fields, similar to the approach we take here, see [74].

and the ultrametric inequality

$$\upsilon_D(a+b) \geqslant \min\{\upsilon_D(a), \upsilon_D(b)\}.$$

We set

$$|a|_D = \|\mathrm{Nrd}_D\, a\| = q^{-\upsilon_D(a)}.$$

The metric space topology on D induced by $|\cdot|_D$ coincides with the natural topology of D as F-vector space.

53.3. The set

$$\mathfrak{O} = \{x \in D : \upsilon_D(x) \geqslant 0\}$$

is thus an open subring of D, and

$$\mathfrak{q} = \{x \in D : \upsilon_D(x) \geqslant 1\}$$

is an open, two-sided ideal of $\mathfrak{O}$: it is the unique maximal left (or right) ideal of $\mathfrak{O}$. If $\varpi_D \in D$ satisfies $\upsilon_D(\varpi_D) = 1$, we have

$$\mathfrak{q} = \varpi_D\mathfrak{O} = \mathfrak{O}\varpi_D\,;$$

one calls ϖ_D a *prime element of* D. Also

$$\varpi_D^n\mathfrak{O} = \mathfrak{q}^n = \{x \in D : \upsilon_D(x) \geqslant n\}, \quad n \in \mathbb{Z}.$$

All the sets $\mathfrak{q}^n$ are stable under conjugation by $D^\times$. In particular, $D^\times$ acts by conjugation on the ring $\boldsymbol{k}_D = \mathfrak{O}/\mathfrak{q}$.

Proposition. *We have:*

(1) $\mathfrak{p}\mathfrak{O} = \mathfrak{q}^2$;
(2) *the ring* $\boldsymbol{k}_D = \mathfrak{O}/\mathfrak{q}$ *is a field of* q^2 *elements;*
(3) *if* ϖ_D *is a prime element of* D, *then*

$$\varpi_D x \varpi_D^{-1} = x^q, \quad x \in \boldsymbol{k}_D.$$

For $n \geqslant 1$, the ring $\mathfrak{O}/\mathfrak{q}^n$ has q^{2n} elements. As in the commutative case of the field F (1.2), we deduce that $\mathfrak{O}$ is the projective limit of the system of finite rings $\mathfrak{O}/\mathfrak{q}^n$, $n \geqslant 1$, and is compact.

It follows that the groups

$$U_D = U_D^0 = \mathfrak{O}^\times \quad \text{and} \quad U_D^n = 1+\mathfrak{q}^n, \quad n \geqslant 1,$$

are compact open normal subgroups of $D^\times$. They provide a fundamental system of open neighbourhoods of the identity in $D^\times$.

The subgroup $F^\times U_D$ has index 2 in $D^\times$; we deduce that $D^\times$ is compact modulo its centre $F^\times$.

53.4. We consider the group $\widehat{D}$ of characters of D. We define the level of $\eta \in \widehat{D}$ to be the least integer k such that $\mathfrak{q}^k \subset \operatorname{Ker}\eta$ (with the understanding that the trivial character has level $-\infty$).

Lemma 1. *For $\eta \in \widehat{D}$ and $a \in D$, the map*

$$a\eta : x \longmapsto \eta(ax)$$

is a character of D. If $\eta \neq 1$, the map $a \mapsto a\eta$ is an isomorphism $D \to \widehat{D}$.

For $\psi \in \widehat{F}$, we set $\psi_D = \psi \circ \operatorname{Trd}_D \in \widehat{D}$.

Lemma 2. *If $\psi \in \widehat{F}$ has level one, the character ψ_D of D has level one.*

Let us now fix $\psi \in \widehat{F}$ of level one. For $a \in D$, we form the function

$$\psi_a^D : x \longmapsto \psi_D(a(x-1)), \quad x \in D.$$

Proposition. *For integers m, n such that $0 \leqslant [n/2] \leqslant m < n$, the map $a \mapsto \psi_a^D$ is an isomorphism of $\mathfrak{q}^{-n}/\mathfrak{q}^{-m}$ with the character group of U_D^{m+1}/U_D^{n+1}.*

53.5. The group of characters of $D^\times$ takes a particularly simple form. If χ is a character of $D^\times$, we define the level of χ to be the least integer $n \geqslant 0$ such that $U_D^{n+1} \subset \operatorname{Ker}\chi$.

Proposition. *The map $\phi \mapsto \phi_D = \phi \circ \operatorname{Nrd}_D$ is an isomorphism of the character group of $F^\times$ with the character group of $D^\times$. If ϕ has level j, then ϕ_D has level $2j$.*

54. Representations

We can give a complete classification of the irreducible smooth representations of $D^\times$, following parts of Chapter IV quite closely. For the purposes of this section, it is convenient to fix a character ψ of F of level one.

54.1. We have noted already (53.3) that the group $D^\times/F^\times$ is compact. So, if π is an irreducible smooth representation of $D^\times$, then $\dim\pi$ is finite (2.7) and $\operatorname{Ker}\pi$ contains an open normal subgroup of $D^\times$. In particular, π is trivial on a unit group U_D^m, for some $m \geqslant 0$. We define the *level* $\ell_D(\pi)$ of π to be the least integer $n \geqslant 0$ such that $U_D^{n+1} \subset \operatorname{Ker}\pi$.

We write $\mathcal{A}_1(D)$ for the set of equivalence classes of irreducible smooth representations of $D^\times$.

If χ is a character of $F^\times$ and $\pi \in \mathcal{A}_1(D)$, we set

$$\chi\pi = \chi_D \otimes \pi,$$

where, as in 53.5, $\chi_D = \chi \circ \operatorname{Nrd}_D$. We say that π is *minimal* if $\ell_D(\pi) \leqslant \ell_D(\chi\pi)$, for all characters χ of $F^\times$.

54.2. From 53.5, we know that the one-dimensional smooth representations of $D^\times$ are the characters χ_D, where χ ranges over the characters of $F^\times$. We let $\mathcal{A}_1^0(D)$ denote the set of classes of representation $\pi \in \mathcal{A}_1(D)$ such that $\dim \pi \neq 1$.

By way of an example, let us consider the elements π of $\mathcal{A}_1^0(D)$ of level zero. We use the notion of admissible pair, as in Chapter 5.

Let $(E/F, \theta)$ be an admissible pair in which θ has level zero. In particular, E/F is unramified. We identify E with a subfield of D, and extend θ, by triviality, to a character Θ of the group ${}_DJ = E^\times U_D^1$. We form the smooth representation

$$\pi_\theta^D = \text{c-Ind}_{{}_DJ}^{D^\times} \Theta.$$

Proposition.

(1) *The representation π_θ^D is irreducible and of level* 0.
(2) *The map $(E/F, \theta) \mapsto \pi_\theta^D$ is a bijection between the set of isomorphism classes of admissible pairs of level zero and the set of $\pi \in \mathcal{A}_1^0(D)$ of level zero.*

54.3. Let $\alpha \in D$, $\alpha \notin F$. The algebra $F[\alpha]$ is then a quadratic field extension. As we noted in 13.4, the notion of α being minimal over F can be formulated purely in terms of the extension $F[\alpha]/F$. Adapting 13.4 Definition to the context of $\alpha \in D$, we get:

Definition. *Let $\alpha \in D^\times$, and set $n = -v_D(\alpha)$. The element α is minimal over F if either*

(1) *n is odd, or*
(2) *n is even and, for a prime element ϖ of F, the reduced characteristic polynomial of $\varpi^{n/2}\alpha$ over F is irreducible modulo $\mathfrak{p}$.*

Observe that the field extension $F[\alpha]/F$ is totally ramified in case (1), unramified in case (2). Just as in 13.4, we have:

Proposition. *Let $\alpha \in D^\times$, with $n = -v_D(\alpha)$.*

(1) *The element α is minimal over F if and only if $\alpha + \mathfrak{q}^{1-n} \cap F = \emptyset$.*
(2) *If α is minimal over F, the coset $\alpha + \mathfrak{q}^{1-n} = \alpha U_D^1$ contains only minimal elements.*

54.4. Let $\alpha \in D$ be minimal over F, and such that $n = -v_D(\alpha) \geqslant 1$. We can then form the character ψ_α^D of the group $U_D^{[n/2]+1}$ (53.4). We get the same intertwining properties as in $G = \mathrm{GL}_2(F)$:

Intertwining Theorem. *Let $\alpha \in D$ be minimal over F, with $n = -v_D(\alpha) \geqslant 1$. Write $E = F[\alpha]$ and let $x \in D^\times$. The following are equivalent:*

(1) *x intertwines the character ψ_α^D of $U_D^{[n/2]+1}$.*
(2) *$x \in E^\times U_D^{[(n+1)/2]}$.*

Since each U_D^m is a normal subgroup of $D^\times$, the element x intertwines $\psi_\alpha^D \mid U_D^{[n/2]+1}$ if and only if it normalizes it. The analogue of the Conjugacy Theorem (15.2), for characters ψ_α^D, is thus trivial: if β is another minimal element of D with $\upsilon_D(\beta) = -n$, the characters ψ_α^D, ψ_β^D of $U_D^{[n/2]+1}$ intertwine in $D^\times$ if and only if they are conjugate.

In the situation of the Intertwining Theorem, we write

$$_DJ_\alpha = E^\times U_D^{[(n+1)/2]}. \tag{54.4.1}$$

Immediately:

Lemma. *Let Λ be an irreducible representation of $_DJ_\alpha$ such that $\Lambda \mid U_D^{[n/2]+1}$ contains the character ψ_α^D; then $\Lambda \mid U_D^{[n/2]+1}$ is a multiple of ψ_α^D.*

We write $C(\psi_\alpha^D)$ for the set of equivalence classes of irreducible representations of $_DJ_\alpha$ such that $\Lambda \mid U_D^{[n/2]+1}$ is a multiple of ψ_α^D.

Proposition.

(1) *Let $\Lambda \in C(\psi_\alpha^D)$. The representation*

$$\pi_\Lambda = c\text{-}\mathrm{Ind}_{J_\alpha}^{D^\times} \Lambda$$

is irreducible.

(2) *Representations $\Lambda_1, \Lambda_2 \in C(\psi_\alpha^D)$ intertwine in $D^\times$ if and only if they are equivalent. In particular, $\pi_{\Lambda_1} \cong \pi_{\Lambda_2}$ if and only if $\Lambda_1 \cong \Lambda_2$.*

54.6. Let $\pi \in \mathcal{A}_1^0(D)$ be minimal of level $n \geqslant 1$. Consider the restriction of π to U_D^n. This is a sum of characters ψ_β^D, where the elements β are all minimal over F. Thus π contains a character ψ_α^D of the group $U_D^{[n/2]+1}$, with α minimal over F. The conjugacy class of this character is determined uniquely by π.

Taking this further, π must contain a unique $\Lambda \in C(\psi_\alpha^D)$. This reduces the classification of the minimal elements of $\mathcal{A}_1^0(F)$ to the description of the elements of the sets $C(\psi_\alpha^D)$.

54.7. Let $\alpha \in D$ be minimal over F, with $n = -\upsilon_D(\alpha) \geqslant 1$. Write $E = F[\alpha]$. Thus $e(E|F) = 2/\gcd(2,n)$.

Lemma 1.

(1) *If n is odd or if $n \equiv 0 \pmod 4$, then $_DJ_\alpha = E^\times U_D^{[n/2]+1}$.*
(2) *If $n \equiv 2 \pmod 4$, then $_DJ_\alpha$ contains $E^\times U_D^{[n/2]+1}$ with index q^2.*

In case (1) of Lemma 1, therefore, the set $C(\psi_\alpha^D)$ consists of the *characters* Λ of $E^\times U_D^{[n/2]+1}$ such that $\Lambda \mid U_D^{[n/2]+1} = \psi_\alpha^D$.

Lemma 2.

Suppose that $n \equiv 2 \pmod 4$. *Let* θ *be a character of the group* ${}_DH^1_\alpha = U^1_E U_D^{[n/2]+1}$ *extending* ψ^D_α.

(1) *There exists a unique irreducible representation* η_θ *of the group* ${}_DJ^1_\alpha = U^1_E U_D^{[(n+1)/2]}$ *such that* $\eta \mid {}_DH^1_\alpha$ *contains* θ.
(2) *The representation* η_θ *satisfies* $\dim \eta_\theta = q$, *and* $\eta_\theta \mid {}_DH^1_\alpha$ *is a multiple of* θ.
(3) *An irreducible representation* Υ *of* ${}_DJ_\alpha$ *lies in* $\mathcal{C}(\psi^D_\alpha)$ *if and only if* $\Upsilon \mid {}_DJ^1_\alpha = \eta_\theta$, *for some character* θ *of* ${}_DH^1_\alpha$ *such that* $\theta \mid U_D^{[n/2]+1} = \psi^D_\alpha$.

54.8. There is an entertaining assymmetry. Let $(\mathfrak{A}, n, \alpha)$, $n \geqslant 1$, be an *unramified* simple stratum in $A = \mathrm{M}_2(F)$. In particular, α is minimal over F. The elements of $\mathcal{C}(\psi_\alpha, \mathfrak{A})$ have dimension 1 if n is odd, and dimension q if n is even. Now choose an F-embedding $F[\alpha] \to D$, and view α as an element of $D^\times$. Thus $\upsilon_D(\alpha) = -2n$. The elements of $\mathcal{C}(\psi^D_\alpha)$ have dimension q if n is odd, and dimension 1 if n is even.

55. Functional Equation

We review the theory of functional equations for irreducible smooth representations of $D^\times$. While the formalism resembles that of §24, the underlying methods are much closer to those of §23.

55.1. Let (π, V) be an irreducible smooth representation of $D^\times$ and, as before, let $\mathcal{C}(\pi)$ denote the space of coefficients of π (defined exactly as in 10.1). We consider integrals of the form

$$\zeta(\Phi, f, s) = \int_{D^\times} \Phi(x)\, f(x)\, \|\mathrm{Nrd}_D\, x\|^s \, d\mu^*(x),$$

for functions $\Phi \in C_c^\infty(D)$, $f \in \mathcal{C}(\pi)$ and a Haar measure μ^* on $D^\times$. This integral can be re-written as a formal Laurent series in q^{-s}. We set

$$\mathcal{Z}(\pi) = \{\zeta(\Phi, f, s{+}\tfrac{1}{2}) : \Phi \in C_c^\infty(D), f \in \mathcal{C}(\pi)\}.$$

Proposition. *There is a unique polynomial* $P_\pi(T) \in \mathbb{C}[T]$ *such that* $P_\pi(0) = 1$ *and*

$$\mathcal{Z}(\pi) = P_\pi(q^{-s})^{-1}\, \mathbb{C}[q^s, q^{-s}].$$

We define

$$L(\pi, s) = P_\pi(q^{-s})^{-1}.$$

Theorem. *Let* (π, V) *be an irreducible smooth representation of* $D^\times$. *We have* $L(\pi, s) = 1$ except *when* $\pi = \chi_D$, *for an unramified character* χ *of* $F^\times$. *In this exceptional case, we have*

$$L(\chi_D, s) = L(\chi, s{+}\tfrac{1}{2}).$$

55.2. We fix a character $\psi \in \widehat{F}$, $\psi \neq 1$. Using the character ψ_D, we can define an operation of Fourier transform $\Phi \mapsto \hat{\Phi}$ on $C_c^\infty(D)$, and self-dual Haar measure on D, exactly as for F (23.1) or A (24.1).

Theorem. *Let $\psi \in \widehat{F}$, $\psi \neq 1$. There is a unique rational function $\gamma(\pi, s, \psi) \in \mathbb{C}(q^{-s})$ such that*

$$\zeta(\hat{\Phi}, \check{f}, \tfrac{3}{2}-s) = \gamma(\pi, s, \psi)\,\zeta(\Phi, f, s+\tfrac{1}{2}),$$

for all $\Phi \in C_c^\infty(D)$, $f \in \mathcal{C}(\pi)$.

The function

$$\varepsilon(\pi, s, \psi) = \gamma(\pi, s, \psi)\,\frac{L(\pi, s)}{L(\check{\pi}, 1-s)}$$

is of the form aq^{bs}, for some $a \in \mathbb{C}^\times$ and $b \in \mathbb{Z}$.

Warning. It is conventional to insert a minus sign in the definition of ε (or γ) in this context. We feel the picture is clearer without it, so we consciously and deliberately omit it.

55.3. The local constant $\varepsilon(\pi, s, \psi)$, for $\pi \in \mathcal{A}_1(D)$, has properties analogous to that for representations of G. In particular,

$$\varepsilon(\pi, s, a\psi) = \omega_\pi(a)\,\|a\|^{2s-1}\,\varepsilon(\pi, s, \psi), \quad a \in F^\times, \tag{55.3.1}$$

where ω_π denotes the central character of π. So, when calculating local constants, we can always reduce to the convenient case where ψ has level one.

Since the irreducible representation (π, V) has finite dimension, there is no difficulty in working with operator-valued integrals

$$\zeta(\Phi, \pi, s) = \int_{D^\times} \Phi(x)\,\pi(x)\,\|\mathrm{Nrd}_D\, x\|^s\, d\mu^*(x),$$

for $\Phi \in C_c^\infty(D)$. Using this formalism, the proofs of the results 55.1, 55.2 are virtually identical to the corresponding arguments in §23.

For $n \in \mathbb{Z}$, write

$$D(n) = \{x \in D : v_D(x) = -n\}.$$

Theorem. *Let $(\pi, V) \in \mathcal{A}_1^0(D)$ have level n, and let $\psi \in \widehat{F}$ have level one. The integral*

$$\int_{D(n)} \check{\pi}(x)\,\|\mathrm{Nrd}_D\, x\|^{\frac{3}{2}-s}\, d\mu^*(x) \tag{55.3.2}$$

is a scalar operator on $\check{V}$ with eigenvalue is $\mu^*(U_D^{n+1})\,\varepsilon(\pi, s, \psi)/q^{(2n+1)}$.

The integral (55.3.2) reduces to a non-abelian Gauss sum exactly like (25.1.1). If π is written in the form c-Ind Λ, as in §55, one can equally express this Gauss sum in terms of Λ, just as in 25.5.

55.4. For the one-dimensional representations, we have:

Proposition. *Let χ be a character of $F^\times$ and $\psi \in \widehat{F}$, $\psi \neq 1$. If χ is unramified, then*

$$\varepsilon(\chi_D, s, \psi) = \varepsilon(\chi, s, \psi), \tag{55.4.1}$$

while otherwise,

$$\varepsilon(\chi_D, s, \psi) = -\varepsilon(\chi, s-\tfrac{1}{2}, \psi)\, \varepsilon(\chi, s+\tfrac{1}{2}). \tag{55.4.2}$$

55.5. The proof of (55.4.1) is straighforward, but that of (55.4.2) relies one an elementary calculation, which is also crucial at later stages of the section.

Let χ be a character of $F^\times$ of level $k \geqslant 1$. Let $c \in \mathfrak{p}^{-k}$ satisfy $\chi(1+x) = \psi(cx)$, $x \in \mathfrak{p}^{[k/2]+1}$. We have various neo-classical species of Gauss sum. Taking a chain order $\mathfrak{A}$ in $A = \mathrm{M}_2(F)$ with $e = e_{\mathfrak{A}}$ (and putting $\chi_A = \chi \circ \det$):

$$\tau(\chi, \psi) = \sum_{x \in U_F/U_F^{k+1}} \chi(cx)^{-1}\psi(cx)\,;$$

$$\tau_{\mathfrak{A}}(\chi, \psi) = \sum_{x \in U_{\mathfrak{A}}/U_{\mathfrak{A}}^{ek+1}} \chi_A(cx)^{-1}\psi_A(cx)\,;$$

$$\tau_D(\chi, \psi) = \sum_{x \in U_D/U_D^{2k+1}} \chi_D(cx)^{-1}\psi_D(cx)\,.$$

Lemma. *There are positive constants $c_{\mathfrak{A}}$, c_D, depending only on k and $e_{\mathfrak{A}}$, such that*

$$\tau(\chi, \psi)^2 = c_{\mathfrak{A}}\, \tau_{\mathfrak{A}}(\chi, \psi) = -c_D\, \tau_D(\chi, \psi).$$

(The first equality is (25.6.1).)

55.6. Following through the development for $\mathrm{GL}_2(F)$, the next step is to prove:

Converse Theorem. *Let π_1, π_2 be irreducible smooth representations of $D^\times$. Suppose that*

$$L(\chi\pi_1, s) = L(\chi\pi_2, s),$$

$$\varepsilon(\chi\pi_1, s, \psi) = \varepsilon(\chi\pi_2, s, \psi),$$

for all characters χ of $F^\times$ and all $\psi \in \widehat{F}$, $\psi \neq 1$. We then have $\pi_1 \cong \pi_2$.

The L-function $L(\chi\pi, s)$ is non-trivial, for some χ, if and only if $\dim \pi = 1$. It follows that, if one π_i is one-dimensional, then so is the other and the result is a straightforward consequence of 55.1 Theorem.

Assume therefore that $\dim \pi_i > 1$, $i = 1, 2$. One reduces, as in §27, to the case where ψ has level one. The key step in the proof is again:

Inversion Formula. *Let* $\pi \in \mathcal{A}_1^0(D)$ *satisfy* $n = \ell(\pi) \leqslant \ell(\chi\pi)$, *for all characters* χ *of* $F^\times$. *Let* $\psi \in \widehat{F}$ *have level one, and set* $r = [n/2]$. *There is a constant* k_D, *depending only on* n, *such that*

$$\operatorname{tr}\check{\pi}(g) = k_D \sum_{\substack{\chi \in \Gamma_{r+1}/\Gamma_0, \\ c \in U_F/U_F^{r+1}}} \varepsilon(\chi\pi, \tfrac{1}{2}, c\psi)\, \chi(\mathrm{Nrd}_D\, g)\, c\psi(-\mathrm{Trd}_D\, g),$$

for all $g \in D(n)$ *which are minimal over* F.

56. Jacquet-Langlands Correspondence

We come to the main point.

56.1. Let $\mathcal{A}_2^\diamond(F)$ denote the set of equivalence classes of irreducible smooth representations of $G = \mathrm{GL}_2(F)$ which are *essentially square-integrable* (see 17.4).

Jacquet-Langlands Correspondence. *Let* D *be a central* F*-division algebra of dimension* 4. *Let* $\psi \in \widehat{F}$, $\psi \neq 1$. *There is a unique map*

$$\begin{aligned} \mathcal{A}_2^\diamond(F) &\longrightarrow \mathcal{A}_1(D), \\ \pi &\longmapsto \pi_D, \end{aligned} \tag{56.1.1}$$

such that

$$L(\chi\pi_D, s) = L(\chi\pi, s), \tag{56.1.2}$$

$$\varepsilon(\chi\pi_D, s, \psi) = -\varepsilon(\chi\pi, s, \psi), \tag{56.1.3}$$

for all $\pi \in \mathcal{A}_2^\diamond(F)$ *and all characters* χ *of* $F^\times$.

The map $\pi \mapsto \pi_D$ *is bijective and* (56.1.3) *holds for all* $\psi \in \widehat{F}$, $\psi \neq 1$. *It further satisfies*

$$\begin{aligned} (\chi\pi)_D &= \chi\pi_D, \\ \omega_{\pi_D} &= \omega_\pi, \\ (\check{\pi})_D &= (\pi_D)^\vee. \end{aligned} \tag{56.1.4}$$

The relation (56.1.3) implies $\ell(\pi_D) = 2\ell(\pi)$, $\pi \in \mathcal{A}_2^0(F)$.

In the remainder of the section, we sketch a proof of the result.

56.2. Let $\pi \in \mathcal{A}_2^\diamond(F) \smallsetminus \mathcal{A}_2^0(F)$. Therefore (17.5 Theorem) π is a special representation $\phi{\cdot}\mathrm{St}_G$, for some character ϕ of $F^\times$. In this case, we set $\pi_D = \phi_D$. This gives a bijection

$$\begin{aligned} \mathcal{A}_2^\diamond(F) \smallsetminus \mathcal{A}_2^0(F) &\longrightarrow \mathcal{A}_1(D) \smallsetminus \mathcal{A}_1^0(D), \\ \phi \cdot \mathrm{St}_G &\longmapsto \phi_D. \end{aligned} \tag{56.2.1}$$

Our calculations of L-functions in 55.1, 26.1 Theorem show that this is the unique map satisfying (56.1.2). The ε-calculations in 26.1 Theorem and 55.4 show that it also satisfies (56.1.3).

56.3. A representation $\pi \in \mathcal{A}_2(F)$ lies in $\pi \in \mathcal{A}_2^0(F)$ if and only if $L(\chi\pi, s) = 1$, for all characters χ of $F^\times$ (27.2 Proposition). The analogous property holds in $\mathcal{A}_1(D)$. So, we have to produce a bijection $\mathcal{A}_2^0(F) \to \mathcal{A}_1^0(D)$ satisfying (56.1.3). We need only do this for one character ψ, arguing exactly as in 27.4. We may therefore assume ψ has level one.

We first define π_D when $\pi \in \mathcal{A}_2^0(F)$ is minimal. We proceed by cases.

56.4. Let $\pi \in \mathcal{A}_2^0(F)$ have level zero. Thus there is an admissible pair $(E/F, \theta)$, in which θ has level zero, such that $\pi \cong \pi_\theta$, in the notation of 19.1. Using the notation of 54.2, we set

$$\pi_D = (\pi_\theta)_D = \pi_\theta^D.$$

The relation $\varepsilon(\pi, s, \psi) = -\varepsilon(\pi_D, s, \psi)$ follows from (25.4.1) and an easy calculation in D.

56.5. Let $\pi \in \mathcal{A}_2^0(F)$ be minimal, with $\ell(\pi) > 0$. There is a simple stratum $(\mathfrak{A}, n, \alpha)$ in A such that π contains the character ψ_α of $U_{\mathfrak{A}}^{[n/2]+1}$. More precisely, π contains a representation $\Lambda \in C(\psi_\alpha, \mathfrak{A})$, and so $\pi = c\text{-Ind}_J^G \Lambda$, where $J = E^\times U_{\mathfrak{A}}^{[(n+1)/2]}$, $E = F[\alpha]$.

Suppose, in this paragraph, that $e_{\mathfrak{A}} = 2$. Thus $\dim \Lambda = 1$.

We choose an F-embedding $E \to D$, and henceforth regard α equally as an element of D. We form the group ${}_DJ = E^\times U_D^{[n/2]+1}$ and define a character Λ_D of ${}_DJ$ by

$$\begin{aligned} \Lambda_D(u) &= \psi_\alpha^D(u), & u &\in U_D^{[n/2]+1}, \\ \Lambda_D(x) &= (-1)^{v_E(x)}\, \Lambda(x), & x &\in E^\times. \end{aligned}$$

We let π_D be the representation of $D^\times$ induced by Λ_D.

Remark. Various choices were made in the definition of π_D. We shall see (56.8 below) that the definition is independent of these, so the matter need not deflect us now.

56.6. In the same initial situation, we assume that $e_{\mathfrak{A}} = 1$ and that n is odd.

Again, π contains a character Λ of $E^\times U_{\mathfrak{A}}^{[n/2]+1}$, $\Lambda \in C(\psi_\alpha, \mathfrak{A})$. Let θ denote the restriction $\Lambda \mid U_E^1 U_{\mathfrak{A}}^{[n/2]+1}$ and set $\omega_\Lambda = \Lambda \mid F^\times$.

Now view E as embedded in D. We get a character θ_D of $U_E^1 U_D^{n+1}$ by

$$\theta_D \mid U_E^1 = \theta \mid U_E^1, \quad \theta_D \mid U_D^{n+1} = \psi_\alpha^D.$$

There is a unique irreducible representation η_D of the group $U_E^1 U_D^n$ such that $\eta_D \mid U_E^1 U_D^{n+1}$ is a multiple of θ_D (54.7 Lemma 2). We define an irreducible representation Λ_D of ${}_DJ = E^\times U_D^n$ by (*cf.* 19.4)

$$\begin{aligned} \Lambda_D \mid U_E^1 U_D^{n+1} &= \eta_D; \\ \Lambda_D \mid F^\times &= \text{a multiple of } \omega_\Lambda; \\ \operatorname{tr} \Lambda_D(\zeta) &= -\Lambda(\zeta), \quad \zeta \in \boldsymbol{\mu}_E \smallsetminus \boldsymbol{\mu}_F. \end{aligned}$$

We let π_D be the representation of $D^\times$ induced by Λ_D.

56.7. Finally, we have the case $e_{\mathfrak{A}} = 1$ and n even. Here, it is easiest to take an admissible pair $(E/F, \chi)$, with E/F unramified and χ of level n, such that $\pi = \pi_\chi$, in the notation of (19.4.2). We define the character Λ_D of ${}_DJ = E^\times U_D^{n+1}$ by

$$\Lambda_D \mid E^\times = \chi, \quad \Lambda_D \mid U_D^{n+1} = \psi_\alpha^D.$$

Again, π_D is defined to be the representation of $D^\times$ induced by Λ_D.

56.8. Taking $\pi \in \mathcal{A}_2^0(F)$ of minimal normalized level, the next step is to use these definitions to verify the relation $\varepsilon(\chi\pi_D, s, \psi) = -\varepsilon(\chi\pi, s, \psi)$, for all characters χ of $F^\times$ and the fixed $\psi \in \widehat{F}$.

This done, the Converse Theorem (55.6) for D shows that the isomorphism class of π_D is determined by that of π, so π_D is independent of the various choices made in its definition.

We now extend the definition of π_D to all $\pi \in \mathcal{A}_2^0(F)$ using the relation $(\chi\pi)_D = \chi\pi_D$. The equation (56.1.3) is then satisfied and it is clear from the classification in §55 that the map $\pi \mapsto \pi_D$ gives a surjection $\mathcal{A}_2^0(F) \to \mathcal{A}_1^0(D)$. It is injective, because of the Converse Theorem (27.1) for $\mathrm{GL}_2(F)$. □

56.9. Comment. Let $g \in G = \mathrm{GL}_2(F)$; one says that g is *regular semisimple* if its characteristic polynomial has distinct roots (in a splitting field), *elliptic* if its characteristic polynomial is irreducible, *elliptic regular* if the characteristic polynomial is also separable. For an elliptic element $g \in G$, there is an element $g_D \in D$ whose minimal polynomial over F coincides with that of g: this condition determines g_D up to conjugacy in $D^\times$. The process $g \mapsto g_D$ gives a bijection between the set of elliptic conjugacy classes in G and the set of $D^\times$-conjugacy classes in $D^\times \setminus F^\times$.

By way of an example, consider a representation $\pi \in \mathcal{A}_2^0(F)$ of minimal normalized level $\ell = \ell(\pi) \in \frac{1}{2}\mathbb{Z}$. Let $(\mathfrak{A}, \Xi)$ be a cuspidal inducing datum in π. Let $g \in \mathcal{K}_{\mathfrak{A}}$ be minimal, with $v_F(\det g) = -n$. Taking more care with positive constants than hitherto, the two inversion formulas and the local constant relation (56.1.3) together yield

$$\operatorname{tr} \Xi(g) = -\operatorname{tr} \pi_D(g_D). \tag{56.9.1}$$

Taking account of the relation $\omega_\pi = \omega_{\pi_D}$, the equation (56.9.1) is valid for all minimal elements $g \in G$ such that

$$v_F(\det g) \equiv 2\ell \pmod 2. \tag{56.9.2}$$

It reflects a deeper reality.

One knows that there is a function θ_π, defined (at least) on the open dense set $G^{\mathrm{ss}}_{\mathrm{reg}}$ of regular semisimple elements of G and locally constant there, such that

$$\operatorname{tr}\pi(f) = \int_G \theta_\pi(g) f(g)\, dg, \quad f \in \mathcal{H}(G).$$

(Observe that, since π is admissible, the operator $\pi(f)$ has finite-dimensional image and its trace is therefore defined.) It is more suggestive to write $\theta_\pi = \operatorname{tr}\pi$.

For minimal elements g satisfying (56.9.2), one can show that

$$\operatorname{tr}\pi(g) = \operatorname{tr}\Xi(g).$$

Thus we have

$$\operatorname{tr}\pi(g) = -\operatorname{tr}\pi_D(g_D).$$

The Jacquet-Langlands correspondence is usually stated in the form:

There is a unique bijection

$$\mathcal{A}_2^\diamond(F) \longrightarrow \mathcal{A}_1(D), \tag{56.9.3}$$

$$\pi \longmapsto \pi_D,$$

such that

$$\operatorname{tr}\pi(g) = -\operatorname{tr}\pi_D(g_D),$$

for all regular elliptic elements $g \in G$.

Our characterization (56.1.2), (56.1.3) is thus weaker than the usual one, but nontheless adequate.

Further reading For a central F-division algebra H of arbitrary finite dimension, the representations of $H^\times$ have been described by Zink [91] and, in a version closely parallel to [19], by Broussous [9]. In characteristic zero, the Jacquet-Langlands correspondence was proved in dimension 4 in [46] and in arbitrary dimension by Rogawski [72]. The case of positive characteristic and arbitrary dimension is due to Badulescu [3]. The compatibility of the classification schemes of [9] and [19] via the Jacquet-Langlands correspondence has not been completely written down, but see [16] for a particularly hard case.

The functional equation is in [46], but is a special instance of [35], which treats $\mathrm{GL}_m(H)$, for any $m \geqslant 1$. An introductory account of the basic theory of characters can be found in an appendix to [13]. The domain of definition of a character is a subtle matter, for which one can consult [59].

References

1. J. Arthur and L. Clozel, *Simple algebras, base change, and the advanced theory of the trace formula*, Annals of Math. Studies, vol. 120, Princeton University Press, 1989.
2. E. Artin and J. Tate, *Class field theory (second edition)*, Addison-Wesley Publishing Company, 1990.
3. A. Badulescu, *Orthogonalité des caractères pour* GL_n *sur un corps local de caractéristique non nulle*, Manuscripta Mathematica **101** (2000), 49–70.
4. J. Bernstein and S. Gelbart (eds.), *An introduction to the Langlands Program*, Birkhauser, 2004.
5. I.N. Bernstein and A.V. Zelevinsky, *Representations of the group* $GL(n, F)$ *where* F *is a local non-archimedean field*, Uspekhi Mat. Nauk **31** (1976), 5–70.
6. ——, *Induced representations of reductive* p*-adic groups*, Ann. Scient. Éc. Norm. Sup. (4) **10** (1977), 441–472.
7. A. Borel, *Admissible representations of a semisimple group over a local field with vectors fixed under an Iwahori subgroup*, Invent. Math. **35** (1976), 233–259.
8. A. Borel and W. Casselman (eds.), *Automorphic forms, representations and* L*-functions*, Proc. Symposia Pure Math., vol. 33, Amer. Math. Soc., Providence RI, 1979.
9. P. Broussous, *Extension du formalisme de Bushnell-Kutzko au cas d'une algèbre à division*, Proc. London Math. Soc. (3) **77** (1998), 292–326.
10. D. Bump, *Automorphic forms and representations*, Cambridge Studies in Advanced Mathematics, vol. 55, Cambridge University Press, 1997.
11. C.J. Bushnell, *Hereditary orders, Gauss sums and supercuspidal representations of* GL_N J. reine angew. Math. **375/376** (1987), 184–210.
12. C.J. Bushnell and A. Fröhlich, *Non-abelian congruence Gauss sums and* p*-adic simple algebras*, Proc. London Math. Soc. (3) **50** (1985), 207–264.
13. C.J. Bushnell and G. Henniart, *Local tame lifting for* $\mathrm{GL}(N)$*, I: simple characters*, Publ. Math. I.H.É.S. **83** (1996), 105–233.
14. ——, *Local tame lifting for* GL_n*, II: wildly ramified supercuspidals*, Astérisque **254** (1999), 1–105.
15. ——, *Davenport-Hasse relations and an explicit Langlands correspondence, II: twisting conjectures*, J. Th. Nombres Bordeaux **12** (2000), 309–347.
16. ——, *Local tame lifting for* $\mathrm{GL}(n)$*, III: explicit base change and Jacquet-Langlands correspondence*, J. reine angew. Math. **508** (2005), 39–100.

17. ——, *The essentially tame local Langlands correspondence, I*, J. Amer. Math. Soc. **18** (2005), 685–710.
18. ——, *The essentially tame local Langlands correspondence, II: totally ramified representations*, Compositio Math. **141** (2005), 979–1011.
19. C.J. Bushnell and P.C. Kutzko, *The admissible dual of $GL(N)$ via compact open subgroups*, Annals of Math. Studies, vol. 129, Princeton University Press, 1993.
20. ——, *Smooth representations of p-adic reductive groups; Structure theory via types*, Proc. London Math. Soc. (3) **77** (1998), 582–634.
21. H. Carayol, *Représentations cuspidales du groupe linéaire*, Ann. Scient. Éc. Norm. Sup. (4) **17** (1984), 191–225.
22. R.W. Carter, *Finite groups of Lie type: conjugacy classes and complex characters*, Wiley-Interscience, 1985.
23. R.W. Carter and M. Geck (eds.), *Representations of reductive groups*, Cambridge University Press, 1998.
24. P. Cartier, *Representations of p-adic groups: a survey*, Automorphic forms, representations and L-functions (A. Borel and W. Casselman, eds.), Proc. Symposia Pure Math., vol. 33(1), Amer. Math. Soc., Providence RI, 1979, pp. 111–156.
25. W. Casselman, *Introduction to the theory of admissible representations of p-adic reductive groups* (1974), Preprint, University of British Columbia.
26. C.W. Curtis and I. Reiner, *Methods of representation theory with applications to finite groups and orders I, II*, Wiley-Interscience, 1981, 1987.
27. P. Deligne, *Les constantes des équations fonctionnelles des fonctions L*, Modular functions of one variable II, Lecture Notes in Math., vol. 349, Springer Verlag, 1973.
28. P. Deligne and G. Henniart, *Sur la variation, par torsion, des constantes locales d'équations fonctionnelles des fonctions L*, Invent. Math. **64** (1981), 89–118.
29. F. Digne and J. Michel, *Representations of finite groups of Lie type*, London Math. Soc. Student Texts, vol. 21, Cambridge University Press, 1991.
30. A. Fröhlich, *Local fields*, Algebraic number theory (J.W.S. Cassels and A. Fröhlich, eds.), Academic Press, London, 1967, pp. 1–41.
31. A. Fröhlich and M.J. Taylor, *Algebraic number theory*, Cambridge Studies in Advanced Mathematics, vol. 27, Cambridge University Press, 1991.
32. S. Gelbart, *Automorphic forms on adèle groups*, Annals of Mathematics Studies, vol. 83, Princeton University Press, 1975.
33. P. Gérardin and J.-P. Labesse, *A solution of the base change problem for* GL(2) *(following Langlands, Saito, Shintani)*, Automorphic forms, representations and L-functions (A. Borel and W. Casselman eds.), Proc. Symposia Pure Math., 33(2), Amer. Math. Soc., Providence RI, 1979, pp. 115–133.
34. P. Gérardin and W.-C. Li, *Degree two monomial representations of local Weil groups*, J. reine angew. Math. **394** (1989), 1–30.
35. R. Godement and H. Jacquet, *Zeta functions of simple algebras*, Lecture Notes in Math., vol. 260, Springer Verlag, 1972.
36. J.A. Green, *The characters of the finite general linear groups*, Trans. Amer. Math. Soc. **80** (1955), 402–447.
37. M. Harris, *Supercuspidal representations in the cohomology of Drinfel'd upper half spaces: elaboration of Carayol's program*, Invent. Math. **129**, (1997), 75–120.
38. M. Harris and R. Taylor, *The geometry and cohomology of some simple Shimura varieties*, Annals of Mathematics Studies, vol. 151, Princeton University Press, 2001.

39. G. Henniart, *Galois ε-factors modulo roots of unity*, Invent. Math. **78** (1984), 117–126.
40. ——, *On the local Langlands conjecture for* GL(n)*: the cyclic case*, Ann. Math. (2) **123** (1986), 145–203.
41. ——, *La conjecture de Langlands locale numérique pour* GL(n), Ann. Scient. Éc. Norm. Sup. (4) **21**, (1988), 497–544.
42. ——, *Caractérisation de la correspondance de Langlands locale par les facteurs ε de paires*, Invent. Math. **113** (1993), 339–356.
43. ——, *Une preuve simple des conjectures de Langlands pour $GL(n)$ sur un corps p-adique*, Invent. Math. **139** (2000), 439–455.
44. G. Henniart and R. Herb, *Automorphic induction for $GL(n)$ (over local non-archimedean fields)*, Duke Math. J. **78** (1995), 131–192.
45. R.E. Howe, *Tamely ramified supercuspidal representations of GL_n*, Pacific J. Math. **73** (1977), 437–460.
46. H. Jacquet and R.P. Langlands, *Automorphic forms on $GL(2)$*, Lecture Notes in Math., vol. 114, Springer Verlag, 1970.
47. H. Jacquet, I. Piatetski-Shapiro and J. Shalika, *Automorphic forms on* GL(3), Ann. of Math. (2) **109** (1979), 169–258.
48. ——, *Rankin-Selberg convolutions*, Amer. J. Math. **105** (1983), 367–483.
49. H. Koch, *Classification of the primitive representations of the Galois group of local fields*, Invent. Math. **40** (1977), 195–216.
50. P.C. Kutzko, *On the supercuspidal representations of GL_2, I*, Amer. J. Math. **100** (1978), 43–60.
51. ——, *On the supercuspidal representations of GL_2, II*, Amer. J. Math. **100** (1978), 705–716.
52. ——, *The Langlands conjecture for GL_2 of a local field*, Annals of Math. **112** (1980), 381–412.
53. ——, *The exceptional representations of GL_2*, Compositio Math. **51** (1984), 3–14.
54. S. Lang, *Algebraic number theory*, Graduate Texts in Math., vol. 110, Springer Verlag, 1994.
55. R.P. Langlands, *Problems in the theory of automorphic forms*, Lectures in modern analysis and applications III, Lecture Notes in Math., vol. 170, Springer Verlag, 1970, pp. 18–86.
56. ——, *On the functional equations of Artin's L-functions*, Unpublished notes (1970).
57. ——, *Base change for $GL(2)$*, Annals of Mathematics Studies, vol. 96, Princeton University Press, 1980.
58. G. Laumon, M. Rapoport and U. Stuhler, *D-elliptic sheaves and the Langlands correspondence*, Invent. Math. **113** (1993), 217–338.
59. B. Lemaire, *Intégrabilité locale des caractères-distributions de* $\mathrm{GL}_n(F)$ *où F est un corps local non-archimédien de caractéristique quelconque*, Compositio Math. **100** (1996), 41–75.
60. I.G. Macdonald, *Symmetric functions and Hall polynomials (second edition)*, Oxford University Press, 1995.
61. J. Martinet, *Character theory and Artin L-functions*, Algebraic number fields: L-functions and Galois properties (A. Fröhlich, ed.), Academic Press, London, 1977, pp. 1–87.
62. C. Mœglin, M-F. Vignéras and J-L. Waldspurger, *Correspondances de Howe sur un corps p-adique*, Lecture Notes in Math., vol. 1291, Springer Verlag, 1987.

63. D. Montgomery and L. Zippin, *Topological transformation groups*, Wiley 1955.
64. L.E. Morris, *Level zero G-types*, Compositio Math. **118** (1999), 135–157.
65. A. Moy, *Local constants and the tame Langlands correspondence*, Amer. J. Math. **108** (1986), 863–930.
66. A. Moy and G. Prasad, *Unrefined minimal K-types for p-adic groups*, Invent. Math. **116** (1994), 393–408.
67. ——, *Jacquet functors and unrefined minimal K-types*, Comment. Math. Helv. **71** (1996), 98–121.
68. J. Neukirch, *Class field theory*, Springer Verlag, 1986; *Algebraic number theory*, Springer Verlag, 1999.
69. I. Reiner, *Maximal orders*, Reissue, Oxford University Press, 2003.
70. F. Rodier, *Whittaker models for admissible representations of reductive p-adic split groups*, Harmonic Analysis on Homogeneous Spaces (C.C. Moore ed.), Proc. Symposia Pure Math. vol. 26, Amer. Math. Soc., Providence RI, 1973, pp. 425–430.
71. ——, *Représentations de $GL(n, k)$ où k est un corps p-adique*, Astérisque **92–93** (1982), 201–218.
72. J. Rogawski, *Representations of $GL(n)$ and division algebras over a local field*, Duke Math. J. **50** (1983), 161–196.
73. V. Sécherre, *Représentations lisses de $GL(m, D)$, I: caractères simples*, Bull. Soc. Math. France **132** (2004), 327–396.
74. J-P. Serre, *Local class field theory*, Algebraic number theory (J.W.S. Cassels and A. Fröhlich, eds.), Academic Press, London, (1967), pp. 128–161.
75. ——, *Galois cohomology*, Springer Verlag, 1994.
76. ——, *Local fields*, Graduate Texts in Math., vol. 67, Springer Verlag, 1995.
77. ——, *Linear representations of finite groups*, Graduate Texts in Math., vol. 42, Springer Verlag, 1996.
78. J-P. Serre and J. Tate, *Good reduction of abelian varieties*, Ann. Math. (2) **88** (1968), 492–517.
79. F. Shahidi, *Fourier transforms of intertwining operators and Plancherel measures for* GL(n), Amer. J. Math. **106** (1984), 67–111.
80. S. Stevens, *Semisimple characters for p-adic classical groups*, Duke Math. J. **127** (2005), 123–173.
81. J. Tate, *Fourier analysis in number fields and Hecke's zeta functions*, Thesis, Princeton University, 1950; J.W.S. Cassels and A. Fröhlich (eds.), Algebraic number theory, Academic Press, 1967, pp. 305–347.
82. ——, *Local constants*, Algebraic number fields: L-functions and Galois properties (A. Fröhlich ed.), Academic Press, London, 1977, pp. 89–131.
83. ——, *Number theoretic background*, Automorphic forms, representations and L-functions (A. Borel and W. Casselman eds.), Proc. Symposia Pure Math., vol. 33(2), Amer. Math. Soc., Providence RI, 1979, pp. 3–26.
84. A. Weil *L'intégration dans les groupes topologiques et ses applications*, Hermann, Paris, 1965.
85. ——, *Sur la théorie du corps de classes*, J. Math. Soc. Japan **3** (1951), 1–35.
86. ——, *Fonctions zêta et distributions*, Séminaire Bourbaki **312** (1965–66).
87. ——, *Basic number theory*, Springer Verlag, 1995.
88. ——, *Exercices dyadiques*, Invent. Math. **27** (1974), 1–22.
89. J.-K. Yu, *Construction of tame supercuspidal representations*, J. Amer. Math. Soc. **14** (2001), 579–622.

90. A.V. Zelevinsky, *Induced representations of reductive p-adic groups II: On irreducible representations of $GL(n)$*, Ann. Scient. Éc. Norm. Sup. (4) **13** (1980), 165–210.
91. E.-W. Zink, *Representation theory of local division algebras*, J. reine angew. Math. **428** (1992), 1–44.

Index

Some Common Symbols

$\boldsymbol{a}_F$:	Artin Reciprocity map
c-Ind:	compact induction functor
δ:	module
$\varepsilon(*, s, \psi)$:	local constant
Ind:	(smooth) induction functor
$\boldsymbol{\iota}$:	normalized induction functor
$L(*, s)$:	L-function
$\ell(*)$:	normalized level
$\lambda_{K/F}(\psi)$:	Langlands constant
ω_π:	central character of π
$\boldsymbol{\pi}_F$:	Langlands correspondence over F
$\tau(*, \psi)$:	Gauss sum

Some Common Abbreviation

If E/F is a finite separable field extension:

$$\begin{aligned}
\chi_E &= \chi \circ \mathrm{N}_{E/F}, && \chi \text{ a character of } F^\times; \\
\psi_E &= \psi \circ \mathrm{Tr}_{E/F}, && \psi \in \widehat{F}; \\
\rho_E &= \rho \mid \mathcal{W}_E, && \rho \in \mathfrak{G}_n^{\mathrm{ss}}(F); \\
\mathrm{Ind}_{E/F}\,\theta &= \mathrm{Ind}_{\mathcal{W}_E}^{\mathcal{W}_F}\,\theta, && \theta \in \mathfrak{G}_m^{\mathrm{ss}}(E); \\
\pi_E &= \mathrm{Lft}_{E/F}\,\pi, && \pi \in \mathcal{A}_2^{\mathrm{wr}}(F).
\end{aligned}$$

The last only applies when E/F is tame and $p = 2$.

Grundlehren der mathematischen Wissenschaften

A Series of Comprehensive Studies in Mathematics

A Selection

246. Naimark/Stern: Theory of Group Representations
247. Suzuki: Group Theory I
248. Suzuki: Group Theory II
249. Chung: Lectures from Markov Processes to Brownian Motion
250. Arnold: Geometrical Methods in the Theory of Ordinary Differential Equations
251. Chow/Hale: Methods of Bifurcation Theory
252. Aubin: Nonlinear Analysis on Manifolds. Monge-Ampère Equations
253. Dwork: Lectures on ρ-adic Differential Equations
254. Freitag: Siegelsche Modulfunktionen
255. Lang: Complex Multiplication
256. Hörmander: The Analysis of Linear Partial Differential Operators I
257. Hörmander: The Analysis of Linear Partial Differential Operators II
258. Smoller: Shock Waves and Reaction-Diffusion Equations
259. Duren: Univalent Functions
260. Freidlin/Wentzell: Random Perturbations of Dynamical Systems
261. Bosch/Güntzer/Remmert: Non Archimedian Analysis – A System Approach to Rigid Analytic Geometry
262. Doob: Classical Potential Theory and Its Probabilistic Counterpart
263. Krasnosel'skiĭ/Zabreĭko: Geometrical Methods of Nonlinear Analysis
264. Aubin/Cellina: Differential Inclusions
265. Grauert/Remmert: Coherent Analytic Sheaves
266. de Rham: Differentiable Manifolds
267. Arbarello/Cornalba/Griffiths/Harris: Geometry of Algebraic Curves, Vol. I
268. Arbarello/Cornalba/Griffiths/Harris: Geometry of Algebraic Curves, Vol. II
269. Schapira: Microdifferential Systems in the Complex Domain
270. Scharlau: Quadratic and Hermitian Forms
271. Ellis: Entropy, Large Deviations, and Statistical Mechanics
272. Elliott: Arithmetic Functions and Integer Products
273. Nikol'skiĭ: Treatise on the shift Operator
274. Hörmander: The Analysis of Linear Partial Differential Operators III
275. Hörmander: The Analysis of Linear Partial Differential Operators IV
276. Liggett: Interacting Particle Systems
277. Fulton/Lang: Riemann-Roch Algebra
278. Barr/Wells: Toposes, Triples and Theories
279. Bishop/Bridges: Constructive Analysis
280. Neukirch: Class Field Theory
281. Chandrasekharan: Elliptic Functions
282. Lelong/Gruman: Entire Functions of Several Complex Variables
283. Kodaira: Complex Manifolds and Deformation of Complex Structures
284. Finn: Equilibrium Capillary Surfaces
285. Burago/Zalgaller: Geometric Inequalities
286. Andrianaov: Quadratic Forms and Hecke Operators
287. Maskit: Kleinian Groups
288. Jacod/Shiryaev: Limit Theorems for Stochastic Processes

289. Manin: Gauge Field Theory and Complex Geometry
290. Conway/Sloane: Sphere Packings, Lattices and Groups
291. Hahn/O'Meara: The Classical Groups and K-Theory
292. Kashiwara/Schapira: Sheaves on Manifolds
293. Revuz/Yor: Continuous Martingales and Brownian Motion
294. Knus: Quadratic and Hermitian Forms over Rings
295. Dierkes/Hildebrandt/Küster/Wohlrab: Minimal Surfaces I
296. Dierkes/Hildebrandt/Küster/Wohlrab: Minimal Surfaces II
297. Pastur/Figotin: Spectra of Random and Almost-Periodic Operators
298. Berline/Getzler/Vergne: Heat Kernels and Dirac Operators
299. Pommerenke: Boundary Behaviour of Conformal Maps
300. Orlik/Terao: Arrangements of Hyperplanes
301. Loday: Cyclic Homology
302. Lange/Birkenhake: Complex Abelian Varieties
303. DeVore/Lorentz: Constructive Approximation
304. Lorentz/v. Golitschek/Makovoz: Construcitve Approximation. Advanced Problems
305. Hiriart-Urruty/Lemaréchal: Convex Analysis and Minimization Algorithms I. Fundamentals
306. Hiriart-Urruty/Lemaréchal: Convex Analysis and Minimization Algorithms II. Advanced Theory and Bundle Methods
307. Schwarz: Quantum Field Theory and Topology
308. Schwarz: Topology for Physicists
309. Adem/Milgram: Cohomology of Finite Groups
310. Giaquinta/Hildebrandt: Calculus of Variations I: The Lagrangian Formalism
311. Giaquinta/Hildebrandt: Calculus of Variations II: The Hamiltonian Formalism
312. Chung/Zhao: From Brownian Motion to Schrödinger's Equation
313. Malliavin: Stochastic Analysis
314. Adams/Hedberg: Function spaces and Potential Theory
315. Bürgisser/Clausen/Shokrollahi: Algebraic Complexity Theory
316. Saff/Totik: Logarithmic Potentials with External Fields
317. Rockafellar/Wets: Variational Analysis
318. Kobayashi: Hyperbolic Complex Spaces
319. Bridson/Haefliger: Metric Spaces of Non-Positive Curvature
320. Kipnis/Landim: Scaling Limits of Interacting Particle Systems
321. Grimmett: Percolation
322. Neukirch: Algebraic Number Theory
323. Neukirch/Schmidt/Wingberg: Cohomology of Number Fields
324. Liggett: Stochastic Interacting Systems: Contact, Voter and Exclusion Processes
325. Dafermos: Hyperbolic Conservation Laws in Continuum Physics
326. Waldschmidt: Diophantine Approximation on Linear Algebraic Groups
327. Martinet: Perfect Lattices in Euclidean Spaces
328. Van der Put/Singer: Galois Theory of Linear Differential Equations
329. Korevaar: Tauberian Theory. A Century of Developments
330. Mordukhovich: Variational Analysis and Generalized Differentiation I: Basic Theory
331. Mordukhovich: Variational Analysis and Generalized Differentiation II: Applications
332. Kashiwara/Schapira: Categories and Sheaves. An Introduction to Ind-Objects and Derived Categories
333. Grimmett: The Random-Cluster Model
334. Sernesi: Deformations of Algebraic Schemes
335. Bushnell/Henniart: The Local Langlands Conjecture for GL(2)

MIX
Papier aus verantwortungsvollen Quellen
Paper from responsible sources
FSC® C105338

If you have any concerns about our products,
you can contact us on
ProductSafety@springernature.com

In case Publisher is established outside the EU,
the EU authorized representative is:
Springer Nature Customer Service Center GmbH
Europaplatz 3, 69115 Heidelberg, Germany

Printed by Libri Plureos GmbH
in Hamburg, Germany